The Biology of Temporary Waters

The Biology of Temporary Waters

D. Dudley Williams
University of Toronto at Scarborough, Canada

OXFORD
UNIVERSITY PRESS

OXFORD
UNIVERSITY PRESS

Great Clarendon Street, Oxford OX2 6DP

Oxford University Press is a department of the University of Oxford.
It furthers the University's objective of excellence in research, scholarship,
and education by publishing worldwide in

Oxford New York

Auckland Cape Town Dar es Salaam Hong Kong Karachi
Kuala Lumpur Madrid Melbourne Mexico City Nairobi
New Delhi Shanghai Taipei Toronto

With offices in

Argentina Austria Brazil Chile Czech Republic France Greece
Guatemala Hungary Italy Japan Poland Portugal Singapore
South Korea Switzerland Thailand Turkey Ukraine Vietnam

Published in the United States
by Oxford University Press Inc., New York

First published 2006

British Library Cataloguing in Publication Data
Data available

Library of Congress Cataloging in Publication Data
Williams, D. Dudley.
The biology of temporary waters / D. Dudley Williams.
p. cm.
Includes bibliographical references and index.
ISBN 0–9–852811–6 (hardback: alk. paper)—ISBN 0–19–852812–4
(paperback : alk. paper)
1. Aquatic ecology. 2. Aquatic habitats. I. Title.
QH541.5.W3W449 2006
577.6—dc22 2005019559

Typeset by Newgen Imaging Systems (P) Ltd., Chennai, India
Printed in Great Britain
on acid-free paper by
Antony Rowe Ltd, Chippenham, Wiltshire

ISBN 0–19–852811–6 978–0–19–852811–1
ISBN 0–19–852812–4 (Pbk.) 978–0–19–852812–8 (Pbk.)

1 3 5 7 9 10 8 6 4 2

This book is dedicated to the memory of my parents
Frank and Bette Williams

Preface

Temporary waters are fascinating venues in which to study the properties of species, as the latter deal with the day-to-day business of living in a highly variable environment. Obligate temporary water species display a remarkable array of adaptations to the periodic loss of their primary medium that largely sets them apart from the inhabitants of permanent water bodies. Survival of individuals frequently depends upon exceptional physiological tolerance or effective migrational abilities, and communities have their own, distinctive hallmarks. Quite apart from their inherent biological interest, however, temporary waters are now in the limelight from a conservation perspective as these habitats come more and more into conflict with human activities. Traditionally, many temporary waters, be they ponds, pools, streams, or wetlands, have been considered to be 'wasted' areas of land, potentially convertible to agriculture once drained. In reality, they are natural features of the global landscape representing distinct and unique habitats for many species—some that are found nowhere else, others that reach their maximum abundance there.

In 1985, the late W.D. Williams lamented '.... The extent of reference to temporary waters is not in accord with their widespread occurrence and abundance, ecological importance, nor limnological interest'. In 1987, I published *'The Ecology of Temporary Waters'* which was a first attempt to gather together the highly scattered literature on these fascinating habitats. Gratifyingly, since Bill Williams' lament there has been a steady increase in the study of temporary waters, and many research articles have been published throughout the world. The latter range from the highly descriptive, to the applied, to the conceptual. Each has contributed something to our knowledge base, yet this information remains largely scattered, in need of collation and synthesis. Several worthwhile attempts have been made through 'special sessions' at recent conferences, however, there remains little integration. Perhaps, the available data remain a little too patchy and the subdiscipline is still a little too immature for a thorough treatment, but hopefully this book is a step in that direction.

I have many people to thank for helping me put this volume together. First and foremost are several of my past and present graduate students and postdoctoral fellows who have shared my interest in temporary waters, and who have allowed me to use some of their unpublished data. In particular, Katarina Magnusson and Oksana Andrushchyshyn have provided information on intermittent pond faunas in southern Ontario (Chapter 4), and Ian Hogg has kindly provided information on the faunal changes along the Murrumbidgee River, Australia (Chapter 10). I am also grateful to those colleagues who have allowed me to reproduce photographs and figures from their published works, and who are acknowledged by name in the figure legends. Several of these same colleagues also provided encouragement to undertake the writing of this book, and I hope that the completed product does not disappoint them. I would also like to formally thank Diane Gradowski and Ken Jones of the University of Toronto at Scarborough graphics department for their help with many of the illustrations, and the Natural Sciences and Engineering Research Council of Canada for financial support of my research. Thanks, too, go to Annette Tavares for the cover design.

On a personal note, I am grateful to my family, Judy, Siân, and Owen, whose support and companionship helped me complete what turned out to be a far bigger project than I had originally thought, and to Bronwen for always being happy to see me.

D. Dudley Williams
Toronto, May 2005

Contents

CHAPTER 1

Introduction

1.1 What are temporary waters?

We have short time to stay, as you,
We have as short a Spring,
As quick a growth to meet decay,
As you or any thing

(Robert Herrick, 1591–1674)

Temporary waters, in general, are to be found throughout the world. Some types, such as the reservoirs of bromeliad leaf axils and turloughs (seasonal limestone lakes), are restricted by factors such as climate or geology. Others, such as temporary ponds and streams, and rain pools are ubiquitous—although these, too, may exhibit some regional differences (e.g. rain pools characterize the high, open grasslands of South Africa). However, all are, for the most part, natural bodies of water which experience a recurrent dry phase. Often, the latter is predictable both in its time of onset and duration. The defining element is the *cyclical* nature of the drought, as some permanent waterbodies may dry up in exceptional years. In the latter case, because most of the biota is not adapted to survive such conditions there will be significant mortality. Temporary water species, on the other hand, are generally well adapted to dealing with water loss. Indeed, many species have spread beyond the boundaries of natural waterbodies to colonize those temporary waters that have been created by human culture, such as bird baths, and rainwater-filled tin-cans and tyre tracks. Although the contents of this book are devoted largely to temporary *fresh* waters, it is important to remember that a great number of inland saline waters also experience drought. Such habitats are widespread and they have received considerable attention elsewhere (e.g. W.D. Williams 1981; Hammer 1986; International Society for Salt Lake Research 2001). Coastal marine habitats, for example salt marsh ponds and supralittoral rock-pools, are also subject to drying, and will be touched on in this volume.

The physicochemical features of temporary waters strongly influence the biotas present, but biological factors may be important also especially with increased duration of the aquatic phase. Insects and crustaceans tend to dominate the fauna, but temporary water communities, as a whole, may comprise bacteria, protoctists, vertebrates, fungi, and an abundance of higher and lower plants. Many species exhibit opportunistic and pioneering traits, and also a range of drought-survival mechanisms, such as diapause and seed formation.

Whereas wetlands, in general, comprise a very important subset of temporary waters, and data from their study will be drawn on throughout this book, the latter is not intended as a comprehensive synthesis of wetland biology *per se*. For such information the reader is directed to, for example: Williams (1993), Finlayson and Van der Valk (1995), Mitsch and Gosselink (2000), Spray and McGlothlin (2004), and the websites of organizations such as the Ramsar Convention on Wetlands (http://www.ramsar.org/), and Wetlands International (http://www.wetlands.org/).

1.2 Biological importance

Despite perhaps being regarded as the cinderellas of aquatic science, temporary waters represent significant components of the global landscape.

From a cultural perspective, cyclically fluctuating water levels often have determined the sustainability and evolution of riparian societies. A prime example is the annual flooding of the River Nile upon which the agricultural activities of both ancient and modern civilizations have depended. Further, from a resource perspective, wetlands, for example, represent a significant store of our planet's freshwaters. However, temporary waters also have considerable significance to Biology *per se*. Blaustein and Schwartz (2001) outlined four reasons for studying temporary pools, specifically, but which well encompass other temporary water types: (1) temporary waters can contribute to our general understanding of 'ephemerality', especially as it relates to life histories, population dynamics, and community organization; (2) these habitats represent convenient systems in which to study ecological concepts, particularly as they are amenable to manipulation experiments, and their abundance allows easy replication. Further, those habitats with simple communities can be mimicked in semi-natural or even artificial set-ups; (3) temporary waters frequently harbour the vectors of disease-causing organisms that afflict mankind; and (4) temporary waters contain many species important to global biodiversity.

To these may be added: (5) that, in a biogeographical context, there is evidence to suggest that temporary ponds may have acted as postglacial dispersal routes for taxa possessing dormant stages capable of 'island-hopping' (e.g. copepods) from glacial refugia (Stemberger 1995); (6) in an evolutionary context, there has been exploration of the idea that life may have evolved on earth more than once, and that an alternative environment-of-origin to the oceans may have been ponds that dry out periodically. In such ponds, chemicals in solution would have been progressively concentrated to a state that enabled the maintenance of protoplasmic systems (Hinton 1968). Further, W.D. Williams (1988) has suggested that there is an alternative explanation to the hypothesis that the biota of permanent, standing freshwaters came from marine ancestors via either the terrestrial environment, rivers or estuaries, and which relegates the biota of temporary freshwaters to a subset of the 'permanent' biota that developed adaptations to resist desiccation and good powers of dispersal (Wiggins *et al.* 1980). The alternative viewpoint considers most permanent lakes to be geologically ephemeral (few are older than 20,000 years) and regards temporary water bodies as being very ancient, not as individuals but as a habitat type. Williams cited Lake George, a temporary freshwater lake near Canberra, Australia, as being 10 million years old. He hypothesized that from rivers and the terrestrial environment, a large contingent of the biota first colonized temporary freshwaters (perhaps temporary floodplain pools), and that a subset of this flora and fauna developed, or regained, the ability to withstand permanent inundation and hence were subsequently able to establish themselves in permanent lakes. Under such a scenario, Williams stated that one would expect to see some of the following properties: the more evolutionary ancient groups of the biota should occur in temporary waters; much of the biota living in permanent waters should retain effective dispersal mechanisms—to counter the geological ephemerality of their habitats and as a reflection of their lineage; many of the 'active migrants' of the permanent lentic biota, should persist in lotic habitats, or have close relatives which do; and overall species richness in temporary waters should be greater than in permanent freshwaters. Evidence from the literature provides some support for each of these properties (Tasch 1969; Elgmork 1980; Fernando 1980; Schram 1986; Fernando and Holcík 1989; Lake *et al.* 1988); (7) with increasing interest in land-water ecotones, the margins of temporary ponds and streams have the potential to be important sites for modelling hydrological processes, nutrient transport and transformation, and the role played by the biota (Bradley and Brown 1997; Giudicelli and Bournard 1997); and (8) there is now evidence to indicate that variations in the physical environment of inland waters impact both molecular and morphological evolution by changing mutation rates and by exposing (through genotype–environment interactions) otherwise cryptic variation. Extreme environments tend to accelerate morphological change, promoting diversification (Hebert 1999).

Temporary waters may be important sites of such altered rates of molecular evolution and, therefore, worth further study.

1.3 Classification of temporary waters

1.3.1 Review of some previous classification schemes

Temporary waters are amazingly diverse in the habitats that they present for the development and sustainment of life. By way of an example, Table 1.1 lists the main types of temporary standing waters to be found in the British Isles. It subdivides these habitats into those of natural origin (e.g. peatland pools and cup fungi; Figure 1.1) and those resulting from human activities (e.g. quarry ponds and saw-pit ponds), and also distinguishes common from rarer, or regional, types. Including lotic temporary waters would swell the list considerably. Faced with such an inventory, it is perhaps little wonder that classification attempts for temporary waters are few and far between. One proposal has been based upon habitat size (micro-, meso-, and macro-; Table 1.2) but this tends to lump habitats that may support quite different communities, for example lowland, floodplain pools and alpine lakes.

Length and intensity of the dry period also have been suggested as criteria, and may be more biologically relevant. Length of the dry phase can be divided simply into seasonal, annual, and greater than annual—but cyclical. Intensity of the drought is important because, for example, two habitats which both remain dry for 4 months of the year might have different moisture-retaining capacities of their substrates, allowing the survival of significantly different biotas. Climatologists have derived a number of indices for drought that may have useful application to temporary waters. One, widely used example is the Palmer Hydrologic Drought Index, which combines precipitation and temperature values with soil water content data—including outflow and storage measures (Heddinghaus and Sabol 1991). Some researchers have found significant correlations between this index and temporary water invertebrate population dynamics (Hershey *et al.* 1999).

As with all systems of classification, there are bound to be exceptions which do not fit any of the categories, for example, Lake Eyre in southern Australia only fills with water every half century or so (Mawson 1950). Can this really be called cyclical? The majority of species that would colonize such a lake would die when it dried up, with

Table 1.1 The main types of temporary standing water found in Britain

Natural origin

Common types: Intermittent and episodic ('seasonal') ponds, lakes; margins of permanent ponds and lakes; floodplain ponds; oxbow ponds; deltaic ponds; tidal wetlands; supralittoral tide pools; peatland pools; woodland pools; vernal ponds (filled only in spring); autumnal ponds (summer-dry); rain pools (both in clay soils and crevices in bedrock); pools associated with uprooted trees and land surface undulations; empty snail shells; water-filled hoof-prints; liquid dung.

Rarer and/or regional types: Ponds associated with glacial activities (e.g. kettle ponds, formed by subsidence resulting from melting of subsurface ice, and moraine ponds, formed in glacially deposited sediments), solution of bedrock, and iron pans; turloughs (water-filled depressions underlain by limestone); plunge pools (formed at the base of dried-up waterfalls); water retained by cup fungi, teasels and mosses; tree holes; water retained in large leaf axils.

Human origin

Quarry pools; pools associated with mining and landscaping; wheel-rut pools; cattle-watering ponds and troughs; pools resulting from peat-digging; waterfowl-decoy channels; fish ponds; ponds associated with ancient rural activities (e.g. dewponds, rainwater-collection ponds, armed/water-distribution ponds, saw-pit ponds, charcoal-burning pits); depressions associated with defence and warfare (e.g. ditches, trenches, moats, bomb craters); rain-filled tyres and plastic sheeting (e.g. silage yards); midden pools; rain barrels; cisterns; ornamental bird baths; pools associated with landfill sites; slurry vats; footprints; water-filled, cattle-trampled depressions (e.g. around gates and feeding areas).

Source: Information taken from Rackham (1986) and Williams (1987), where greater description of these various habitats can be found.

Figure 1.1 Photograph of a rainfilled cup fungus; note the accumulated particles of detritus that may serve as food for the biota, and the presence of several semi-aquatic oligochaetes.

Table 1.2 Classification of temporary water habitats based on size

Microhabitats: Axils of plant leaves (e.g. bromeliads); tree holes; rain-filled rockpools; tin cans; broken bottles and other containers; foot prints; tyre tracks; cisterns; empty shells (e.g. molluscs and coconuts)
Mesohabitats: Temporary streams and ponds; snow-melt pools; monsoon rain pools; floodplain pools; dewponds; wetland pools
Macrohabitats: Periodically flooded, large old river beds; shallow oxbow lakes, drying lakes, drying lakeshores, alpine lakes; sloughs; turloughs

Source: Adapted from Decksbach (1929).

perhaps only a few, highly specialized forms being able to span the 50 or so years between fill ups. The occurrence of some waterbodies in the United Kingdom affords another example of a misfit. *Triops* (a notostracan, or tadpole shrimp), was first recorded in 1738 from a 'temporary' pond, it was next recorded in 1837 and again in 1948, after a lapse of 111 years (Schmitt 1971). Although it did not appear, naturally, during all that time, it could be hatched from dried mud taken from the pond and rehydrated. This animal requires a period of desiccation prior to hatching and the wet British climate did not create suitable conditions very often. Whether such a pond could be truly termed 'temporary', or whether it was simply a permanent pond that dried up infrequently is debatable. Clearly it contained at least one species normally characterized as being indicative of temporary waters.

Classifications based on indicator species, or species groups also are not infallible. For example, Klimowicz (1959) attempted to classify small ponds in Poland on the basis of their molluscan faunas. Granted, some snail and bivalve species are very resistant to water loss and may be usefully assigned to different habitat types, however some, such as *Musculium partumeium*, are known to occur in both temporary and permanent ponds (Way *et al.* 1980). Colless (1957) modified a classification scheme of Laird's (1956) based on breeding habitats for mosquitoes. In it, he created two main categories—*Surface Water*, and *Containers*—the former were then subdivided into Lakes and ponds; Swamps and marshes; Transient pools; Obstructed streams; and Flowing streams; and the latter subdivided according to whether they were Large, simple containers (e.g. kerosene tins); Small simple containers (e.g. cans and bottles); or specialized containers (e.g. pitcher plants, plant axils, and crab holes). Laird later, (1988), summarized a number of other classification schemes based on larval mosquito habitats, many of them temporary.

Pichler (1939) proposed that water temperature could be used to establish a classification scheme for small waterbodies, as follows:

1. puddles—very small waterbodies up to 20 cm deep with the bottom strongly heated by the sun; practically no stratification in the summer, when, daily, the variation may be as much as 25°C (see Figure 1.2);
2. pools—waterbodies up to 60 cm deep, consequently less heat reaches the bottom; thermal stratification is upset daily by a turnover and the summer temperature variation may be up to 15°C at the surface and 5°C at the bottom;
3. small ponds—up to 100 cm deep with very little heat reaching the substrate; stratification is more stable but can be upset daily, summer temperature variation is up to 10°C at the surface and 2°C near the bottom.

All these characteristics were based on open ponds, thus shading by emergent vegetation would make an important difference to the scheme. In addition, differences would be expected to occur between temporary pools and streams, as water in the latter is in motion and may run through shaded and non-shaded reaches.

Other schemes have been proposed by special interest groups, such as those interested in saline lakes, where the term 'athalassohaline waters' (non-marine waters with a significant salt content; Bayly 1967; W.D. Williams 1981) has been coined to distinguish them from 'thalassohaline waters' (NaCl-rich waters with a similar concentration to sea water, or slightly diluted; Cole 1979). Poff and Ward (1990) put forward a proposal to classify streams based on their hydrological patterns. Two of their study streams were temporary and

Figure 1.2 An example of rainfilled puddles, in eastern Utah; such habitats are typically less than 20 cm deep and may experience daily temperature variations as much as 25°C (Scale: diameter of the pool in the foreground is approximately 3 m; water depth is a maximum of 12 cm).

produced streamflow histories that were termed 'intermittent flashy' (Sycamore Creek, Arizona) and 'harsh intermittent' (Dry Creek, Oklahoma), respectively (Figure 1.3). The authors concluded that these radically different flow regimes, together with climatic and substratum differences, must strongly influence the respective biotas present. Applying a similar argument, Puckridge *et al.* (1998) identified 11 relatively independent measures of hydrological variability that, primarily for large rivers, could categorize river types and were related to the properties of the resident fish faunas. For example, protracted zero flows were associated with greater numbers of small and carnivorous species, and dominance of physiologically tolerant forms. Rivers with high-amplitude variation, in contrast, were dominated by small, omnivorous species, and greater numbers of 'colonizing' species. Keeley and Zedler (1998) showed that vascular plants in California could be organized according to average water duration in vernal pools (Figure 1.4). Those species experiencing little or no inundation were typically annual grasses and forbs characteristic of the surrounding grasslands, whereas those inundated for longer periods were typically those restricted to vernal pools.

There are also many temporary waters that are included within the term 'wetlands'. These transitional areas between terrestrial and aquatic systems are associated with many other names that, over the years, have become either synonymous with wetlands, or regarded as a subset of habitats, such as peatlands, swamps, marshes, bogs, fens, and floodplains. In the United States, the extent of these areas is believed to be around 42 million hectares, and so their significance is considerable (Dahl and Johnson 1991). However, such has been the subdivision of these habitats that there are now more than 70 categories described in Canada alone (Warner and Rubec 1997). The terms 'seasonal wetlands' and 'seasonal ponds' also crop up in the literature, generally with reference to habitats in temperate parts of North America and Europe. These have limited use in climate zones that are more typically non-seasonal.

While many of these schemes are helpful, the persistence of other habitat names (such as salt

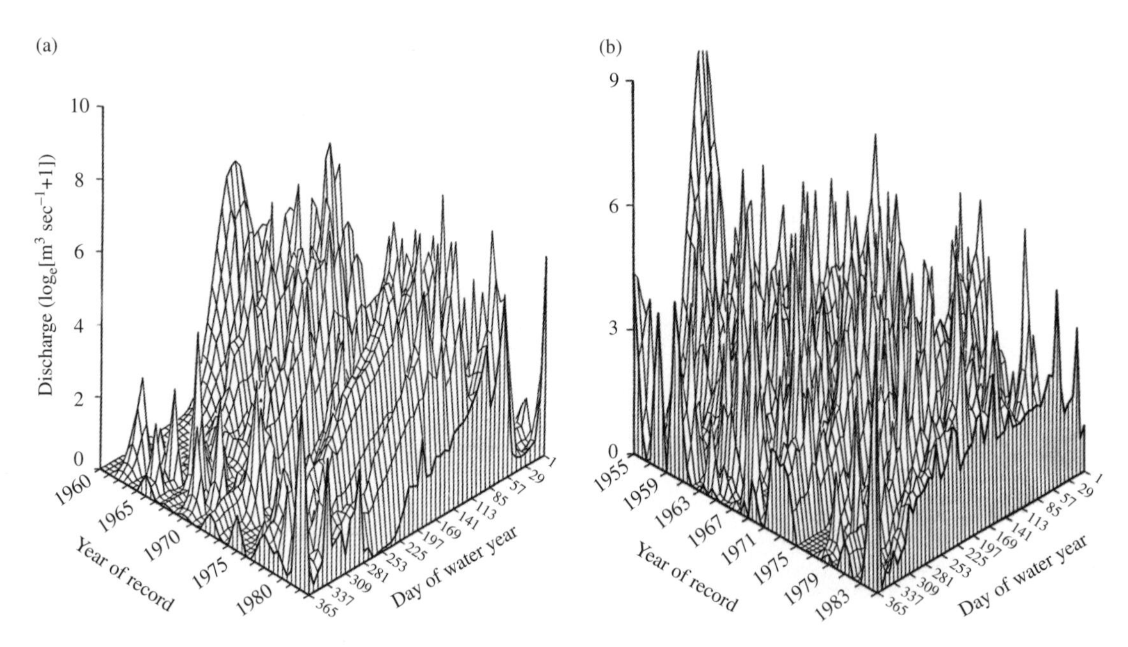

Figure 1.3 Streamflow patterns for two temporary streams, based on long-term, daily mean discharge records over several years: (a) Sycamore Creek, Arizona; (b) Dry Creek, Oklahoma (redrawn from Poff and Ward 1990).

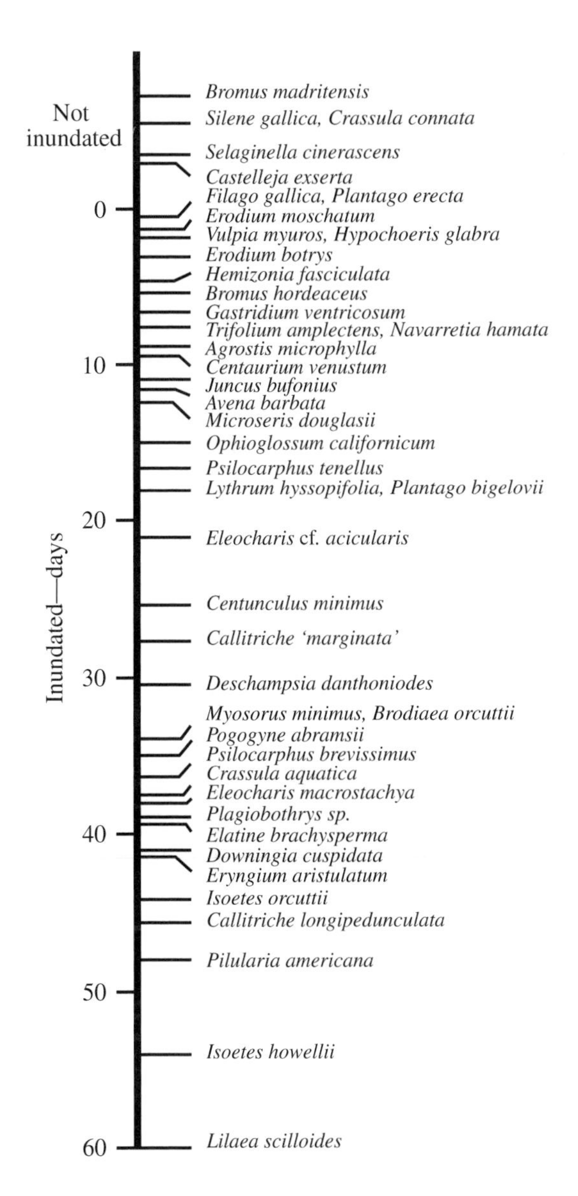

Figure 1.4 Common vernal pool plants from southern California arranged along an axis of inundation tolerance (redrawn from Keeley and Zedler 1998).

pans, playas, and astatic ponds) together with adoption of regional names (such as claypans, gnammas, and vegetated pans, Australia; vleis, southern Africa; dayas, North Africa; prairie potholes and tinajas, North America; ramblas, eastern Spain), has created a confusing plethora of terms (Comín and Williams 1994). Some regional names are clearly still useful in a local context. The term 'ephemeral', in particular, has been used loosely, and often interchangeably with 'intermittent' and 'temporary'. Its derivation (from the Greek 'ephemeros'—living but a day) suggests that it should be abandoned as a biological term to describe temporary waters. Hydrologists, however, define ephemeral waters as those having basins or channels which are above the water table at all times (Gordon *et al.* 1993). Some (e.g. Comín and Williams 1994) have suggested that the term 'temporary waters' itself has become confused (it is sometimes used to refer to 'intermittent' waters—see below); herein it is used to encompass *all* waters which experience cyclical drought, and also to refer to temporary waterbodies in which the precise nature of the hydroperiod is unknown.

A potentially useful classification scheme for temporary wetlands has been proposed by Boulton and Brock (1999). Based on Australian waters, it focuses on the predictability and duration of flooding events (Figure 1.5). Although it recycles terms such as 'ephemeral' and 'seasonal', it defines them quite precisely.

1.3.2 A suggested classification framework

Based on the experience of those who have attempted to classify permanent bodies of water, a desirable approach to creating an overarching framework would be to keep any scheme relatively simple and not overly concerned with attempting to account for every single habitat variation. Table 1.3 proposes a classification into which most temporary waters will fit. Habitat typing beyond this level can be left to the specifics of individual studies or localities (e.g. vernal pools versus autumnal pools, which differ in their time of filling; Wiggins *et al.* 1980). The proposal initially assigns a temporary water body to one of the existing *global biomes*. These are accepted major regional groupings of plants and animals discernible at a global scale. They represent distribution patterns that are strongly correlated with regional climate patterns and are identified according to the climax vegetation type with affiliated characteristics of successional communities, faunas,

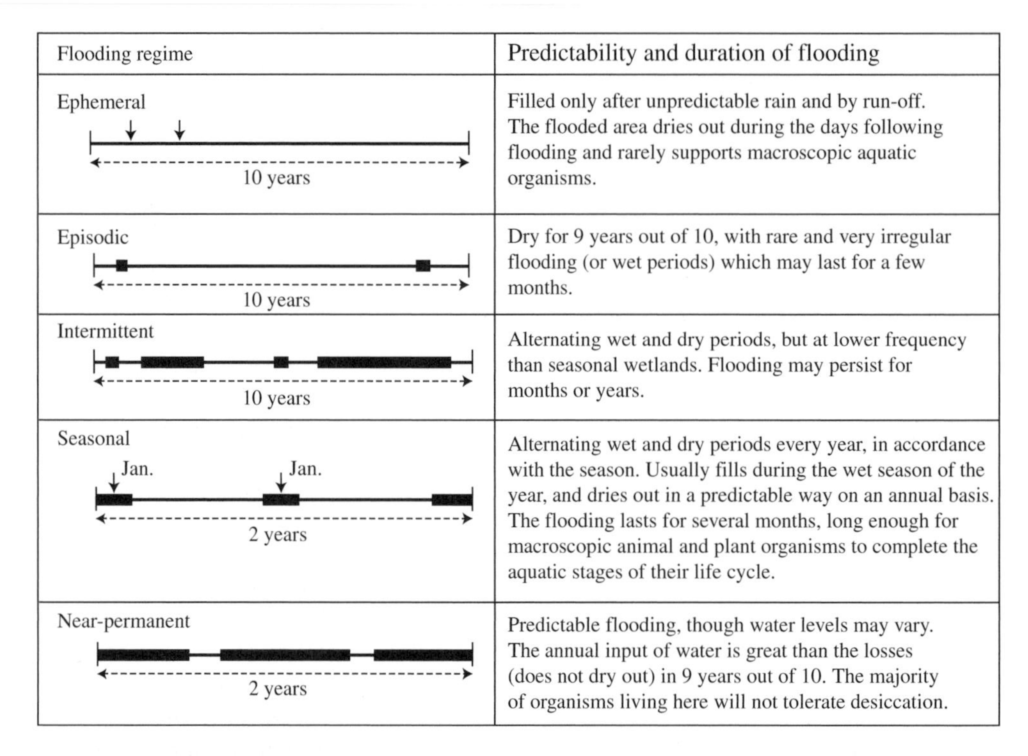

Flooding regime	Predictability and duration of flooding
Ephemeral (10 years)	Filled only after unpredictable rain and by run-off. The flooded area dries out during the days following flooding and rarely supports macroscopic aquatic organisms.
Episodic (10 years)	Dry for 9 years out of 10, with rare and very irregular flooding (or wet periods) which may last for a few months.
Intermittent (10 years)	Alternating wet and dry periods, but at lower frequency than seasonal wetlands. Flooding may persist for months or years.
Seasonal (Jan. Jan. 2 years)	Alternating wet and dry periods every year, in accordance with the season. Usually fills during the wet season of the year, and dries out in a predictable way on an annual basis. The flooding lasts for several months, long enough for macroscopic animal and plant organisms to complete the aquatic stages of their life cycle.
Near-permanent (2 years)	Predictable flooding, though water levels may vary. The annual input of water is great than the losses (does not dry out) in 9 years out of 10. The majority of organisms living here will not tolerate desiccation.

Figure 1.5 Simplified classification scheme for temporary waters (vertical arrows indicate times of water input; thick horizontal bars indicate the relative duration of the hydroperiod; based on Boulton and Brock 1999, but as modified by Yavercovski *et al.* 2004).

Table 1.3 Suggested classification framework for temporary water habitats

Biome	Hydrological character	Relative size	Chemical state
Tundra		Micro-	
Boreal forest (Taiga)			
Temperate broadleaf deciduous forest	Intermittent		Freshwater
Temperate grassland			
Tropical broadleaf evergreen forest		Meso-	
Tropical savanna	Episodic		Saline
Desert scrub			
Mediterranean scrub		Macro-	
Icefield zone			

Note: 'Intermittent' refers to waterbodies which contain water or are dry at more or less predictable times in a cycle, while 'episodic' refers to waterbodies which only contain water more or less unpredictably, and tend to be confined to arid regions.
Source: After Comín and Williams (1994). The boundary between saline and freshwater is considered to be 3 g l^{-1}; after W.D. Williams (1964).

and soils. Keeley and Zedler (1998), working on vernal pools in California, have concluded that there is a close association between these specific pools and the Mediterranean climate in general, giving credence to the use of biomes as a classification parameter.

Within these biomes, waterbodies are assigned to either *intermittent*, or *episodic* categories (*sensu*

Comín and Williams 1994), and within these to *macro-*, *meso-*, and *microhabitats* (*sensu* Decksbach 1929). As many temporary lentic waters are high in dissolved minerals, a distinction needs to be made between *saline* and *freshwater* habitats, and the boundary concentration of $3\,g\,l^{-1}$ salinity (established by W.D. Williams 1964) is a useful one. So, for example, the water-filled leaves of the pitcher plant *Nepenthes* would be classed as intermittent, freshwater, microhabitats from the Tropical Broadleaf Evergreen Forest; the large, shallow sloughs of the Canadian prairies would be classed as intermittent, saline, macrohabitats from the Temperate Grasslands; unpredictably flowing, headwater streams in arid regions would be classed as episodic, freshwater, mesohabitats from the Desert Scrub, or Mediterranean Scrub; unpredictable rainwater pools in east Africa would be classed as episodic, freshwater, mesohabitats from the Tropical Savanna; and rainfilled tyres would be classed as episodic, freshwater, microhabitats from whichever biome they were located in. Irregularly occurring, meltwater rivulets arising in the Antarctic (see Vincent and Howard-Williams 1986) and high Arctic (and perhaps glacial margins in high alpine regions should also be included here) would necessitate the addition of another 'biome', resulting in episodic, freshwater, mesohabitats from the Icefield Zone. Of course, once formal description has been assigned to a particular water body, for example, in the 'habitat' section of a publication, it could thereafter be shortened for stylistic convenience (e.g. intermittent saline stream; episodic rainwater pool; intermittent woodland pond).

A lesson learned very early on by limnologists was that it is not easy to assign waterbodies within strict classification schemes. Pearsall (1921) demonstrated that lakes in the English Lake District fell into a continuum of habitats. Although he was able to assign lakes at the extreme ends of this range to 'oligotrophic' and 'eutrophic' types, overall he found it difficult to establish logical boundaries in-between. The classification scheme for temporary waters outlined above, should therefore be viewed only as a reference framework within which there are likely to be many continua.

1.4 Importance of temporary waters in the landscape

Exactly how extensive temporary waters are in the global landscape is difficult to assess as few surveys of these habitats, *per se*, have been made. However, wetlands are better known and can reasonably be used as a model for examining the extent of regions of the world where conditions likely to support some types of temporary waters occur. Table 1.4 provides an overview of major wetland areas. Clearly, temporary waters are highly varied, widely distributed, and a dominant feature of our planet, occurring across virtually all continents and in all climatic zones. Further, some, such as the floodplains of tropical South America are regarded as major sites of speciation—as plants and animals respond to the flood pulse via morphological, physiological, and other adaptations (Junk 1993).

Despite such importance, the environs that support temporary waters have been, and continue to be, under threat from human activities. Agriculture, urban sprawl, drainage, pollution, deforestation, and many other processes have taken their toll, worldwide. In Europe, for example, temporary ponds were in the past a more common feature of the landscape than today. Although precise numbers are difficult to obtain, the loss of small ponds, in general, from the United Kingdom during the period 1984–90 has been estimated at between 4 and 9%. Encouragingly, in that country, there currently seems to be some restoration of their numbers as a result of ponds created for wildlife, and by altered farming practices (Duigan and Jones 1997), but most of these are likely to be permanent waters. Only comprehensive conservation programmes will restore temporary waters (P. Williams *et al.* 2001). These issues will be discussed further in Chapter 10.

In another, more applied, landscape sense, Mozley (1944) drew attention to the fact that temporary ponds are a neglected natural resource. In temperate regions, when a pond dries up in early summer the bed becomes part of the terrestrial habitat. This habitat is well fertilized, due to the excrement and debris (e.g. exoskeletons) left by the

Table 1.4 Some areas of the globe that support major wetlands

Africa
Swamps of the Upper Nile; the Rift and High altitude lakes of Eastern Africa; the Niger and its floodplains; the Lower Senegal Valley; coastal lagoons of the Ivory Coast; Lake Chad (West Africa); the vast floodplains of the south, including the Pongolo River floodplains, the Mkuze Wetland System, the Nyl River floodplains, and various pans and dambos; the internal deltas of rivers (e.g. Timbuktu in Mali and the Lorian Swamp in Kenya). There are also temporary habitats associated with various man-made lakes.

Mediterranean (southern Europe and North Africa)
A large array of geomorphological formations that support wetlands, including river deltas (widely distributed from Spain to Greece); coastal lagoons (extensive around the Mediterranean Sea and along the Atlantic coast of Morocco); riverine floodplains (various oxbow lakes, for example the Rhône area); floodplain marshes (e.g. the River Tejo, Portugal, the Languedoc and Crau regions of France; and flooded woodlands, for example the Moraca River, Yugoslavia and the River Strymon, Greece); freshwater lakes (e.g. those of glacial origin in the Sierra Nevada, the Pyrenees, Apennines and Alps, and also Morocco; those associated with volcanic activity, such as the calderas of Italy; and those of karstic origin, such as those found in Albania, Yugoslavia, northwestern Greece, southern France, Spain, Algeria and Tunisia); man-made reservoirs (e.g. on the rivers Guardiana and Tejo, western Spain, the Esla Reservoir in central Spain, and Lake Boughzoud in Algeria); salt basins (restricted to the Maghreb and central Spain); intertidal regions (localized along the Atlantic coast); and seasonally flooded channels (very extensive: 75% of first order streams in southern France are believed to be seasonal and in Morocco and Tunisia the incidence for first- and second- order streams is 97%, and 80% for third and fourth order streams).

Australia (northern Australia)
Along the Queensland coast, upland areas contain seasonal wetlands: floodplain lakes, billabongs (oxbow lakes), swamps, waterholes, and river flats subject to flooding; there are also extensive mangroves and tidal flats. Lowlands along the Gulf of Carpenteria support intermittent swamps in shallow pans and seasonal billabongs. Waterholes, seasonal swamps and floodplain lakes skirt the Arnhem Land Plateau, and the coastal plains east of Darwin have extensive floodplains. In Western Australia (Pilbara Area), there are numerous waterholes along river channels and also intermittently flooded lakes. The large inland arid region, which occupies almost 50% of the continent, is characterized by saline intermittently flooded and episodic wetlands.

Papua New Guinea
This predominantly high-rainfall country supports the following wetland types: saline and brackishwater swamps (including mangrove); freshwater swamps (including seasonal swamp forest and woodland, swamp savanna and herbaceous swamps). These habitats occupy around 7.5% of the land area.

South Asia (India and southeast Asia)
All climatic zones support wetlands, a large percentage of which are seasonal due to the long dry summer. Many are saline (mangroves), and among the world's largest such habitats. Freshwater wetlands are dominated by shallow lakes, ponds and temporary waters. Major and medium-sized rivers all support extensive floodplain wetlands, many of which have been converted into paddy fields and fish ponds.

Canada and Greenland
Wetlands are estimated to comprise some 14% of the land area of Canada, with peatlands accounting for 88% of that figure. They variously include marshes (both freshwater and saline), shallow open water, bogs, fens and swamps, ranging through seven bioclimatic zones: arctic, subarctic, boreal, temperate, prairie, mountain and coastal. Greenland supports shallow open water, salt marshes, bogs and fens.

United States
The United States supports a vast array of wetland types, which range from those found in tropical rain forests (Hawaii), to those of wet tundra (Alaska), and those found in deserts (Southwest). Wetlands have been subdivided into five ecological groups: marine (e.g. pools on rocky shores); estuarine (e.g. salt and brackish marshes, mangrove areas); riverine (e.g. intermittent streams and shallow rivers); lacustrine (e.g. temporary ponds of various types, shallow lakes, reservoirs); and palustrine (e.g. inland marshes, wet meadows, bogs, swamps, flooded forests).

Mexico
Estuarine and marine wetlands are the most extensive, ranging along the 10,000 km long coastline. The major watershed formed by the rivers Grijalva and Usumacinta also supports temporary aquatic habitats. Palustrine habitats include flooded marshes and savannas, together with forested wetlands, palm thickets, and inundated lowland forests of the Yucatan Peninsula. Lacustrine wetlands are restricted largely to inland mountainous regions.

Table 1.4 (*Continued*)

Central Asia (Russia and the newly independent states, northern China)
Many areas within this region, particularly to the north and east, support a similar diversity of wetlands to that seen in the nearctic (marshes, bogs, swamps, fens, wet tundra, intermittent streams, riverine floodplains, flooded forests, salt and brackish marshes, and coastal rockpools). However, some areas are water scarce, for example, Uzbekistan, Turkmenistan and southern Kazakhstan have largely desert climates and have only two principal rivers, the Amu Darya and Syr Darya. As a consequence, irrigation has been practiced for millenia and has altered the water balance, with significant loss of natural wetlands (e.g. the surface area of the Aral Sea has decreased over 50% since 1960). In addition, shallow groundwaters have become highly saline due to the mobilization of vast quantities of salt, whereas excessive irrigation has resulted in waterlogged soils. Extensive dam building has also had an impact.

Tropical South America
Hydrologically, South America is dominated by large rivers, such as the Amazon, Orinoco, and Magdalena, which result from high annual rainfall (up to 5 m per year). Marked seasonality in rainfall produces intermittent flooding of vast areas of forest and savanna, creating many types of temporary waters. In total, these habitats are estimated to cover 20% (2,000,000 km^2) of the country, with the Pantanal of the Mato Grosso (Brazil) being considered the largest wetland in the world. The floodpulse is predictable and monomodal in the savannas and along the floodplains of large rivers, but it is unpredictable and polymodal in the floodplains of small streams. The floodplains of tropical South America are regarded as regions of high speciation. They are also areas of high interchange between permanent and temporary waters. Salt marshes and mangroves occur on the Atlantic coast. The wet Paramos of the high Andes supports reed swamps, cushion bogs, and peat bogs.

Source: Information taken largely from Whigham *et al.* (1993).

aquatic organisms, and thus supports a considerable biomass of land plants during the terrestrial phase. The terrestrial community, in turn, leaves a legacy of organic matter (e.g. decaying leaves, stems, and roots) which can be used the following spring by the aquatic community. Mozley pointed out that it should be possible to use temporary ponds for the rotational harvesting of stocked fish fry during the aquatic phase and a field crop such as oats during the terrestrial phase; each community would be nourished by the remains of the other. Such a practice has in fact been in operation in France since the fourteenth century, and the growing of rice in flooded (paddy) fields alongside nutrient-generating invertebrates and fishes has been practiced in many tropical and subtropical countries for millenia. Details will be given in Chapter 8.

Unfortunately, temporary waters also have deleterious aspects. Many temporary waters, especially in the tropics and subtropics, are breeding places for the vectors of disease organisms. For example, tree holes are the ancestral habitat of *Aedes aegypti*, the yellow fever mosquito, that now breeds in many man-made water containers, such as discarded tin cans and tyres. Intermittent ponds and ditches, irrigation canals, marshes, and periodically flooded areas support large numbers of mosquitoes and also aquatic snails (Styczynska-Jurewicz 1966). The latter are, for example, host to the blood trematode *Schistosoma*, a debilitating and eventually fatal parasite of humans and cattle, and the liver fluke *Fasciola hepatica*. As well as yellow fever, mosquitoes transmit malaria, dengue and viral encephalitis, while sucking the blood of humans and domestic animals. Such diseases are not, however, restricted to the tropics and, presently, the inhabitants of some temperate regions (e.g. Europe and North America) are being increasingly affected. The presence of suitable vector species, existence of a pool of affected individuals, and the availability of suitable local aquatic habitats for the vectors are all factors in the equation. The increasing trend of global warming is likely to escalate this spread—perhaps through creation of more, and warmer, temporary ponds. The past, too, has seen distributional shifts. For example, a 'touch of the ague' was a common complaint in Londoners in the 1600s, where residents living in the newly developed and fashionable districts of St James' Park, Piccadilly, and Haymarket were infected by malaria-carrying mosquitoes breeding in the nearby Pimlico marshes (Johansson 1999). These, negative, aspects of temporary waters will be revisited in Chapter 9.

CHAPTER 2

The physical environment

2.1 Hydrological considerations

2.1.1 Temporary streams and ponds, and the run-off cycle

Hydrological characteristics vary in different regions of the globe as a result of many factors, foremost among which are local climate, near-surface geomorphology, vegetation, and land use. However, at the base of the formation of most waterbodies, both permanent and temporary, is the run-off cycle.

Precipitation is the most important source of water (Figure 2.1), but before this even touches the ground surface it can be intercepted several times by trees and other vegetation. Water trapped on these exposed surfaces is very quickly evaporated by wind. The water that reaches the soil surface is taken up by infiltration and the rate at which this occurs depends on the type of soil and its aggregation. At this stage, in some exceptional clayey soils, water may collect on the surface in small depressions and form puddles and even small trickles. Both tend to be short-lived, as the water they contain is usually absorbed quite quickly by soil cracks and patches of more permeable soil over which it may run. Such waterbodies would fall under the definition of 'episodic' temporary waters, as their temporal occurrence, and sometimes also their precise physical location, are unpredictable. Unless such waters are close to sources of rapidly colonizing biota, they are unlikely to support many species. Pools and rivulets resulting from storms or snow melt are examples, as are shallow depressions in bedrock outcrops where, of course, there is no infiltration.

Where there is soil, infiltrated water near the surface is subject to direct evaporation back into the atmosphere due to air currents and uptake and subsequent transpiration by surface vegetation such as grasses. If the intensity of precipitation at the soil surface is greater than the infiltration capacity of the soil, and if all the puddles have been filled, then overland flow begins. When this reaches a stream channel it becomes surface run-off. If the topography is such that the water cannot flow away in a channel, then it collects in a low point and forms a pond. Some of the water that penetrated the now-saturated soil will reach the stream channel or pond as interflow, usually where a relatively impervious layer is found close to the soil surface. The rest of the infiltrated water, which has penetrated as far as the groundwater table, eventually will also reach the stream or pond as baseflow or groundwater flow.

The water flowing in a stream or collecting in a pond is thus derived from the following sources: overland flow, interflow, groundwater flow, and direct precipitation on the water body itself. Infiltration is perhaps the single most important factor in the regulation of temporary waters, for it determines how precipitation will be partitioned into the categories of overland, inter-, and groundwater flow. Horton (1933) defined infiltration capacity as the maximum rate at which a given soil can absorb precipitation in a given condition. In the initial phase of infiltration the attraction of water by capillary forces of the soil is of great importance, although the effect of these forces in medium- to coarse-grained soils is only minor after the infiltration front has penetrated more than a metre or so. These capillary forces are greatest within fine-grained soils which have low initial moisture (Davis and DeWeist 1966). Air

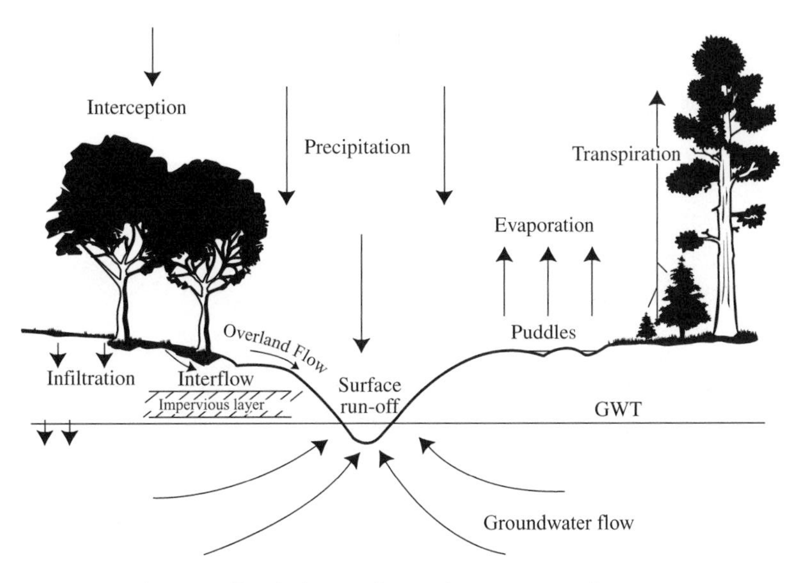

Figure 2.1 The basic components of the Run-off Cycle that contribute to the water in a pond or stream.

trapped between the soil particles may have an effect opposite to that of soil structure as at first the infiltration rate will be slowed down, as the advancing front of infiltrating water will have a tendency to expel any air it meets and this may result in the formation of pockets of dry soil which will form barriers to water movement. However, as the front continues, some air may be dissolved and the rate of advance will speed up.

The condition of the soil, particularly its texture and structure, is also of great importance as, for example, a bare soil surface will be directly exposed to rain which will tend to compact the soil and also wash small particles into open cracks and holes. This in turn will have the effect of reducing infiltration as the rain continues, and may lead to overland flow. Conversely, a dense cover of vegetation will protect the soil surface so that compaction and the filling up of cracks will be less. Further, the roots of these plants may hold the soil open and increase the normal infiltration rate. Infiltration rate is also affected by surface crusting which may block the larger pore spaces and persist until disrupted by vegetation growth, soil fauna activity, erosion, cultivation, or freeze-thaw action. Stone cover may have positive or negative impacts on infiltration, as rocks and coarse gravel may both protect the surface from rainsplash and crusting, tending to increase infiltration rate, and reduce the surface area available for infiltration (Bull and Kirby 2002).

Taking the above factors into consideration, we could speculate that, in temperate regions, a large number of temporary streams and ponds would be supported on areas of land with clay-loam soil under heavy cultivation, where many of the large stands of trees and much of the bush have been cleared, where bare soil is more common than pasture and where wire fences are preferred to earth-bank hedgerows (Figure 2.2). Were this land to be left uncultivated it would probably support a much smaller number of permanent streams and ponds instead of a large number of temporary ones. However, such predictions may be confounded by recent evidence that shows a relationship between forest growth cycles and stream discharge. In managed forests, pre-planting drainage often results in an increase in low flows, if greater than 25% of the catchment is drained. In all but the driest years, forest growth decreases low flows, whereas clear-cutting increases low flows initially, but thereafter they decline according to the rate at which the vegetation regrows (Johnson 1998).

Where the soil has a high sand content, infiltration will be high and retention of water in the

(a)

(b)

Figure 2.2 Kirkland Creek, an intermittent stream in southern Ontario, Canada during late spring (left photograph) and late summer (right photograph).

stream or pond bed will be influenced by the position of the groundwater table. In many arid and semi-arid regions, the water table is depressed, and this combined with low precipitation and high evaporation rates typically produces a landscape rich in temporary waters (Nanson *et al.* 2002). Related to the regularity of rainfall, these habitats will be either intermittent or episodic. The examples of Sycamore Creek in Arizona, and Dry Creek in Oklahoma, given in Chapter 1 (Figure 1.3), demonstrate that a range of hydrological types exists in such streams.

In the humid tropics, soils are often fully saturated for much of the year and here the lifespan of surface waters is controlled largely by evaporation, for example, around two thirds of the water which falls on the river basins of Surinam is returned back into the atmosphere (Amatali 1993). The lifespan of floodplain pools will therefore be influenced by such factors as the degree of exposure to winds and direct sunlight. For vernal forest pools in New England, Brooks and Hayashi (2002) studied predictive relationships between depth–area–volume and hydroperiod. The strongest relationship they detected was between hydroperiod and maximum pool volume, but in general, pools with a maximum depth greater than 50 cm, a maximum surface area larger than 1,000 m^2, or a maximum volume greater than 100 m^3 contained water on more than 80% of the occasions on which they were visited. These authors concluded that the weak relationships detected between pool morphometry and hydroperiod indicate that other factors, such as temporal patterns of precipitation and evapotranspiration and groundwater exchange, are likely to significantly influence the hydrology and hydroperiodicity of such ponds.

2.1.2 Components of subsurface water

The factors that contribute to and influence the regime of water in temporary ponds and streams were summarized above, but what happens when the water disappears and the habitat becomes part of the terrestrial environment? This is best approached by looking at the way in which

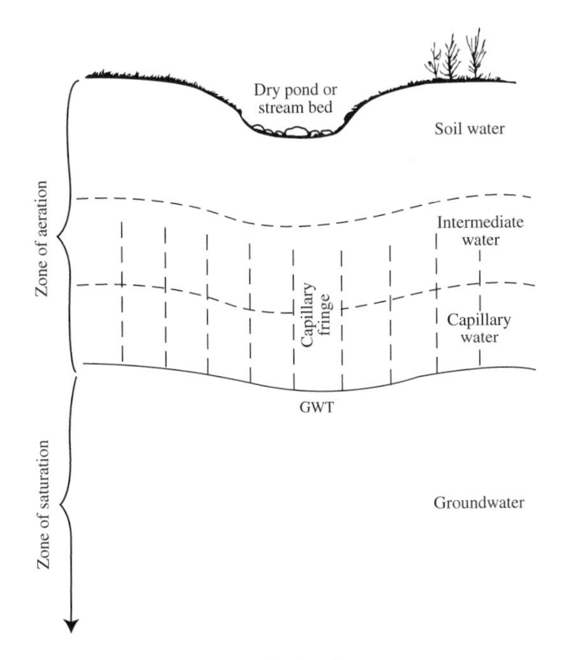

Figure 2.3 The components of subsurface water.

subsurface waters are classified. Figure 2.3 shows the components of subsurface water beneath a dry pond or stream bed to consist of four zones known as soil water, intermediate water, capillary water, and ground- or phreatic water. There are other zones beneath that of the groundwater, such as internal water, but they include water in unconnected pores and water in chemical combination with rocks. As the latter are beyond the access of most aquatic organisms they are largely irrelevant to the present discussion.

Soil water is subject to large fluctuations in amount as a response to transpiration and direct evaporation, and it is this feature which separates it from the other unsaturated zones. The intermediate water zone lies beneath that of the soil water and separates it from the saturated zone. The water here is sometimes referred to as suspended water since although it can move downwards in response to gravity, it also can move upwards into the soil water zone should the latter become very dry. This zone is variable in size, being greatest in arid regions and absent in moist areas. The lower limit of the intermediate zone is continuous with the capillary fringe. This fringe is irregular in outline and consists of water moving up through the soil (by capillary action) from the lower parts of the capillary water zone which may be as fully saturated as the groundwater. In areas of fine-grained soils, where recharge is active, the capillary fringe may extend well into the intermediate zone.

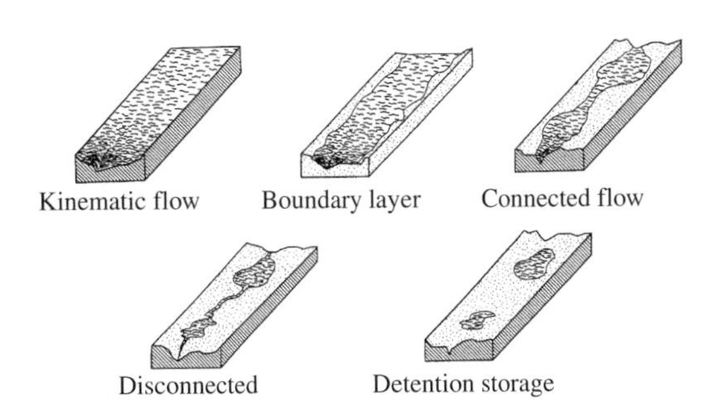

Figure 2.4 The progressive hydrological phases that occur as stream channels drain (modified from Shannon *et al.* 2002).

The groundwater table (GWT) separates the capillary fringe from the next water zone, the groundwater, where all the material is saturated. The GWT is commonly approximated to the level to which water will rise in a well, and it can move up or down in response to recharge. When, for example, a stream is flowing, the GWT will be near the level of the stream surface and, as it recedes, it will lower the level of the stream (or pond) unless it is offset by sudden precipitation and consequent overland flow and interflow. If it continues to recede, the stream will cease to flow and only a few pools will remain in the channel (Figure 2.4). Further recession will result in disconnection of the GWT from the pools which may soon vanish due to evaporation.[1] If the GWT remains fairly close to the ground surface, the capillary fringe may extend up to the surface of the stream or pond bed and thus provide a moist environment for those aquatic organisms that have sought refuge

[1] According to strict *hydrological* definition, an intermittent stream that feeds the underlying groundwater is known as 'influent'. However, when channel flow decreases to the point where groundwater feeds the stream it becomes known as 'effluent' (Gordon *et al.* 1993). If a receding GWT disconnects from the channel, these definitions become fuzzy and a state akin to *intermittent ephemerality* is reached—as 'ephemeral' streams are defined as permanently above the GWT; see Chapter 1.

by burrowing. It is in situations such as this, where the GWT is close to the soil surface and soil moisture is high that a relatively small imput of moisture from rain, is sufficient to produce a substantial relaxation of moisture tension within the soil pores (Carson and Sutton 1971). The result is that the GWT may rise quickly and cause water to reappear in the pond basin. It is for this reason, too, that some temporary streams may restart for short periods after quite small amounts of rainfall.

Any further drop in the GWT will result in the setting up of the other subsurface water zones outlined above, and hence subject organisms to the large fluctuations of water content that are characteristic of the soil water zone. Deeper burrowing of active forms to reach the saturated zone, or the coordination of a special drought-resistant stage in the life cycle will be required if these taxa are to survive. In some cases, it is possible that a few species may obtain sufficient moisture from condensation of dew on the ground surface to remain in the soil water zone.

Figure 2.4 also illustrates the progressive hydrological phases that occur as ephemeral stream channels drain. Although the term 'ephemeral' was dimissed in Chapter 1 as being of limited biological use, it is, as already noted, a term still used by hydrogeologists to encompass running waters that are permanently disconnected from the GWT, for example, dryland rivers that flow only for a short period during and after rainstorms (Bull and Kirkby 2002). These phases have some relevance to the flow patterns seen in many intermittent and episodic (?unpredictably ephemeral) streams and are summarized in Table 2.1. The kinematic flow phase is seen as being relatively infrequent in dryland streams, such as those of the 'ramblas' of the Spanish Mediterranean region, but may be an annual event in temporary streams in wetter climates. Boundary layer flow, again, is commonplace in temperate and humid tropical regions. Connected flow is a very dominant phase in many temporary streams and proceeds alongside GWT recession and reduced rainfall towards disconnected flow and detention storage. Although aquatic organisms are present throughout all five phases, the latter three are often times of intense biological activity (Williams and Hynes 1976a).

Table 2.1 Characteristics of the progressive hydrological phases of temporary streams

Flow domain	Frequency	Type of flow	Comments
Kinematic flow	Low sometimes measured in yrs; requires many, cumulative rainfall events	Bankfull flow that just submerges a channel reach; friction from bed does not influence hydrograph propagation	Major reworking of bed materials; sediment transport and deposition as waves; transmission losses reflect channel dimensions and infiltration characteristics
Boundary layer flow	Low; reflects channel width; requires cumulative rain	Same as above, but more shallow; friction from bed can affect hydrograph propagation speeds	Same as for kinematic flow, but transmission losses tend to be lower as over-bank flow is less likely
Connected flow	Depends on antecedent conditions; less rainfall amount and duration; (1–3 times/yr)	Low flow condition that may occupy braided channels; moderate water depths	Hydrographic propagation speeds moderate; flow and transmission loss controlled by bed morphology; some reworking of fine sediments
Disconnected flow	Largely same as connected flow but even fewer rainfall events and duration; (usually 1–3/yr)	Low flow largely a result of detention storage pools overflowing	Hydrographic propagation downstream decreases; little reworking of bed; transmission losses minimal and related to local bed conditions
Detention storage	Will occur for most rainfall events; storage duration will reflect volume of rain and antecedent channel conditions	No flow, only ponded water in channel depressions	No flow or reworking of bed materials; ponding results in channel abstracting all of the surface water; channel bed morphology determines location of detention storage

Source: Modified from Shannon *et al.* (2002).

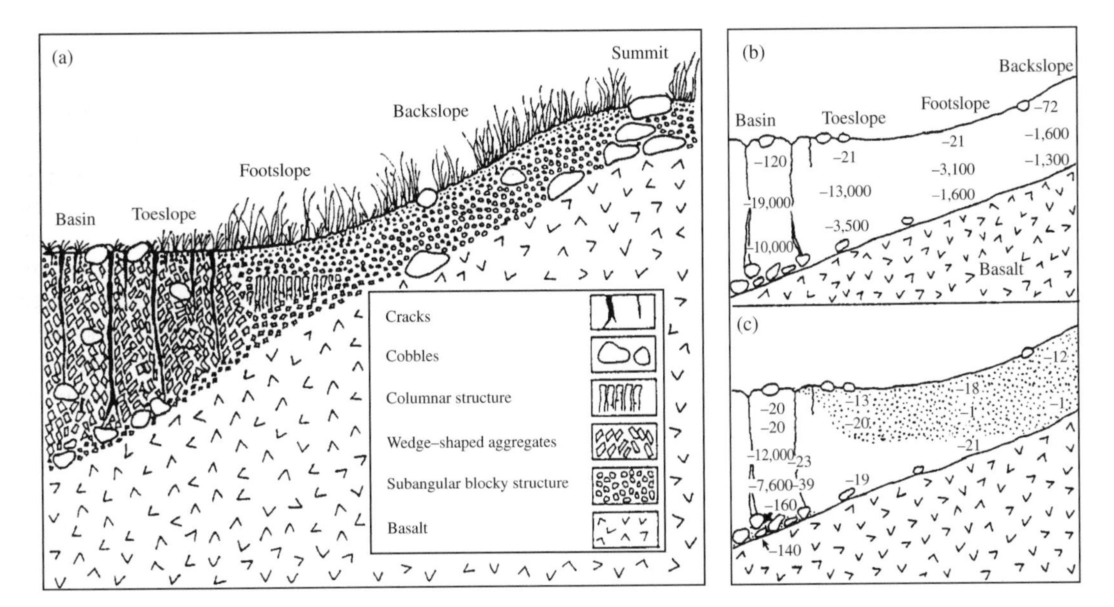

Figure 2.5 Cross sections of a vernal pool basin in southern California showing: (a) soil morphology (horizontal axis is 30 m, basin soil is ~1 m thick); (b) water potentials (in J kg^{-1}) after 12 mm of rain; and (c) after 90 mm of rain. Stippled area indicates where free water was observed on the soil surface or in the soil matrix. Open and closed cracks are indicated in the basin soil. (redrawn from Weitkamp *et al.* 1996).

Details of how water begins to accumulate in a basin after drought have been provided for a Californian vernal pool (hydroperiod February to April) by Weitkamp *et al.* (1996). The study revealed a major relationship between water movement and basin substrate (soil) morphology. Figure 2.5(a) shows a 30-m cross-section through the pool catena,[2] where soil depth varied from 20 cm at the summit to around 100 cm in the basin; the region is underlain by bedrock. Soil morphology varied from a weak and moderate subangular blocky structure (with <20% clay) near the summit, to strong, very coarse angular blocky structure in the basin. In addition, the latter had wedge-shaped aggregates, slickensides, and shrinkage cracks, with >40% clay. Movement of water into the basin was followed throughout two wet seasons. After 12 mm of rainfall, infiltration occurred through the shallow upper-slope soils and flowed down towards the basin along the surface of the bedrock and within the weathered, vesicular basalt. At the footslope, further downward movement was slowed due to a marked change in soil texture. At this stage, water potentials were highest at the surface and lowest in the mid-section of the profile. They were also somewhat higher at bedrock contact points (Figure 2.5(b)) where, in a previous wet season, water was seen flowing from the soil matrix over bedrock exposed by an observation pit.

After 90 mm of rain, the upper 27 cm of footslope soil became saturated (Figure 2.5(c)) and water began to accumulate on the surface such that the perimeter of the basin exhibited standing water before the centre. Water from the ponded foot- and toe-slope soils tended to flow overland and contributed to the wetting of the basin soil. At this stage, some of the cracks in the basin vertisol (areas of dark soil rich in clay) swelled up and closed, whereas others remained open, or closed only to a depth of 5–10 cm. Later in the season, some of the cracks that were open to the basin

[2] *Catena* refers to a series of related soils of about the same age, derived from similar parental material, and occurring under similar climatic conditions, arranged in a sequence of increasing wetness.

surface filled with free water. Interestingly, in the subsurface horizon, open and closed cracks were detected as close as 15 cm to one another (indicated by the different water potentials, bottom left of Figure 2.5(c)). Such differences could be important to the survival of pond organisms seeking areas of highest substrate saturation. So, too, could Weitkamp *et al.*'s observation that soil materials surrounding cobbles at the bedrock contact were moister than the materials above. Upon removal of this material from the bedrock surface, via an excavated pit, water rose up from the weathered basalt.

2.1.3 Small temporary waters

At the microhabitat level, the hydrological characteristics of such waterbodies as rockpools, empty shells, tree holes, and 'container' habitats in general, are largely controlled by rainfall and evaporation. Typically, they will have a longer hydroperiod during wet seasons and where the effects of drying winds are minimal. There will also be some influence of basin shape; deeper bodies with restricted openings (e.g. pitcher plants and tin cans) tending to retain their water longer. In terms of phytotelmata, there will be the added factor of growing season and the availability of leaf axils and pitchers. In the case of the North American pitcher plant *Sarracenia purpurea*, for example, the pitchers are available for colonization only during the late spring, summer, and early autumn. In contrast, the many species within the tropical genus *Nepenthes* produce pitchers virtually continually, thus providing a fairly regular and somewhat more predictable supply of new habitats, within a restricted area, year round (Beaver 1983). However, even in the tropics, the flowering seasons of certain plants, for example, *Heliconia*, may dictate a restricted hydroperiod.

2.2 Origins: basin and channel formation

2.2.1 Pond basins

Pond basins may be formed by natural, geological processes or by human activities. Naturally produced ponds frequently result from glacial activity, involving both erosional and depositional processes. Ice scour of flat bedrock areas typically creates basins of varying size and depth. These ponds receive all their water either directly from precipitation or by surface run-off. Other types of erosion-formed lakes and ponds are cirques, which are amphitheatre-shaped basins carved at the heads of glaciated valleys, and paternoster lakes, which are chains of ponds formed in the bottoms of these valleys. Different pond types result from the deposition of glacial debris. For example, retreating glaciers leave large deposits of ground moraine over wide areas. Where these overlie impermeable till, vast areas of wetlands and very shallow lakes and ponds have been created. Similarly, kettle ponds have been left in many areas by the melting of ice masses buried in the moraine. Deposits left at the bottoms of valleys may confine melting ice and so form basins. Besides these erosionally and depositionally formed basins, glaciers produce depressions through alternate freezing and thawing of the ground surface, resulting in subsidence. Shallow arctic and antarctic ponds are formed in this way (Reid 1961). Where many of these ponds merge the result is a thermokarst lake (Rex 1961).

Solution ponds are formed in regions where soluble rock has been dissolved by water. Infiltrating surface water freshly charged with carbon dioxide (forming weak carbonic acid) is particularly effective as a solvent.

Small, deep ponds may result from plunge-pools at the base of dried-up waterfalls. Large, shallow ponds may be formed in oxbow fashion (called billabongs in Australia) in any wide valley through which a river meanders, and shallow deltaic ponds are formed by sediment deposition at a river mouth.

Percolating waters may deposit an insoluble iron-pan layer in sandy, permeable soils and above this a pond may form. There are many examples of this type of pond in Surrey, in England, as well as in Sologne, France (Bowen 1982).

Meteor impact is known to have created both large and small basins but these appear to be rare, or are perhaps rarely recognized as such.

Figure 2.6 Small intermittent pool basins created by uprooted trees.

Uprooting of trees by storms commonly creates small shallow basins that drain readily in sandy soils but which may contain water for several months in clay soils (Figure 2.6).

Man-made basins result from industrial activities such as mining, quarrying, landscaping, etc., and also from ancient rural activities (especially in the Old World). Examples of these basins include: watering holes; peat-digging holes; moats (not just around castles, as, in the thirteenth century, moats were common features added to the houses of gentry and farmers alike, as status symbols; Taylor 1972); fish ponds (see Chapter 8); decoys (long, reed-lined, shallow channels leading from a lake and used for luring wild ducks into a trapnet); dewponds (shallow, nineteenth century, clay-lined ponds for collecting rainwater and run-off); armed ponds (watering holes with several spreading arms used for sharing water between several fields); saw pits (a practice begun in the fourteenth century, where one of two men sawing a log lengthwise stood in a pit beneath the log); charcoal pits (an ancient process producing charcoal by burning wood buried in a pit) (Rackham 1986); and temporary ponding areas created to collect sediment-laden runoff from various construction projects.

In both ancient and modern times, ponds, ditches, trenches and craters have been created during warfare. Most notably, in Vietnam's Xieng Khuang Province, the onslaught of bombs dropped continually from 1964 to 1973 has created a unique, pockmarked landscape. Bombs, each up to 900 kg, have created thousands of craters many of which are now water-filled. Some are used as fish ponds, and many dry up regularly.

2.2.2 How many ponds?

As many temporary ponds are small they do not appear on most topographical maps, and it is therefore difficult to estimate their abundance. By careful study of fine-scale Ordnance Survey maps and applying a correction factor for small ponds (less than 6 m in diameter) not marked, Rackham (1986) estimated the total number of ponds in England and Wales to be 800,000 (or 5.4 ponds km^{-2}) around 1880. He considered this period to have been when the total number of natural and man-made basins was at a maximum, and his estimate included both basins that were permanently and intermittently filled. The frequency of occurrence of these basins was not uniform across the country (Figure 2.7), being least dense in mountainous areas (e.g. 0.12 km^{-2}) and most dense in areas of ancient agriculture and ancient woodlands (e.g. 115 km^{-2}). Undoubtedly, this analysis will have missed many of the smallest

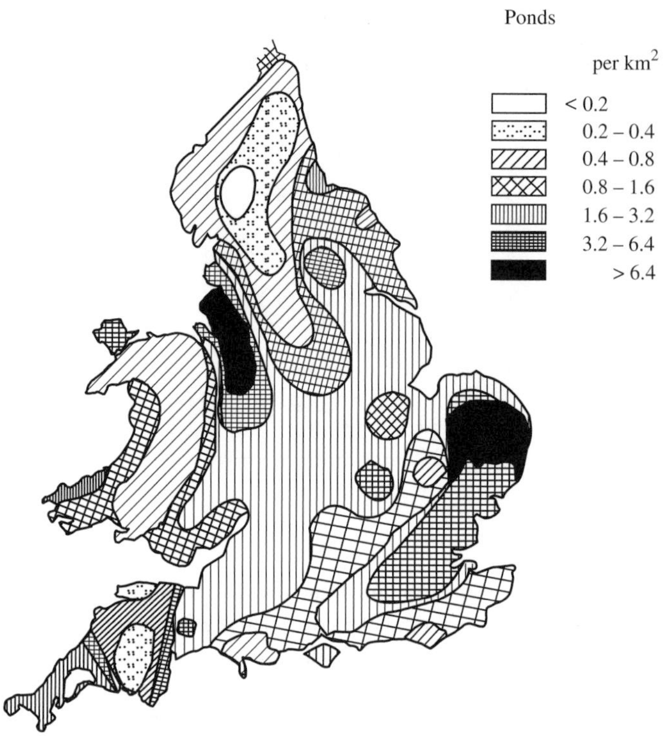

Figure 2.7 Map of England and Wales showing the distribution and approximate densities of all pond types (permanent and temporary) in the 1920s. The data were derived from fine-scale Ordinance Survey maps and a correction factor for ponds smaller than 6 m in diameter has been applied (redrawn from Rackham 1986).

temporary pond basins and so the nineteenth century total for England and Wales may well have peaked at more than one million. In a more recent survey, Everard *et al.* (1999) have estimated the number of ponds, of all sorts, in England and Wales to be between 650,000 and 750,000. Further, data from a Countryside Survey in 1996, in which temporary ponds were, for the first time, identified as a separate habitat type, showed that almost 40% of all ponds in lowland Britain were temporary; more than 82,000 in total, and representing a density of around 0.7 ponds km^{-2} (Biggs *et al.* 1996; Williams *et al.* 2001).

2.2.3 Stream and river channels

Most river valleys and stream channels are formed by erosion, so the processes of formation are much less diverse than for pond basins. However, in their lower reaches these channels become much modified by depositional processes that may create a variety of features (e.g. meanders, braided channels, and deltas) some of which are associated with intermittent water residency.

Intermittency in general, however, is considered to be more a characteristic of small headwater or tributary streams. Nevertheless, many large rivers can also become dry, for example, all of the 12 major rivers that flow through northwestern Namibia (some with catchments over 400 km long) are dry, sandy channels for much of each year (Jacobson *et al.* 1995). Similarly, rivers in most of the world's drylands are subject to prolonged drought, for example on the west coasts of Southern Hemisphere continents (Australia, South America, southern Africa), the deserts of the American southwest, central Eurasia, and the Tibetan plateau (Nanson *et al.* 2002).

Irrigation ditches represent man-made temporary streams and their frequency varies according to regional agricultural practices, soil characteristics, and climate. Other examples include spillways and fish migration bypass channels that may hold water only at certain times of the year.

Table 2.2 Frequency of occurrence of intermittent streams in southern and central Ontario, Canada

Area	Map scale	Map area (km^2)	No. intermittent	No. permanent	% intermittent
Rural and woodland					
Brampton	1 : 50,000	540	570	95	85.7
Alliston	1 : 50,000	540	43	8	84.3
Bolton	1 : 50,000	540	582	117	83.3
Barrie	1 : 50,000	540	87	10	89.7
Markham	1 : 50,000	540	363	73	83.3
Forested					
Haliburton	1 : 50,000	540	29	78	24.8
Kawagama	1 : 50,000	540	27	86	23.9
Algonquin	1 : 50,000	540	25	125	16.7
Coe Hill	1 : 50,000	540	0	208	0[a]
Gravenhurst	1 : 50,000	540	0	326	0[b]

[a] 64%, and

[b] 83% of the streams were marsh-fed.

2.2.4 How many streams?

Temporary streams are sometimes marked on fine-scale topographical maps as broken lines surrounded by symbols indicating a bog or marsh. In contrast to the many inventories made of permanent running waters, very few have been done for temporary ones. However, a preliminary survey (Table 2.2) of some randomly selected 1 : 50,000 topographical maps of mixed rural and woodland areas in southern and central Ontario, Canada, shows densities of intermittent streams ranging from 43 to 582 per 540 km^2 (0.08 to 1.08 km^{-2}). On average, 85.3% of all first-order streams shown on these maps were marked as being intermittent at their source. This value dropped to 21.8% in more heavily wooded areas. Regions differing in geomorphology and precipitation pattern would naturally deviate from these estimates, but these figures may serve as a useful snapshot.

2.3 Climate, seasonality, and habitat persistence

Regional climate is clearly a major factor in determining the nature of the local precipitation/evaporation balance, and hence the propensity for temporary waters to form. Rainfall frequency, amount and pattern, are all important, as are the various mechanisms of water loss. As already noted, this balance is also influenced by near-surface geomorphology (basin/channel availability), vegetation (density and type), and land use. The various permutations of these factors together with climate produce temporary water habitats on all continents, whether they be episodic streams in arid regions, or intermittent, vernal or autumnal ponds in the temperate zone.

The pattern of water availability is particularly important as it often characterizes the type of habitat and the physicochemical signature that its biota will have to endure. For example, in India and much of the Oriental tropics there are the annual monsoons. Here, warm and moist air from the ocean is blown north where it encounters either high mountain ranges or cool, dry air moving south. The result is heavy rainfall from June to September resulting in the establishment of a host of temporary water habitats, ranging from extensive wetlands, to floodplain lakes, to rain-filled rockpools. The latter are likely to be characterized by water that is low in nutrients and electrolytes, reflecting a direct rainwater origin; the former, in contrast, are more likely to have turbid, more nutrient-rich water resulting from overspill from adjacent, swollen rivers.

Seasonality is also a strong feature of temporary ponds and streams formed by snow melt at higher latitudes, and of early summer glacier melt at high altitudes. The influence of growth season on various phytotelmata has already been discussed. Even the availability of man-made habitats may reflect seasonal activities within cultures, for example, the flooding patterns of irrigation canals based on crop growth, and the summer rejuvenation of ornamental water fountains.

An aspect of climate that will likely impact temporary waters markedly in the future is global climate change. The earth's climate is being warmed, as a result of the accumulation of greenhouse gases, such that within 100 years not only will the earth be warmer than it has been for a million years, but the rate of increase will have been greater than any on record (Schneider 1989). A range of climate-related changes are expected to accompany a doubling of atmospheric carbon dioxide, including melting of the polar ice sheets and permafrost, and rising sea level (Poiani and Johnson 1991). The mean summer air temperature of the Great Plains of North America, as an example, are predicted to rise by 1–2°C, with an even greater rise in winter (3–4°C). Although less certain, models foresee increases in mean global precipitation of 3–11%, but accompanied by faster return to the atmosphere due to the warmer temperatures (Bradley *et al.* 1987). Ironically, continental interiors may become more arid as a result of a shift in current mid-latitude rain belts. All of these changes will impact the numbers, types, and properties of temporary water habitats. To appreciate this, Figure 2.8 illustrates the hydrological budget of a prairie wetland pond. In reality, this pond was classed as being 'semipermanent' (i.e. holding water throughout the growing season during most years) (Poiani and Johnson 1991). This type of pond thus sits on a hydrological knife-edge between being permanent or temporary. It is easy to see how it's balance, and that of many other types of temporary waters, could be tipped by changes in local climate: for example, more precipitation and it will never dry out; less precipitation, accompanied by an increase in seepage outflow by virtue of a lowered GWT, and it will become very short-lived. Further, Manabe and Wetherald (1986) predicted a 30–50% decrease in the summer moisture levels of Great Plains' soils, which could adversely affect the survival of the dormant stages of pond bed biota (see Chapter 5). In contrast, rising sea levels would flood many coastal temporary waterbodies, rendering them

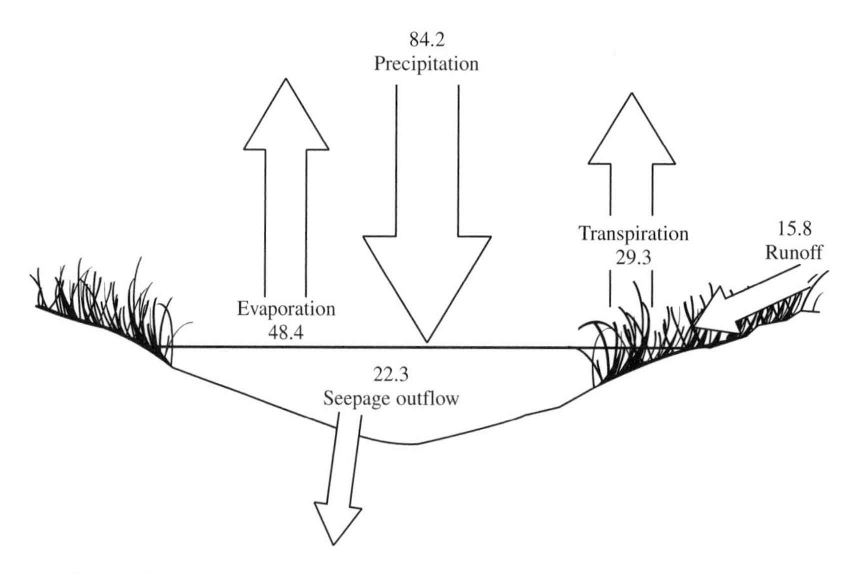

Figure 2.8 Hydrological budget for a prairie wetland pond (water loss/gain units are expressed in percentages; modified from Poiani and Carter 1991).

permanent with the loss of drought-adapted species.

In another modelling exercise, Larson (1995) looked at the effects of climate-related variables on the number of wetland basins occurring in the northern prairies. Her models which included information on spring and autumn temperatures, annual precipitation, the previous year's basin count, and the previous autumn's precipitation—accounted for 63–65% of the variation in the number of wet basins. She also demonstrated that wetlands in more wooded areas ('parkland') were more vulnerable to increased temperatures than those in grasslands.

Some of the predicted, as well as the more complex, effects of global warming are already being seen in North American lakes (Schindler *et al.* 1996). For example, Yan *et al.* (1996) observed that in a boreal lake that experienced a drought-related drop in water level the increased exposure caused a re-oxidation of sediment sulphur, which resulted in a re-mobilization of acid into the lake water. This produced a decrease in dissolved organic carbon concentrations that was sufficient to increase, by three times, the depth to which ultraviolet radiation could penetrate.

Another climatic phenomenon that is likely to significantly affect temporary water habitats is El Niño. Under normal conditions, easterly winds blow across the equatorial Pacific pushing sea surface water westwards. The water thus displaced from the west coast of South America is replaced by upwelling of cold, deep-ocean water that lowers sea surface temperatures in the eastern Pacific. This promotes a cool, dense air mass that fails to rise sufficiently high to form rain clouds, thus creating a high pressure system, with cool dry conditions. Low pressure systems, producing monsoon rains, are therefore confined to the warmer, western Pacific. During an El Niño event, winds shift so that surface water is no longer pushed away from the west coast of South America. Deep-water upwelling diminishes, or ceases, and seas surface water and air temperatures rise in the eastern Pacific, producing low atmospheric pressures. This results in the monsoon rain belt extending eastwards into the central Pacific, creating drought in the western Pacific and flooding in the east. In addition to these regional changes, El Niño produces changes in the circulation of the upper atmosphere as dense, tropical rain clouds rise, altering the positions of both global rain belts and jet streams. The net effect is to produce unseasonal weather in many parts of the world, including shifts in the frequency and duration of floods and droughts

Habitat persistence is another important aspect of temporary waters, although its strict definition 'being permanent' (Oxford Concise Dictionary), seems at odds with the basic nature of temporary waters. But it is necessary to distinguish between the persistence of an individual temporary water body and the persistence of that type of temporary water body within a given locality; the former may be short-lived (e.g. a rain puddle created in a tyre track), whereas the latter may persist over a much longer timespan (e.g. pools on the floodplain of a major river). Such floodplain pools may be short-lived, individually, but, collectively, they allow establishment of metapopulations of suitably adapted temporary water biota on the floodplain. Indeed, a hypothesis was aired in Chapter 1: that temporary waters represent very ancient habitats that may have played a significant role in the origins of life (W.D. Williams 1988).

Many equate habitat persistence/permanence with habitat stability, for example, the stability of ancient Lake Baikal is often cited as the main reason for the preservation of its endemic flora and fauna. In terms of stability, then, how do temporary waters compare with, for example, freshwater springs, which, with their classically constant discharge and uniform water temperatures, would seem to be at the opposite end of a stability/permanence axis? Although the field data are limited, some tentative comparisons can be made at the regional level. In southern Ontario, permanent cold-water springs are dominated by nemourid stoneflies, chironomids, caddisflies, mites, copepods, ostracods, and amphipods (Figure 2.9). With the exception of the mites and chironomids (e.g. 20 species and 10 genera, respectively, present in Valley Spring, Ontario; Williams and Hogg 1988), diversity within these taxa is not high. For example,

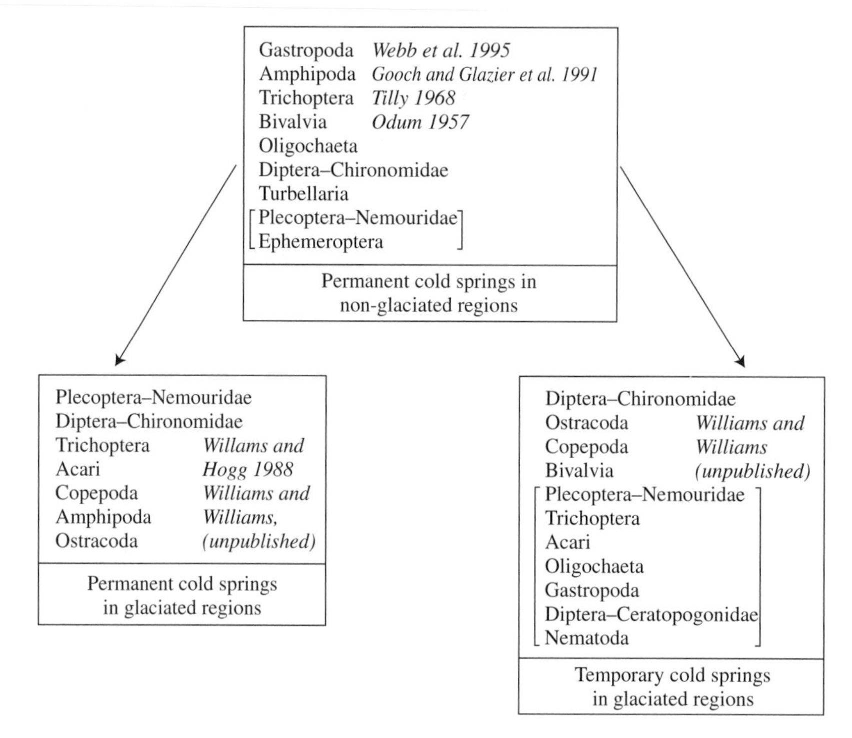

Figure 2.9 Comparison of the dominant invertebrate taxa known to occur in temporary and permanent, cold-water springs in southern Ontario, Canada, together with those found in permanent springs in non-glaciated regions. In each box, the taxa are ranked from top to bottom in the approximate order of numerical dominance seen in that spring type. Taxa enclosed within the square brackets are less common but usually present. It is possible that, due to non-standard collecting mesh size, the microcrustaceans may have been underrepresented in some of the studies from which the data were gathered.

there are usually only one or two species of stonefly and amphipod per spring, although population densities can be quite high. Intermittently flowing, cold-water springs in southern Ontario have a fauna that is taxonomically similar to that found in permanent springs, however, there is a predominance of chironomids, ostracods, copepods and sphaeriid clams. Although moderate numbers of caddisfly larvae are sometimes present, they do not match the dense populations recorded from a temporary spring in California in which trichopterans dominated the fauna (Resh 1982).

Figure 2.9 also compares the faunas of these southern Ontario springs with those from adjacent regions of eastern North America that were largely unaffected by recent glacial activity (see Matthews 1979). The latter, geologically more persistant habitats, are dominated, numerically, by gastropod and bivalve molluscs, triclads, amphipods, oligochaetes, and trichopteran and chironomid larvae. Nymphs of nemourid stoneflies and mayflies are present often, also. The diversity of species within some of these taxa may be very high, for example, 17 species of Oligochaeta in Old Driver Spring, southern Illinois (Webb *et al.* 1995). The insect/non-insect dominance seen between permanent, cold-water springs in Ontario and those to the south may well be related to respective glacial histories, with the more vagile insects predominating in less persistant habitats in regions that have been subject to recent glacial activity.

CHAPTER 3

Influential environmental factors

3.1 Introduction

As was pointed out in Chapter 1, there are indications that variations in the abiotic environments of inland waters strongly influence molecular and morphological evolution by altering mutation rates and exposing cryptic variation. Extreme environments, such as temporary waters, are likely to be particularly influential, and have been hypothesized as foci of biological diversification. Major parameters in the physical and chemical environments of temporary freshwater ponds and streams are summarized in Figure 3.1. Some are dealt with below under separate headings, but others, such as light and pH, are interactive and are discussed in multiple contexts. Biological influences are introduced at the end of the chapter.

3.2 Water balance

Of fundamental importance to the existence of any body of water is the result of the balance between water gain and loss. For a permanent stream or pond, water input equals water loss. However, as noted in Chapter 2, there are several sources of input (surface run-off, groundwater flow, precipitation, etc.) and several forms of output (loss downstream, absorption by the soil, evaporation, uptake by plant roots, etc.), the magnitudes of which vary in space and time causing water levels to fluctuate. Waterbodies in which input and output rates are highly variable are frequently temporary.

The exact length of the aquatic phase (the hydroperiod) varies according to both geographic location and local hydrological conditions, and is also related to water depth and surface area. For example, in the tropics, temporary pools and streams contain water immediately after the monsoons but may soon lose water due to evaporation by the sun; in areas of extended high rainfall, the life of the water body will likely be extended until the dry season. At high latitudes, although evaporation may not be as severe, a pond may dry up for a short period in late summer but may also 'lose' its water in winter due to freezing solidly to its bed. By way of illustration, Figure 3.2 shows the water level and some associated variables in Sunfish Pond, Canada. This temperate zone, intermittent pond receives most of its water in spring from snow-melt. At this time, loss to surrounding soil is negligible as the pond margins are either saturated or still frozen. Evaporation is also small because of diminished solar radiation. As time passes, solar radiation intensifies, air and water temperatures rise and evaporation increases. Simultaneously, the groundwater table drops and the water level in the pond decreases. Eventually, in mid-summer, the pond becomes dry. Occasionally, very heavy summer rainfall may cause the pond to fill temporarily (a few days), but usually the pond remains dry until the following spring. In years with low snowfall or an early, warm spring, the hydroperiod may be shortened. Conversely, in years with high snowfall or a late, cold spring, it will be extended. Wiggins *et al.* (1980) termed this kind of pond 'vernal', to distinguish it from intermittent 'autumnal' ponds which flood in the autumn and remain wet until the following summer.

Temporary streams, as noted, experience several distinct phases in their hydroperiod: surface flow connected to the GWT, reduced flow disconnected

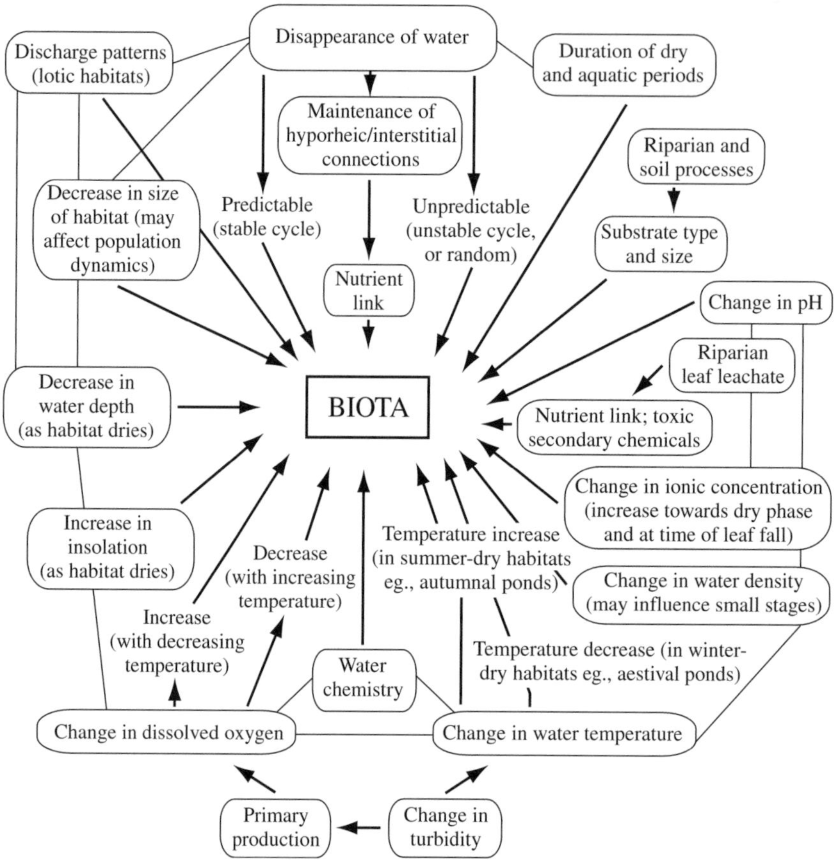

Figure 3.1 Summary of the major physical and chemical factors that influence the biotas of temporary freshwaters (from Williams 1996; an arrow indicates that the 'boxed' factor has been shown, or is likely, to have an effect on organisms—sometimes indirectly through another boxed factor; interactions between factors are shown by the lines joining the factor boxes).

from the GWT, isolated pools (detention storage), and dry bed (Figure 2.4). During the latter, heavy rains may again allow brief surface flow. A number of discharge-related factors, shown in Figure 3.1 (e.g. temporal pattern, maxima–minima, maintenance of connection with the hyporheic zone), are likely to influence the biota both directly and indirectly; the latter through correlation with several of the other physicochemical factors (such as water depth and dissolved oxygen content) shown in the model. The pattern of water loss is important particularly in terms of whether it is predictable (i.e. part of a stable cycle) or unpredictable (i.e. random, or linked to vagaries of local climate). One might assume that adaptations to deal with predictable disappearance of the water (intermittent waters) evolve more readily than those required to deal with random disappearance (episodic waters). However, a strategy such as 'bet-hedging' (Stearns 1976; Baird *et al.* 1987; see Chapter 5) would seem suited to both situations—although it might optimize rather than maximize reproductive effort in episodic waters.

Comparison of the invertebrates living in three water-filled ditches in southern Ontario (Table 3.1) indicates a decrease in species richness with decreasing length of the hydroperiod. A similar relationship between species richness and flow duration has been shown for some streams in Australia (Boulton and Suter 1986). However, in a

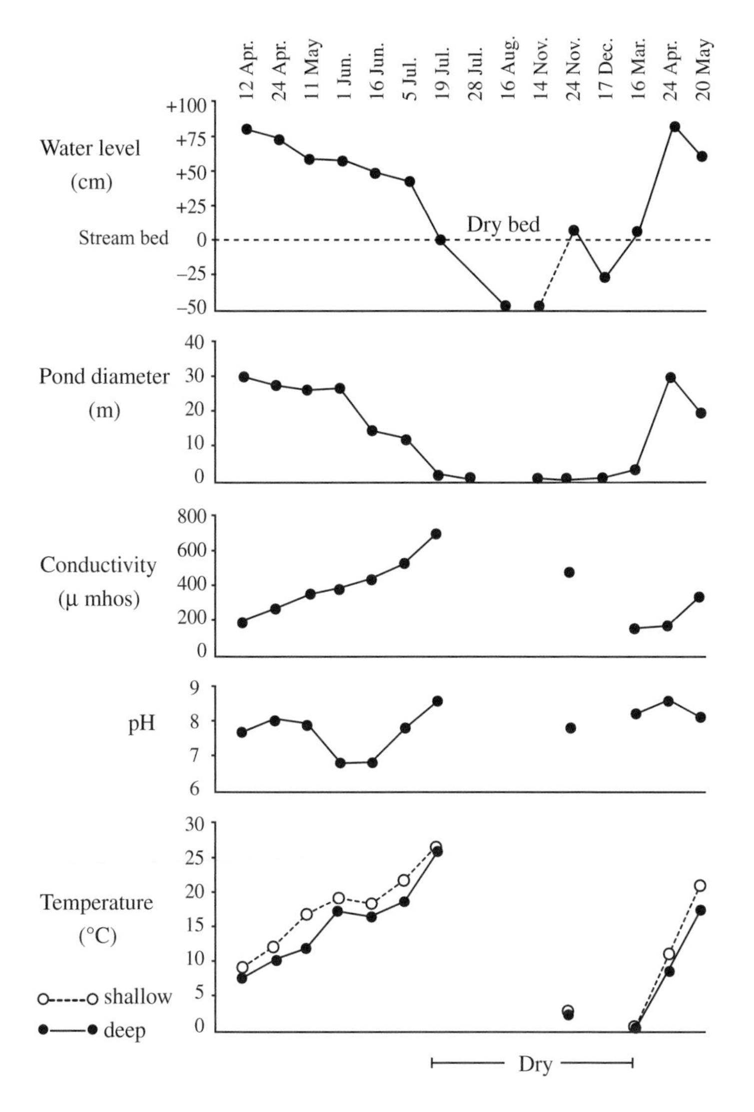

Figure 3.2 Characteristics of the hydroperiod and some associated parameters in Sunfish Pond, an intermittent, vernal pond.

study of six, small upland streams in Alabama, Feminella (1996) found that invertebrate assemblages differed only slightly, despite large differences in flow permanence: 75% of species (primarily insects) were common to all of the streams, or showed no relationship to permanence; only 7% were exclusive to the normally intermittent streams. Indeed, inter-year differences in assemblages within single streams were as great as within-year differences among streams of contrasting permanence.

In a comparison of crustacean diversity and hydroperiod in standing waters in North America (Table 3.2), Williams (2002) concluded that: (1) maximum taxon richness occurred in habitats with a hydroperiod of between 150 and 250 days; (2) some additional taxa occurred in habitats with hydroperiods of between 70 and 150 days, but no additions (to the genus level) occurred in habitats wet for less than 70 days; and (3) richness of the classic, three, large temporary water branchiopod groups (Anostraca, Notostraca, and Conchostraca) fell dramatically in habitats containing water for more than 250 days per year.

In floodplain ponds of the Mississippi River, species richness and abundance of predatory

Table 3.1 Comparison of the insect and non-insect faunas of three Ontario water-filled ditches with different lengths of hydroperiod, shown as the number of months in the year in which water is present. Absence of a particular taxon is shown by a dash

Taxonomic group	Length of wet period (months)		
	11	8	6
Collembola	—	*Isotomurus* sp.	*Isotomurus* sp.
Hemiptera	—	*Gerris buenoi*	—
	Sigara sp.	*Sigara* sp.	—
Trichoptera	*Ironoquia punctatissima*	—	—
	Hydropsyche betteni	—	—
	—	*Limnephilus rhombicus*	—
Coleoptera			
Hydrophilidae	—	—	*Helophorus orientalis*
	—	—	*H. grandis*
	—	—	*H. lacustris*
Elmidae	*Stenelmis* sp.	—	—
	Optioservus sp.	—	—
Diptera			
Chironomidae			
Tanypodinae	—	*Natarsia* sp.	—
	Thienemannimyia sp.	—	—
Diamesinae	*Diamesa* sp. A	—	—
	—	*Diamesa* sp. B	—
Orthocladiinae	*Cricotopus/Orthocladius* sp.	—	—
	Orthocladius sp.	—	—
	—	*Cricotopus* sp. A	—
	Cricotopus sp. B	—	—
	—	—	*Pseudosmittia* sp.
Chironominae Chironomini	*Endochironomus* sp.	—	—
	Micropsectra sp.	*Micropsectra* sp.	—
	—	*Polypedilum* sp.	—
	—	*Paratendipes* sp.	—
Tipulidae	*Antocha* sp.	—	—
Ceratopogonidae	*Bezzia/Probezzia* grp	*Bezzia/Probezzia* grp	—
Simuliidae	*Simulium venustum*	—	—
Dolichopodidae	—	—	Unident
Syrphidae	*Tubifera* sp.	—	—
Total insect taxa	16	11	6
Other invertebrates	15	10	8

invertebrates increased with increasing hydroperiod, whereas overall invertebrate abundance and richness decreased (Corti *et al.* 1997). Boix *et al.* (2001) similarly found a positive relationship between hydroperiod duration (and also flooded surface area) and species richness in Espolla Pond, Spain. Hershey *et al.* (1999) found that, in the prairie wetlands of Minnesota, for some groups (especially insects) species counts decreased with shortening hydroperiod, however, for others (e.g. molluscs) there was no relationship. In ponds in central Italy, Bazzanti *et al.* (1997) have shown that in those with a long hydroperiod (but also lower oxygen content), the chironomid assemblage was dominated by members of the Chironominae and Tanypodinae. A pond with a shorter hydroperiod,

Table 3.2 Occurrence of crustaceans in North American temporary waters in relation to length of the hydroperiod

	Mean length of hydroperiod (days)					
	<10	10–40	40–70	70–150	150–250	250–330
Number of studies	2	2	4	10	11	17
Taxa						
Branchiopoda						
Anostraca						
Artemina					*	*
Chirocephalopsis				A		
Branchinecta					*	
Eubranchipus		*	*	*	*	
Streptocephalus					C	
Thamnocephalus					*	
Notostraca						
Triops					*	
Conchostraca						
Caenestheriella					*	
Eocyzicus					*	
Eulimnadia					*	
Leptestheria					*	
Limnadia					C	
Lynceus			*	A	*	*
Cladocera						
Alona					C	
Alonella				*	*	*
Bosmina						R
Chydorus				C	*	*
Chydoridae				C		
Ceriodaphnia				C		
Daphnia		*	*	*	C	*
Diaphanosoma					C	
Macrothrix					C	C
Moina					C	
Pleuroxus						*
Polyphemus					C	
Pseudosida					C	
Scapholeberis	*	*		*	C	*
Sididae						R
Simocephalus			R	C	C	*
Ostracoda						
Candona				A	*	
Cypria				A		
Cypricercus				C		
Cypridopsis						*
Cyprinotus					*	
Cyclocypris					*	
Eucypris				C		
Limnocythere					*	
Megalocypris					C	
Pleocypris					*	
Potamocypris					*	

Table 3.2 (*Continued*)

	Mean length of hydroperiod (days)					
	<10	10–40	40–70	70–150	150–250	250–330
Copepoda						
Calanoida				*	*	*
Diaptomus				C		
Leptodiaptomus			C	C	C	
Harpacticoida					*	*
Canthocamptus				A		
Cyclopoida	*	*	*	*	*	*
Acanthocyclops				C		
Cyclops				A		
Siphonostomatoida						
Argulus					*	*
Malacostraca						
Isopoda						
Caecidotea				C	*	*
Oniscus				R		
Amphipoda						
Crangonyx					*	C
Gammarus						C
Hyallela					*	C
Decapoda						
Palaemonetes					C	*
Cambarus				C	*	R
Procambarus					*	C

Abbrevations: A = Abundant, C = Common, R = Rare, * = Present (not quantified).
Source: Based on data in Batzer *et al.* (1999).

but also higher oxygen, was dominated by Orthocladiinae. Suter *et al.* (1994) found that chironomids responded to flooding and drying events in the floodplain wetlands of the Lower Murray River in South Australia. Notably, species of *Chironomus* responded rapidly to flooding, and species of *Procladius* (Tanypodinae) became more abundant as drying progressed. However, these authors cautioned that while, for some species, the responses were consistent across several locations, for others they were not. Timing of flooding, salinity variations, and colonization sources were all thought to have some influence.

Temporary coastal rockpools in Jamaica exhibited low variability in community structure, and were dominated by 'weedy' species with good powers of dispersal and colonization (Therriault and Kolasa 2001). Schneider (1999) concluded that the invertebrate communities of short hydroperiod, snowmelt ponds in Wisconsin were structured primarily by species adaptations to the threat of drying (i.e. to abiotic factors). On the other hand, the communities in ponds with longer hydroperiods were structured more by biotic interactions, particularly predation and competition. Greater consideration of such r- and K-selected species will be left until Chapter 5.

Working on the largely crustacean-dominated pans of South Africa, Meintjes (1996) found that larger pans supported more species than smaller pans. Decreasing water volume in tree hole habitats has been shown to induce metamorphosis at smaller sizes in females of the mosquito *Aedes triseriatus*, indicating that larvae can detect changes in the hydroperiod and take appropriate steps to escape a deteriorating environment (Juliano and Stoffregen 1994). As to the mechanism involved, a laboratory simulation experiment using the frog *Rana temporaria* demonstrated that tadpoles were able to respond to pond drying by speeding up their development through behavioural mediation, irrespective of any change in water temperature (Laurila and Kujasalo 1999).

3.3 Water temperature and turbidity

Temperature is a very important environmental variable, and not just seasonally—as it also fluctuates on a daily, or even hourly, basis. Because temporary waters are typically shallow, they are highly susceptible to rapid heating (from solar radiation) and cooling (at night and also from wind). Temperature inversions together with kinetic energy transfer from wind blowing over the water surface set the water column in motion, which in turn stirs up bottom materials.

Turbidity is thus a variable closely linked with temperature. Figure 3.3, shows that a section of a pond with low turbidity absorbs significantly more heat close to the bed, especially if the latter is darkly coloured, than a more turbid section. Whereas low turbidity leads to a more uniform temperature with depth, high turbidity leads to greater absorbance of heat at the surface (typically the upper 5–6 cm; Figure 3.4). The source of turbidity varies. In the Oklahoma pond it was suspended clay particles as well as microscopic organisms. Chromogenic bacteria are responsible for the red colour of many saline desert ponds, and purple, sulphur bacteria often occur in shallow, stratified waters (Cole 1968).

The annual temperature regime of Sunfish Pond is shown in Figure 3.2. Temperature increases from a post-snowmelt value of around 8°C to a maximum of 27°C in mid-summer. There is little evidence of stratification in this clear-water pond, with surface water only ever being a degree or two warmer than that near the bed. Some shallow waterbodies experience a daily temperature turnover similar to that seen annually in permanent

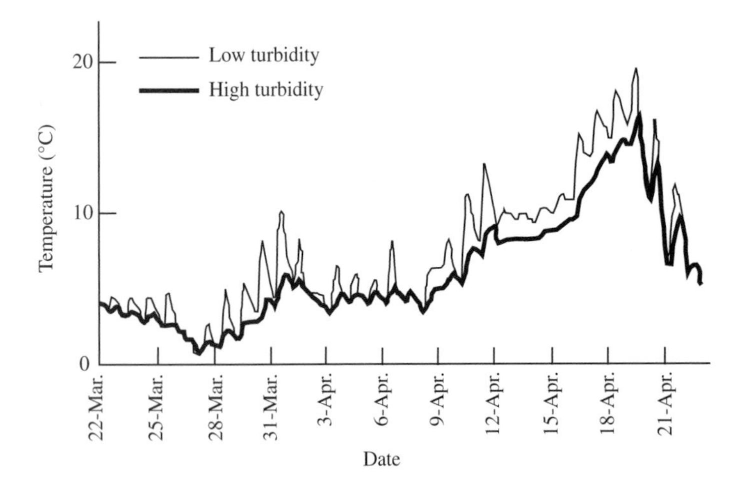

Figure 3.3 Temperature profiles for two sections of an intermittent pond (Vandorf Pond II, Ontario) that differ in turbidity level (low = 13 FTU; high = 32 FTU). Temperatures were measured close to the bed, thus readings were higher in the low turbidity section (deeper light penetration) compared with the high turbidity section. (Data from Magnusson and Williams, unpublished).

lakes (Eriksen 1966). Figure 3.5 shows the broad seasonal temperature profile of an intermittent, temperate pond, together with the diel variation measured at three points in its hydroperiod. While mean temperature from early April to late June ranged from around 4 to 19°C, the diel fluctuations were 2–20°C (early April), 10–28°C (mid-May), and 17–24°C (mid-June). The biological consequences of such massive, short-term temperature changes are poorly understood. To take an extreme example, the surface water of shallow ponds in temperate regions may, on occasion, approach 40 °C in mid-afternoon in summer. This is very near the thermal death point of most insects. Young and Zimmerman (1956) found many aquatic beetles active in such ponds in Florida even though the predominant aquatic vegetation (*Chara* spp.) was dead. Typically, however, these insects do not stay near the water surface, but congregate under debris or burrow in the bottom mud. They return to the surface at night and in the early morning, to forage, when the water is cooler. Among dytiscid beetles, species composition is known to change as pond temperatures rise (Nilsson and Svensson 1994). Eisenberg *et al.* (1995) have suggested that populations of the mosquito *Culex tarsalis* from different ponds in southern California have evolved separately to maximize survival in their respective temperature regimes by adapting to different optimal larval survival temperatures and egg-development rates.

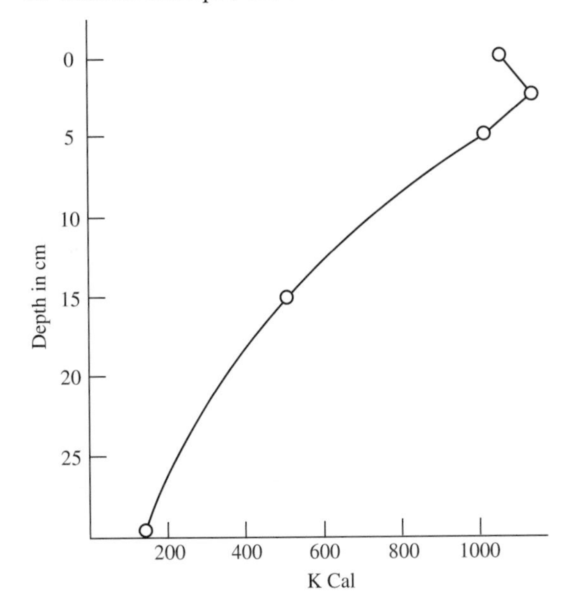

Figure 3.4 Heat absorption by 1-cm strata in a shallow intermittent pond (180 ppm (~80 FTU) turbidity) in Oklahoma (redrawn from Butler 1963).

In arctic and alpine regions, the effect of temperature is also marked but the conditions produced are quite different. Daborn and Clifford (1974) found that the first persistent ice cover in autumn, in a pond in western Canada, significantly reduced the normal close correlation between water and air temperatures. Initially, diurnal changes disappeared but subsequently reappeared as soon as the pond had frozen to the bottom. This was thought to be a function of lower thermal conduction across an ice–water interface than through ice alone, and of variation in snow

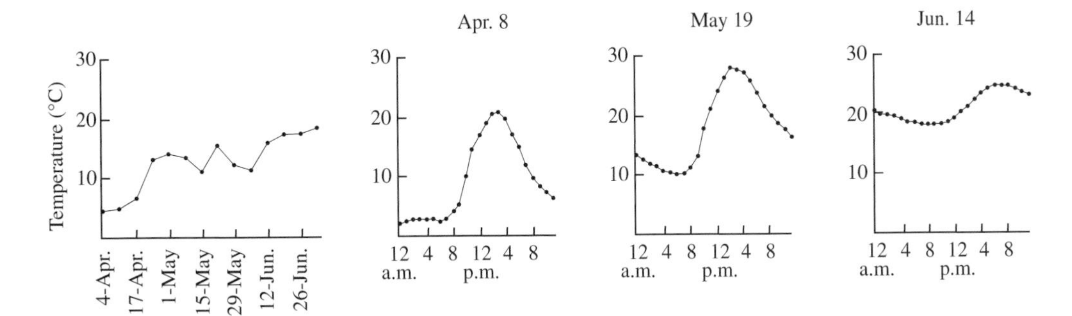

Figure 3.5 Comparison of seasonal (mean values) vs. diel (measured on 3 separate days) temperature fluctuations in an intermittent pond (Vandorf Pond I, Ontario) (data from Magnusson and Williams, unpublished).

depth on top of the ice. Ice temperatures as low as −8°C were recorded and aquatic invertebrates, such as damselfly nymphs, survived embedded in the ice.

3.4 Dissolved oxygen and carbon dioxide

Dissolved oxygen in temporary waters may fluctuate diurnally as a result of photosynthesis and respiration. Whitney (1942) found this oxygen pulse to be at a maximum just after dark, when the day's photosynthesis had ended, but thereafter it fell gradually due to overnight respiration. He concluded that, in many cases, absorption of oxygen from the air was of relatively minor importance, as often absorption values were far below the air saturation value for a particular temperature. Further, the oxygen content of the water frequently changed during a period when a uniform temperature prevailed. Schneller (1955) found that during the low flow stages of Salt Creek, Indiana, large quantities of decaying leaf matter were sufficient to cause an oxygen depletion combined with an increase in free carbon dioxide from the activities of decomposers. Surprisingly, few mortalities among the fish species present were reported. In contrast, George (1961), in a study of the diurnal variation of oxygen in two shallow ponds in India, found that, in one, the levels varied between 4.5 and 9.9 ppm, being maximal at 5:30 p.m. and minimal at 5:30 a.m.; these times agree closely with the 6:00 a.m. minima and 4:00 p.m. maxima measured in a pond on the Ivory Coast (Guiral *et al.* 1994). In the other Indian pond, the corresponding variation was from 0.1 to 28.2 ppm. These increases were attributed to photosynthesis, and the quantity of oxygen removed at night by community respiration (mostly due to the blue-green alga *Microcystis aeruginosa*) was sufficient to cause a fish kill. Podrabsky *et al.* (1998) similarly measured oxygen maxima (e.g. 256% saturation) in mid- to late-afternoon, with minima (e.g. 2%) between 9 and 10 a.m., in rainwater pools in Venezuela. Annual killifish (*Austrofundulus* spp., *Pterolebias* spp., and *Rachovia* spp.) embryos living in these pools were thought to be exposed, both intermittently and chronically, to hypoxic or anoxic conditions.

Decreased oxygen, alongside increased water temperature, has been shown to increase the level of brooding behaviour in the amphipod *Crangonyx pseudogracilis*. This species is a common inhabitant of temporary waters and its response to oxygen stress consists of increasing embryo ventilation within the brood pouch together with selective ejection of non-viable eggs. Some other amphipods that live in harsh environments show similar behaviours (Dick *et al.* 1998).

Cerny (cited in Vaas and Sachlan 1955) found that in some small ponds in Europe, the amount of daily photosynthesis could completely exhaust all of the available carbon dioxide. pH may rise as a result of this depletion—although the magnitude of pH change would depend not only on the intensity of the photosynthesis, but also on the degree of buffering available, for example, from surrounding alkaline soils. Eriksen (1966) showed that turbidity could cause stratification in the above parameters, as suspended material limits the penetration depth of sunlight thus restricting photosynthesis to the upper layers. This, in turn, might lead to oxygen, carbon dioxide, and pH stratification. Decreasing depth as the hydroperiod progresses may increase insolation which, again through changes in primary production, but also wind-induced aeration, may change the concentrations of dissolved gases in the water.

In various temporary waters, oxygen levels are also know to become depleted rapidly soon after inundation, as basin sediments and soils become flooded. Renewed microbial activity removes the oxygen, creating a reduced redox state in the bed (Sposito 1989).

In a study of the physicochemical properties of four intermittent ponds in southern Ontario, Magnusson and Williams (2006) determined that, overall, dissolved oxygen, along with pH, turbidity, and nutrients (nitrate, ammonia, and total phosphorus) showed large fluctuations seasonally, among ponds, and between years. However, these parameters were more stable in ponds III and IV, which had longer hydroperiods (Figure 3.6).

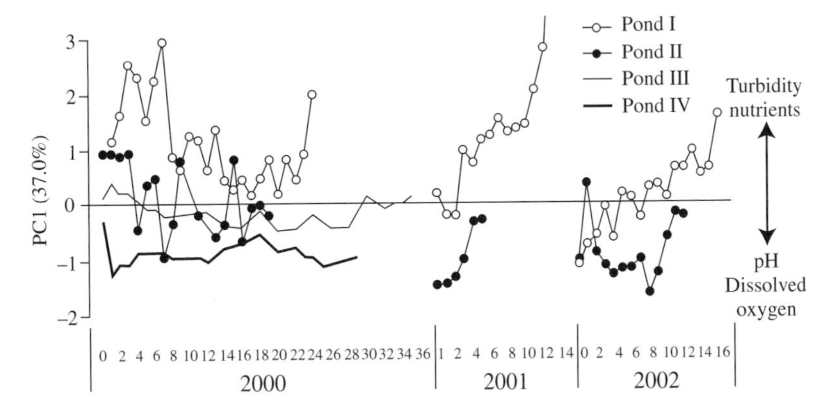

Figure 3.6 Plot of Principal Components Analysis-scores for the physicochemical parameters in four intermittent ponds in southern Ontario, over the period 2000–2002 (the most important eigenvector coefficients are given on the right-hand side of the graph; ponds III and IV show the smallest fluctuations along the axis, indicating more stable conditions in these two ponds; data from Magnusson and Williams 2006).

3.5 Other chemical parameters

The concentrations of dissolved substances in temporary waters vary more than in most permanent waters. This is due, largely, to three physical processes to which temporary waters are subjected: drying out, refilling, and freezing.

Taking Sunfish Pond as an example (Figure 3.2), from the time that it fills in early spring to just before it dries up the conductivity of the water typically increases four-fold. This concentration of chemical ions is due primarily to evaporation. Cole (1968) described the changes in water chemistry that occur as this proceeds—the normal trend is from carbonate to sulphatochloride to chloride water as calcium carbonate and, subsequently, calcium sulphate and possibly sodium sulphate are precipitated. The loss of calcium ions is accompanied by relative increases in sodium and magnesium ions. Some calcium remains however, as calcium chloride—as the latter is extremely soluble. Even this, though, may precipitate out of solution in ponds in extreme climates. Cole cites the example of Don Juan Pond in Antarctica. This is a calcium chloride pond, only 10 cm deep, situated in a very dry, cold region. Needle-like crystals of calcium chloride hexahydrate form on the pond bottom and in the water. Species richness in such ponds is likely to be very low. Recent work on the ciliates of a salt pan in Spain has provided some information on how species composition is influenced by such strong chemical variables (Esteban *et al.* 2002). At its normal concentration of 8.1% sodium chloride (~2.5 times that of sea water) the Spanish pan water contained only seven species. However, upon dilution with filtered fresh water in the laboratory, more species began to appear from cysts, ultimately producing a total of 34 species. Species appearance was clearly controlled by water chemistry rather than by habitat accessibility and, additionally, provided evidence to support the contemporary belief that most microorganisms are cosmopolitan in nature and, globally, are not represented by vast numbers of species.

When temporary waters first fill after the dry phase, the water is likely to have a chemical signature that changes soon thereafter—although this has not been well studied. McLachlan *et al.* (1972) recorded the events as Lake Chilwa, a shallow, saline lake in central Africa, filled after having been virtually dry for over a year. Refilling took 5 months during which time there was a rapid dilution of initially high levels of dissolved salts, and a high suspended sediment load resulting from erosion of the lake bottom; this turbidity gradually decreased over a further 2-year period. Increased turbidity is often evident also during the spring thaw of temporary ponds and streams in temperate regions. Initial nutrient release from the mineralization of dead tissues accumulated during the dry phase takes place largely at the sediment surface. Thus, at first and where light levels are suitable, near-bottom-dwelling autotrophs are likely to be favoured, especially green and blue-green algae. In intermittent ponds, this may be followed, as nutrients circulate into the water column, by an increase in phytoplankton which may accelerate into blooms that, in eutrophic waters, may compromise subsequent periphyton production through reduced light penetration (Brönmark and Hanson 1998). Light penetration may also be affected by the release of tannins from leaves that have accumulated on the bed, especially in the autumn in temperate regions. In addition, Cameron and LaPoint (1978) showed that although not directly toxic, the tannins from Chinese tallow (*Sapium sebiferum*) inhibited feeding in the isopod *Asellus militaris* and the amphipod *Crangonyx shoemackerii*, causing high mortality in both.

In intermittent streams with permeable substrates, below-bed flow has been shown to be an important source of nutrients, particularly nitrogen. Grimm *et al.* (1981) proposed that biological production in a stream should be greatest at points on the bed where subsurface water rises to the surface. This model was subsequently validated by Valett *et al.* (1994) who showed that hyporheic upwelling could locally stimulate the productivity of surface algae, and probably bacteria (Dent *et al.* 2000), with benefits to grazing invertebrates. In other streams, however, hyporheic sediments have been shown to remove nutrients, for example, through denitrification (Duff and Triska 2000). Dieterich and Anderson (1998) have pointed to the potential, applied importance of some summer-dry streams for the efficient removal of nutrients and pollutants.

Concentration of chemicals in temporary waters may occur also as the result of freezing. Daborn and Clifford (1974), in a study of a shallow aestival pond in western Canada, found that, as the winter ice cover thickened and the volume of the pond decreased, there was a rapid rise in conductivity, alkalinity, water hardness (calcium and magnesium ions), sulphate, and orthophosphate. They found this cryogenic 'salting out' to be a consequence of the stable and selective nature of ice crystals—only a few chemical elements or compounds have an appropriate configuration that allows them to become incorporated into the crystal structure of ice. They further observed that the quantity of dissolved and particulate matter in the top layer of ice influenced the pattern of ice break up in the spring—primarily through control of the penetration of sunlight.

The chemical composition of temporary waters may also be influenced by rainwater. Paradise and Dunson (1998) monitored tree holes in three regions of Pennsylvania subject to high, but different inputs of hydrogen and sulphate ions. Although no regional trends were detected, tree hole insect densities and species richness were related to water volume, the concentrations of sulphate and sodium ions, and to dissolved organic carbon. More complex chemicals are known to influence the behaviour of temporary water inhabitants. For example, oviposition by the mosquitoes *Aedes aegypti* and *Ae. albopictus* has been shown to be significantly higher when females were exposed to water in which larvae had been previously reared (Allan and Kline 1998). Also, the percentage of eggs hatching of another mosquito, *Anopheles diluvialis*, was largest in infusions made from intermittently flooded swamp soil, and in hexane extracts of swamp water, suggesting that the hatching factor is a chemically stable organic compound (Jensen *et al.* 1999).

Inland, saline waters of course have significant chemical signatures that need to be considered. They are included in this book because not only do many of the small ones dry completely, annually, but larger, more permanent ones experience substantial changes in water level, thus producing intermittently wet habitats in their littoral zones. Salinity is thought to be an important ecological determinant, and has been highly correlated with species richness and composition. However, W.D. Williams *et al.* (1990) found that although this relationship held over the salinity range of 0.3–343 $g\,l^{-1}$ in Australian salt lakes, it broke down over intermediate ranges. Many taxa found in the latter had very broad tolerances, suggesting that other environmental factors (both abiotic and biotic), together with chance (stochasticity), determined their distribution. These authors cited Herbst's (1988) study of the population dynamics of *Ephydra hians*, the brine fly, as a good example of how population dynamics are affected by different factors at different salinity levels. Larvae of this species in North American lakes are found over the range 20–200 $g\,l^{-1}$. At the lower end of this scale, larval densities seem largely to be limited by biotic influences, such as predation and competition. Near the upper end, however, they are constrained by physiology. Salinity *per se* did not seem to be an important determinant of this species' abundance.

Tibby *et al.* (2000) outlined the problem of determining whether whole community response in hydrologically closed systems, such as lakes and wetlands, is driven by the direct physiological effects of salinity stress, or by the restructuring of microhabitats that results from changes in water level and salinity. They extracted a core from the bed of a shallow, fluctuating lake in Kenya and analysed the fossil macroflora, diatoms, invertebrates, and sediment, representing the period 1870–1991. The data showed distinct species-specific responses to water level, salinity, and the development of papyrus swamp. Significantly, the latter two explained 51% of the observed historical variation in benthic community composition. The authors concluded that a significant proportion of the published correlation between salinity and invertebrate community structure along the full gradient of inland aquatic ecosystems may be an indirect effect of broad, but diffuse, relationships between salinity and the distribution of various types of benthic microhabitat. Thus local populations may well be regulated by the fluctuating availability of specific habitat and associated resources rather than by the limit of their osmoregulatory capacity. Details of the communities found in saline temporary waters will be given in Chapter 4.

3.6 Substrate

In both lotic and lentic temporary waterbodies, the size and type of bed substrate particles are likely to exert influences ranging from suitability as attachment sites during the wet phase to protection from desiccation during drought (Boulton 1989). As an example of the latter, Tabacchi *et al.* (1993) demonstrated that nymphs of the mayfly *Thraulus bellus* seek out the water-filled substrate interstices of temporary floodplain ponds as a refuge from fluctuating water levels. They also suggested that this behaviour may prevent overlap with populations of two other mayfly species, *Caenis horaria*, which lives in silt, and *Cloeon* gr. *simile*, which lives among macrophytes. Substrate composition also may be important to species survival. For example, the organic cases of larvae of the caddisfly *Limnephilus coenosus* proved to be better at holding water, and hence reducing mortality, in drying pools than the mineral cases of *L. vittatus* (Zamoramunoz and Svensson 1996).

In Chapter 2, it was described how various physical and chemical properties of vernal pool soils control the manner in which water drains from and returns to the basin (Weitkamp *et al.* 1996). Although, intuitively, the presence of a wetted substrate both prior to and at the end of the true hydroperiod would seem beneficial to the survival of early and late colonizers, little empirical evidence is available to support this. There is evidence, however, that soil type in flooded meadows influences microbial activity. Alternate flooding and draining has been shown to enhance nitrogen removal through stimulation of both nitrification and denitrification (Busnardo *et al.* 1992), with sandy soil removing more nitrogen than peaty soil. Infiltrating water may increase leakage of dissolved forms of both organic and inorganic nitrogen from the soil. Further, this water will also bring electron acceptors (nitrate-N, and dissolved oxygen) into the soil to promote mineralization (Davidsson and Leonardson 1998). Soil moisture and organic content in woodland floodplains has been shown to influence the abundance and distribution of larval tipulids (Merritt and Lawson 1981).

Apart from the role of microbes, there is a huge gap in our knowledge of other biological processes that must be presumed to take place in the wetted phases of temporary water basins. It is known that, in soils in general, besides bacteria, protoctists,

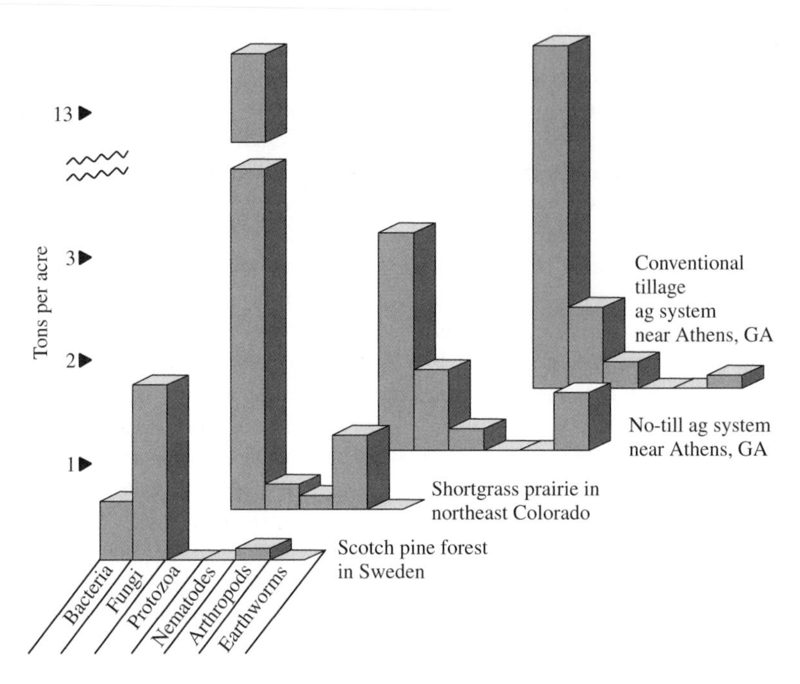

Figure 3.7 Data indicating that different soil types support different proportions of litter processing organisms (redrawn after Tugel and Lewandowski 1999).

and fungi, a number of metazoan groups are crucial to organic matter processing, particularly nematodes, earthworms, and arthropods. Among the latter, millipedes, insects, and a huge diversity of mites shred dead plant material, which begins the cycling of carbon, nitrogen and other nutrients by making available a large surface area for micro-organism colonization. These biological communities tend to be concentrated in the top several centimetres of substrate and around growing plant roots—as the rhizosphere attracts bacteria to sloughed-off plant cells and plant-released sugars and proteins. Fungal hyphae are common decomposers of plant debris and, unlike bacteria, can transport (via their spreading hyphae) nitrogen from the underlying soil to the litter layer. Fungi are also important as they are often the only organisms able to secrete enzymes capable of degrading complex compounds. Although nutrient release, which can benefit both terrestrial plants during the wetted phases and aquatic species when the water returns, is an important process, so also is the production of humus. Humus, or humidified organic matter, is what remains because it is not readily decomposed (it may be chemically too complex to be used by soil organisms, or it may be physically protected inside aggregates). Nevertheless, it is important in binding tiny soil aggregates and plays an essential role in improving the water and nutrient-holding capacities of the substrate. Different soil types support different proportions of litter-processing taxa (Tugel and Lewandowski 1999; Figure 3.7), which may well influence the organic matter breakdown rates and efficiencies in different temporary water basins, and hence perhaps the community composition and productivity of the aquatic phase. Further information on the organic component of bed substrates is given in Section 3.8.

3.7 Light

Light is of course an important factor in regulating primary production. Plant biomass lies at the base of most temporary water food webs, either as autochthonous algal and macrophyte tissue, or as the breakdown products of riparian leaf litter. Shoreline vegetation also may be influential in terms of shading, as this has been shown to contribute to the structure of the invertebrate communities of intermittent streams in Swedish meadows (J. Herrmann, personal communication). Photoperiod is a key aspect of light that is crucial to the survival of many temporary water species, as it provides essential cues that regulate the timing of life cycles, emergence from, and entry into, diapause, flight periodicities, and colonization dynamics. Responsiveness to photoperiod may vary with latitude, as has been shown for the pitcher-plant mosquito *Wyeomia smithii* (Wegis *et al.* 1997). More specific aspects of photoperiod will be covered in Chapter 5.

3.8 Biological factors

Brief reference, above, to the role of organisms in basin substrate processing and nutrient release brings us to consider some of the *biological* factors that influence the biotas of temporary freshwaters—summarized in Figure 3.8. Whereas it is not difficult to identify the influence of single biological factors, assessing their relative importance and linkage is less easy owing to lack of research specifically on temporary waters. Strong biological features include the succession of species so characteristic of temporary waters (e.g. Williams 1983; Jeffries 1994), seasonal influx of aerial colonizers (Fernando 1958; Nilsson and Svensson 1994), and temporal changes in the food base—as surplus bed nutrients from the dry phase are used up, aquatic-phase fungal and other decomposers continue to process allochthonous bed materials, and phyto- and zooplankton become established (Bärlocher *et al.* 1978; Wiggins *et al.* 1980; Maher and Carpenter 1984). As a result of changes in the food base, together with variation in the numbers of predators (both as aquatic species succeed each other and as terrestrial ones invade the shrinking habitat), the structure of the food web must be subject to change (see later discussion). So, too, will the extent and intensity of inter- and intraspecific competition (Morin *et al.* 1988; Johansson 1993), although it has been argued that, in general, biotic interactions may contribute relatively little to community structure given the

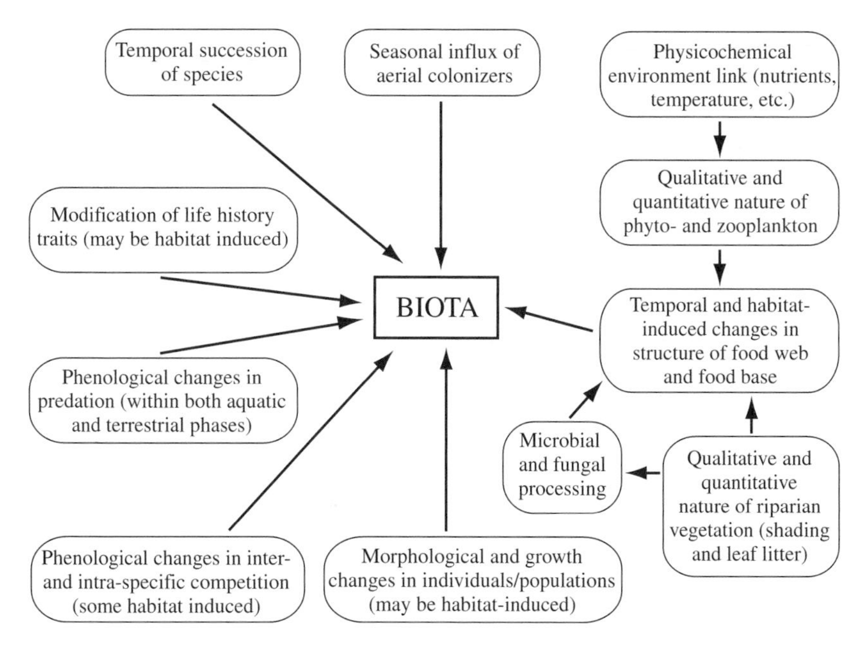

Figure 3.8 Summary of biological factors that influence the biotas of temporary fresh waters (from Williams 1996; see Figure 3.1 for explanation of arrows; note that interaction lines between factors have not been added due to uncertainty stemming from the lack of research on these factors specifically in temporary waters).

demanding requirements of the physicochemical environment (Poff and Ward 1989). In addition, a number of changes affect individual insects, such as modification of growth, morphology, and life history (Harrison 1980; Landin 1980; Juliano and Stoffregen 1994), that may be classed as genotype–environment interactions. More detailed coverage of these largely population- and community-level factors is left until Chapters 5 and 6.

A biological aspect that has received relatively little study, apart from the shading effect mentioned above, is the link between forested temporary waters and the surrounding vegetation. While plant biomass is known to lie at the base of such communities, the qualitative and quantitative influence of leaf-litter input is poorly known, relative, for example, to permanent waters (e.g. Hynes 1975; Lake 1995; Webster *et al.* 1995).

When a winter-frozen, intermittent pond begins to fill in the spring ('vernal' pond), degradation of both autochthonous (basin grasses and weeds) and allochthonous (riparian trees) leaf-litter begins; in the case of 'autumnal' ponds, this process may have begun, briefly, before winter. Degradation processes in intermittent waters can be inferred using flooded riparian woodlands as models. Merritt and Lawson (1992) identified various abiotic and biotic factors in the floodplain environment that influence litter processing (Figure 3.9), and have emphasized the importance of leaf 'conditioning'. Initially, autumn-shed leaves that have accumulated in the pond basin are subject to physical fragmentation via wet–dry rain cycles, winter freeze-thaw cycles, and wind. Moisture levels and ambient temperature and oxygen levels are important as the basin begins to fill with water, and microbes and fungi colonize and begin to degrade the leaf pieces. At this stage, soluble organic compounds (e.g. sugars, polyphenols, and amino acids) are leached into the surrounding water (Suberkropp and Klug 1976). Leaf chemical composition is important in controlling the rate of these various processes, as some species (e.g. oak)

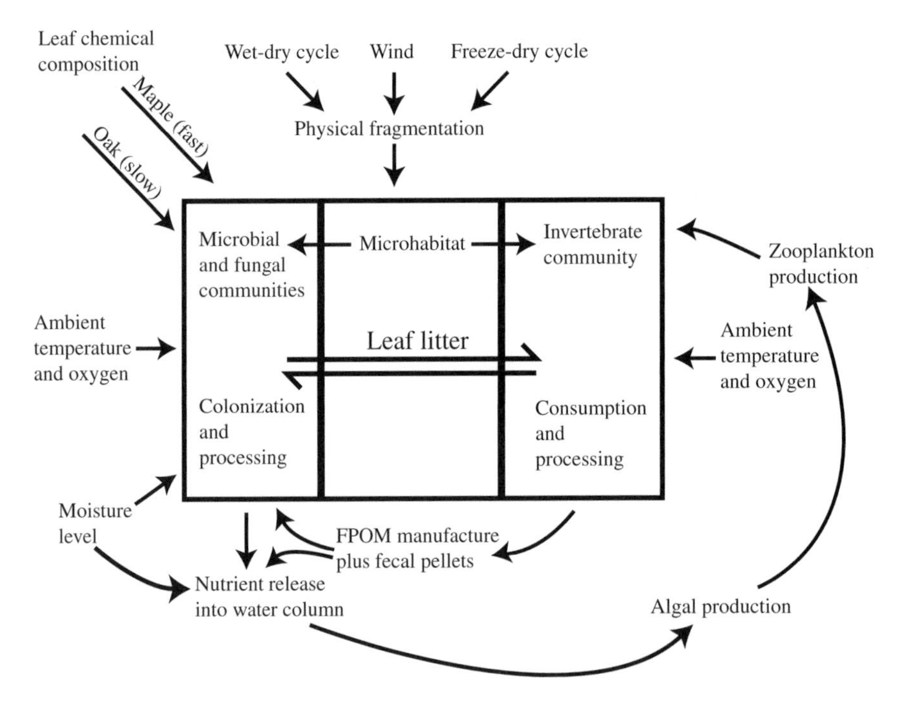

Figure 3.9 Schematic of the main abiotic and biotic factors associated with leaf-litter processing in forested temporary waters (based on Merritt and Lawson 1992).

are more resistant to conditioning than others (e.g. maple) (Kaushik and Hynes 1968). Release of leachates into the water column, together with other nutrients released by fungal and bacterial activity, contribute to an increase in benthic and planktonic algae, which in turn promotes the development of zooplankton populations. Leaf conditioning directly stimulates the growth of benthic invertebrate species, as aquatic hyphomycetes are know to be an important food source for many benthic invertebrates (Kaushik and Hynes 1971). As invertebrates consume and process leaf fragments, they manufacture FPOM (crumbs) and fecal pellets that are cycled back into the food web for consumption by others—frequently after recolonization by fungi and bacteria. In addition, whole leaves and fragments serve as microhabitats for pond inhabitants (e.g. as a physical substrate for microorganisms, and as shelter and attachment sites for invertebrates).

CHAPTER 4

The biota

4.1 Introduction

The global temporary water biota is best known in terms of its higher plants and metazoans—although current studies are beginning to catalogue and understand the roles of bacteria, protoctists, and even fungi. However, it is the insects and crustaceans that dominate the fauna. The latter are well represented by micro-and macro-forms from the following major taxonomic groups: Branchiopoda, Ostracoda, Copepoda, Decapoda, Peracarida, and many rare species are present. Among the insects, it is the Hemiptera, Coleoptera, Trichoptera, and Chironomidae (Diptera) that are best represented. Aquatic mites (Hydracarina) are also common. Although typically absent from standing temporary waters, several fish species migrate in and out of temporary streams, rivers, and floodplains. Several types of amphibian live in temporary waters, some of them very rare, and temporary waters provide important feeding areas for migratory birds. Tropical waters are more likely to have reptiles and mammals associated with them. This chapter will present summaries of each taxonomic group, providing examples of habitat-specific requirements and adaptations. The dearth of studies on the 'minor' taxa of temporary waters makes it difficult to discuss these groups in any other context than on the local scale. Case history studies of predominant temporary water types will be highlighted, ranging from the very large (floodplains) to the very small (rain-filled rockpools). Commonality in community composition, structure, and function will be sought. Whether any of the environmental influences discussed in Chapter 3 actually constrain the biota will be discussed.

4.2 Taxa

4.2.1 Prokaryotes

There are few inventories of named bacteria from temporary waters. Table 4.1 compares those found in an intermittent, snow-melt pool in northern Quebec with those in an episodic puddle in Singapore (Laird 1988). Although it is difficult to draw many conclusions from such a limited dataset, there are some similarities in community composition, despite the huge geographical separation and differences in predictability and lengths of their respective hydroperiods (the pool more than 3 weeks, the puddle no longer than 9 days).

Felton *et al.* (1967) found nine physiological groups of bacteria in a temporary pond in Louisiana, USA. Sulphur-, ammonia-, and nitrite-oxidizing autotrophs were notably absent. It was felt that bacteria did not make a major contribution to the pond ecosystem as primary producers. However, counts of aerobic nitrogen-fixers and urea-using forms ranged from 10^6 to 10^8 per gram of bottom mud, suggesting a significant role in the nitrogen cycle in the pond. Bacteria capable of decomposing cellulose were found both in the water column and in the mud. Mud samples taken in the spring contained fewer cellulose-decomposers than samples taken in the autumn when, presumably, terrestrial plant material growing in the basin during the summer dry period would be starting to die and decay. Heterotrophs (those species requiring a supply of organic material from their environment) were the largest physiological group found in the pond, with anaerobes being more abundant than aerobes. Counts of both types decreased as

Table 4.1 Comparison of the prokaryotes found in an intermittent, snow-melt pool in northern Quebec with those in an episodic puddle in Singapore

Snow-melt pool	Tropical puddle
Oxyphotobacteria	
Cyanobacteria:	Cyanobacteria
Anabaena sp.	*Lyngbya aerugineo-caerulea*
Aphanizomenon flos-aquae	*Oscillatoria nigra*
Chroococcus turgidus	*Oscillatoria princeps*
Gloeocapsa granosa	
Nostoc commune	
Stigonema mamillosum	
Stigonema ocellatum	
Eubacteria	
Pseudomonads:	Pseudomonads
Pseudomonas fluorescens grp.	*Pseudomonas* sp.
Facultatively anaerobic rods	Facultatively anaerobic rods
Unident. coliforms	Unident. coliforms
Spirillas:	
Spirillum volutans	
Sheathed bacteria:	Sheathed bacteria
Sphaerotilus natans	*Sphaerotilus natans*
Leptothrix ochracea	
Chemolithotrophic bacteria	
Rhodobacterial detritus	
Gliding bacteria:	
Saprospira sp.	
Endospore-forming bacteria:	
Bacillus sp.	
Uncertain standing:	
Achromobacter guttatus	Gram-positive asporogenous rods and cocci
Undetermined bacteria	*Aerobacter* sp.

Source: (From Laird 1988; original classification has been retained)

the pond dried up but later increased after the basin was dry. Heat-shock experiments revealed that about 1.5% of the aerobic heterotroph population consisted of heat-resistant spores. The authors concluded that the bacteria functioned as decomposers and transformers in the nitrogen, carbon, and energy cycles of the pond as well as acting as a source of nutrients and as a primary source of food for protoctists and plankton. Walker *et al.* (1991) have further highlighted the importance of bacteria as a direct food source for mosquito larvae. In field and laboratory microcosms, they were able to show that feeding by *Aedes triseriatus* reduced microbial densities, and that selective removal of larvae from tree holes and tyres was followed by rapid increases in microbial densities (Morgan and Merritt 1992). Kaufman *et al.* (1999) have shown that predation by *Ae. triseriatus* may also cause a shift in bacterial community composition in container habitats, from Pseudomonaceae to Enterobacteriaceae. Larvae additionally affected microbial nitrogen metabolism, either through grazing on the bacterial taxa responsible for nitrification–denitrification, or through modification of the physicochemistry of the water.

Ectoenzymatic activities of cyanobacteria in the biofilm on Mediterranean stream substrates were found to resume significantly after only 2 h of re-immersion after drought. Romani and Sabater (1997) suggested that such rapid recovery was due to the stromatolitic (organosedimentary) structure of the biofilm, which acted as an organic matter reserve, alongside rapid rehydration permitted by the cell sheaths. This implies that such biofilms are capable of exploiting even short rainfall events during dry summers. The rapid re-establishment of the biofilm when streams resume flow ensures an almost immediate supply of food for grazers. Castillo (2000) has shown that populations and productivity of bacteria in neotropical floodplain lakes reach their highest values during low water, when they significantly exceed those of bacteria in the adjacent river water. In contrast, in blackwater lakes of Central Amazonia, bacterial abundance peaked during rising water—primarily due to input of allochthonous bacteria from the river at the beginning of the rainy season (Rai and Hill 1984).

Bacteria are well known for their tolerance extremes. In Don Juan Pond in the Antarctic, for example, where the freezing point of water is −40°C, bacteria have been observed growing in water temperatures ranging from −3 to −24°C (Hinton 1968). Many are also highly resistant to drying: of 2,724 bacterial strains tested, 83% of cultures dried in a vacuum for periods of up to 14 years, proved to be still viable (Rhodes 1950).

In general, functional studies of bacteria have been largely laboratory- or microcosm-based, and those that have been field-oriented have tended to concentrate on either purely aquatic ecosystems or terrestrial ones, although past methodologies have limited progress. Bacterial studies in habitats that experience a transition (e.g. from aquatic to terrestrial), such as temporary waters are rare, although rice fields are an exception (Conrad 1993). Aragno and Ulehlova (1997) have pointed out that transient ecosystems are likely to provide a multiplicity of ecological niches for a range of functional groups of bacteria, and are ripe for investigation with modern techniques. One such approach is terminal restriction fragment length polymorphism (T-RFLP), a phylogenetic method for the rapid analysis of complex microbial communities that has recently been applied to a freshwater ecotone (Sliva and Williams 2005).

A strictly numerical analysis of total bacteria in two adjacent intermittent ponds in Ontario has shown substantial differences both between ponds and within ponds in successive years. Mean densities in Pond I varied almost two orders of magnitude in the second year of the study, whereas in the same year, variation in Pond II was the least (Figure 4.1). Examination of environmental variables indicated that few were correlated with bacterial abundance in Pond I, but that the latter was strongly, and positively, correlated with turbidity in Pond II.

A final aspect of temporary water bacteria worth mentioning is their role as pathogens. Again, the data are limited, but a vibrio bacterium has been found infecting the larvae of several

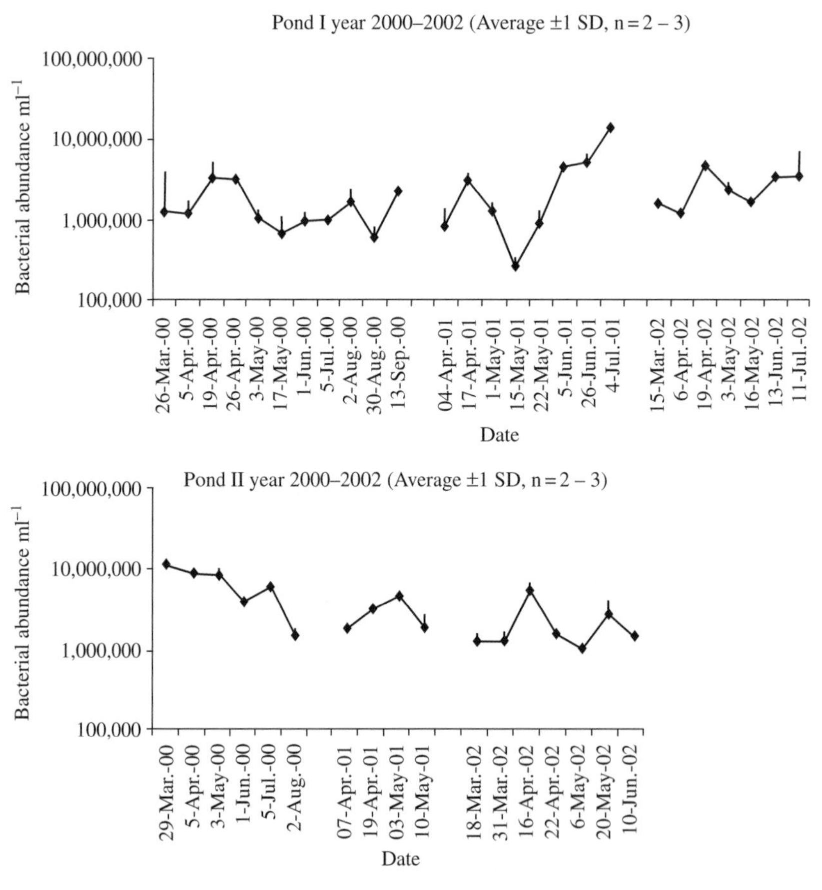

Figure 4.1 Comparison of the mean abundances (±1 SD) of bacteria in two intermittent ponds in southern Ontario throughout their entire hydroperiods, and over three years.

container-inhabiting mosquito species in Florida (Fukuda *et al.* 1997). In addition, a baculovirus has been recorded from populations of the mosquito *Wyeomyia smithii* living in the pitcher plant *Sarracenia purpurea*, in Massachusetts (Hall and Fish 1974).

4.2.2 Protoctists

Membership of this group follows that proposed by Margulis and Schwartz (1988), and includes both those phyla known as *protozoans*, and those grouped as *algae*. Some information on cyanobacteria is also included here, as the latter are often still included in general studies of algal communities.

Algae

Sheath and Hellebust (1978) studied the various algal communities in an arctic tundra pond and also reviewed the literature on these very abundant habitats. Tundra ponds develop where vegetation has been injured resulting in an increased thawing of permafrost, where there is a loss of volume as ice is melted, and where subsidence of sediments occurs. These ponds held water during the brief summer (June to August) but froze solid under 2 m of ice each winter. The plankton community showed two peaks in biomass and primary production during the short ice-free season. Only one peak in biomass was recorded for the periphyton community but primary productivity showed one or two peaks. Laird (1988) recorded a total of 114 species of diatom and 24 species of green alga from his snowmelt pond in northern Quebec.

Antarctic streams, which periodically lose their free water by virtue of freezing solid in winter, have been shown to support rich epilithic communities (87 taxa; Pizarro *et al.* 1996) of high biomass (>20 μg chlorophyll *a* cm^{-2}, or >20 mg C cm^{-2}), yet production rates are low (Vincent and Howard-Williams 1986). Nutrient supply and light were not thought to be limiting factors in these streams but water temperatures seldom rose above 5.0°C and, more often, lay between 0 and 2.0°C. Metabolism in these algae thus occurs at a low rate and the high biomass observed represents accumulation over several seasons of growth (as theoretical turnover times were of the order of several hundred days yet each annual growing season lasted less than 80 days). Despite this, the overwintering community retained a high metabolic capacity and responded rapidly to hydration at the beginning of summer. Figure 4.2 shows that photosynthetic capacity of the cyanobacterium-dominated epilithon rose as a log function of time over the first 6 h, and then at a faster rate over the subsequent two days. Thus, simple rehydration allowed immediate resumption of some photosynthesis but full recovery necessitated longer-term biosynthesis and repair. Vincent and Howard-Williams (1986) likened this resurrection to the response of desert plants in warmer regions, each community inhabiting a seasonally arid environment. They pointed

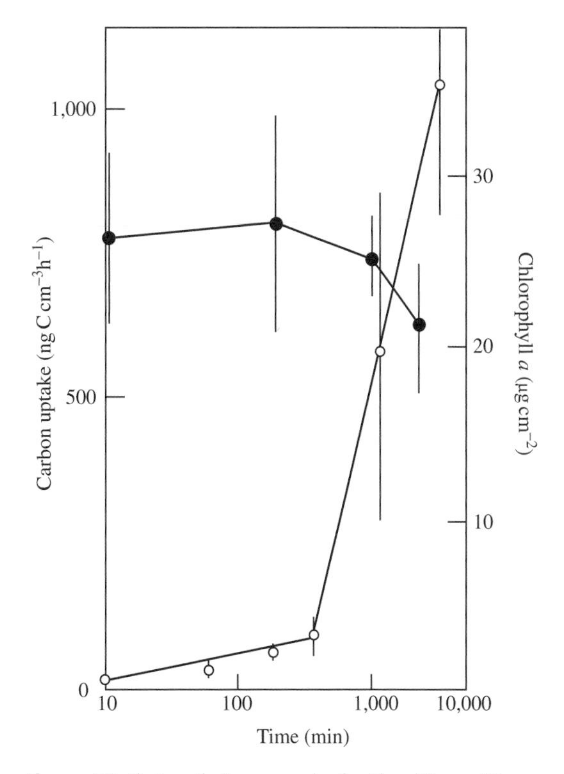

Figure 4.2 Photosynthetic recovery by the *Phormidium* epilithon as a log function of hydration time (closed circles represent chlorophyll *a* content; open circles represent photosynthesis; each point is the mean of three samples ±2 SE; redrawn from Vincent and Howard-Williams 1986).

out, however, that in addition to loss of water, Antarctic epilithon must contend with continuous darkness in winter and a harsh freeze-thaw cycle. Physiological resilience to freezing must therefore be an essential property of the microflora in the Antarctic. Fumanti *et al.* (1995) found a similarly rich microflora in Lake Gondwana, Northern Victoria Land, Antarctica. Although the phytoplankton was species poor (5 taxa), there was a rich benthic community in the form of shoreline mats comprising 34 taxa (8 species of green algae, 7 of diatoms, and 19 of cyanobacteria). *Phormidium frigidium* predominated in the lower mat layer, whereas the upper mat layer was dominated by *Pleurococcus antarcticus* (a green alga) alongside several other taxa all of which were characterized by gelatinous sheaths.

In many types of temporary waters, the formation of algal mats, especially those formed from the drying and felted remains of filamentous species, can be crucial to the survival of other organisms that may take refuge under them during drought. Reynolds (1983) has recorded sheets of this 'algal paper' as large as 50 × 100 m on the beds of Irish turloughs (temporary lakes that fill and drain through underlying limestone karst).

Survival of the algae, themselves, to prolonged exposure to drying is typically via modified vegetative cells with thickened walls, mucilage sheaths, and an accumulation of oils. The ability to resist exposure by such means is the major factor controlling zonation of algae at pond margins. Taxa with a gelatinous structure predominated in an intermittently dry section of Speirs' Pond in southern Ontario, and proved to be a subset (22) of those genera (30) found in permanently inundated sections (Williams *et al.* 2005). In the temporary site, green algae (*Oocystis*, followed by the filamentous *Oedogonium*) were dominant early in the hydroperiod, although diatoms (*Pinnularia*) were also present (Figure 4.3). For much of May, diatoms (*Pinnularia*, *Navicula*, and *Synedra*) predominated, including in the pre-summer-drought community—along with the Chlorophyta (*Chlorococcum*), Euglenophyta (*Euglena*), and some

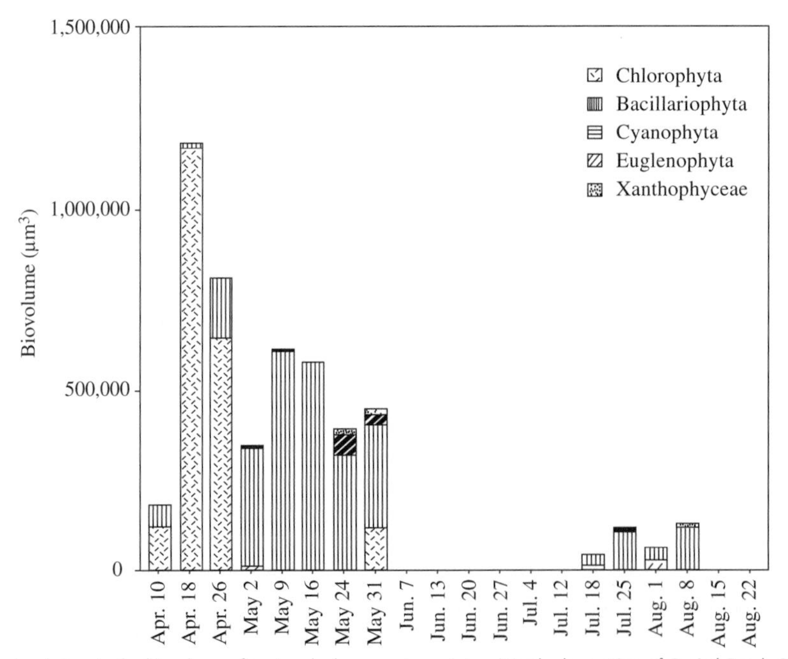

Figure 4.3 Temporal variation in the biovolume of major algal groups in an intermittently dry section of Speirs' Pond, Ontario. The site was dry from June 7 to July 12, and again from August 15 to 22.

Xanthophyta (*Tribonema*). The community found in the mid-drought-respite period of late July (which resulted from heavy rainfall), consisted primarily of diatoms (*Navicula, Diatoma*), with some *Tribonema* (Xanthophyta).

Many species of alga in temporary ponds and streams appear to be opportunists. Many pass through predictable life cycle phases with maximum zygospore germination ocurring when water levels are highest. *Vaucheria*, a typical temporary pool alga, survives drought as the thick-walled zygotes discussed earlier, but it also has a 'back-up' system. The latter involves forming hypnospores in response to rapid desiccation. These are specialized structures that release amoeboid cells capable of movement to areas where water is more abundant and where they give rise to new filaments (Sands 1981). Such properties are clearly those of algal species adapted to temporary waters, as Benenati *et al.* (1998) found that the recovery and maintenance of the phytobenthos in a permanent, but regulated, river in Arizona were compromised by erratic flow management. Repeated desiccation of the algal community had major effects on the bottom-up interactions of the Colorado River ecosystem.

A number of algal species live epizootically on the branchiopod crustaceans found in temporary ponds. In Moroccan ponds, Thiéry and Cazaubon (1992) found that filamentous greens, such as *Stigeoclonium* and *Oedogonium*, predominated on the shells of conchostracans (Spinicaudata), whereas small species of Chlorococcales and Tetrasporales colonized anostracan bodies. All of the algae were common species with wide holarctic distributions, such that no special adaptations were evident, and colonization site was thought to be determined by the microhabitat and swimming behaviour of the hosts.

Protozoans

Few studies have tackled the protozoan component of temporary water communities. Fenchel (1975) found densities of around 10^6 m^{-2} and biomass of 20–40 mg m^{-2} for ciliates in an arctic tundra pond between June and August. Some fed on algae or bacteria, whereas others were carnivores feeding on zooflagellates and other ciliates. Stout (1984) studied the protozoans of seasonally inundated soil under grassland in New Zealand and found a resemblance to the biota of a shallow temporary pond: an autotrophic community of phytoflagellates and diatoms, and a largely bacteriovorous community of zooflagellates, sarcodines (33 species), ciliates (57 species), and meiofauna. Of the 94 species of protozoan recorded, 45% could be described as freshwater forms and 35% as aquatic-terrestrial, the former dominating during winter when the soil was covered by several centimetres of water.

In an experimental ricefield, in Italy, Madoni (1996) found that the ciliate community was strongly influenced by environmental factors generated by the growing rice plants. Initially, when phytoplankton productivity was high both in the water column and at the sediment–water interface, grazing benthic species dominated the system. However, as the rice plants grew, algae decreased (likely due to shading and competition for nutrients) and decomposition processes became more intense. In the water column, this led to a predominance of bacteriovorous ciliate species, whereas the shift towards reducing conditions at the sediment–water interface resulted in the disappearance of many species and the relocation of others. In the final phase of rice cultivation, general ciliate diversity and production decreased markedly. Interestingly, despite these changes, one species, the prorodontid *Coleps hirtus*, dominated the system throughout the 4-year study—this was attributed to its extreme flexibility in feeding behaviour which allowed it to ingest algae, bacteria, other protozoans, and even small metazoans.

Coleps hirtus was also a dominant species in Laird's snow-melt pond in northern Quebec. Together with nine other common species, it formed the first phase of this pond's ciliate community. Using the 'Saprobiensystem' of classifying freshwaters by their degree of organic enrichment, Laird found that the post snow-melt, oligosaprobic phase of the pond was quickly replaced by beta-mesosaprobic conditions which lasted about one week. Subsequent high bacterial production resulting from decaying grasses produced alpha-mesosaprobic conditions which, after another week, gave way to polysaprobic conditions. Ciliates found in these three enrichment phases of the pond are listed in Table 4.2.

Few studies have addressed inter-pond comparisons. However, Andrushchyshyn *et al.* (2003) showed that the relative abundance of ciliates in two adjacent intermittent ponds in southern Ontario differed significantly from one another: that in Pond II rose rapidly from day 1 increasing two orders of magnitude by day 7. In contrast, abundance in Pond I began at the same level but increased much more slowly, reached a plateau of around 500 individuals l^{-1}, and increased again late in the hydroperiod (Figure 4.4(a)). The two ponds were also fairly dissimilar in terms of their species richness and species composition. Pond I contained 50 species compared with 70 species for Pond II, with only 24 species shared. Variation in ciliate abundance in Pond I could be explained by the number of days after filling (39%) and enclosure treatment (23%). These two parameters also explained 72% of the variation in species richness in Pond I. Sixty-five per cent of the variation in abundance in Pond II could be explained by the number of days after filling (27%), pH (19%), and nitrate levels (12%). Fifty-two per cent of the variation in species richness was explained by environmental parameters, of which pH was the most influential. Species succession was a strong feature of both ponds. Pond II contained more mid-sized ciliates (50–200 μm), whereas Pond I was dominated by smaller ciliates, especially in mid-May and early June (Figure 4.4(b)). There were more algivorous species in Pond II, although their abundance was greater in Pond I. In Pond II, bacteriovore relative abundance increased as the pond dried up; in Pond I there was a similar but more variable trend. Facultative algal feeders were most numerous in Pond II early in the season (e.g. 38% of the relative abundance on day 7 was due to *Uroleptus gallina* alone) but thereafter decreased rapidly. Pond I contained only one, though very abundant, facultative algivore, *Pelagohalteria cirrifera*. Omnivorous ciliates were moderately important parts (21%) of the communities of both ponds early in the season but in Pond I thereafter declined (Figure 4.4(c)). Although 13–14% of all species in the two ponds were predators, feeding chiefly on other ciliates, their abundance never exceeded 5% (Pond I) and 8% (Pond II) of the total ciliate community. Predator peaks occurred at around the same time (days 13–21) in both ponds. Experimental addition of invertebrate predators to these ponds resulted in higher ciliate abundance and species richness for a limited time in one of the ponds—suggesting that differences in food-web dynamics may influence ciliate community composition.

Stout (1984) put forward the idea that the biotas of moist soils and temporary ponds lie on a transition from freshwater to edaphic habitats—indeed, in a review of the world soil ciliate fauna, Foissner (1998) concluded that of the, at least, 1,000 known species about 25% are also known from freshwater habitats. Protozoan faunal elements from each can persist in the restricted

Table 4.2 Succession of the most common ciliate species throughout three enrichment phases in an intermittent, snow-melt pool in northern Quebec

Alpha-mesosaprobic phase [2 June]	Beta-mesosaprobic phase [9 June]	Polysaprobic phase [16 June]
Coleps hirtus	*Coleps hirtus*	*Caenomorpha medusula*
Cothurnia marina	*Colpoda cucullus*	*Epistylis lacustris*
Epistylis lacustris	*Epistylis lacustris*	*Euplotes patella*
Frontonia acuminata	*Euplotes patella*	*Intrastylum invaginatum*
Halteria grandinella	*Frontonia acuminata*	*Metopus es*
Intrastylum invaginatum	*Intrastylum invaginatum*	*Tetrahymena pyriformis*
Ophrydium versatile	*Stylonychia mytilus*	*Vorticella convallaria*
Pleuronema crassum	*Tetrahymena pyriformis*	*Vorticella microstoma*
Stylonychia mytilus	*Vorticella convallaria*	
Vorticella striata ssp. *octava*	*Vorticella striata* ssp. *octava*	

Source: From Laird (1988)

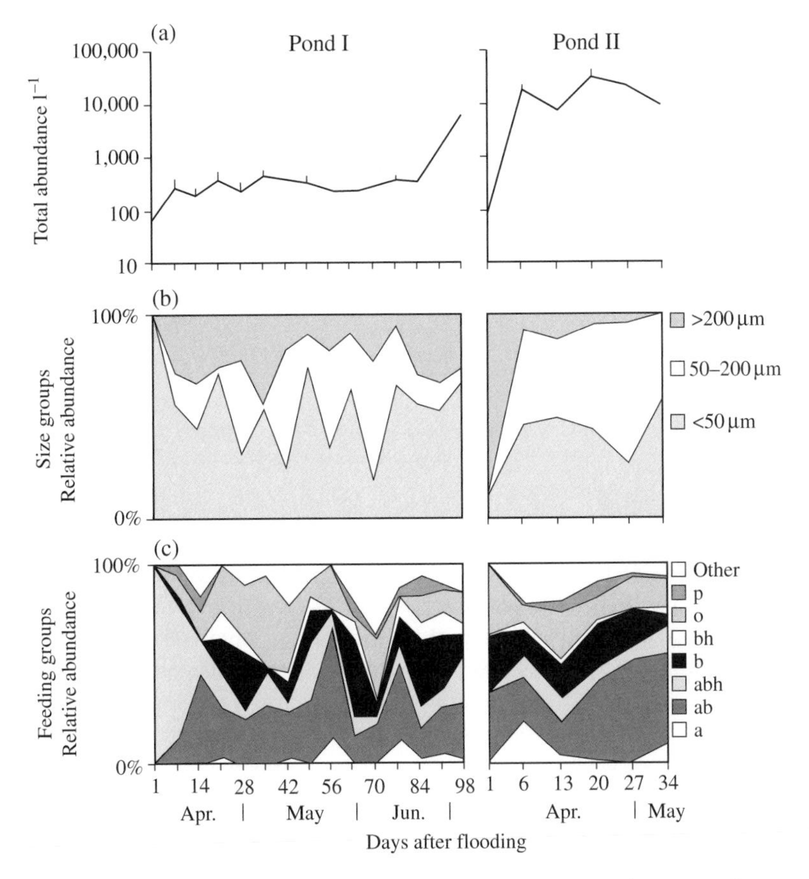

Figure 4.4 Seasonal development of (a) total ciliate abundance in two intermittent ponds in southern Ontario (means ±1 SE); (b) three size groups within the two communities; (c) relative abundance of the dominant ciliate feeding categories (a = algivores; b = bacteriovores; h = predators on heterotrophic flagellates; p = predators on other protozoans, including ciliates; o = omnivores; other = unknown prey).

moisture films around soil particles in pond bottoms, and extend into forest litter where physical contact between fallen, decaying leaves provides an unbroken moisture film. From here, Stout hypothesized that protozoans and some meiofauna have invaded soils, root zones, and vegetation surfaces, and have even been provided with colonization pathways to aerial habitats such as tree holes, leaf axils, and pitcher plant chambers (Figure 4.5).

Fukada *et al.*'s (1997) survey of Floridian container habitats revealed four protozoan parasites of mosquito larvae, the most common being *Ascogregarina taiwanensis* (Apicomplexa) and the microsporidium *Vavraia culicis*.

4.2.3 Fungi

In running waters, aquatic hyphomycetes are known to be critical to the breakdown of riparian leaf litter, as well as serving as an important food source for many benthic invertebrates—indeed they are fundamental to the base of lotic food webs (Kaushik and Hynes 1971). It is likely that they are also important contributors to nutrient release and dynamics in temporary waters, especially, for example, those nutrients contained within the terrestrial summer vegetation that grows in dry pond basins and stream channels. Unfortunately, little work has been done on these habitats. However, in a study of an intermittent vernal pond in southern

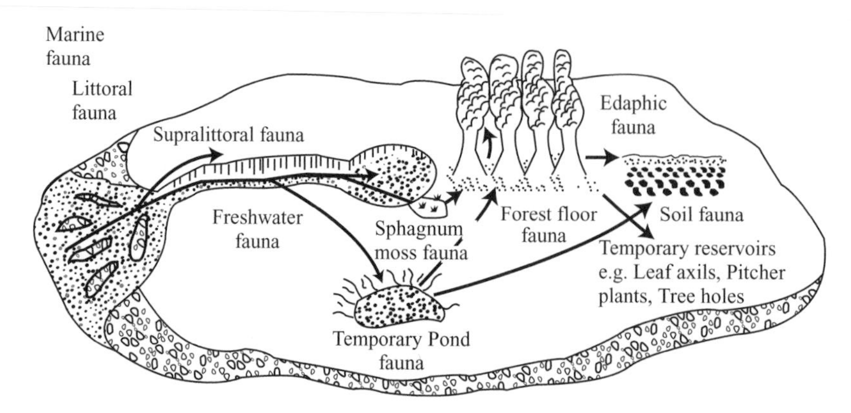

Figure 4.5 Hypothesized evolutionary pathway of terrestrial protozoans and meiofauna. Temporary ponds and sphagnum moss habitats provide reservoirs from which soil organisms disperse from their freshwater origins to colonize new edaphic habitats and recolonize seasonal habitats (redrawn after Stout 1984).

Ontario, Barlocher *et al.* (1978) found the fungal flora on bottom detritus to be very different from that found in nearby permanent ponds. Terrestrial fungi predominated during the waterless period and a seasonal succession was evident. In running waters, fungi actually add protein to decomposing leaves, a process that has since been described as 'conditioning' (Barlocher 1985). Detritivorous insects, when given a choice, select conditioned leaves over those that are fresh, because of their enhanced nutritional value. The mechanisms used by fungi to degrade leaves have been examined by Suberkropp and Klug (1980), by initially isolating five species of aquatic hyphomycete (*Alatospora acuminata, Clavariopsis aquatica, Flagellospora curvula, Lemonniera aquatica*, and *Tetracladium marchalianum*) and growing them in pure culture with hickory leaves. Enzymatic activity of each fungal species resulted in the skeletonization of leaves via maceration of the leaf matrix and the subsequent release of leaf cells as FPOM. After incubation, further fractionation and analysis of leaf material indicated that all fungal species metabolized (degraded) cellulose, and that two species (*T. marchalianum* and *F. curvula*) metabolized hemicelluloses. In cultures treated with *T. marchalianum*, the release of fine particulates coincided with increases in fungal biomass (as measured by ATP) and enzymatic activity in the supernatant, which degraded carboxymethycellulose, xylan, and polygalacturonic acid. Macerating activity increased with pH, indicating that pectin transeliminase was involved in the softening of leaf tissue by *T. marchalianum*. This finding prompted Suberkropp and Klug to suggest that transeliminase is more important in the release of leaf cells than hydrolytic enzymes with lower pH optima.

Aquatic insects show a positive selection for conditioned food. Bacteria also can condition detrital material, but evidence suggests that fungi may be more effective in enhancing its nutritional value. For example, when Mackay and Kalff (1973) presented leaf discs from fungal and bacterial cultures to the caddisflies *Pycnopsyche gentilis* and *P. luculenta*, it was the fungal discs that were preferred.

4.2.4 Higher plants

As previously indicated, most temporary ponds represent the penultimate stage of a sere, the climax of which is terrestrial. In a study of plant succession in temporary ponds in Oregon, USA, Lippert and Jameson (1964) found that the species present were characteristically those found in wet places, for example, the cat-tail (*Typha latifolia*) and the spike rush (*Eleocharis palustris*). In northern Germany, Caspers and Heckman (1981) found *T. latifolia* and the grass *Glycera maxima* in large numbers in Stage 5 (the final stage) of their ditches. In this locality, these latter two species are eventually responsible for the total disappearance (terrestrialization) of the aquatic habitat. The margins of temporary waters are very susceptible to invasion by terrestrial species, the roots of which may proliferate through the basin substrate. Litter from all species provides a rich substrate for microorganisms and subsequently a source of food for aquatic invertebrates. Barlocher *et al.* (1978) measured an average of 132.8 $g\,m^{-2}$ of leaf litter (ash-free dry weight) entering Ontario pond basins in the autumn.

Many temporary waters, especially those with longer hydroperiods and low- or no-flow, support heavy growths of aquatic macrophytes. In certain localities, free-floating annual macrophytes such as *Azolla* may contribute significantly to the productivity of temporary ponds. The benthic invertebrate communities of shallow lakes on the floodplain of the River Parana, Argentina, experience shifts in composition related to floating macrophyte cover and changes in depth. During high-water periods, communities are similar among lakes and are dominated by ostracods and oligochaetes. During low water, when biomass is highest, chironomid larvae (*Chironomus*) and burrowing mayflies (*Campsurus*) have been found to dominate lakes with 60% cover of *Eichhornia crassipes*, whereas lakes without floating plants were dominated by sphaeriid clams (*Pisidium* and *Eupera*), snails (*Littoridina*), and burrowing mayflies (Bechara 1996). Macrophytes provide an important structural 3-dimensional framework for the activities of many invertebrates, extending the benthic realm into the water column, and perhaps allowing greater interactions between bottom-dwellers and the plankton. However, some invertebrates also eat plant parts, especially the leaves. Cronin *et al.* (1998) found that aquatic insects consumed 0.2–1.7% of the leaf surface of water lilies (*Nuphar variegata* and *Nymphaea odorata*) per day in a Michigan lake, with some leaves suffering more than 60% damage. Submerged leaves appeared to be more susceptible. Interestingly, their data also provided support for the hypothesis that herbivorous insect species belonging to primarily aquatic groups have a much wider diet than those derived from terrestrial groups.

In a study of the microdistribution of macrophytes of vernal pools in southern California, Zedler (1981) related water level to a succession of species groups each with characteristic germination times, and flowering and seed maturation stages. Species were assigned to weighted average water duration classes (WADC), calculated from duration-frequency data, ranging from 1 to 16. Species falling into classes 1 to 4 are those rarely found within the inundated areas of pools but which grow commonly near the margins (e.g. *Bromus* spp., bromegrass; and *Erodium* spp., filaree). Species with WADC values of between 4.5 and 7.0 (e.g. *Juncus bufonius*, toad rush; and *Agrostis microphylla*, bentgrass) possess a physiology capable of tolerating inundation but fall short of being true marsh or aquatic plants. Species in classes 7.0–9.5 tolerate water-cover for long periods, but do not thrive under long submergence—they are able to germinate and grow underwater but require a considerable period out of water in order to mature (e.g. *Callitriche marginata*, water-starwort; and *Anagallis minimus*, false pimpernel). Characteristic temporary pool species fall into the class range 9.5–11.5. These are species possessing the morphological and physiological plasticity that allows them to withstand prolonged submergence (e.g. *Downingia cuspidate; Pogogyne abramsii*, pogogyne; and *Eryngium aristulatum*, marsh eryngo). All three species produce submerged leaves that are very different from the emergent foliage of the mature plant. The final group of plants (classes 11.5–16) contains species that are almost true aquatics, yet they can withstand some dry period (e.g. *Pilularia americana*, pillwort; *Callitriche longipedunculata*, water-starwort; and *Lilaea scilloides*, flowering quillwort).

On the basis of this study, Zedler proposed several hypotheses: The first was that the distribution of standing water in time and space is the single most important factor influencing temporary pool macrophytes. Physical stress of inundation is the primary cause of the distinctive assemblages of species but pattern of soil moisture may be important also. That waterlevel fluctuation stress differentially affects semi-emergent plants, can be illustrated by the water buttercup, *Ranunculus*. In a comparison of four species living on floodplains, He *et al.* (1999) showed that each exhibited different adaptations to inundation: *R. repens*, a species from lower, frequently inundated floodplains, proved to be very tolerant of prolonged waterlogging and submergence, as it was able to use oxygen generated by underwater photosynthesis for root respiration; *R. sceleratus*, common in low-lying mudflats, responded by ameliorating flooding stress through high root porosity and an ability to rapidly elongate its petioles under water; *R. acris*, from less-frequently flooded areas, responded in a manner similar to *R. repens*, although it had a lower resistance to submergence in the dark; and *R. bulbosus*, a species from seldom-flooded river levées, was generally intolerant to both waterlogged soil and complete submergence. Coxon (1987) similarly found that flooding depth was pivotal in the development of vegetation in Irish turloughs. Floor vegetation became increasingly dominated by marsh plants as the depth increased to around 3 m. Surprisingly, some of the deepest turloughs, along with those with variable depth due to undulating beds, had short hydroperiods, and a floor vegetation dominated by dry land ruderals (e.g. *Rumex*). Vegetation was, in addition, related to deposit characteristics, although the latter were also correlated with hydroperiod—peat and marl were associated with longer water residence, whereas sand, silt, clay, and diamicton were more common in turloughs with shorter hydroperiods. Linhart and Grant (1996) pointed out that soils are especially strong agents of selection in short-lived plant species. Consequently, pools located in regions with different soil characteristics (e.g. mineral content, pH) would be likely to harbour genetically differentiated populations of a variety of species.

Zedler's second hypothesis was that local extinctions of temporary pond species are rare despite considerable year-to-year variations in rainfall and habitat availability. Several characteristics of temporary pool species lessen the probability of extinction, namely small minimum plant size; small seed size and thus many seeds/unit biomass; and high vegetative and reproductive plasticity typical of annuals. Holland and Jain (1981), however, in a survey of over 250 vernal pools in central California, showed that species composition varied significantly among years and that species richness varied nearly twofold over sites, apparently in response to regional differences in rainfall among sites and edaphic conditions, and was less in a drought year. However, Gafny and Gaisith (1999) demonstrated that even though the historical appearance of macrophytes in the littoral zone of Lake Kinneret, Israel was sporadic (due to water level fluctuations), species stands were always located in exactly the same places, and that this was highly dependent on sediment structure. Husband and Barrett (1998) found that in the arid region of northeastern Brazil, the annual *Eichhornia paniculata* occurs in discrete temporary waters, such as pools, wet ditches, and flooded pastureland. Population size averaged 86 plants, with 64% of the populations persisting from one year to the next, and, on average, 21.6% of suitable habitat patches were occupied. No populations occurred in areas where the density of patches was less than around 2 patches km^{-1}, suggesting the existence of a habitat threshold for persistence.

The third hypothesis was that despite the fundamental influence of inundation on community structure, competition is also a factor. Most temporary pool species are restricted at the upper (drier) end of the water duration-elevation gradient by competition from species that are better at exploiting drier conditions. At the lower (wetter) end, however, morphological and physiological tolerance to inundation are probably more influential. Holland and Jain showed that several grassland species grew in vernal pools during a year of severe drought but they were excluded in years of average rainfall. Further, several taxa characteristic of the vernal pool flora in this region were not evident during the drought although they were common and widespread in subsequent years. They showed that the number of species specialized for a given pool depth is proportional to the relative area in a pool at that depth; thus more species are adapted to the margins than to the centre. They concluded that species richness is determined chiefly by certain physiographic niche properties whereas competitive niche-partitioning factors influence congeneric sympatry and zonation within a pool.

Elam (1998) has reviewed the limited population genetics data of vernal pool plants and concluded that species often form 'races' or biotypes, at the geographical (among-population) scale, but also show intrapopulation differentiation at the local scale. The latter may involve changes over distances of as little as 2–5 m. In *Veronica peregrina*, for example, the environmental conditions experienced by individual plants from a pool's centre to its margin are known to change with respect to competition regime, moisture availability, and environmental predictability (Linhart 1988). At the centre, individual plants grow among a high density of conspecifics, but under a relatively predictable, wet environment. In contrast, individuals growing at the margin come under heavy interspecific competition from grassland species, as well as having to endure a low moisture, less predictable environment. The resulting differentiation occurs over distances not considered a barrier to gene flow by seed and pollen, and suggests that the homogenizing process of gene flow can be offset by the effects of strong disruptive selection (which favours extreme traits in a population when sudden changes occur in the environment).

Local climate change can influence the macrophytes associated with shallow ponds. For example, the mid-1990s in Britain were considered to be drought years in which the vegetation of a pond in Epping Forest was invaded by creeping bent grass (*Agrostis stolonifera*), with isolated clumps of Yorkshire fog (*Holcus lanatus*) becoming established on the margins. In prior, wetter years (1989 and 1991), the dominant vegetation was floating sweet-grass (*Glyceria fluitans*), together with soft rush (*Juncus effusus*), reed mace (*Typha latifolia*), and yellow flag (*Iris pseudocorus*). Panter and May (1997) concluded that should the drying trend persist, the pond would be likely to continue to develop a terrestrial succession of plants. Welling *et al.* (1988) working on prairie wetlands showed, similarly, that if drought conditions were extreme this favoured the germination of annual species over emergents. Reflooding, on the other hand, eliminated annuals and stimulated the vegetative growth of emergent species, unless the latter's water depth tolerance was exceeded (Poiani and Johnson 1991). In the coastal wetlands of southeastern Alaska, Pollock *et al.* (1998) showed that sites with intermediate flood frequencies and high spatial variation of flood frequencies (SVFFs) were plant species-rich, whereas sites frequently, rarely, or permanently flooded and with low SVFF were species-poor (Figure 4.6). Such changes in plant species composition are likely to affect faunal abundance and composition. For example, Wrubleski (1987) demonstrated that the microdistributions of insects (Chironomidae) in Delta Marsh, Canada varied among major vegetation communities along a gradient from deeper water to dry land (Figure 4.7).

The Pantanal region of Brazil supports the largest wetland complex in the world, part of which includes vegetation well adapted to seasonal cycles of flooding and drought (Por 1995). Several vegetation zones have been identified, and those most associated with the network of water channels include: (1) *Amphibious herbaceous vegetation* that grows under permanently wet, temporarily

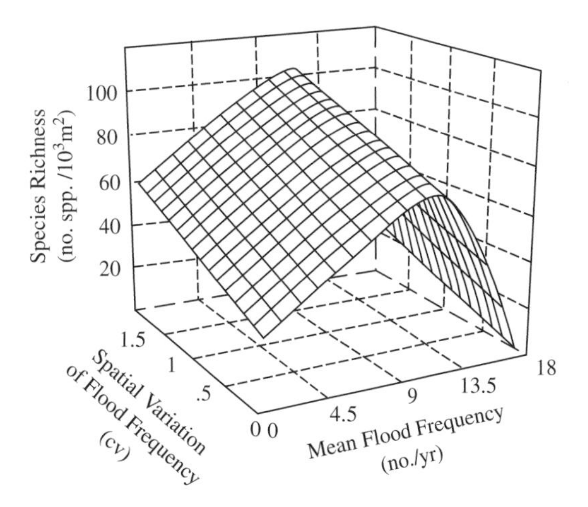

Figure 4.6 Three-dimensional representation of the relationships among macrophyte species richness, the spatial variation in flood frequencies, and mean flood frequency for 16 wetland sites on Chichagof Island, southeastern Alaska (redrawn after Pollock *et al.* 1998).

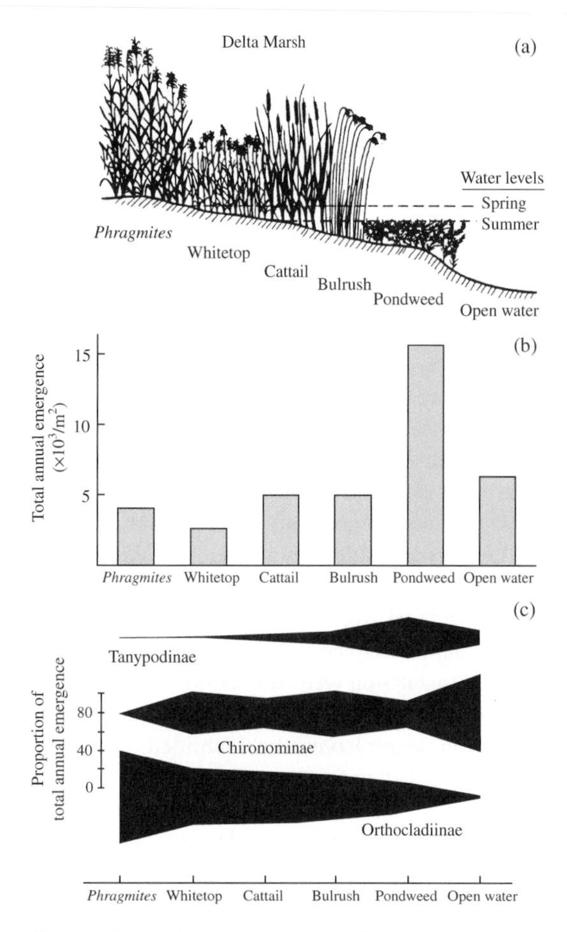

Figure 4.7 Microdistribution of chironomids along a moisture and vegetation gradient in Delta Marsh, Canada: (a) elevational and vegetation gradients; (b) total annual chironomid emergence from each microhabitat; (c) proportion of the total emergence represented by each chironomid subfamily (redrawn after Wrubleski 1987).

submerged conditions, such as are found associated with shallow lakes (baías), and includes grasses (*Cyperus giganteus, Scirpus validus*, and *Paspalum repens*), the reed *Typha dominguensis*, and the maranthaceans *Thalia geniculata* and *Polygonum hispidum*; (2) *Herbaceous campo vegetation* that represents seasonally flooded grasslands, comprising species of *Panicum, Paratheria, Oryza*, and *Setaria*, together with some exotic herbs, such as *Brachiaria*, introduced by cattle ranchers; (3) *Gallery forests* that cover the higher banks of the major rivers and comprise *Triplaris formicosa, Rheedia brasiliensis*, several species of *Inga, Vochysia*, and *Ficus*, including *Ficus elliotiana* a species endemic to the Pantanal and especially well adapted to surviving drought; and (4) *Cerrado-type forest* (shrub savannah) that may often also cover seasonally flooded forest floors. However, whereas the reliance of these plant communities on inundation cycles is quite well understood, comparatively little is known of their influences, energetic or otherwise, on the aquatic communities.

In a review of the adaptations of plants to flooding stress, Blom (1999) emphasized that, for woodlands, zonation is largely determined by local hydrology. For example, successful establishment of softwood species, such as those belonging to the Salicaceae, on river banks depends on the interaction between water levels and the timing of seed dispersal—they are well adapted to irregular, high, and prolonged inundation. In contrast, hardwoods, such as those belonging to the genera *Quercus, Fraxinus, Ulmus*, and *Acer*, are flood-sensitive, are found higher on the floodplain, and have heavy seeds that germinate in shade. The most shade-tolerant hardwoods are the least well adapted to inundation. These properties will influence which species are likely to have their roots in and around temporary water bodies, and hence the quantity and quality of allochthonous leaf litter that will fall into the basin and upon which the aquatic food web will, in part, be based. Further, inundation-tolerant species survive as a result of having aerenchymatous roots together with shoot parts that elongate upon submergence. Most importantly, as a result of high porosity in the roots of inundation-tolerant species, radial oxygen loss greatly influences nitrification and denitrification processes in the flooded basin soil. Root-derived oxygen thus has the capacity to restore nutrient cycles and 'detoxify' the oxygenated rhizosphere.

Forestry practices, especially the stage in forest cycles, have been shown to influence low flow conditions in streams. For example, pre-planting drainage may increase low flow if more than 25% of the watershed is drained. In contrast, forest growth decreases low flow levels (in all but the driest years). Clearfelling increases low flows initially, but this may be followed by a gradual decrease depending on the rate of vegetation regrowth (Johnson 1998). As streams dry up, the degree of retreat of the groundwater table can have a profound effect on the structure of riparian plant communites. For example, in the San Pedro River floodplain, Arizona, Stromberg *et al.* (1996) observed that as the depth to groundwater ranged from 0 to 4 m, their wetland indicator score (an index based on cover of plants within wetland indicator groups and frequency of indicator species) changed sharply. In particular, the abundance of obligate wetland herbs declined significantly at groundwater depths below 25 cm. Other impacts of retreating groundwater included reduced establishment of *Populus fremontii–Salix gooddingii* forests, and reduced cover of herbaceous species associated with fine-textured soils and the shady conditions of floodplain terraces.

Forestry practices may also influence the structure of temporary water communities through alteration of shading. For example, woodland ponds are heavily shaded, have little emergent vegetation, and receive much of their energy input from fallen tree leaves—the communities that have evolved under these conditions are thus largely detritus-based. Timber harvesting removes this cover, opens the ponds to sunlight, reduces allochthonous energy input, and promotes a shift to alga/herbivore-based communities.

The presence of temporary waters in arid regions can have a profound influence on the local riparian flora. For example, in the western United States, where riparian areas occupy less than 1% of the landscape, Zimmerman *et al.* (1999) found that variation in vegetation distribution and composition in Arizona was best explained by a complex temperature/moisture/substrate gradient, with understory diversity being related to changes in slope and sand/gravel substrate. Most of these environmental properties are of course running-water-induced. Further, each tributary canyon supported unique and oftentimes rare plant species. In southern California, Bauder *et al.* (1998) identified five vernal pool plant species as being endangered, a consequence of a historical 95–97% loss in pool habitat.

4.2.5 Invertebrates

Sponges

Sponges, although fundamentally a marine group of simple multicellular animals, are nevertheless also quite well represented in a variety of freshwater habitats, including temporary waters. In permanent waters, their natural life cycle includes a dormancy phase, typically during winter. At this time, the active tissues of the sponge body become transformed into gemmules, which represent regression of cells into masses that then become surrounded by a protective coat comprising collagen layers embedded with spicules (Frost 1991). Dormancy seems to be initiated by a number of factors, including changes in water temperature, declining water level, or increasing environmental stress. Species living in tropical regions enter dormancy prior to high summer temperatures, and this is also the case for temporary water forms. Gemmules are particularly resistant to water loss and are able to withstand passive dispersal episodes to other habitats. Both properties are well suited to living in temporary waters. Gemmule hatching seems to be primarily cued by rehydration and a return to normal water temperatures, although photoperiod may also be involved. In some species, the gemmules undergo a true diapause which may add safeguards to prevent premature hatching under temporarily improved environmental conditions (Simpson and Fell 1974).

The role of sponges in freshwater systems is poorly known, although they are believed to be a relatively minor component of the benthic community. However, a study of *Spongilla lacustris* in Mud Pond, New Hampshire, showed maximum biomass to be over 3.5 g dry weight m^{-2}. At this density, the sponge population would, in theory, have been capable of filtering the entire volume of water in the pond every seven days, and, in the process, could have removed most of the bacteria and phytoplankton from the water column (Frost 1978). Similarly, Lake *et al.* (1989) found *Radiospongilla sceptroides* to be a 'conspicuous' component of the fauna of an intermittent pond near Victoria, Australia, where it grew in masses up to

8 cm across on the stems of dead terrestrial vegetation. Clearly then, under certain circumstances sponges can be significant members of the benthos, although quantitative studies in temporary waters are very rare.

Coelenterates

Coelenterates are another group considered to be of minor importance in most freshwaters. Again, they have clear saltwater origins, and are represented in freshwaters by only a few forms whose ecology is poorly understood. The hydras are perhaps the best recognized, represented by simple, solitary polyps, and are often found in temporary ponds or slow-flowing streams. They exhibit both asexual (budding) and sexual reproduction, with the latter producing fertilized eggs surrounded by a thecal coat. The latter appear to have some resistance to adverse environmental conditions. *Craspedacusta* and *Limnocodium* are freshwater genera in which the dominant form is a medusa which is budded off from minute, larval polyps ('microhydra'). *Craspedacusta* is known to inhabit shallow pools in the Yang-tse River valley, China, where it is subject to large fluctuations in water level and temperature. Severe temperature changes result in the polyp contracting into a cellular ball which becomes encased in a protective chitin-like membrane (Slobokin and Bossert 1991). Interestingly, mass release of the hydra *Chlorohydra viridissima* has been tried as a possible control mechanism for mosquito larvae in temporary ponds in California, but with limited success (Yu *et al.* 1974).

Flatworms

Micro- and macroturbellarians (triclads) are known to live in temporary ponds and streams as well as in some other extreme aquatic habitats including ponds in the high Arctic, alpine meltwater ponds, the leaves of bromeliads, and on wet moss and leaves at water margins (Kenk 1944; Kolasa 1991). However, in general, turbellarians tend to get short shrift in ecological studies, although they clearly are often a component of the biota of temporary waters. Frequently, they are listed as unidentified, and species richness typically seems low (e.g. a single species of *Dugesia* from an intermittent prairie stream in Kansas, Fritz and Dodds (2002); unidentified Rhabdocoela from a temporary pond in central Italy, Bazzanti *et al.* (1996); unidentified 'Turbellaria' from ephemeral pans in South Africa, Meintjes (1996)), although occasionally it is higher (e.g. unidentified Neorhabdocoela and Typhloplanidae, together with *Rhynchodemus* cf. *sylvaticus* from Espolla Pond, Spain, Boix *et al.* (2001); nine species, from six families, from Namibian wetlands, Curtis (1991)).

In a unique review of the triclads of temporary waters in eastern North America, Ball *et al.* (1981) determined that only two species occurred regularly in this region: *Hymanella retenuova* and *Phagocata velata*. These authors stated that the typical turbellarian life cycle is simple and without resistant stages; moreover adults are fragile and susceptible to high temperatures and desiccation. *P. velata* has become adapted to withstand desiccation by a modification of its asexual cycle in which adults undergo multiple fission at unfavourable times followed by each fragment secreting a layer of slime which hardens into a resistant cyst. The same behaviour has been observed in *Phagocata fawcetti*, an inhabitant of intermittent streams in California. *Hymanella retenuova*, which appears to be almost entirely restricted to vernal ponds, does not reproduce asexually, however adults are capable of producing egg cocoons that are thick-shelled and capable of resisting drought.

Densities of both micro- and macroturbellarians may be high. For example, Collins and Washino (1979) recorded more than 1,000 individuals m^{-2} of *Mesostoma lingua* in Californian rice fields. At such high densities, turbellarians have been cited as both regulating the population dynamics of zooplankton in ponds (Maly *et al.* 1980), and engaging in intense predation on other turbellarians (e.g. *Dalyellia* have been observed to consume *Mesostoma* and *Rhynchomesostoma* in temporary pools; Heitkamp (1982)). As a result, some microturbellarians (Typhoplanidae) have been explored as a means of reducing populations of *Culex tarsalis* larvae in ricefields in California (Case and Washino 1979). Mass release of *Dugesia dorocephala* has also been tried (Tsai and Legner 1977).

Gastrotrichs

The gastrotrichs represent a phylum of pseudocoelomate animals (aschelminthes) closely allied to nematode worms. They have both marine and freshwater representatives, although the latter are understudied. Freshwater species live in sediments and in slimy coatings on submerged stones, vegetation, and twigs. They may occur at very high densities (10,000–100,000 m^{-2}) yet their ecological roles are unclear—although they are known to feed on protozoans, algae, bacteria, and fine particulate detritus. They have been recorded from temporary ponds and boggy pools in Poland (Kisielewska 1982; Szkutnik 1986). It is likely that they survive drought in one of the two types of parthenogenetically produced egg: opsiblastic eggs are thick-shelled and are highly resistant to freezing and drying, they also seem to be the principal dispersal phase (Strayer and Hummon 1991). Smock (1999) reported that specimens of *Chaetonotus* appeared within 3 days of flooding of floodplain soils in the southeastern United States.

Nematodes

Nematodes belong to the largest of the pseudocoelomate phyla, comprising about 80,000 described species, although some authorities believe that there are upwards of 1 million species yet to be described. There are both free-living and parasitic species, the former being found in the sea, in freshwater, and in the soil, often in huge numbers. Some nematode species are known to occupy very specific niches; for example, hotsprings, meltwater pools on glaciers, and intermittently soaked felt beermats in German pubs. Many species are renowned for withstanding severe environmental conditions, and some have survived immersion in liquid nitrogen. Many live in environments with very low oxygen tensions (e.g. internal parasites and those that live in anoxic muds), yet they are able to maintain their body functions. Under extreme stress, they may enter cryptobiosis, a trait shared by many other lower invertebrates (e.g. rotifers and tardigrades). The sex ratio has been shown to be influenced by prevailing environmental conditions, such that when food is limited males predominate. However, changes in environmental temperature, ions and carbon dioxide content may produce similar results. Females can contain millions of eggs and can produce them at a rate of up to 200,000 per day. Eggs may also be produced as a result of self fertilization (hermaphroditism) or from unfertilized eggs (parthenogenesis). It is clear, then that nematodes possess several physiological and life history traits that would serve them well in temporary waters.

Unfortunately, whereas the category 'Nematoda' frequently shows up in inventories of temporary water communities, it is rare to see species listed. Because nematodes are known to be very abundant in freshwaters (e.g. an estimate of 235,000 individuals of *Tobrilus grandipapellatus* m^{-2} was made for the benthos of an Austrian alpine lake, in winter; Bretschko (1973)), they are believed to be important consumers of bacteria, fungi, algae, and higher plants (Nicholas 1984). However, their significance in the general economy of freshwater ecosystems is not known (Poinar 1991). By way of comparison, on an estuary mudflat in the United Kingdom, nematode annual production was calculated to be 6.6 $g\,C\,m^{-2}\,yr^{-1}$ (Warwick and Price 1979).

From the limited knowledge of species from temporary waters, the following facts are notable: Bazzanti *et al.* (1996) recorded *Monhystera* sp., *Dorylaimus stagnalis, Laimydorus centrocercus*, and unidentified Dorylaimida as being prominent in a pond in central Italy. Laird (1988) listed free-living nematodes (Adenophorea) as being 'always abundant' in his snow-melt pool in northern Quebec, and 'quite common' in his episodic puddle in Singapore. Taylor *et al.* (1999) reported nematodes to be common and diverse in shallow wetland ponds on the Atlantic Coastal Plain of North America, however, only one genus, *Dorylaimus*, was identified. Smock (1999) recorded the genera *Miconchus, Labronema, Prismatolaimus*, and *Dorylaimus* from rehydrated riverine floodplain forest soils in Virginia. Also from Virginia, although not strictly from a truly temporary habitat—but from an intermittently freshwater one, Yozzo and Diaz (1999) identified at least 16 species (representing 5 orders and 11 families) from the tidal freshwater marshes of the James River. These included the genera *Monhystera*, *Prismatolaimus*, *Tylenchus*, and *Dorylaimus*. This latter genus, together with *Mesodorylaimus*, *?Rhabdolaimus*, and a species from the parasitic family Mermithidae have been identified from intermittent ponds in southern Ontario (A.K. Magnusson, Personal communication). For North America, in general, Poinar (1991) lists at least 21 families of freeliving nematodes as containing species which during all or a large part of their life cycles are to be found in freshwaters.

Nematomorphs

Nematomorphs are all aquatic or semiaquatic, with most species living in freshwater. The phylum, which comprises about 320 known species, has a worldwide distribution. Freshwater species live in ponds, lakes, rivers, ditches, and a variety of man-made container habitats. Indeed, the common name 'horsehair worms' was derived both from their general appearance and colour, and the fact that they sometimes live in livestock-watering troughs—where, in past times, superstition had them spontaneously come to life from shed horse-hairs. The eggs are laid in long strings in water or very damp soil, with a single female being capable of producing several million eggs. Nematomorph larvae hatch after 15–80 days and each has a protrusible proboscis that is covered with spines. After hatching, each larva must seek out and enter a host, either by direct penetration of the host's body wall, or by being eaten. Arthropods seem to be the only viable hosts known, although other animals (leeches, snails, amphibians, fishes, and even humans) may serve as temporary, or 'accidental' hosts. If a larva is eaten by an inappropriate host, the larva may become encysted and thus remains dormant until its host is eaten by a more suitable host. In many instances this does not happen and the life cycle is never completed. There is still considerable uncertainty about the exact processes involved in host acquisition. After several weeks or months of metamorphosis, the sexually immature, adult worm emerges from the host's body either when the host is near water or during rain. Maturation occurs in the free-living stage, and some species are capable of encysting as adults if suitable environmental conditions are not available. As with several other aschelminth phyla, the Nematomorpha seem to possess traits that pre-adapt them to life in temporary waters. Poinar (1991) lists four genera known from freshwaters in North America, together with another 12 thought likely to occur there but unconfirmed due to lack of study.

Rotifers

Approximately 2,000 species of rotifer have been described, the vast majority of which live in freshwaters. Habitats include lakes, ponds, bogs, streams, rivers, and puddles, but some live on the moisture-covered surfaces of mosses, lichens, tree bark, or in damp soil. In these latter, temporary water, habitats species survive periodic desiccation by secreting a protective layer of gel. Rotifers can reproduce at prodigious rates with up to 40 generations per year being common. Some species undergo a seasonal change in body shape, known as cyclomorphosis, which is associated with temperature changes and/or changes in predation pressure.

Many species produce two types of egg. 'Summer' eggs are diploid and result in only females. In autumn, heavy-shelled, haploid 'winter' eggs are produced that require fertilization before they can develop. In the spring they give rise to parthenogenetic females. Males develop in the population when haploid eggs fail to be fertilized. 'Winter' eggs are also highly drought-resistant, and may remain dormant for as long as 4 years. In a study of the invertebrates of prairie wetland marshes in Minnesota, Hershey *et al.* (1999) reported that recovery of rotifer populations after drought was more rapid than in insects, as recruitment of the former was from egg banks in the soil, whereas most insects were believed to recruit from nearby permanent water bodies. These authors also found that rotifers, together with cladocerans, were more abundant and species-rich during a regional drought—with abundance being significantly correlated with the Palmer Hydrologic Drought Index for the previous month. Some 35 rotifer taxa were recorded from these wetlands, with 6 genera being common to all sites sampled (*Lepadella*, *Monstyla*, *Mytilina*, *Platyias*, *Testudinella*, and *Trichocera*).

In two English ponds that were freshly dug out after having been dry for 20 and 40 years, respectively, the earliest colonizers, after only a few days, were the rotifers *Keratella valga* and *Brachionus urceolaris* (along with cyclopoid copepods and *Daphnia obtusa*) (Pontin 1989). In both ponds, the rotifers rapidly produced large numbers of parthenogenetic females, but males and resting eggs also appeared soon. Herbivorous species were followed by the omnivorous species *Asplanchna brightwelli* which was observed to prey on *K. valga*. Initial colonizers persisted in the ponds for several years after, despite the diversification of the fauna—although the time of their appearance shifted. The 20-year-dry pond subsequently developed a greater diversity of species that may have been related to a higher abundance of macrophytes there.

Bonecker and Lansactoha (1996) related changes in the community structure of rotifers in standing and running water sections of the upper Parana River floodplain in Brazil to environmental factors. Densities of the most abundant species were strongly correlated with chlorophyll *a*, dissolved oxygen, water temperature, and water level. In contrast, rotifer diversity was mainly related to water level, with species groups being associated with different phases of the hydrological cycle. Hydroperiod was also found to be a controlling influence in a survey of 18 temporary ponds in the Donana National Park, southwestern Spain (Fahd *et al.* 2000). The ponds were divided into 'seasonal', 'intermediate', and 'ephemeral' according to the length of their hydroperiod. The total numbers of rotifer taxa (and also of crustaceans) were highest in the intermediate-hydroperiod ponds (32 and 26 taxa, respectively). However, even the 'ephemeral' ponds supported a comparatively rich rotifer fauna (21 taxa; plus 20 crustacean species) despite their small size and short wet phase.

Alkins-Koo (1989/90) found the colonial rotifer *Conochilus* mainly in the slower-flowing, deeper sections of two intermittent streams in southwestern Trinidad. In another Caribbean study, Janetzky *et al.* (1995) surveyed the rotifers of inland waters on Jamaica, including those species found in phytotelmata and gastrotelmata. The former, primarily in bromeliads, are a typical feature of the understory of the Wet Limestone Forests of the Cockpit Country and represent the only persistent waterbodies there. The latter normally hold water only in the rainy season (September–November) and develop water quality characteristics based on water source (direct rainfall or throughfall via vegetation), the nature of debris in the shell, and by materials dissolved from the shells themselves. The survey produced a list of 205 species (179 monogononts and 26 bdelloids), dominated by three genera: *Lecane* (25%), *Cephalodella* and *Lepadella* (10% each). Ten species were exclusive to aquatic microhabitats: six in gastrotelmata (belonging to the genera *Cephalodella, Habrotrocha* (also known from the northern pitcher plant *S. purpurea*), *Macrotrachela*, and *Rotaria*); three in phytotelmata (*Collotheca, Lecane*, and *Macrotrechela*); and one in rock pools (*Lecane*). Overall, the Jamaican rotifer fauna resembles that of South and Central America.

Tardigrades

These are very small coelomate animals, ranging from 50 μm up to 1.2 mm in length. There are about 600 species which are found in terrestrial, freshwater, and marine habitats. There are four body segments, each bearing a pair of short legs which end in claws. The body is covered by a cuticle which is shed periodically throughout the animal's life, and which may be either smooth, sculptured, and/or armoured. Although the cuticle contains chitin, it has not developed into the rigid exoskeleton typical of arthropods. It is believed that rigidity would prevent anhydrobiosis (dormancy induced by loss of body water), specifically the formation of the 'tun'stage. This form of the animal requires that the body length be reduced to about 50% through loss of water from the tissues, folding of the dorsal intersegmental cuticle, and invagination of the legs (resulting in reduced surface area and volume). During this process, trehalose (a protective, non-reducing sugar) is synthesized. Again, unlike many arthropods, the tardigrade cuticle is permeable, although it has special lipid layers and air-filled spaces which allow the animal to control cuticular transpiration.

In many terrestrial and freshwater species, parthenogenesis is common and, in some species, there appear to be no males. Parthenogenesis is thought to be linked to the development of anhydrobiosis which is rare in marine species. The eggs of terrestrial species are protected by a thick, ornamented shell. The eggs of aquatic tardigrades are either attached singly to suitable substrates or are laid inside the female's recently shed cuticle. There is no larval stage and a small version of the adult animal emerges from the egg after from 5 to 40 days, depending on species and environmental conditions. In most tardigrades, growth is achieved by an increase in cell size rather than in cell number, however mitosis has been observed in some eutardigrades.

Anhydrobiosis represents one of five types of latency identified by Crowe (1975), the others being: encystment, anoxybiosis, cryobiosis, and osmobiosis. When in one of these states, metabolism, growth, reproduction, and aging are reduced or suspended, and resistance to environmental extremes such as drought, heat, cold, chemicals, and radiation increases (Nelson 1991). For example, tardigrades can tolerate immersion in liquid helium at −272°C, temperatures as high as 340°C, and can also survive exposure to 570,000 roentgens of radiation (1,140 times the lethal dose for humans). Non-marine tardigrades are incredibly resistant to desiccation (doubtless an adaptation to the intermittently wet habitats in which many of them live). Both adult tardigrades and their eggs can enter a state of deep hibernation which can last for at least 100 years. These dormant stages may be dispersed by the wind. Transition from marine to terrestrial environments has been accompanied by a shift from predominantly striated muscle cells in the tardigrade body towards predominantly smooth muscles. Again, this may be related to the development of anhydrobiosis.

Many tardigrades live in moist, semiterrestrial habitats, such as in damp soil and leaf litter, and among lichens, mosses, and liverworts. Occasionally, they are found in temporary waters and in interstitial environments such as the hyporheic zone. Ramazzotti and Maucci (1983) classified moss-dwelling species into three groups: wet, moist, and dry mosses. Wet mosses, such as those found around lake and stream margins, seldom dry completely. These contrast with mosses growing on trees, rocks, roofs, and walls, which frequently dry out and are dependent on rain for rehydration. *Echiniscus molluscorum* has been found living in the moist faeces of the land snail *Bulimulus exilis* (Fox and Garcia-Mol 1962).

Few ecological studies exist on tardigrades, as is exemplified by the fact that out of 39 recent studies of regional wetlands in North America (Batzer *et al.* 1999), only one includes tardigrades in its faunal inventory. Yozzo and Diaz (1999) list the following seven species from tidal freshwater marshes on the James River, Virginia: *Isohypsibius saltursus*, *Macrobiotus richtersii*, *M. dispar*, *M. furcatus*, *M. hufelandii*, *Hypsibius* sp. (Eutardigrada: Macrobiotidae), and *Echiniscus* sp. (Heterotardigrada: Scutechiniscidae). In a rare study of the population dynamics of two species living in damp roof moss in Wales, Morgan (1977) found population densities of up to 823 individuals g^{-1} of moss. Temporal variation in the numbers of both *Echiniscus testudo* and *M. hufelandii* appeared to be cyclical and positively correlated with temperature and the number of daylight hours. Further, increases in humidity and rainfall, 10–20 days prior to sampling, produced a decrease in densities, but an increase in the size of individuals. In an attempt to expand on the moss 'wetness/dryness' scale (Petersen 1951) for categorizing moss-dwelling tardigrades, Kathman and Cross (1991) explored correlations with moss species, degree of exposure, and an altitudinal gradient, but none was found. They concluded that species composition was probably related to microenvironmental conditions within individual moss tufts.

Annelids

Annelid worms known from temporary waters primarily comprise species from the classes Oligochaeta and Hirudinea, although freshwater Polychaeta (Sabellidae: *Manayunkia*) have occasionally been found (Yozzo and Diaz 1999). Species belonging to the Branchiopdellida might also be expected to occur in these habitats by virtue of their ectosymbiotic relationship with freshwater crustaceans, in particular astacid crayfishes.

Leeches tend to live in lakes, ponds, and the slower-flowing reaches of streams and rivers, and are often an important component of the benthic community; in small ponds they often represent the top predators (Davies 1991). Several species of leech are known to tolerate degraded water conditions, including anoxia (for up to 60 days). Their distribution is related, at least in part, to the total dissolved solids (TDS) content of the water. For example, the mortality of *Erpobdella punctata*, a common wetland species, has been shown to be strongly influenced by an interaction among water temperature, ionic content, and TDS, whereas *Nephelopsis obscura* was relatively unaffected. For both species, cocoon production was most influenced by temperature. Interestingly, cocoon production and viability in *E. punctata* increased in waters with high TDS (Linton *et al*. 1983), the latter often signalling the end of the hydroperiod. Some leech species (such as those belonging to the Erpobdellidae) seem preadapted to living in temporary waters as their thick-walled cocoons have a spongy outer layer thought to reduce water loss. Further, some species actually deposit their cocoons out of water, in damp places.

Although it is likely that all leeches are genetically iteroparous (i.e. they reproduce several times in their life), Davies (1991) believes that they tend to exhibit semelparity (single reproductive event in their life) under most conditions. He suggested that such flexibility would promote the long-term persistence of a variety of genotypes in aquatic environments that are highly variable. *Helobdella stagnalis*, a species commonly found in temporary ponds, for example, can produce either one or two generations in a year, depending on water temperature (Davies and Reynoldson 1976). Growth of another common wetland leech, *N. obscura*, was found to be more sensitive to prey variation between years than to variations in temperature (Davies 1991). In turn, this leech has been shown to strongly influence the spatial distributions of the benthic chironomid larvae upon which it feeds (Rasmussen and Downing 1988).

Species of leech that occur in temporary waters seem to have cosmopolitan distributions, probably related to the ease with which some are transported, attached to waterfowl and/or macrophytes, and also to their reproductive flexibility. Some are also capable of surviving droughts as adults in mucus-lined cysts.

Many of the oligochaete species that live in temporary waters are also cosmopolitan, and frequently occur in permanent waters. Also, there appears to be less separation of species between running- and standing-water habitats than is seen in many other animal groups. Further, most species seem tolerant of a range of sediment sizes (e.g. sand to mud), and niche separation may depend more on the organic and microbial contents of sediments, together with the intensity of interspecific competition (Brinkhurst and Gelder 1991). Temporary waters support species chiefly from the families Lumbriculidae, Enchytraeidae, Naididae, and Tubificidae.

Lumbriculids typically live among vegetation at pond and stream margins, and commonly appear in temporary water faunal lists. Adult *Lumbriculus variegatus* have the ability to form cysts, but, under experimental conditions, survival without water has not lasted longer than 4 days—although cysts protected by mineral particles may fare better. A different type of cyst is developed in response to freezing (Olsson 1981), and some lumbriculids have been observed to undergo fission while encysted. *Tenagodrilus musculus*, a new genus and species of lumbriculid has recently been described from a temporary pond in Alabama (Eckroth and Brinkhurst 1996), which perhaps underscores their lack of study in such habitats.

The Enchytraeidae is generally regarded to be a terrestrial family, however there are numerous records of these worms occurring in both saturated and moist sediments, including pond basins, stream beds, and marshes (e.g. Williams 1993; Yozzo and Diaz 1999).

Naidids occur in a wide variety of aquatic habitats but reach their greatest abundance in running waters with fairly coarse substrates. In standing waters they are more common in the littoral zone, especially on macrophytes with finely divided leaves; populations tend to be reduced on mud and silt substrates. Reproduction is primarily asexual, via budding and fragmentation, and may be controlled by food availability and water temperature. Among those populations that produce mature individuals, there appears to be only one sexual generation per year, with adults dying after cocoon production (Whitley 1982). There is some evidence to suggest that adults are more resistant to environmental stress and hence sexual reproduction is not used to survive adverse conditions (Brinkhurst and Gelder 1991). Genera that commonly appear in the wetland literature include *Chaetogaster, Dero, Nais, Stylaria,* and *Pristina* (e.g. Gathman *et al.* 1999; Taylor *et al.* 1999).

Tubificid worms represent a large heterogeneous family whose species often possess respiratory pigments that allow them to survive in oxygen-poor conditions such as the terminal stages of many temporary waters. Species often have interesting physiological and behavioural properties such that species mixtures may exhibit reduced respiration rates together with increased growth rates and assimilation efficiencies, compared with monocultures (Brinkhurst and Austin 1978). Because they often occur at very high densities, and through feeding on basin deposits, tubificids can exert a significant effect on sediment properties. Reproduction is typically by sexual means, but parthenogenesis is known in some species. Development can be rapid, depending on the food supply and water temperature both of which may promote considerable local variation in life cycles. Once mature, species frequently breed several times each year. Cosmopolitan tubificid species that commonly appear in the wetland literature include *L. hoffmeisteri, Tubifex tubifex,* and *Branchiura sowerbyi* (e.g. Euliss *et al.* 1999; Hall *et al.* 1999). Adults are capable of surviving drought by forming cysts. As such, *T. tubifex*, for example, has been shown to be able to survive for 14 days in dry sediment, but for up to 70 days if occasionally moistened. Once moistened, individuals can resume normal activities within 20 h (Kaster and Bushnell 1981). Presumably, the cocoons have some desiccation resistance, too.

Molluscs

Both gastropods and bivalves are represented in temporary waters. The bivalve fauna of North America, to take an example, is the most species-rich in the world, with most of them belonging to a single superfamily, the Unionacea (227 native species), which contains many endemics. Typical habitats include permanent lakes and river systems. In contrast, the family Sphaeriidae contains only 33 native (plus 4 introduced) species, but many of these are found in small, frequently temporary, ponds, lakes, and streams (McMahon 1991). Further, most sphaeriids have broad distributions, often extending from the Atlantic to the Pacific coasts. These result from effective dispersal mechanisms that include: juveniles clamping their shells onto the limbs of aquatic insects and salamanders, and onto the feathers of waterfowl; and survival of some individuals ingested and regurgitated by ducks, which feed on them. In addition, sphaeriids are self-fertilizing hermaphrodites which enables a single individual to found a new population—a property favouring successful colonization of isolated drainage systems and temporary ponds. Sphaeriid genera appear to have different substrate preferences that are perhaps associated with different organic matter contents. There are also physiological differences with, for example, the genera *Sphaerium* and *Musculium* being relatively intolerant of low oxygen levels, but many species of *Pisidium* surviving well under hypoxia (Burky 1983). Many sphaeriids (and a few unionaceans) are very tolerant of exposure to air and are thus able to exist at the margins of fluctuating permanent waters, or in truly temporary waters.

The adaptations that allow sphaeriids to survive exposure for periods of up to several months have been studied in some detail, and to some extent seem to be stage-specific (e.g. adults versus newly hatched individuals). Individuals frequently burrow into the basin sediments during the last part of the hydroperiod, and oxygen consumption and rate of metabolism drop. For *Sphaerium occidentale*, Collins (1967) showed that, in air, oxygen requirement was 20% of that in water, exchange in air taking place across specialized pyramidal cells that extended through minute pores in the shell. The latter allows gas exchange while keeping the valves closed to conserve water. The problem of dealing with metabolite detoxification while exposed appears to be solved through suspension of protein catabolism combined with greater dependence on carbohydrates. McKee and Mackie (1980) showed that tolerance of desiccation may involve prior physiological and biochemical adjustments, in that individuals of both *S. occidentale* and *M. securis* taken from populations aestivating in an already dry pond were more tolerant than individuals taken from a permanent pond.

The gastropod fauna of temporary waters comprises species belonging to two of the molluscan subclasses. The Prosobranchia (which are gill-breathers) are primarily marine and only a few species, from the order Mesogastropoda, are able to live in moist, semi-terrestrial habitats. On the other hand, the Pulmonata contains many species that have successfully adapted to live in both permanent and temporary freshwaters, and on land—in fact the freshwater forms are descendants of species that returned to water during the Mesozoic era (~70–250 million years ago) (Kozloff 1990). Foremost among these adaptations is possession of a lung-like structure formed from part of the mantle and which communicates with the outside environment through a closable aperture known as the pneumostome. The lung has spongy walls and a rich blood supply which, along with a slight positive pressure when the pneumostome closes, facilitate gas exchange when the lung is filled. Carbon dioxide-rich air is expelled when the pneumostome opens and the lung floor is raised. Some pulmonates fill their lungs with water and use them as gills.

Other adaptations to living in variable environments include the shell (and also the closable operculum in prosobranchs), production of slime to prevent drying (including production of a mucous epiphragm to seal the shell aperture in pulmonates), and direct development, which eliminates free-living larval stages in the life cycle. Alongside these largely morphological adaptations are a number of physiological ones such as extreme temperature tolerance (0–40°C for the group as a whole), and excretion of nitrogen as urea, a relatively non-toxic form that can be safely stored in the blood during drought hibernation. Pulmonates are also capable of withstanding low oxygen tension by means of breathing at the water surface and switching to anaerobic respiration (McMahon 1983; Brown 1991). The fact that many pulmonate species successfully survive being passively dispersed on plants, birds, and insects (as eggs or adults), enables them to colonize new and isolated habitats, and results in broad distributions (e.g. *Lymnaea stagnalis, Physa gyrina, P. integra, Gyraulus parvus*). A measure of this success has been provided by Davis (1982) who recorded immigration rates to ponds as high as nine species in a single year. Successful establishment in new locations may require habitat features known to influence gastropod populations, such as suitable substratum particle sizes, the presence of macrophytes, moderate-to-high water hardness and pH, and the absence of molluscivorous fishes.

Although pulmonates tend to dominate shallow and temporary waters, a number of prosobranch families are also found in these habitats, for example, Hydrobiidae, Viviparidae, Valvatidae, and Pomatiopsidae. The most abundant pulmonates in similar habitats belong to the Lymnaeidae, Physidae, and Planorbidae. Apart from the adaptations already listed, pulmonates also exhibit considerable adaptive plasticity in their life cycles. For example, Thomas and McClintock (1996) found that the physid *Physella cubensis* survived habitat drying in warm-water temporary ponds in Alabama by burrowing into the bed sediments. In addition, when water was present, it exhibited several 'opportunistic' strategies, such as rapid juvenile growth and attainment of maturity, high fecundity, and continuous reproduction. Other, temperate, pulmonate species have been observed reproducing in cold water, allowing early breeding in the spring followed by rapid maturation before the hydroperiod ends.

Crustaceans

Crustaceans are an ancient group dating back to marine environments of the lower Cambrian era. Of the approximately 40,000 extant species known, about 10% live in inland aquatic habitats. The latter include some extreme habitats, such as polar ponds and streams, hotsprings, highly saline lakes, as well as a wide range of intermittent and episodic waters. For convenience, coverage here will subdivide crustaceans into three main groups: the large, but primitive Branchiopoda; the large, more advanced Malacostraca; and the microcrustaceans.

In terms of diversity, temporary ponds are frequently highly species-rich, especially in their zooplankton. For example, ponds on the upper coastal plain of South Carolina support 44 species of cladoceran and 7 species of calanoid copepod (Mahoney *et al.* 1990). In a single temporary pond in western Morocco, Thiery (1991) found six species of Anostraca, two of Notostraca, and two of Conchostraca (Spinicaudata). Petrov and Cvetkovic (1997) similarly found up to seven species of large branchiopod in a single pond in Yugoslavia. Eder *et al.* (1997) provided evidence for the temporal segregation (related to hydrology, temperature, and water chemistry requirements) of branchiopods in Austrian waters: the anostracans *Chirocephalus shadini* and *Eubranchipus grubii*, and *Lepidurus apus* (Notostraca) occurred in late winter and spring; *Branchinecta ferox*, *B. orientalis* (Anostraca), *Cyzicus tetracerus* (Conchostraca), and *Chirocephalus carnuntanus* occurred only in spring; the anostracans *Streptocephalus torvicornis* and *Tanymastix stagnalis*, and *Eoleptestheria ticinensis* (Conchostraca) were present in spring and summer; and *Branchipus schaefferi* (Anostraca) was found only in summer and autumn.

Despite there being limited comparative data on temporary water crustaceans, properties of the Branchiopoda, and one of its orders in particular—the Anostraca (brine shrimp and fairy shrimp), are better known. Anostracans are important members of the communities of temporary lentic waters throughout the world (Table 4.3). The majority (17) of present-day genera (23) appear either to have

Table 4.3 World distribution of the 23 genera of Anostraca

	Eurasia	Africa	N. America	S. America	Australia	Antarctica
Branchinecta	*	*	*	*		*
Artemia	*	*	*	*	*	
Branchinella	*	*	*	*	*	
Streptocephalus	*	*	*			
Linderiella	*	*	*			
Branchinectella	*	*	*			
Branchipodopsis	*	*				
Branchipus	*	*				
Tanymastix	*	*				
Chirocephalus	*	*				
Artemiopsis	*		*			
Drepanosurus	*					
Polyartemia	*					
Siphonophanes	*					
Tanymastigites		*				
Metabranchipus		*				
Dexteria			*			
Eubranchipus			*			
Polyartemiella			*			
Thamnocephalus			*	*		
Dendrocephalus				*		
Phallocryptus				*		
Parartemia					*	

Source: After Belk (1981)

evolved since the formation of the modern continents or represent relict groups. Resolution of these alternatives is hampered by the paucity of their fossil record (Tasch 1969). Modern distribution patterns that suggest extensive distribution in Pangea (the single super-continent that existed in the Paleozoic era, some 200 million years ago) are seen in only three genera—*Artemia, Branchinella,* and *Branchinecta*. Three other genera, *Artemiopsis, Linderiella,* and *Streptocephalus,* have current distributions that suggest that they were widely distributed across Laurasia (the northern landmass that was derived from the breakup of Pangea in the late Triassic era, some 180 million years ago; Figure 4.8). Belk (1981) proposed that many of the presumptive Pangea genera do not now occur on all of the modern continental fragments of Pangea because of ecological factors. *Branchinecta,* for example, is a genus of cold-water adapted species that is particularly common in the Canadian Arctic and Alaska. It is absent from Australia which, when part of the east coast of Pangea, had a warm climate (Bambach *et al.* 1980). Similarly, *Artemiopsis,* a presumptive Laurasian cold-water genus, is restricted to cold northern regions of Eurasia and North America (e.g. the Northwest Territories of Canada). Conversely, *Branchinella* and *Artemia,* which are warm-water forms, are absent from modern-day Antarctica. The limited distribution of *Artemia* in Australia may be due, in part, to competition from Australia's native brine shrimp, *Parartemia* (Geddes 1980). The absence of streptocephalids from Australia and South America suggests that *Streptocephalus* (another presumptive Laurasian genus) colonized Africa after the breakup of Gondwanaland and has since undergone a major adaptive radiation there—though mostly in the temperate zone (Brtek 1974; Belk 1981). *Streptocephalus sealii* is the most widely distributed anostracan in the United States (27 out of 48 states). In contrast, 18 of the 43 species known from the United States have been reported from only a single state each (Jaas and Klausmeier 2000).

Anostracan species richness, in general, is always highest in temperate parts of the globe. In an attempt to account for climate being an important controlling factor in their distribution, Belk (1981) compared the anostracan faunas of Arizona and South India, two climatically different regions. Arizona is a region in the temperate zone that experiences marked seasonal differences in climate and includes environments that range from lowland desert to high alpine. The anostracan fauna is represented by 13 species from five genera (*Artemia, Branchinecta, Eubranchipus, Streptocephalus,* and *Thamnocephalus*). South India is a region in the tropical zone that experiences only slight seasonal changes in climate. The area of the Western Ghats mountains and to the west is humid and anostracans are not found there, but east of the mountains is a semi-arid area where temporary ponds are very common. Here the anostracan fauna is represented by only six species from three genera (*Artemia, Branchinella,* and *Streptocephalus*). The variation in thermal regime associated with refilling of temporary ponds is much less in southern

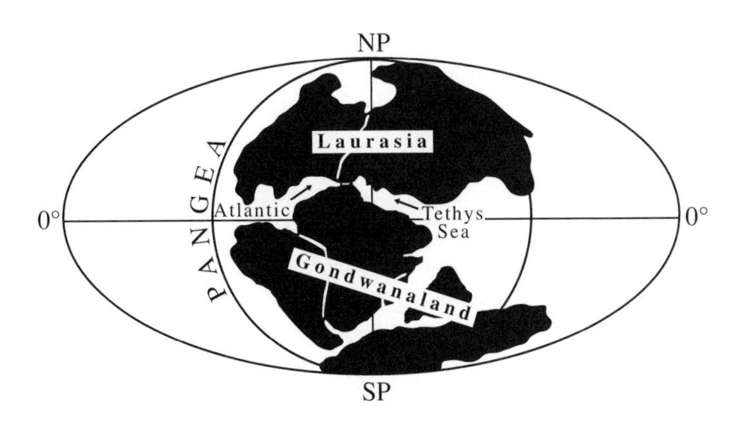

Figure 4.8 Map showing the break-up of Pangea in the late Triassic era (180 million years BP).

India than in Arizona, and Belk maintained that because there are significant differences between the temperature requirements for egg hatching between species, a lower diversity is to be expected in the less variable environment. Southern Africa is particularly rich in anostracans, with 80% (38 species) being endemic. However, the fauna is under threat from agricultural practices, urbanization, pollution, and pesticides (Hamer and Brendonck 1997). In the Sahel, Senegal, downwind drift from routine application of a variety of insecticides used to control Desert Locusts has been shown to be detrimental to anostracans and cladocerans living in local ponds (Lahr 1998). Vekhoff (1997) recorded a comparatively rich branchiopod fauna from the high arctic islands and archipelagoes of the Barents Region of Russia (four anostracans, one notostracan, and two conchostracans). This high diversity was thought to be the result of the fauna comprising both widespread tundra species and species that had dispersed into the region from the nearby mainland. One species, *Branchinecta paludosa*, occurred in 45–50% of the crustacean communities, and reached its most northerly distribution (77°N) at Ivanov Bay.

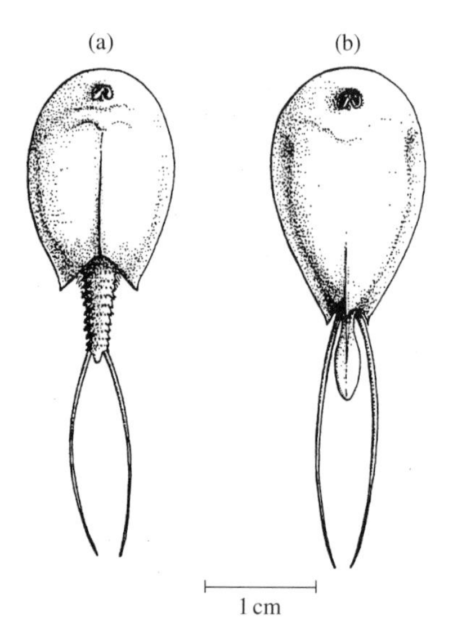

Figure 4.9 Notostraca: (a)—*Lepidurus arcticus* Pallas 1793; (b) —*Lepidurus couesii* Packard 1875.

There are also global and regional differences in the distribution of the Notostraca (tadpole shrimps). Rzoska (1984) stated that the genus *Lepidurus* is northern and may occur in Arctic conditions, whereas the genus *Triops* (=*Apus*) is confined to warmer waters, especially in arid regions and hot climates. In the United States, *Triops longicaudatus* seems to be confined to the drier western half of the country, and its range extends down into Mexico, the West Indies, the Galapagos Islands, Argentina, and the Hawaiian Islands. There is one record for Canada, in Alberta (Figure 4.9). Recent electrophoretic evidence, however, suggests that *T. longicaudatus* is actually a mixture of at least two reproductively isolated species (*T. longicaudatus* and *T. newberryi*)—the former typical of central, prairie pools, the latter more characteristic of large, southwestern, playa pools (Sassaman *et al.* 1997).

The genus *Lepidurus* is represented by several species in North America and these again occur chiefly in the western states but also in central and arctic Canada. *Lepidurus arcticus*, (Figure 4.10), for example, is found only in Alaska, the Northwest Territories and Labrador, plus in Greenland (Figure 4.9). *Lepidurus couesi* (Figure 4.10) is found in Montana, North Dakota, Oregon, Idaho, and Utah in the United States; Alberta, Saskatchewan, and Manitoba in Canada; and in Russia, Northern Siberia and Turkestan in the Palearctic (Linder 1959; National Museum of Canada records). In Australia, both *Lepidurus* and *Triops* occur in virtually all states except Tasmania, where only *Lepidurus* is known; only *Triops* has been found in the Northern Territories. However, *Lepidurus* is largely confined to the more temperate southeastern and southwestern corners of the continent, while *Triops* tends to be more common in the dry interior, although it also occurs on the coast in a few regions (Figure 4.11). These distribution patterns seem to be correlated with regional differences in climate particularly in terms of mean annual temperature and evaporation (Figure 4.12). Even though the ranges of these two genera overlap in the southeast, they are never found coexisting in the same body of water and this appears to be true for much of the rest of the world also (Williams 1968).

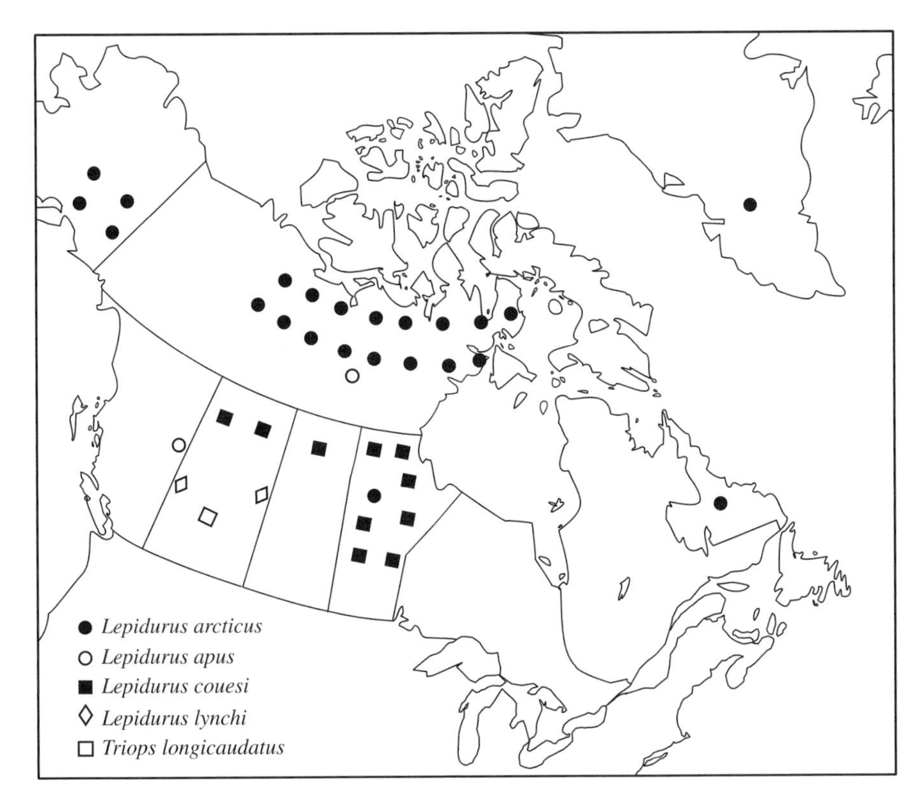

Figure 4.10 Records of anostracans from Canada, Alaska, and Greenland (dots represent single collections made in a particular province/state—they do not accurately depict where the collections were made. Based on records in the Canadian Museum of Nature, Ottawa, and in Linder 1959).

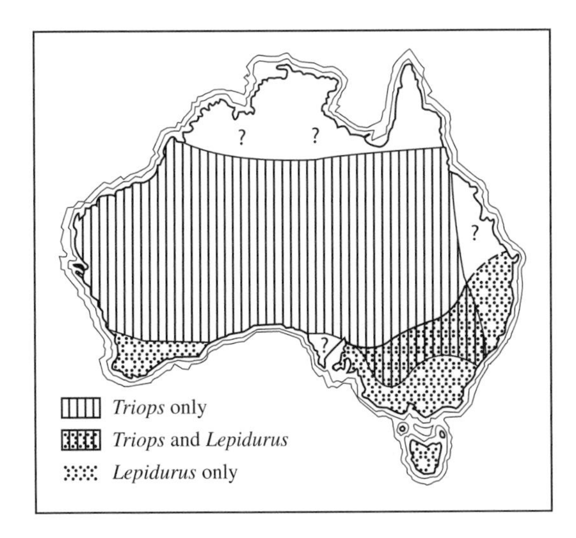

Figure 4.11 Geographical distribution of *Triops* and *Lepidurus* (Notostraca) in Australia (redrawn after W.D. Williams 1968).

The Conchostraca (clam shrimps) have a wide geographic distribution primarily in temporary waters but also in the littoral zone of lakes. In general, they are found in warmer waters than most anostracans. In common with other large branchiopods, their preference for temporary waters is believed to be a stratagem for avoiding predatory fishes against which they seem relatively defenceless. Some species have very extensive distributions while others are known only from their type locality. For example, in North America, *Lynceus brachyurus* is found across the United States as well as in most Canadian provinces (except for the Maritimes), the Northwest Territories and Alaska (Figure 4.13). It is also found in Europe and Asia. *Cyzicus mexicanus* occurs across most of United States and Mexico, and also in Alberta and Manitoba (Mattox 1959;

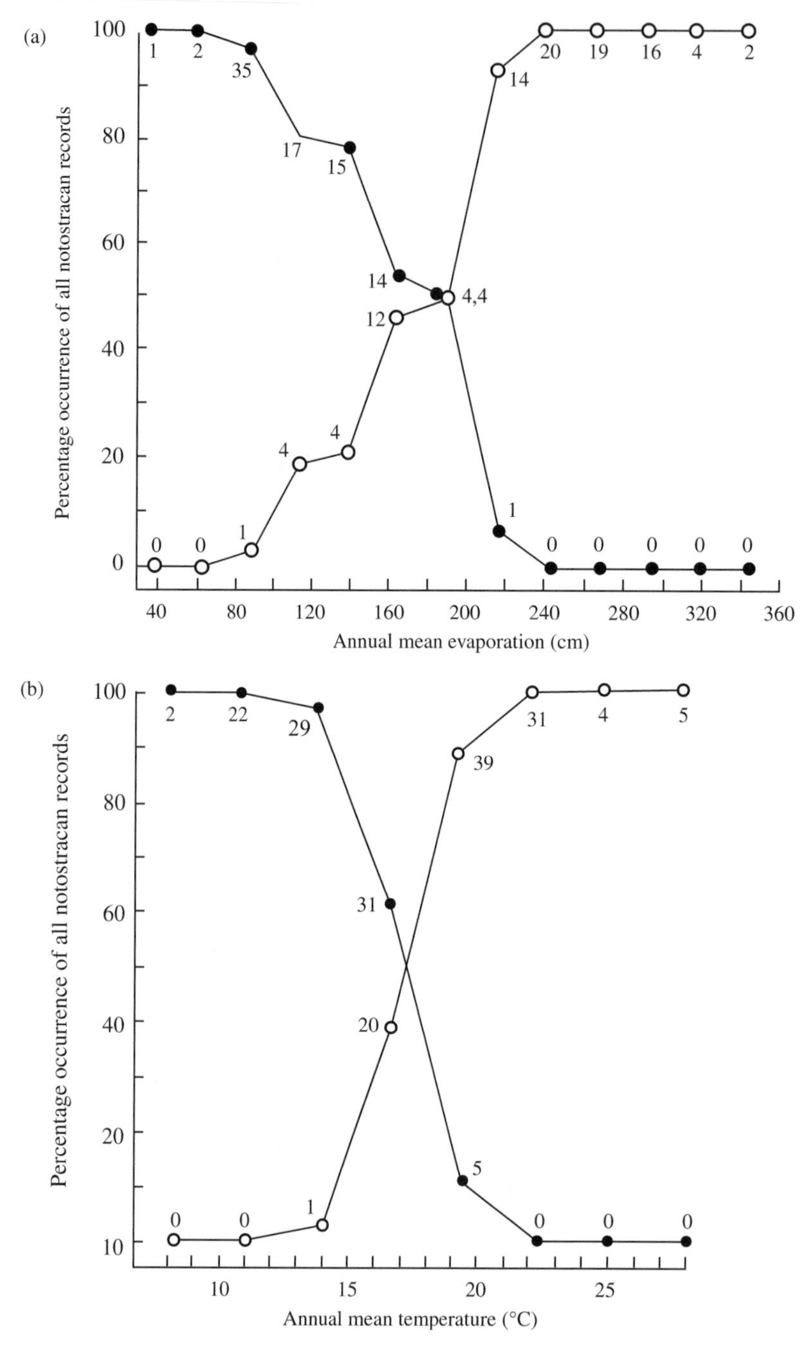

Figure 4.12 Distribution of *Triops* (open circles) and *Lepidurus* (closed circles) (Notostraca) with respect to: (a)—isoclines of annual mean evaporation; (b)—isotherms of annual mean temperature (numbers of records are given; redrawn after W.D. Williams 1968).

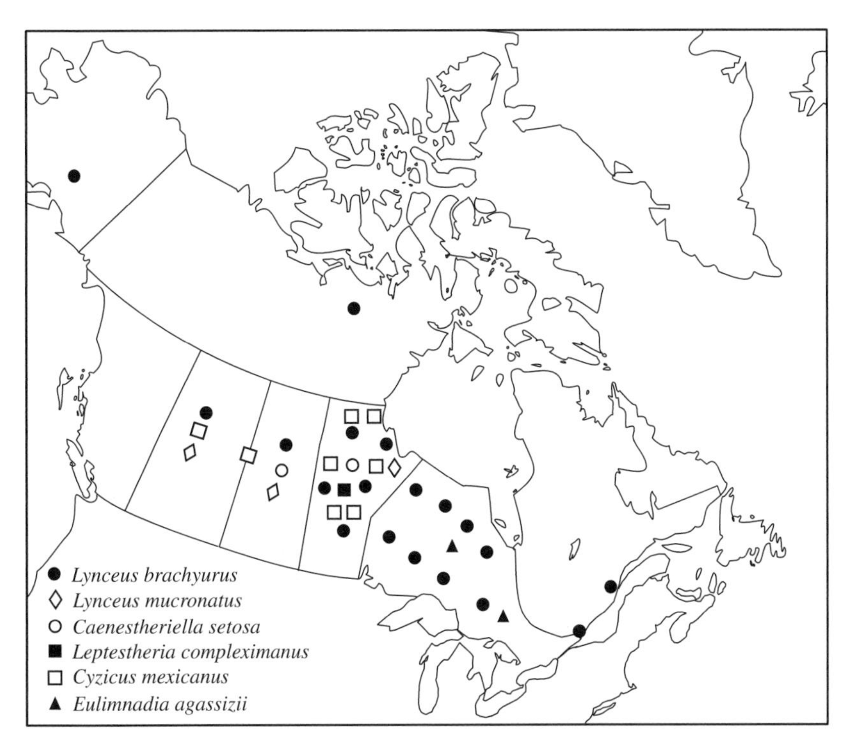

Figure 4.13 Records of conchostracans from Canada and Alaska (dots represent single collections made in a particular province/state—they do not accurately depict where the collections were made. Based on records in the Canadian Museum of Nature, Ottawa, and in Maddox 1959).

National Museum of Canada records). Some species of *Eulimnadia*, however, have very restricted distributions. *Eulimnadia alineata* and *E. oryzae*, for example, have been recorded only from rice fields at Stuttgart, Arkansas (Mattox 1959). In Colombia, *Eulimnadia magdalenensis* seems highly adapted to very short-hydroperiod ponds in arid regions, whereas *E. colombiensis* prefers longer-lasting ponds in cooler locations (Roessler 1995). Three species of *Eulimnadia* have also been recorded from Australia (from the Northern Territories, central and Western Australia, Queensland, and Tasmania). In total, there are some 23 species of Conchostraca in Australia which are all, with the exception of *Cyclestheria hislopi*, endemic (Williams 1980). *Cy. hislopi* has a circumtropical distribution and is one of the few conchostracans capable of existing in permanent waters. Here, it demonstrates effective anti-fish predator behaviours including: hiding among vegetation; production of a special, protective mucous capsule; and direct development, in which larvae leave the safety of the maternal carapace only when they are capable of secreting their own capsule (Roessler 1995).

Most members of the Malacostraca (Amphipoda, Isopoda, Mysidacea, and Decapoda) lack the ability to deal with desiccation and have no capability for diapause, adaptations that would suit them to life in temporary waters. There are, however, a limited number of exceptions. The amphipod *Hyallela azteca*, for example, is a widespread species in permanent waters in North America, but is occasionally found in intermittent ponds and streams where it appears to survive the summer in small, isolated pools. In the wetlands of western Canada, *H. azteca* along with *Gammarus lacustris* occur at such high densities that they influence the abundance and distribution of waterfowl, specifically the lesser scaup (*Aythya affinis*) (Lindeman and Clark 1999). Two species of *Crangonyx*, *C. setodactylus* and *C. minor*, are common in intermittent streams in eastern North America where they reproduce and grow during the lotic phase, but retreat into moist substrate or share the burrows of crayfishes as the hydroperiod ends (Williams and Hynes 1977). There are very few records of isopods from temporary waters. Apart from terrestrial forms, such as species of *Oniscus* that forage on bed detritus during the dry phase, only *Caecidotea* is recorded in the North American literature (Batzer *et al.* 1999). Paucity of peracarids in temporary waters is perhaps surprising given that many species occur in subterranean and interstitial habitats—there are around 116 species of amphipod that live in groundwaters in North America, alone (Covich and Thorp 1991).

In contrast, many decapod species are associated with temporary waters, in particular those belonging to the Astacidae (crayfishes) and Brachyura (crabs), and occasionally the Palaemonidae (shrimps). They are able to live in these habitats because of a range of adaptations that include: physiological adaptations—such as an ability to regulate their oxygen consumption (seen in some cambarid crayfishes); life cycle modifications—such as brooding of their young until an advanced stage; and behavioural adaptations—such as burrowing into the substratum to follow retreating groundwater tables. Burrowing behaviour in North American crayfishes has been categorized into: (1) primary burrowers (e.g. *Fallicambarus devastator*), which typically live most of their lives within complex tunnels that may be some distance from open water (Figures 4.14(a,d)); (2) secondary burrowers (e.g. *F. hedgpethi*), which also tend to live mostly within their burrows, but seek out open water during rain events (Figure 4.14(b)); and (3) tertiary burrowers (e.g. *Procambarus acutus*), which live in open water when it is available, but retreat into burrows at other times (e.g. to brood their eggs, to avoid desiccation, and to move below the frost line in winter; Figure 4.14(c,e)). Within their burrows, individuals do not always immerse themselves in water, as they have the ability to respire in humid air. Some non-burrowing species are also capable of digging simple shafts in bed materials to avoid drought (Hobbs 1991). During drought, a variety of other species are known to share the refugial water of crayfish burrows (see Chapter 5).

Crayfishes are known to have a significant impact on temporary water environments. For example, when *Procambarus clarkii* was introduced into a freshwater marsh in Spain, it altered the structure and function of the marsh community—primarily through impact on the pre-existing food web (Gutierrezyurrita *et al.* 1998).

Several species of true crab (Brachyura) live in brackish- or freshwater wetlands. Indeed, such habitats may have played a significant role in the evolution to terrestrial life seen in, for example, some of the grapsid crabs of Jamaica. There are nine endemic species of land crab on this island, some of which show active brood-care for their larva and juveniles, a phenomenon not known elsewhere in this group. These species demonstrate a range of dependence on water along with specific adaptations to a variety of habitats. The most specialized of these are *Metopaulius depressus*, which raises its larvae in water-filled bromeliad leaf axils, that it defends against predators, and *Sesarma jarvisi*, which breeds in empty snail shells that it keeps topped-up with rainwater (Schubart *et al.* 1998).

Microcrustaceans are represented in temporary waters by the smaller branchiopods (Cladocera), copepods, and ostracods. Cladocerans are typically found in lentic habitats, and many species have cosmopolitan distributions—although some of the latter may be confounded by taxonomic inadequacies. Many species appear to be capable of producing haemoglobin facultatively, when faced with low oxygen levels. Reproduction is parthenogenetic and rapid (often just 1 or 2 days) for much of the hydroperiod, and the number of instars varies according to both genetic and environmental factors. Males appear close to the end of favourable habitat conditions (pre-winter, or pre-drought) and sexual, resting eggs are produced thereafter. These eggs are protected by an ephippium and can withstand freezing, drying, and dispersal to new habitats. Many cladocerans exhibit a wide tolerance of environmental conditions such as pH, temperature, and dissolved oxygen, yet others are restricted to very narrow

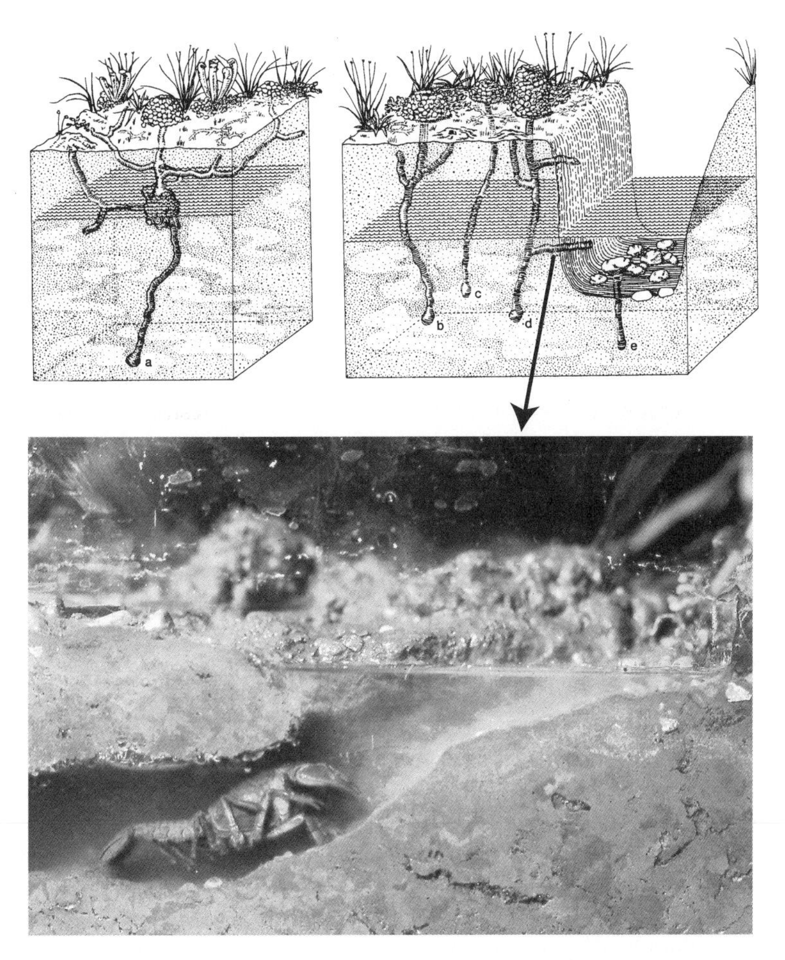

Figure 4.14 Different forms of North American crayfish burrows in relation to ground- and surface water levels (a and d—primary burrowers [see photo of *Fallicambarus fodiens*]; b—secondary burrowers; c and e—tertiary burrowers; redrawn after Hobbs 1981).

ranges of water chemistry, especially dissolved calcium and salinity. In permanent waters, planktivorous fishes are believed to be influential in structuring cladoceran communities; in temporary waters they may be affected by invertebrate predators such as odonates and beetles.

Most of the 12,000 known species of copepod are marine. However, species belonging to the three main orders, Calanoida, Cyclopoida, and Harpacticoida have successfully colonized most freshwater habitats, in all climatic zones. Temporary waters in which they live include snow-melt water pools, desert pools, phytotelmata, damp moss, and wet leaf litter on forest floors and in compost bins. Reproduction is largely by sexual means (except in a few harpacticoids), with fertilized eggs being carried, and possibly still nourished, by the female for several days until their release. There then follow six naupliar stages, followed by six copepodite stages, the last being the adult. Development time can take as little as 1 week, depending on habitat conditions. Diapause is a dominant feature in the life cycle of many species. Harpacticoids and calanoids are capable of producing two types of eggs: subitaneous eggs that hatch within a few days of being laid, and resting eggs capable of long periods of dormancy. Production of egg type seems to depend, at least partly, on environmental cues. Cyclopoids produce only subitaneous eggs, however individual animals may enter a resting stage as late copepodites. The latter may be either a simple developmental arrest, or full diapause (Williamson 1991). Harpacticoids are also capable of near-adult diapause under unfavourable conditions. More information on the adaptations of crustaceans, in general, to temporary waters will be given in Chapter 5.

Ostracods are very common in temporary waters but records often lack detail due to difficult taxonomy—although this is now improving. Fossil records show that the earliest freshwater forms lived in coal-forming swamps, ponds, and streams of early Pennsylvanian age (Benson 1961). Reproduction may be either sexual or asexual, and eggs may hatch soon after being laid, or may be staggered over time. Eight moult stages are passed through before the mature adult emerges. The eggs of freshwater ostracods have a structure (a double wall of chitin impregnated with calcium carbonate, with a fluid-filled space in between) that endows great resilience to environmental adversity, especially desiccation and freezing (Delorme 1991). Drought resistance was one of the criteria listed by Marmonier *et al.* (1994) as characterizing species found in temporary ponds on the floodplain of the Upper Rhone River—the others being: high reproductive rates, short life span, spherical shape, long swimming bristles, and low thigmotaxis. For North American temporary ponds, Delorme (1991) cited the following species association as being typical: *Candona renoensis, Cypricercus deltoidea, Megalocypris alba*, and *Cyclocypris laevis*. In some habitats, such as the semi-arid and arid regions of southern Africa, diversity and endemicity are remarkably high, comparable only to the ostracod faunas of very stable habitats, for example, ancient Lake Tanganyika (Martens 1996).

As shown in Chapter 3, from North American data (Table 3.2), the length of the hydroperiod seems to significantly influence both macro- and microcrustacean diversity (see also Ebert and Balko 1987), with the greatest number of species occurring in habitats that contain water for between 150 and 250 days in a year. Apart from the anostracan *Eubranchipus* and the conchostracan *Lynceus*, microcrustaceans (comprising three cladocerans—*Daphnia, Scapholeberis*, and *Simocephalus*, the calanoid *Leptodiaptomus*, and several cyclopoid copepods) persisted in habitats containing water for less than 70 days. Further, the cyclopoids and *Scapholeberis* survived in hydroperiods of less than 10 days per year. In a study of English wetlands, Jeffries (2001) showed that, at the species level, the relationship between length of the dry period and the subsequent presence of microcrustaceans was negative for some but positive for others, probably contingent on the characteristics of individual ponds. Laird (1988) found that cladocerans (*Chydorus sphaericus* and *Daphnia middendorffiana*), an ostracod (*Cyclocypris globosa*), a cyclopoid (*Acanthocyclops vernalis*), and unidentified harpacticoids were the only crustaceans present in his northern Quebec snow-melt pool (hydroperiod more than 3 weeks). Additionally, a species of the cladoceran genus *Moina*, and two cyclopoids (*Mesocyclops thermocyclopoides* and *Microcyclops varicans*) were the only two crustaceans in his episodic puddle (hydroperiod less than 9 days) in Singapore. Mature specimens of *Moina dubia* were recorded within 2 days of refilling in a short-lived tropical rainpool in the Sudan (Rzoska 1961). Numerically, Lahr *et al.* (1999) found cladocerans to be the dominant taxon present early on in the 7-month hydroperiod of a temporary pond in the arid Sahel region of Senegal. Later, the ponds were dominated by 'more slowly establishing' crustaceans, particularly copepods, which then gave way to insects.

Insects

A low-resolution view of the major groups of aquatic insects (including the Collembola) known to occur in temporary running waters is presented in Figure 4.15. The horizontal axis in this figure represents a relative scale of habitat permanence. Habitats such as stable, coldwater springs are placed at the high permanence end, while temporary streams are placed at the low permanence end. Extending this logic, episodic streams would be placed at a lower permanence level than intermittent streams. However, as there has been little distinction between these two latter

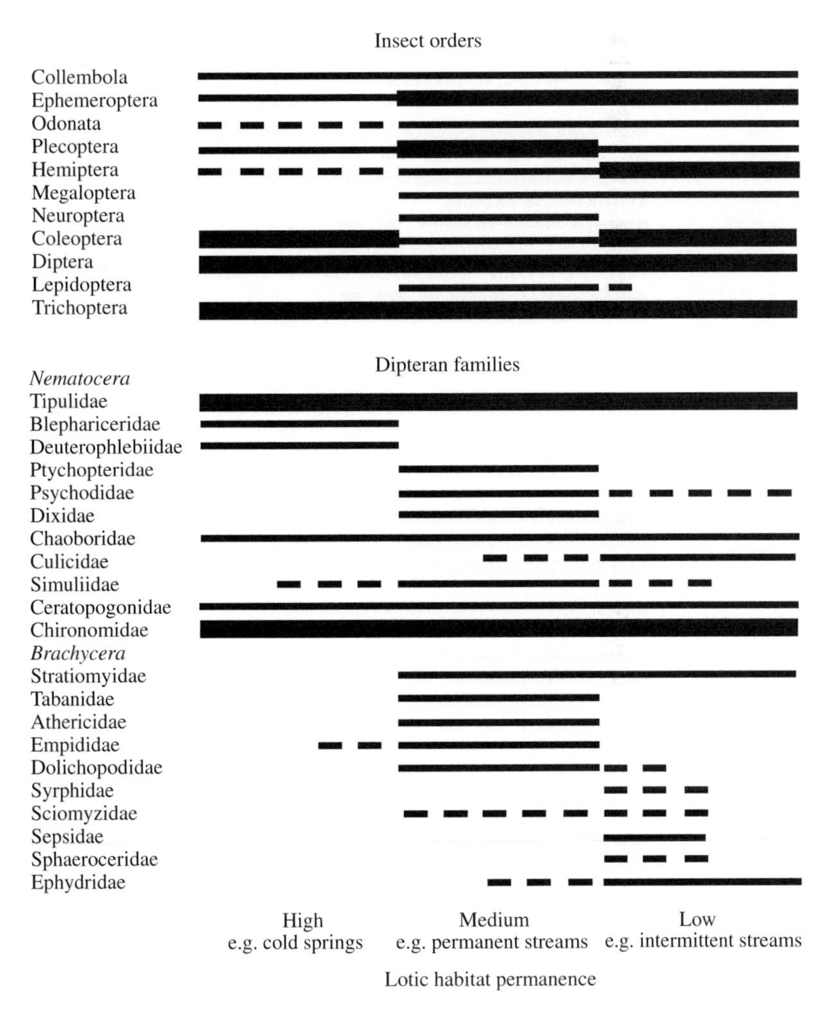

Figure 4.15 Overview of the relative importance of different aquatic insect orders and dipteran families across a gradient of running water habitat permanence (thick line indicates 'well represented', thin line indicates 'moderately well represented', broken line indicates 'present but not significantly so'; information taken mainly from Merritt and Cummins 1984, and Williams and Feltmate 1992).

habitat types in the past literature, the present analysis necessitates combining distributional records under the more general category of 'temporary' streams. In temporary lotic waters, there appear to be no records for the Neuroptera and Lepidoptera (although pyralid larvae have been collected from a stream in southern Australia; A.J. Boulton, Personal Communication), and the Plecoptera and Odonata are poorly represented. Particularly well-suited to temporary streams are the Ephemeroptera, Hemiptera, Coleoptera, Trichoptera, and Diptera. Higher resolution (Figure 4.15) shows that the Tipulidae and Chironomidae (Nematocera) dominate the temporary stream dipteran fauna (Williams and Feltmate 1992), whereas brachycerans are comparatively rare.

With the exception of mayflies (although a few species may attain large populations in vernal pools, Wiggins *et al.* 1980), all of the orders commonly found in temporary running waters also are common in temporary ponds (Figure 4.16). In addition, the odonates are well-represented. The most common nematoceran Diptera in

Figure 4.16 Overview of the relative importance of different aquatic insect orders and dipteran families across a gradient of standing water habitat permanence (thick line indicates 'well represented', thin line indicates 'moderately well represented', broken line indicates 'present but not significantly so'; information taken mainly from Merritt and Cummins 1984, and Williams and Feltmate 1992).

temporary ponds are the Tipulidae, Culicidae, Ceratopogonidae, and Chironomidae, together with the brachyceran Ephydridae. Odonates and dipterans also are well-represented in highly specialized habitats such as phytotelmata (Corbet 1983; Beaver 1985).

A number of observations can be made from these data. First, they indicate that, at the level of orders and families, temporary waters do not support diverse insect faunas as are found in permanent waters. Second, some very general inferences can be made about the adaptive traits and trade-offs demonstrated by different taxa. For example, it appears that Plecoptera seem less able to deal with standing water conditions than with drought, and survive drought only in running water habitats. Conversely, several dipteran families (e.g. Ptychopteridae, Tabanidae, and Empididae) are capable of living in temporary standing waters but not apparently in temporary running waters (although there are records of empidids from a semi-permanent spring in Germany and from two temporary streams in Australia; F.O. Gathmann, Personal Communication, and

Boulton and Lake 1992a, respectively). Despite their considerable powers of flight, Trichoptera and aquatic Hemiptera and Coleoptera do not colonize phytotelmata significantly. Neuropterans seem not to have developed the capability of living in temporary habitats even though they are found in a range of permanent waters.

Although some of these conclusions are tentative, due to inadequate study of temporary waters, they nevertheless point to some interesting avenues for future research into evolutionary trends in habitat selection among aquatic insects. Particularly useful would be determination of why the Plecoptera, Odonata, and brachyceran Diptera are poorly represented in temporary streams, yet the Tipulidae and Chironomidae are successful; why mayflies are not widespread in temporary ponds despite being occasionally abundant locally; what mechanisms, physiological or otherwise, result in decrease in species diversity with decrease in the length of the aquatic stage of the habitat; and why the Trichoptera, Hemiptera, and Coleoptera are poorly represented in phytotelmata. In addition, there needs to be more extensive surveying of habitats in order to confirm the global nature of temporary water communities and to extend/confirm the known ranges of taxa along the permanent–temporary water habitat gradient.

The relationship between a habitat permanence gradient and community structure can best be examined when a dataset contains a high level of taxonomic resolution. Some examples of this relationship, for insects, have already been given in Chapter 3 (e.g. Table 3.1). Another, intruiging insect dataset is provided by Wissinger *et al.* (1999), who compared the insect species found in 41 subalpine wetland basins in Colorado. These basins were divided into three groups, based on their number of open-water days: 40–64; 74–116; and greater than 130 (Table 4.4). The taxon counts were 10, 26, and 50, respectively. Clearly, numerically, these data support the trend of decreasing species richness with decreasing length of the hydroperiod. However, examination of the taxa in the >130-day hydroperiod group, which was designated 'permanent' by the authors, reveals quite a number of taxa frequently seen in temporary waters, such as the hemipterans; the beetle genera *Helophorus, Agabus, Acilius, Dytiscus, Rhantus*; the dipterans *Bezzia* and *Stratiomyia*, and several chironomid genera. The interesting point that emerges about the >130-day hydroperiod group is that although none of these basins has dried out completely in the past 50 years, they are only free of ice and snow cover for about 4 months of each year, and most are frozen to within just a few centimetres of their beds. It could be argued, therefore, that they do not truly represent permanent water habitats, and that their short window of opportunity for colonization attracts pioneering taxa, resulting in a community with distinct *temporary* water elements.

In addition to the largely aquatic taxa described above, there are a number of other insects associated with temporary waters, especially at the end of the hydroperiod, when there is decaying aquatic vegetation and moribund aquatic prey to be harvested. Little attention has been paid to these, and other arthropod taxa (see below), thus their roles in temporary water ecosystems are poorly understood, but perhaps important. Groups of note are the grylloblattids (rock crawlers), ants, wasps, and various terrestrial beetles.

Lude *et al.* (1999) found that at least nine species of ant were able to survive frequent inundation on Alpine floodplains. In particular, *Formica selysi* regularly colonized relatively young, unvegetated gravel islands and bars, and survived flooding by forming swimming rafts when its nest entrances were compromised. Each raft consisted of several dozen workers, a queen, and brood, and remained intact until it reached the shoreline. Clearly, this ant species found suitable foraging opportunities on these periodically flooded areas. Other hymenopterans associated with seasonally flooded sites include wasps and bees. For example, Visscher *et al.* (1994) recorded the ground-nesting bee *Calliopsis pugionis* emerging from sites that had been underwater for more than 3 months in the floodplain of the San Jacinto River. The authors suggested that the flooding regime influenced the sex ratios of bees as they emerged from diapause.

Carabid beetles are known to be commonly associated with damp environments, including

Table 4.4 Comparison of the dominant insect taxa in 41 subalpine wetland basins in Colorado with different degrees of permanence (shown as the number of open-water days in the year)

Taxonomic group	**Number of open-water days**		
	40–64 days	***74–116 days***	***>130 days***
Ephemeroptera			
Baetidae	—	—	*Callibaetis* sp.
Odonata			
Corduliidae	—	*Somatochlora* sp.	*Somatochlora* sp.
Lestidae	—	—	*Lestes* sp.
Coenagrionidae	—	—	*Coenagrion* sp.
	—	—	*Enallagma* sp.
Aeshnidae	—	—	*Aeshna* sp.
Hemiptera			
Corixidae	*Arctocorixa* sp.	*Arctocorixa* sp.	*Arctocorixa* sp.
	—	*Callicorixa* sp.	*Callicorixa* sp.
	—	—	*Coenocorixa* sp.
Notonectidae	—	—	*Notonecta* sp.
Gerridae	—	*Gerris* sp.	*Gerris* sp.
Trichoptera			
Phryganeidae	—	—	*Agrypnia* sp.
Limnephilidae	*Asynarchus* sp.	*Asynarchus* sp.	*Asynarchus* sp.
	Limnephilus coloradensis	*Limnephilus coloradensis*	*L. coloradensis*
	—	*L. picturatus*	*L. picturatus*
		L. externus	*L. externus*
	—	*Hesperophylax* sp.	—
Coleoptera			
Hydrophilidae	—	—	*Helophorus* sp.
Haliplidae	—	—	*Haliplus* sp.
Gyrinidae	—	—	*Gyrinus* sp.
Dytiscidae	*Stictotarsus* sp.	*Stictotarsus* sp.	*Stictotarsus* sp.
	—	*Sanfilippodytes* sp.	*Sanfilippodytes* sp.
	—	*Agabus kootenai*	*Agabus kootenai*
	—	*A. tristus*	*A. tristus*
	—	*A. strigulosus*	*A. strigulosus*
	—	*Acilius* sp.	*Acilius* sp.
	—	*Dytiscus* sp.	*Dytiscus* sp.
	—	—	*Ilybius* sp.
	—	—	*Rhantus* sp.
Chrysomelidae	—	—	*Plateumaris* sp.
Staphylinidae	—	—	*Stenus* sp.
Diptera			
Chironomidae			
Tanypodinae	—	*Procladius* sp.	*Procladius* sp.
	—	—	*Ablabesmyia* sp.
Diamesinae	—	—	*Pseudodiamesa* sp.
Orthocladiinae	*Psectrocladius* sp.	*Psectrocladius* sp.	*Psectrocladius* sp.
	—	*Cricotopus* sp.	*Cricotopus* sp.
	—	—	*Corynoneura* sp.
	—	—	*Eukiefferella* sp.

Table 4.4 (*Continued*)

Taxonomic group	Number of open-water days		
	40–64 days	***74–116 days***	***> 130 days***
Chironominae			
Chironomini	*Chironomus riparius*	*Chironomus riparius*	*C. riparius*
	—	*C. salinarius*	*C. salinarius*
	—	—	*Cladopelma* sp.
	—	—	*Dicrotendipes* sp.
	—	—	*Endochironomus* sp.
	—	—	*Microtendipes* sp.
	—	—	*Pagastiella* sp.
Tanytarsini	*Tanytarsus* sp.	*Tanytarsus* sp.	*Tanytarsus* sp.
	Paratanytarsus sp.	*Paratanytarsus* sp.	*Paratanytarsus* sp.
	—	—	*Cladotanytarsus* sp.
Tipulidae	*Limnophila* sp.	*Limnophila* sp.	—
Ceratopogonidae	—	*Bezzia* sp.	*Bezzia* sp.
	—	—	*Culicoides* sp.
Culicidae	*Aedes* spp.	*Aedes* spp.	—
Stratiomyidae	—	—	*Stratiomyia* sp.

Note: Absence of a particular taxon is shown by a dash (data from Wissinger *et al.* 1999).

wetlands, floodplains, etc. Basta (1998) recorded 94 species from a single marsh in the Czech Republic, where population densities of some of the most abundant species seemed tied to fluctuations in water level. Carabids were also identified as a major component of the riparian fauna on four alpine floodplains in Bavaria, where most of their prey were aquatic species. In particular, on the Isar floodplain, river-derived invertebrates represented 89% of carabid prey, primarily emerging chironomids (fed upon by small species of *Bembidion*) and stoneflies (favoured by *Nebria picicornis*) (Hering and Plachter 1997). Juliano (1985) found a variety of carabids of the genus *Brachinus* associated with different pond types in Arizona. *Brachinus lateralis* dominated the margins of more permanent ponds, whereas high-elevation temporary ponds were dominated by *B. mexicanus*. *Brachinus javalinopsis* and *B. lateralis* co-dominated the margins of a low-elevation temporary pond, but only *B. mexicanus* was found in dry pond basins. All three species were believed to share at least two potentially limiting resources: food for the adults (carrion and other arthropods), and water beetle pupae (required as hosts for their ectoparasitoid larvae). Lott (2001) reported that temporary ponds in lowland England support a rich beetle fauna, comprising chiefly carabids and staphylinids. Some of the species are those associated with the margins of larger, pemanent water bodies and river floodplains. Assemblages differ among ponds, and at least part of this can be attributed to bed substrate composition (e.g. mineral versus peat). A number of interesting adaptations to being occasionally submerged by water have been observed (e.g. retreating to air pockets in the bed litter, retreating up the bank prior to a flood advance, or skimming over the water surface after secretion of a water repellant), however, the ecology and behaviour of most species are largely unknown.

Arachnids and other arthropods

The arachnids most commonly found associated with temporary waters can be divided into three informal groups: water mites, which are primarily associated with the hydroperiod; soil mites, which are more likely to be associated with the moist sediments of the basin after the water has evaporated; and spiders, which, apart from some aquatic and semi-aquatic forms, live in the riparian zone, but also may move onto the drying bed to scavenge.

Water mites are considered as those belonging to five somewhat unrelated subgroups of small arachnids. The Hydrachnida is the most familiar, but some forms belonging to the Oribatida, Halacaridae, Mesostigmata, and Acaridida (the latter three being of minor importance) have invaded freshwaters, giving rise to species that are now fully adapted to an aquatic existence. Water mites may be extremely abundant in ponds and the littoral zones of shallow lakes, for example, 2,000 individuals m^{-2}, representing as many as 75 species. Many mites represent important micropredators in temporary waters, and some have coevolved with major insect groups (especially the Diptera), both parasitizing their bodies and using their adult stages as vessels of dispersal to new habitats (Smith and Cook 1991). In North America, families/subfamilies of hydrachnids commonly found in temporary waters include the: Hydrachnidae, Eylaidae, Piersigiinae, Hydryphantinae, Thyadinae, Tiphyinae, Pioninae, and Arrenurinae. In general, hydrachnids are able to survive in temporary waters by one of two means: physiological endurance, or avoidance—for example, larvae of the Eylaidae and Hydrachnidae can remain attached to their adult insect hosts for the entire duration of the dry phase (Wiggins *et al.* 1980; Smith and Cook 1991).

The ecology of oribatid mites is not well known, partially due to problematic taxonomy. However, they are extremely abundant in most forested ecosystems and often comprise around 50% of the total microarthropod fauna. As such, oribatids are believed to play important roles in the decomposition of organic materials, modification of the physical and chemical textures of soil, the cycling of nutrients, and the conservation of healthy soil environments (Wallwork 1983). Habitat features thought to influence their population biology include habitat complexity and soil micropore size, soil humidity and organic content, soil temperature, surface vegetation, precipitation, and the activity of other soil microfauna (Wauthy *et al.* 1989). Because forest soils are often contiguous with temporary pond and stream beds, it is likely that oribatid mites contribute to biological processes (largely unknown) in these sediments. That there are characteristic assemblages of these mites in temporary waters is illustrated by the fact that oribatids have been found useful in distinguishing between different stages of degradation of freshwater mires in the area of Berlin (Kehl 1997). In coastal freshwater habitats in Antarctica, two species of terrestrial oribatids, *Edwardzetes elongatus* and *Trimaloconothrus flagelliformis*, have become adapted to survive prolonged submergence on aquatic mosses—apparently in response to niches unoccupied by truly aquatic forms (Pugh 1996).

There is only one species of truly aquatic spider. *Argyroneta aquatica* (Agelenidae), found in ponds of central and northern Europe, builds an underwater air-store from silk fashioned in the form of an inverted vase. It supplies this store with air brought from the water surface, and uses it as a medium in which to externally digest prey captured under water. Many other spiders, however, frequent the margins of both permanent and temporary ponds and streams. All members of the Pisauridae, for example, depend on being close to water, and some, such as *Dolomedes* dive while hunting for tadpoles, aquatic insects, and small fishes. There is a rich diversity of pisaurids in Australia, for example, including: *Dendrolycosa* (which also extends from New Guinea to India), *Dolomedes* (cosmopolitan), *Hygropoda* (also in southeast Asia, Venezuela, New Guinea, Madagascar, and central Africa), *Perenthis* (also in India, Burma, Papua New Guinea, and Japan), and the endemics *Inola* and *Megalodolomedes*. (Main *et al.* 1985). Another family of common riparian spider is the Tetragnathidae—the long-jawed orb weavers. The genus *Tetragnatha* has a wide global distribution, with *T. versicolor* and *T. elongata* being circumboreal. Williams *et al.* (1995) calculated that, as a minimum estimate, individuals of these two species captured 0.2% of the total number of insects (particularly chironomids and mayflies) emerging from a small river in Canada. Tetragnathid webs are common on any vegetation overhanging temporary ponds and streams, where their prey also includes emerging mosquitoes and other long-legged dipterans. The diversity of spiders associated with temporary waters can be very high, for example, van Helsdingen (1996) recorded a total of 63 species from two Irish floodplains.

As introduced above, a number of other, non-insect arthropods are associated with the end of the hydroperiod. Whereas zoogeography will dictate local composition, this 'clean-up crew' may include millipedes, centipedes, symphylans, pauropods, pseudoscorpions, harvestmen (Opiliones), diplurans, bristletails (Archaeognatha), and silverfishes (Thysanura).

4.2.6 Vertebrates

Fishes

Apart from highly specialized forms, such as lungfishes which can aestivate in the bottom mud during the dry phase, fishes tend to be absent from temporary ponds. Temporary streams, on the other hand, may support fish populations of considerable size and diversity. This is the result of either some streams becoming intermittently connected to permanent waters from which fishes can migrate, or fishes surviving in permanent pools left in some drying streambeds. Studies of the ecology of fishes in temporary streams are, however, few in number. Possibly, this is because although large rivers do occasionally dry up, particularly in the tropics, intermittency, particularly in temperate regions, is more a characteristic of smaller bodies of water that usually support populations of fishes of little economic or recreational importance. There does exist, however, a more substantial literature on those fish species that use temporary pools on the floodplains of large rivers during part of their life cycles; these will be discussed in more detail in Chapter 8.

Three species of fish are known from the Kalahari Desert. *Clarias gariepinus* is a catfish that possesses suprabranchial respiratory organs. The other two species are both cichilds, *Tilapia sparrmanni* and *Hemihaplochromis philander*; all three species can survive in very little water as stunted individuals (Cole 1968).

In North America, out of a total of 50 species of fish found by Williams and Coad (1979) in the Grand River watershed, Ontario, only 12 were collected from three intermittent tributaries. Species in the latter were largely members of the families Cyprinidae (minnows) and Percidae (perches), with one species from each of the Catostomidae (suckers) and Gasterosteidae (sticklebacks). The intermittent streams entirely lacked catfishes, sunfishes, and salmonids. The brook stickleback *Culaea inconstans* and the cyprinids *Pimphales notatus* and *P. promelas* showed physiological tolerance in that they survived poor water quality, high temperatures, and crowding in shrinking summer pools. *Catostomus commersoni* (white sucker) and *Semotilus atromaculatus* (creek chub) on the other hand, moved into the tributaries, spawned, and then left. The main advantages to fish species colonizing intermittent streams appear to be plentiful food, earlier spring breeding (as the water is often warmer than in adjacent permanent streams), and reduced predation by large fishes.

A problem faced by fishes moving into intermittent streams is that they may become stranded and die if the pools dry up completely. The longfin dace, *Agosia chrysogaster*, possesses behavioural adaptations that contribute to its success as the only species to consistently use intermittent streams in the Sonoran Desert of Arizona. These streams dry up to form pools separated by lengths of dry streambed. The fishes position themselves in the current and this minimizes the chances of them becoming stranded by falling water levels. This species quickly invades new habitats during wet periods when flow is continuous and is capable of existing for at least 14 days in areas where there is no free water, provided that there is moisture beneath mats of algae (Minckley and Barber 1971; Bushdosh 1981). Avoidance of the stream edges and shallows reduces predation from birds and mammals. In contrast, high predation pressure from large aquatic and terrestrial predators has been put forward to explain the coexistence of three catfish species in swamps in Suriname. Mol (1996) suggested that as drying swamp pools become more restricted, predators, such as caiman and birds, exert so much pressure on populations of *Hoplosternum littorale*, *H. thoracatum*, and *Callichthys callichthys* that none can dominate.

Another fish species that is adept at surviving in intermittent streams is *Poecilia reticulata*, the guppy. Although more frequently studied in the clear water, upland permanent streams of the Northern Range of its native Trinidad, populations thrive in the lowland rivers of the southwestern part of the island. During the dry season, hundreds of individuals survive in shallow (5–10 cm), highly turbid, streambed pools no more than 1 m or so in diameter, where they appear to survive by respiring the thin surface layer of oxygen-rich water (Alkins-Koo 1989/90). Alkins-Koo's study focused on two intermittent streams on the Chatham Peninsula: the Carlisle and Quarahoon rivers, both of which range from 1 to 8 m wide and up to 3 m deep in the rainy season, to isolated or chains of pools in the dry season. From 1980 to 82, she collected 31 species from these two rivers (Table 4.5), which, despite the relatively small drainage area and the interrupted flow regime, represented almost half of the freshwater fish fauna of the island. Alkins-Koo suggested that this high diversity may have resulted from several habitat features: (1) an intermediate level of disturbance, resulting from the annual flood–drought cycle (see Townsend *et al.* 1997); (2) the presence of extensive pool refugia; and (3) the dynamic state of the local fauna on the peninsula, due to continued colonization from the nearby mainland of Venezuela. In terms of adaptations to surviving in these two streams, Alkins-Koo pointed out that many of the species are known to be able to endure fluctuating environmental conditions, especially stagnation and hypoxia, increased predation, and crowding. Several species are capable of breathing air, and *Rivulus hartii*, *C. callichthys*, and *Synbranchus marmoratum*, for example, can move overland, allowing them to colonize new habitats. In particular, *S. marmoratum* can survive without free water, in burrow systems (Kramer *et al.* 1978).

Survival and between-pool migration of *Clarias anguillaris* on the floodplain of the Sokoto-Rima River, in Nigeria, was also possible by this species' ability to breathe air. However, Hyslop (1987) proposed that its highly varied diet (from higher plants and algae to invertebrates and other fishes) also contributed to its success in temporary waters. Ouboter (1993) summarized some of the other adaptations that allow fishes to survive in freshwater swamps in Suriname; these are summarized in Table 4.6.

The capacity of temporary water fishes to airbreath is currently causing some concern in the United States. An established population of airbreathing snakeheads (Channidae: *Channa marulius*) was discovered in Broward County, Florida, in 2000, and introduced populations have been captured in several other states (Courtenay and Williams 2005). Arising from Asia, these fish are voracious feeders with a diet that, in the United States, now includes native fish species. Snakeheads possess suprabranchial chambers for breathing air, combined with a ventral aorta that is divided into two parts that permit bimodal (air and water) respiration. Airbreathing allows channids to be very adept at dispersing among ponds via overland travel, creating the potential for displacement of native fish species.

Amphibians

Frogs and salamanders are common inhabitants of temporary ponds and cosmopolitan genera include *Rana*, *Hyla*, and *Ambystoma*. Species richness has been found to be highest in habitats that exhibit an intermediate level of spatial-temporal variability. In the Rhône Valley, for example, these habitats are mesotrophic temporary ponds, and species align themselves along a variability gradient (Joly and Morand 1997). The most unstable sites are colonized by the yellow-bellied toad, *Bombina variegata*. *Bufo calamita*, *Hyla arborea*, and *Pelodytes punctatus* (the parsley frog) are characteristic of ponds of intermediate stability. Where water levels are more predictable, *Bufo bufo*, *Rana temporaria*, and *R. lessonae* occur. *Salamandra salamandra* and *Alytes obstetricans* (the midwife toad) are absent from the floodplain, probably because the adults cannot tolerate submersion during floods. Species such as *R. dalmatina*,

Table 4.5 Teleost fish species collected from two intermittent streams on Trinidad, West Indies

Taxonomic group	Sampling station			
	Shallow pool (dry sometimes)	***Deeper pool (never dry)***	***Deep pool (watering hole)***	***Downstream pool (brackish water)***
Characiformes				
Erythrinidae				
Hoplias malabaricus	X	X	X	
Erythrinus erythrinus			X	
Gasteropelecidae				
Gasteropelecus sternicla	X	X		
Characidae				
Brycon siebenthalae		X		
Triportheus elongatus			X	X
Corynopoma riisei	X	X	X	
Astyanax bimaculatus	X	X	X	
Moenkhausia bondi	[elsewhere in main river]			
Hemigrammus unilineatus	X	X	X	X
Siluriformes				
Pimelodidae				
Rhamdia sebae	X	X	X	
Callichthyidae				
Callichthys callichthys			X	
Corydoras aeneus	X	X	X	
Gymnotiformes				
Gymnotidae				
Gymnotus carapo		X	X	
Cyprinodontiformes				
Aplocheilidae				
Rivulus hartiix	X	X		
Poeciliidae				
Poecilia reticulata	X	X	X	
P. picta				X
P. vivipara				X
Atheriniformes				
Atherinidae				X
Syngnathiformes				
Syngnathidae				X
Synbranchiformes				
Synbranchidae				
Synbranchus marmoratus		X	X	
Perciformes				
Centropomidae				
Centropomus parallelus				X
Serranidae				
Epinephelus itajara				X
Gerreidae				
Diapterus rhombeus				X

Table 4.5 (*Continued*)

Taxonomic group	Sampling station			
	Shallow pool (dry sometimes)	***Deeper pool (never dry)***	***Deep pool (watering hole)***	***Downstream pool (brackish water)***
Haemulidae				
Pomadasys sp.				X
Nandidae				
Polycentrus schomburgkii			X	
Cichlidae				
Cichlasoma bimaculatum		X	X	
Crenicichla alta	X	X		
Mugilidae				
Mugil curema				X
Gobiidae				
Sicydium punctatum				X
Pleuronectiformes				
Bothidae				
Citharichthys sp.				X
Soleidae				
Trinectes sp.			X	

Source: Data from Alkins-Koo (1989/90)

Table 4.6 Summary of some of the adaptations that allow fishes to survive in the freshwater swamps of Suriname

Taxonomic group	Adaptations
Electrophoridae	
Electrophorus electricus	Well-vascularized buccal cavity for air-breathing (gills have degenerated)
Callichthyidae	
Callichthys spp.	Air is swallowed and oxygen taken up through specialized part of the intestine + overland migration in *C. callichthys*
Hoplosternum spp.	Air is swallowed and oxygen taken up through specialized part of the intestine
Synbranchidae	
Synbranchus marmoratus	Supplementary pair of lung-like sacks in gill pouches + overland migration
Erythrinidae	
Hoplerythrinus unitaeniatus	Oxygen is taken up through highly vascularized portion of swim bladder + overland migration
Erythrinus erythrinus	Oxygen is taken up through highly vascularized portion of swim bladder
Hemiodontidae	
Hemiodopsis spp.	Respiration of oxygenated thin surface water layer
Lebiasinidae	
Copella spp.	Respiration of oxygenated thin surface water layer
Cichlidae	
Pterophyllum spp.	Respiration of oxygenated thin surface water layer
Aplocheilidae	
Rivulus spp.	Respiration of oxygenated thin surface water layer + overland migration in *R. urophthalmus*

Source: Data from Ouboter (1993)

R. ridibunda, and *Triturus helveticus* are ubiquitous, able to breed in ponds spanning the gradient.

Clearly, the aquatic larvae of all these species are severely threatened by any untimely onset of the dry phase of the habitat. Wilbur and Collins (1973) have suggested that an endocrine-controlled, metabolic-feedback mechanism exists in temporary pond species. Should the rate of larval growth be slow, metamorphosis to the adult stage is initiated once a certain minimal larval size is attained. Although the resulting small adult may face disadvantages in the terrestrial environment, these are less than those facing the larva if the pond dries up prematurely. If, however, the larval body size is small but its rate of growth is fast, metamorphosis is delayed so as to maximize the animal's growth potential in the pond. Control of metamorphosis is thus related to the stability of the habitat and species with a fixed size for metamorphosis are therefore excluded from temporary waters. In Western Australia, all the species of *Heleioporus* (burrowing frogs) breed in the winter and lay their eggs in a frothy mass in a burrow dug by the male. The site chosen is always one which will be later flooded by heavy winter rains.

Species of *Pseudophryne* (toadlet frogs) have similar egg-laying habits. As the rain raises the level of the water table in the burrow, the larvae break free of the egg mass and develop to metamorphosis often in no more than 0.5 l of water. In the two western species of *Pseudophryne*, larval development takes slightly longer than 40 days. The aquatic phase of their ponds seldom lasts more than 50 days so, potentially, there is little leeway. This is offset, however, by the ability of the eggs to complete embryonic development (6–8 days) in the absence of free water, so that the larvae are ready to hatch as soon as the rains come. In the event of a delay in rainfall, hatching can be postponed for several weeks (Harrison 1922; Main *et al.* 1959).

Adults of the five species of Australian *Heleioporus*, together with those of the genus *Neobatrachus* (spadefoot and other frogs), can withstand a drop in the water content of their bodies of up to 45% of their body weight. Rehydration rates in species of *Neobatrachus* vary according to the severity of water loss in their particular habitats. For example, species that live in the arid interior rehydrate faster than those from the wet coastal regions of the southwest. In contrast, *Heleioporus* shows no difference in rehydration rates among species spanning a wide spectrum of aridity. It is thought that because all species of *Heleioporus* are superior burrowers, selective pressures for increasing the speed of rehydration may not operate (Bentley 1966).

Burrowing seems to be a common method of surviving droughts in amphibians, as even the ability to get just a few centimetres below ground level places the animal away from the drying effects of sun and wind, and into a more moist environment. Even in deserts, moisture from past rains can remain trapped for years in sand at depths of only 20–30 cm (Bagnold 1954). *Scaphiopus couchi*, the spade-foot toad of California aestivates in its burrow surrounded by a layer of dried, skin-like material. This may help to limit evaporation of moisture from its body in much the same way as the cocoon-like covering of the African lungfish. Another species, *S. hammondi*, lines its burrow with a gelatinous substance that presumably slows down water loss. In this genus, aggregations of tadpoles have been found shortly before metamorphosis and subsequent emergence from temporary ponds (Bragg 1944). It has been suggested that these dense aggregations may conserve water, as the combined rapid beating of many tadpole tails tends to deepen that part of the pond basin and water from shallower parts of the pond will drain into it.

That amphibians, particularly frogs and toads, can successfully contend with the intermittent availability of water (though many people think of them as being associated with cool, moist environments) is evidenced by phenomena, such as those that occur in the western deserts of Australia. Here, after rain, the number of frogs emerging from burrows is so large as to interfere with rail transportation, as thousands of frogs are crushed as they attempt to cross railway lines thus making traction impossible (Bentley 1966). Colonization of new ponds by the alpine newt, *Triturus alpestris*, has been shown to be by both adults and juveniles (Joly and Grolet 1997).

Mayhew (1968) summarized the general adaptations of amphibians found in dry areas as follows:

— no definite breeding season;
— use of temporary waterbodies for reproduction;
— breeding behaviour initialized by rainfall;
— loud voice in male attracts both females and other males, resulting in rapid congregation of breeding animals;
— rapid development of eggs and larvae;
— omnivorous feeding habits of tadpoles;
— production of inhibiting substances by tadpoles which influence the growth of other tadpoles;
— high tolerance of heat by tadpoles;
— adults have metatarsal 'spades'for digging;
— ability to withstand considerable dehydration, compared with other anurans;
— nocturnal activity.

A particularly noteworthy observation on amphibian populations in temporary waters is that there exists a great deal of variation in reproductive characteristics (especially in egg and clutch size) among individuals. Kaplan (1981) has examined this variation in the light of theory on 'adaptive coin-flipping' or natural selection for random individual variation. Given two genotypes with equal mean fitnesses, the genotype with less variance in fitness will eventually outcompete the genotype with more variable fitness (Felsenstein 1976). Kaplan argues that natural selection for random individual variation has been overlooked because, in general, neo-Darwinian theory is a theory of genes and not a theory of development and, consequently, population geneticists often do not take into consideration influences of the environment on development. A well-buffered phenotype may be advantageous in many cases, but a less-well-buffered developmental system also might be of advantage to an individual, particularly in environments that are temporally variable. This is supported by the observations made on variation in size for the onset of metamorphosis, discussed earlier. Further evidence for the selection of phenotypic plasticity in amphibian populations living in highly variable environments has been provided by Van Buskirk (2002), although he concluded that species-specific attributes may sometimes obscure this relationship. Van Buskirk also suggested that *behavioural* solutions to surviving in such environments are likely to evolve under different scales of variation than morphological responses.

In a large-scale experiment, Wilbur (1987) demonstrated that when several anuran species coexist, assemblage structure is determined by an interaction between biological and environmental factors. In an assemblage comprising *Rana utricularia*, *Scaphiopus holbrooki*, *Bufo americanus*, and *Hyla chrysoscelis*, neither predation nor competition proved to be the single unifying force, instead they interacted in determining the consequences of the date of pond drying to the emergence success of each species. In another experiment, Blaustein *et al.* (1996) found that *Hyla savignyi* and *Bufo viridis* were heavily preyed upon in ponds containing larvae of the fire salamander *S. salamandra infraimaculata*. Interestingly, the tadpoles tended to be larger in the pools with salamanders. Not only did this predator affect the anuran populations, but it had an impact on most of the other pond inhabitants. For example, it reduced invertebrate species richness by 53%, eliminated the large cladoceran, *Simocephalus expinosus*, and reduced populations of the numerically dominant calanoid copepod, *Arctodiaptomus similis*, and a species of *Chironomus*. As a consequence of reduced invertebrate grazing pressure together with increased nutrient input from the salamander's excreta, periphyton and bacterial populations rose. Laurila and Aho (1997) examined whether adult behaviour of *R. temporaria* might play a role in larval survival through selection of ponds that are predator free. However, there was no evidence to support this, leading these authors to conclude that competition, in combination with pool desiccation, was the main factor affecting fitness, perhaps favouring females that primarily selected vacant pools, or pools with low densities of competitors.

Reptiles

A variety of terrestrial reptiles use temporary waters as sources of drinking water. In addition to these, there are several types of reptile that live in close association with both permanent and temporary waterbodies, for example, crocodiles, monitor lizards, turtles, and iguanas. Alkins-Koo (1989/90) recorded four species of reptile from two intermittent streams in southwestern Trinidad: the turtles *Phrynops gibbus*, *Rhinoclemmys p. punctularia*, and *Kinosternon s. scorpioides*, and the caiman *Caiman crocodilus*. In the Amazon basin, the breeding cycle of the latter appears to be closely linked to fluctuations in water level, with nests constructed in the grass mats that cover the margins of large lakes (Lang 1989). Three species of caiman have been recorded in the Pantanal wetlands of Brazil: *Paleosuchus palpebrosus* and *Caiman latirostris* (though both may be now locally extinct), and *C. yacare*, the jacaré, which is still abundant. The jacaré frequents shallow bays, where it hauls out during the day. Its food consists primarily of aquatic snails and small fishes, but aquatic insects, especially beetles and hemipterans, are important in the diet of juveniles (Por 1995). The Nile crocodile, *Crocodylus niloticus*, is known to use temporary streams that flow in the rainy season as dispersal routes to new permanent water bodies.

Ehrenfeld (1970) found that alligator holes (temporary ponds excavated by the reptiles) in the Florida Everglades serve as collecting ponds and biological reservoirs for the surrounding aquatic life—both vertebrate and invertebrate—in the dry season. Rich growths of algae and higher aquatic plants are nourished by the reptiles'droppings and these, in turn, maintain a variety of animal life. At the end of the drought, the survivors move out to colonize the glades anew. The Florida Everglades also support a large number of snakes and several lizards that commonly use temporary pools and marshes as habitats in which to feed, drink, or regulate their body temperature; these include the following: *Eumeces inexpectatus* (southeastern five-lined skink), *Ophisaurus ventralis* (eastern glass lizard), *O. compressus* (island glass lizard), *Nerodia floridana* (Florida green water snake), *N. taxispilota* (brown water snake), *N. fasciata* (Florida water snake), *N. clarkii* (mangrove salt marsh snake), *Seminatrix pygaea* (South Florida swamp snake), *Thamnophis sauritus* (peninsula ribbon snake), *Regina alleni* (striped crayfish snake), *Farancia abacura* (eastern mud snake), *Opheodrys aestivus* (rough green snake), *Drymarchon corais* (eastern indigo), *Elaphe obsoleta* (Everglades rat snake), and *Aghistrodon piscivorus* (Florida cottonmouth) (Steiner and Loftus 1997). Such a high diversity of predators must have a significant effect on the wetland communities. In the water hyacinth-choked savanna wetlands of Venezuela, anacondas (*Eunectes murinus*) prey on capybaras and wading birds that frequent these waters. These reptiles also are highly dependent on these habitats for other life processes, including mating (breeding balls), dispersal, and thermoregulation (Figure 4.17).

Birds

Wetlands are crucial habitats, in terms of both reproduction and feeding, for many bird species

Figure 4.17 A female anaconda (~4.5 m long) living in the water hyacinth-choked savanna wetlands of Venezuela (photograph provided and copyrighted by A. Chartier).

throughout the world. Indeed, a number of studies have demonstrated a close correlation between the annual production of waterfowl and the number of wetlands. For example, breeding densities and brood numbers of the mallard (*Anas platyrhynchos*) were found to be highest in years when the densities of ponds holding water were highest (Leitch and Kaminski 1985). Poiani and Johnson (1991) have pointed out that the quality of habitat for waterfowl is highly dependent on the mix of permanence types found in wetland complexes. For example, temporary pools provide an abundant food source early in the season, and tend to be heavily used by dabbling ducks in the spring. Seasonally flooded wetlands also provide nesting habitat and sites for brood-rearing for both dabbling and diving ducks, especially in years of high water. As wetland areas dry, birds capable of re-nesting tend to relocate to open-water areas (Swanson 1988).

In Australia, extensive breeding of waterfowl occurs in floodplain areas adjacent to rivers. In lightly wooded and treeless plains, vast areas of new waterfowl habitat are created when rivers overflow their banks. These shallow waters are soon colonized by huge numbers of grey teal, pink-eared ducks and shovelers, and by lesser numbers of black duck and white-eyed duck (Frith 1959; Timms 1997). Piscivorous bird species also have life cycles closely linked to floods such that, in Africa, fledglings are produced just when small fishes appear in floodplain pools. There is also heavy predation as the floodwaters recede and fishes become stranded in temporary pools and channels (Bonetto 1975). Foraging success of bald eagles (*Haliaeetus leucocephalus*) has been shown to be strongly dependent on fluctuating water levels of the Colorado River, Arizona, being higher at low flow, particularly in isolated pools (Brown *et al.* 1998).

In Asia, wetlands are extremely productive in terms of invertebrates, fishes, and amphibians, and this food base draws in both resident species, such as storks and herons, and migrants, such as waders and egrets. The wetlands of northern Asia, in general, and Siberia, in particular, are among the most extensive in the world, and regions, such as the Ob floodplain, are especially important breeding grounds for water birds. Not all wetlands are natural, however. In the nineteenth century, for example, the Maharaja of Bharatpur created what has become one of the finest water-bird sanctuaries in the world. Originally intended as a shooting preserve, its construction involved building small dykes and dams, and diverting water from an irrigation canal. Within a few years, this 29 km^2 low lying area became surrounded by marginal forests, and the complex now supports over 300 species of bird. Among these is the Siberian crane (*Grus leucogeranus*) which flies over 6,400 km from its summer retreat to get there.

Of course, birds also provide some benefits to wetlands, primarily in the form of nutrient input. For example, Lesser snow geese were measured as increasing the nutrient loading rates in wetland ponds in New Mexico by up to 40% for total nitrogen and 75% for total phosphorus; nitrogen proved to be consistently limiting to algal primary production in these ponds (Kitchell *et al.* 1999). Many bird species are also the primary dispersal mechanisms for temporary water invertebrates and plants, which sometimes show distribution patterns that coincide with migration routes.

Mammals

Temperate wetlands tend to support relatively few species of mammal. Those present either eat fishes and amphibians, like otters and mink, or feed on wetland vegetation, like mice, rats, beavers, and water-voles. Unfortunately, information on mammals frequenting isolated waterbodies, *per se*, seems rare. However, in a study of the use of a vernal pool by small mammals in chaparral and coastal sage scrub communities in southern California, Winfield *et al.* (1981) found that it did not appear to be used heavily. This was despite the fact that it provided a potential source of food in the form of protracted growth of riparian vegetation and semi-aquatic, ground-dwelling insects. Of seven common species of mouse, rat, and rabbit, only one, *Reithrodontomys megalotis*, the western harvest mouse, had a higher estimated population at the pool than elsewhere, but these results were considered tentative.

Table 4.7 Mammal species associated with a variety of Asian wetlands

Wetland	Location	No. species	Notable species
The Sundarbans (~6,500 km^2)	Bangladesh/India	49	*Panthera t. tigris* [Bengal tiger], *Felis chaus* [Jungle cat], *F. bengalensis* [Leopard cat], *Prionailurus viverrinus* [Fishing cat], *Axis axis* [Spotted deer], *Muntiacus muntiac* [Barking deer], *Sus scrofa* [Wild pig], *Macaca mulatta* [Rhesus Macaque]
Dongdongtinghu (~1,328 km^2)	Hunan Province, China	31	*Lipotes vexillifer* [Chinese river dolphin], *Neophocaena phocaenoides* [Finless porpoise], *Neofelis nebulosa* [Clouded leopard]
Bharatpur (~29 km^2)	Keoladeo National Park, India	29	*Vulpes bengalensis* [Bengal fox], *P. viverrinus* [Fishing cat], *Lutra perspicillata* [Smooth Indian otter]
Kushiro Marsh (~183 km^2)	Hokkaido, Japan	26	*Vulpes schencki* [Red fox], *Cervus Nippon yesoensis* [Sika deer]
Azraq Oasis (~60 km^2)	Amman, Jordan	?	*Canis lupus* [Grey wolf], *Gazella gazella* [Mountain gazelle], *Caracal caracal* [Caracal lynx]—may all be now locally extinct

Source: Based on data in Hails (2003)

In contrast, wetlands in southern Asia have a high diversity which includes some rare species restricted to these habitats, such as *Lutra sumatrana* (hairy-nosed otter), and the wholly aquatic dugong and Chinese river dolphin. Table 4.7 lists some other mammal species commonly associated with Asian wetlands.

In Africa, many species of wildlife move onto the floodplains of rivers during the dry season in search of grazing and prey. Some antelopes, such as the bushbuck, *Tragelaphus scriptus*, and the lechwe, *Kobus leche*, migrate back and forth across swampy ground as the floods rise and fall. Their life cycles are aptly timed such that they drop their young as the floodwaters recede and new pasture is exposed (Welcomme 1979). Other African antelope that live in wetlands include the sitatunga (*Tragelaphus spekii*) and the water chevrotain (*Hyemoschus aquaticus*), both of which are often seen almost fully submerged in water. In addition, the waterbuck (*Kobus ellipsiprymnus*) is a shaggy antelope with oily fur that feeds around wetlands.

Hippopotamus are important transporters of nutrients through their habits of grazing on floodplains at night and depositing large quantities of nutrient-rich dung in water as they bathe during the day. The capybara, *Hydrochoerus*, of South America similarly inhabits floodplains where it feeds on grasses and aquatic plants; it is generally associated with permanently wet areas (Gonzales-Jimerez 1977).

In Kenya, some water holes are formed from the erosion of termitaria by wildlife such as rhinoceros and hartebeest rubbing against the mounds. When subsequently weathered below soil level, water collects in them and elephants, warthogs, buffalo, and other animals use them as sources of drinking water, thus accelerating the drying-out process (Ayeni 1977). In many arid or semi-arid regions of the world, scattered water holes provide drinking water for animals that can move long distances. Most large mammals in areas such as the grasslands of Africa generally drink every one or two days, in hot weather, and are thus very dependent on finding waterbodies regularly. Only a few species of ungulate are capable of going without water for longer periods (e.g. camels), although there are a number of rodents (e.g. kangaroo rats, gerbils, pocket mice, jerboas) that seem capable of existing on water obtained from food alone.

4.3 The temporary water community—global scale comparisons

4.3.1 Comparison of the communities of intermittent ponds

The aim of this section is to identify common elements among intermittent pond invertebrate communities from across the globe. Table 4.8 lists the taxonomic groups recorded in pond studies on four continents: North America (northeastern),

Table 4.8 Comparison of the invertebrate faunas recorded in intermittent ponds on four continents

Taxon	Pant-y-Llyn (Wales) +Oxfordshire ponds	NE North America	NW Australia +SE Australia	Singapore
Cnidaria	—	—	—	Hydra
Turbellaria	atbr	(3)	—	*Mesostoma*
Gastrotricha	atbr	atbr	atbr	*Chaetonotus*
Rotifera	atbr	atbr	*Asplanchna/Keratella*	*Conochilus*
Nematoda	atbr	atbr	indet.	indet.
Gastropoda	*Lymnaea* (2)	*Lymnaea*	[*Lymnaea*]*	—
	Planorbis (+1)	*Planorbula* (+3)	*Gyraulus*	*Gyraulus* (2)
	—	—	Viviparidae	—
	—	Physidae	—	—
Bryozoa	—	*Fredericella*	—	—
Bivalvia	*Pisidium*	*Pisidium/Sphaerium*	—	—
Oligochaeta	Naididae	Naididae (3)	[Naididae]*	[Naididae]*
	[Enchytraeidae]*	Enchytraeidae	[Enchytraeidae]*	[Enchytraeidae]*
	—	Lumbricidae	[Lumbricidae]*	[Lumbricidae]*
	—	Lumbriculidae	[Lumbriculidae]*	[Lumbriculidae]*
	[*Limnodrilus hoffmeisteri*]*	*Limnodrilus hoffmeisteri*	[*Limnodrilus hoffmeisteri*]*	*L. hoffmeisteri*
Hirudinea	Erpobdellidae	Glossiphoniidae (2)	Glossiphoniidae	—
Anostraca	[*Chirocephalus*]*	*Chirocephalopsis*	[*Parartemia*]*	—
Conchostraca	—	[*Caenestheriella*]*	*Cyclestheria* (+3)	*Cyclestheria*
Notostraca	[*Triops*]*	[*Triops/Lepidurus*]*	[*Triops*]* *Lepidurus*	—
Cladocera	*Chydorus*	*Chydorus* + *Alona*	*Chydorus* + *Alona* (+5)	—
	Daphnia pulex/D. obtusa	*Daphnia pulex*	*Daphnia carinata*	—
	—	*Ceriodaphnia*	*Ceriodaphnia*	—
	Simocephalus expinosus	?*S. expinosus*	—	—
	S. vetulus	?*S. vetulus*	*S. vetulus*	—
	—	*Scapholeberis kingi*	—	—
	—	*Moina*	*Moina*	*Moina*
	—	—	Macrothricidae (3)	—
	—	—	Sididae (2)	—
Ostracoda	*Cypricercus*	*Cypricercus* (3)	*Cypricercus/Cypretta*	*Cypretta*
	Eucypris	*Eucypris* (2)	*Eucypris*	—
Copepoda				
Calanoida	*Diaptomus*	*Diaptomus*	*Diaptomus*	*Tropodiaptomus*
Cyclopoida	*Cyclops* (2)	*Cyclops*	*Mesocyclops*	*Mesocyclops*
	Acanthocyclops	*Acanthocyclops*	indet.	—
Harpacticoida	*Canthocamptus* (2)	*Canthocamptus* (+2 genera)	[*Canthocamptus*]*	indet.
Amphipoda	*Crangonyx pseudogracilis*	*C. pseudogracilis*	*Austrochiltonia*	—
Isopoda	*Asellus*	*Asellus*	—	—
Decapoda	—	*Cambarus*	*Holthuisana*	—
Acari	Hydrachnida	Hydrachnida	Hydrachnida (5)	Hydrachnida (4)
Aranaea	*Tetragnatha*	*Tetragnatha*/Lycosidae	*Lycosa*	—
Collembola	*Isotomurus*	*Isotomurus*	*Cryptopygus*	—
	Sminthurides	*Sminthurides*	—	—
Ephemeroptera	*Cloeon*	*Cloeon*	[*Cloeon*]*	*Cloeon*
	[Siphlonuridae/Leptophlebiid]*	Siphlonuridae/Leptophlebiidae	*Atalophlebioides*	—

Table 4.8 (*Continued*)

Taxon	Pant-y-Llyn (Wales) +Oxfordshire ponds	NE North America	NW Australia +SE Australia	Singapore
Odonata				
Zygoptera	*Coenagrion/Enallagma*	*Lestes*	*Ischnura*	*Agriocnemis*
Anisoptera	*Sympetrum*	*Sympetrum*	*Orthetrum* (2)	*Orthetrum*
	[*Pantala*]*	? *Pantala*	*Pantala/Diplacodes* (2)	—
	Cordulegaster	? *Anax*	*Anax/Hemianax*	[*Anax*]*
	(1)	(2)	(2)	—
Plecoptera	*Nemoura cinerea*	—	*Dinotoperla bassae*	—
Hemiptera				
Corixidae	*Callicorixa/Corixa*	*Callicorixa* (+ 2 genera)	*Agraptocorixa/Micronecta*	*Micronecta* (2)
Gerridae	*Gerris*	*Gerris*	*Limnogonus*	[*Gerris*]*
Veliidae	*Microvelia*	*Microvelia*	[*Microvelia*]*	*Microvelia*
Notonectidae	*Notonecta*	*Notonecta*	*Nychia/Enithares/Anisops*	*Anisops* (2)
Belostomatidae	—	*Belostoma*	*Diplonychus*	*Diplonychus*
Pleidae	—	*Plea*	*Plea*	—
Nepidae	[*Ranatra*]*	*Ranatra*	—	—
Hydrometridae	[*Hydrometra*]*	*Hydrometra*	*Hydrometra*	*Hydrometra*
Saldidae	—	*Saldula*	—	—
Coleoptera				
Dytiscidae	*Agabus*	*Agabus*	*Rhantaticus*	[*Agabus*]*
	Colymbetes	*Colymbetes*	*Guignotus* (2)	—
	Dytiscus	*Dytiscus*	*Laccophilus* (2)	*Laccophilus*
	Hydroporus (3)	*Hydroporus*	*Hyphydrus*	—
	Ilybius	—	*Liodessus*	—
	—	—	*Copelatus*	—
Gyrinidae	*Gyrinus* (2)	*Gyrinus*	*Dineutus*	—
Hydrophilidae	*Helophorus* (4)	*Helophorus*	—	[*Helophorus*]*
	Anacaena limbata (+ 1)	*Anacaena limbata*	*Regimbartia*	—
	Hydrobius	*Hydrobius*	*Paracymus*	—
	[*Berosus*]*	*Berosus*	*Berosus* (2)	—
Hydraenidae	*Limnebius* (2)	—	*Hydraena*	—
Haliplidae	*Haliplus*	*Haliplus/Peltodytes*	*Haliplus* (3)	—
Helodidae/Scirtidae	[*Cyphon*]*	*Cyphon*	[*Cyphon*]*	—
Noteridae	—	—	—	*Canthydrus*
Trichoptera				
Limnephilidae	*Limnephilus* (2)	*Limnephilus* (2)	—	—
	Glyphotaelius	*Anabolia*	—	—
	—	*Ironoquia*	—	—
Phryganeidae	—	*Ptilostomis*	—	—
Diptera				
Tipulidae	[*Tipula*]*	*Tipula* spp.	indet.	—
Culicidae	[*Aedes/Culiseta*]*	*Aedes/Culiseta/Psorophora*	*Aedes/Anopheles/Culex*	*Anopheles/Culex*
Tabanidae	[*Tabanus*]*	*Tabanus*	indet.	—
Ceratopogonidae	—	*Bezzia/ProbezziaPalpomyia*	*Palpomyia*	*Bezzia*
Stratiomyidae	[*Odontomyia*]*	*Odontomyia*	indet.	—
Chaoboridae	—	(2)	—	—
Sciomyzidae	—	(8)	indet.	—

Table 4.8 (*Continued*)

Taxon	Pant-y-Llyn (Wales) +Oxfordshire ponds	NE North America	NW Australia +SE Australia	Singapore
Chironomidae				
Diamesinae	[*Prodiamesa*]*	*Prodiamesa*	—	—
Tanypodinae	*Zavrelimyia*	*Procladius*	*Procladius/Pentaneura*	—
Orthocladiinae	*Psectrocladius*	*Psectrocladius*	—	—
	Allopsectrocladius	—	—	—
	—	*Cricotopus/Eukiefferiella*	*Cricotopus/Eukiefferiella*	—
Chironominae	*Chironomus*	*Chironomus*	*Chironomus*	*Chironomus*
	—	*Dicrotendipes/Tanytarsus*	*Dicrotendipes/Tanytarsus*	—

Note: Families, genera and species in common are indicated, as are ecologically similar taxa that may be tentatively regarded as ecological equivalents; the bracketed numbers indicate the number of species, above one, recorded in the preceding taxonomic group; 'indet.' indicates that specific identification was not done; 'atbr' indicates ubiquitous taxa, not recorded in the study but 'assumed to be represented'.

Source: (The Pant-y-Llyn/Oxfordshire records were obtained from Blackstock *et al.* 1993, and Collinson *et al.* 1995; the northeastern North American records from Wiggins *et al.* 1980, and Williams 1983; the northwestern Australian records from Watson *et al.* 1995; the southeastern Australia records from Lake *et al.* (1989); and the Singapore records from Laird 1988) (*although these genera were not recorded in these specific pond studies, they are known in these geographical locations from other studies).

Europe (Wales and England), Asia (Singapore), and Australasia (northwestern Australia). Genera and species that overlap in two or more of these locations are named.

Despite wide geographical separation, large differences in climate, endemism, and local pond characteristics (e.g. differences in length of hydroperiod, water chemistry,or riparian vegetation), there is similarity among the faunas. Characteristically present are snails (typically Planorbidae); oligochaetes; usually one species of Anostraca, Notostraca, and/or Conchostraca; microcrustaceans (especially chydorid cladocerans, ostracods, and copepods); aquatic mites; springtails; odonates (Zygoptera and Anisoptera); chironomid, culicid, and ceratopogonid dipterans; and a high diversity of Hemiptera and aquatic Coleoptera (although only two genera were present in the Singapore pond). In addition, there are several ubiquitous meiofaunal taxa that were most probably overlooked in these studies, for example, the protozoans, microturbellarians, gastrotrichs, rotifers, and nematodes. Remarkably, there are a number of genera common to three or more of ponds/regions: the cladoceran *Moina*, the tadpole shrimp *Triops*, the copepods *Diaptomus* and *Canthocamptus*, the mayfly *Cloeon*, the odonate *Anax*, the hemipterans *Gerris*, *Hydrometra*, and *Microvelia*, the beetles *Agabus*, *Berosus*, *Haliplus*, *Cyphon*, and *Helophorus*, and the midge *Chironomus*. In addition, the tubificid oligochaete *Limnodrilus hoffmeisteri*, is common to all four ponds.

Further, the U.K. ponds share at least 33 genera and three (possibly five) species (a cladoceran, an amphipod, and a beetle) with their North American counterparts. Were the datasets to be more extensive, even greater similarities would doubtless be evident. For example, the ostracods *Cypria ophthalmica*, *Cypricercus ovum*, and *Cypridopsis vidua*, the oligochaete *Lumbricus variegatus*, and the leech *Helobdella stagnalis* are known from intermittent ponds in both North America and Germany (Caspers and Heckman 1981), and all species occur in the United Kingdom. By the same token, the Australian pond shares five genera with the pond in Singapore: the dragonfly *Orthetrum*, the corixid *Micronecta*, the notonectid *Anisops*, the belostomatid *Diplonychus*, and the dytiscid beetle *Laccophilus* (the latter is also known from Europe and North America). Such taxa invariably show special characteristics either of their physiology or life cycle which make them successful in temporary waters, as well as allowing them the means to colonize them. Further coverage of these adaptations is left until Chapter 5.

In contrast, certain freshwater invertebrate groups are notably absent from Table 4.8, suggesting that they likely do not occur in intermittent ponds. These include sponges (although these possess suitable adaptations—see section 4.2.5), megalopterans, stoneflies (but see below), and the caseless caddisflies (Hydropsychoidea). In the case of the latter two groups, however, this may be the result of the habitat being a standing water one rather than an intermittent one, as both have been recorded from intermittent streams. Taxa, such as bryozoans and tardigrades, may be present, but tend to be overlooked. This may be true also for aquatic lepidopterans (e.g. the Pyralidae), hydrophilous orthopterans, such as certain species of katydid and cricket, and some neuropterans, such as the semi-aquatic Osmylidae. Surprisingly, the four ponds in Table 4.8 also lack representatives of a number of dipteran groups known to be associated with pond edges and other moist substrates, for example, the Ptychopteridae (craneflies), Psychodidae (moth flies), Dixidae (midges), Rhagionidae (snipe flies), Empididae (dance flies), Dolichopodidae, Syrphidae (rat-tail maggots), Sepsidae (black scavenger flies), Sphaeroceridae (small dung flies), Scathophagidae (dung flies), Ephydridae (shore flies), Anthomyiidae (root maggot flies), and Sarcophagidae (flesh flies). Difficulties in identifying the larvae of many of these dipterans may be a contributing factor.

An attempt to characterize the global intermittent pond fauna reveals three components. The first consists of genera, or occasionally species, with broad distributions (see Table 4.8), some of which may be found in permanent waters, either during part (e.g. *Microvelia*) or the whole of their life cycles (e.g. *Daphnia pulex*). The cosmopolitan distribution of species such as *D. pulex*, together with uncertainty as to the latter's status as a single species (e.g. Innes 1991), calls into question whether populations from permanent ponds are indeed genetically the same as those from temporary ponds—given the different environmental selection pressures that probably operate in these two pond types.

A second component of the global fauna consists of species that occur only, or predominantly, in temporary waters. Foremost among these are the larger Branchiopoda. The fairy shrimp, *Chirocephalus diaphanus*, the tadpole shrimp, *T. cancriformis*, and species of the clam shrimp genus, *Cyclestheria*, for example, are restricted to temporary waters because of their physiological requirement of a dry-phase in their life cycles. Often, such species are both geographically and temporally rare. For example, in Britain, *C. diaphanus* is known from only one site in Wales and about 12 in England (from ponds in the New Forest, the southwest, Cambridgeshire and Sussex). The most northern record is from near York, in 1862, but most sightings lie south of a line from the Severn Estuary to the Wash (Bratton and Fryer 1990). *T. cancriformis* has been recorded from only about 10 localities over the past 200 years and is currently known only from a single pool in the New Forest (Bratton 1990). Populations are able to persist by means of a rapid life cycle and eggs that are both drought-resistant and viable over many years, and that hatch within hours of a pond basin filling.

Forming a third component of the intermittent pond fauna are species, which although rare in a particular geographical region, occur there primarily in temporary waters. Whereas there is some obvious overlap with the previous component, the emphasis is on occurrence in a particular locality rather than exclusivity in intermittent ponds. For example, from their extensive survey of Oxfordshire ponds, Collinson *et al.* (1995) found that although intermittent and permanent ponds yielded similar species rarity indices, four of the five highest rarity index scores were from temporary or 'semi-permanent' sites. Notable and Red Data Book species included the snail *Lymnaea glabra*, the damselfly *Lestes dryas*, and the waterbeetles *Graptodytes flavipes*, *Agabus uliginosus*, *Haliplus furcatus*, *Dryops similaris*, *Helophorus strigifrons*, *H. nanus* and *H. longitarsus*. A somewhat surprising occurrence in the Pant-y-Llyn pond was the stonefly *Nemoura cinerea*, as the Plecoptera are rarely found in standing waters, and especially intermittent ponds—although, in North America, species of Capniidae and Taeniopterygidae are known from temporary streams where they survive the drought as diapausing nymphs (Harper and Hynes 1970). *N. cinerea* has also been recorded in Irish turloughs (intermittent lakes) (Byrne 1981), and in temporary waters in Sweden (Brinck 1949). The stonefly *Dinotoperla bassae* has been recorded from temporary ponds in Victoria and Tasmania (Hynes 1982).

In summary, and bearing in mind the limited number of comparable, full faunal inventories available, it would appear that intermittent ponds throughout the world provide very similar niches for colonizing animals. In many instances, these niches are filled by the same genera, lending credence to the theory that the taxonomic unit of 'genus' is an ecological as well as a morphological entity (Wiggins and Mackay 1978). As noted above, in the case of cosmopolitan, readily disseminated forms, such as *D. pulex*, these niches are filled by the same species, or members of a species complex. Where dispersal powers are weak and do not allow a species to colonize habitats far afield, locally endemic species of the same major taxon fill the gap.

4.3.2 Comparison of the communities of intermittent streams

As was attempted for intermittent ponds, Table 4.9 compares common elements among the invertebrate communities of intermittent streams from four widely separated regions of the world: western North America (California), eastern North America (Ontario), the Caribbean/South America (Trinidad and Brazil), and southeastern Australia.

Again, despite significant geographical separation, differences in climate, endemism, and local hydrological characteristics, there are common elements among the faunas. Groups characteristically present (defined as occurring at three or more locations) include turbellarians; nematodes; gastropods (typically including planorbids); sphaeriid clams; naidid, tubificid, and enchytraeid worms; cyclopoid and harpacticoid copepods; amphipods; decapods; water mites; leptophlebiid mayflies; stoneflies (although the families represented vary considerably; also this group is particularly well represented in Australia); corixid, gerrid, and veliid hemipterans; elmid, dytiscid (highly diverse in the Australian streams), hydrophilid, and psephenid beetles; hydropsychid caddisflies; tipulid, ceratopogonid, simuliid, and chironomid dipterans. The high species richness of hemipterans and dytiscid and hydrophilid beetles is likely due to their colonization of pools after flow ceases.

In contrast to the global comparison of intermittent ponds, there are no instances where genera are common to three or more locations, although some spanned two, for example: the clam *Sphaerium* (Canada and Australia); the mayfly *Paraleptophebia* (Ontario and California); the beetles *Helochares, Berosus*, and *Enochrus* (South America and Australia); the caddisfly *Cheumatopsyche* (Canada and Australia); and the dipterans *Tipula, Aedes, Bezzia, Eukiefferiella, Cricotopus, Chironomus*, and *Polypedilum* (Canada and Australia). Greater taxonomic resolution of some groups may have improved this comparison. Taxa better represented in these streams than in their lentic counterparts are, as might be expected, the Ephemeroptera, Plecoptera, Trichoptera, elmid and psephenid Coleoptera, and certain Diptera. Taxa not well represented, again as might be predicted, include planktonic microcrustaceans; and large branchiopod crustaceans are absent.

Of course the precision of the kind of comparison made above will also be affected by differences in length of the hydroperiods, as well as hydroperiodicity. In a rare study, Dieterich and Anderson (2000) examined the differences in invertebrate communities between 'temporary' (intermittent) and 'ephemeral' (episodic) streams in western Oregon. They recorded 125 or more species in intermittent forest streams (more than that [100] found in a local permanent stream), but no more than 35 species in episodic streams. The intermittent stream communities were dominated by insects, and characterized by the mayflies *Paraleptophlebia gregalis* and *Ameletus andersoni*, the stoneflies *Soyedina interrupta, Sweltsa fidelis, Calliperla luctuosa*, and *Ostrocerca foersteri*, and the caddisfly *Rhyacophila fenderi*. Chironomids were not included in the analysis. In contrast, the

Table 4.9 Comparison of the invertebrate faunas recorded in intermittent streams in four widely separated regions of the world

Taxon	Cronan Creek (California)	Southern Ontario streams (Canada)	Trinidad and Pantanal (Brazil)	Southeastern Australia
Porifera	—	—	*Drulia*	—
Turbellaria	—	*Fonticola velata*	Microstomidae (2)	*Cura/?Mesostoma*
Nematoda	indet.	indet.	indet.	indet.
Rotifera	—	indet.	29 spp.	—
Gastrotricha	—	—	*Polymerurus*	—
Ectoprocta	—	—	*Plumatella*	—
Gastropoda	—	*Lymnaea/Gyraulus/Physa*	Planorbidae/Ampullariidae	Planorbidae/Hydrobiidae
			Ancylidae/Pilidae	Ancylidae
Bivalvia	—	*Sphaerium*	Sphaeriidae (2 + 4)	*Sphaerium*
Oligochaeta	indet.	Naididae	*Stylaria*	Naididae
		Enchytraeidae	Enchytraeidae	Enchytraeidae
		Limnodrilus hoffmeisteri (+3)	Tubificidae	Tubificidae
		Tubifex tubifex (+3)		
		Lumbriculus		Lumbriculidae
Hirudinea	—	*Helobdella stagnalis/Erpobdella*	*Helobdella/Placobdella*	—
	—	*Glossiphonia/Mollibdella*	*Glossiphonia/Oligobdella*	—
Ostracoda	—	*Candona/Cypridopsis/Cypricercus*	—	—
Cladocera	—	—	Daphniidae/Sididae	—
			Chydoridae	—
Copepoda				
Cyclopoida	indet.	*Cyclops vernalis/Eucyclops agilis*	*Mesocyclops/Thermocyclops*	—
Harpacticoida	indet.	*Attheyella nordenskioldii* (+1)	indet.	—
Amphipoda	—	*Crangonyx/Hyalella*	indet.	*Afrochiltonia/Niphargus*
Decapoda	—	*Fallicambarus fodiens*	Penaeidae/Palaeomonidae	Parastacidae/Atyidae
	—	—	Trichodactylidae	—
Acari	Hydrachnida	*Hydrachna/Lebertia/*Oribatida	Hydrachnida	Hydrachnida (17)
Aranaea	—	*Tetragnatha*	*Tetragnatha*	—
Collembola	—	*Isotomurus/Sminthurides*	*Dicranocentrus*	—
Ephemeroptera				
Baetidae	*Baetis*	—	—	*Pseudocloeon*
Heptageniidae	*Cinygmula*	—	—	
Leptophlebiidae	*Paraleptophlebia*	*Paraleptophlebia/Leptophlebia*	Leptophlebiidae	Leptophlebiidae (3)
Siphlonuridae	—	*Siphlonurus*	—	—
Odonata	—	—	Coenagrionidae/Gomphidae	Corduliidae/Synthemidae
	—	—	Calopterygidae/Aeshnidae	Aeshnidae/Lestidae
	—	—	Libellulidae	—
Plecoptera				
Chloroperlidae	*Sweltsa*	—	—	—
Leuctridae	*Despaxia*	—	—	—
Capniidae	—	*Allocapnia vivipara*	—	—
Nemouridae	*Nemoura/Melenka*	—	—	Notonemouridae
Peltoperlidae	*Soliperla*	—	—	Gripopterygidae (12)
Perlidae	*Calineuria*	—	—	Austroperlidae (2)
Hemiptera				
Corixidae	—	*Sigara*	*Tenagobia*	*Sigara/Micronecta*
Gelastocoridae	—	—	*Gelastocoris*	—
Notonectidae	—	—	*Bueno*	*Anisops* (2)

Table 4.9 (*Continued*)

Taxon	Cronan Creek (California)	Southern Ontario streams (Canada)	Trinidad and Pantanal (Brazil)	Southeastern Australia
Gerridae	—	*Gerris*	Gerridae (5)	Gerridae (2)
Veliidae	—	*Mesovelia*	*Rhagovelia*	*Microvelia* (2)
Belostomatidae	—	—	*Belostoma* (2 + 1)	—
Nepidae	—	—	*Ranatra/Curicta*	—
Pleidae	—	—	*Neoplea*	—
Naucoridae	—	—	*Pelocoris*	—
Hydrometridae	—	—	*Hydrometra* (2)	—
Megaloptera	*Sialis*	—	—	*Protochauliodes*
Coleoptera				
Elmidae	*Ordobrevia*	*Stenelmis/Optioservus*	—	*Austrolimnius* (5 + 5)
Dytiscidae	*Hydrovatus*	*Hydroporus/Hydrobius/Agabus*	*Laccophilus* (+6)	*Rhantus* (+26)
Gyrinidae	—	—	*Gyretes*	(3)
Noteridae	—	—	*Suphisellus*	indet.
Hydraenidae	*Octhebius*	—	—	*Octhebius/Hydraena*
Hydrophilidae	*Cymbiodyta*	*Anacaena/Helophorus*	*Helochares/Tropisternus* *Berosus/Enochrus*	*Helochares/Paracymus* *Berosus/Enochrus*
Psephenidae	*Eubianax/Acneus*	*Psephenus*	—	*Sclerocyphon*
Haliplidae	—	*Peltodytes*	—	—
Staphylinidae	*Stenus*	indet.	—	—
Carabidae	—	indet.	*?Omophron*	—
Trichoptera				
Apataniidae	*Apatania*	—	—	—
Calamoceratidae	*Heteroplectron*	—	—	*Anisocentropus*
Glossosomatidae	*Glossosoma/Anagapetus*	—	—	*Agapetus*
Hydropsychidae	*Hydropsyche*	*Hydropsyche/Cheumatopsyche*	—	*Cheumatopsyche*
Hydroptilidae	*Ochrotrichia*	—	—	*Oxyethira* (+2)
Lepidostomatidae	*Lepidostoma*	—	—	—
Phryganidae	—	*Ptilostomis*	—	—
Limnephilidae	—	*Ironoquia/Limnephilus/Anabolia*	—	*?Archaeophylax*
Odontoceridae	*Parthina*	—	—	—
Polycentropodidae	*Polycentropus*	—	—	*Plectrocnemia*
Rhyacophilidae	*Rhyacophila*	*Rhyacophila*	—	—
Uenoidae	*Neophylax*	—	—	—
Helicopsychidae	—	*Helicopsyche borealis*	—	*?Helicopsyche*
Leptoceridae	—	—	—	(9)
Hydrobiosidae	—	—	—	(4)
Diptera				
Tipulidae	indet.	*Tipula/Limnophila/Helobia*	—	*Tipula/Limnophila* (+14)
Culicidae	—	*Aedes*	—	*Aedes/Anopheles/Culex*
Ceratopogonidae	indet.	*Bezzia/Probezzia*	—	*Bezzia/Nilobezzia* (+6)
Dixidae	indet.	—	—	*Dixa*
Stratiomyidae	—	*Stratiomyia/Odontomyia*	—	indet.
Sciomyzidae	—	—	indet.	indet.
Sphaeroceridae	—	*Leptocera* (3)	*Leptocera*	—
Empididae	indet.	—	—	indet. (2)
Psychodidae	*Maruina*	—	—	*Maruina*
Simuliidae	indet.	*Simulium* (2)	—	*Austrosimulium* (4)
Tabanidae	—	*Tabanus/Chrysops*	—	indet. (2)

Table 4.9 (*Continued*)

Taxon	Cronan Creek (California)	Southern Ontario streams (Canada)	Trinidad and Pantanal (Brazil)	Southeastern Australia
Chironomidae	indet.		indet.	
Diamesinae		*Diamesa*		*?Monodiamesa*
Tanypodinae		*Psectrotanypus/Natarsia*		*Pentaneura/Procladius*
Orthocladiinae		*Trissocladius/Orthocladius*		*Corynoneura* (+13)
		Eukiefferiella/Cricotopus		*Eukiefferiella/Cricotopus*
Chironominae		*Chironomus/Polypedilum*		*Chironomus/Polypedilum*
		Micropsectra		*Tanytarsus* (+8)

Note: Families, genera and species in common are indicated, as are ecologically similar taxa that may be tentatively regarded as ecological equivalents; the bracketed numbers indicate the number of species, above one, recorded in the preceding taxonomic group; 'indet.' indicates that specific identification was not done.

Source: Williams, D.D. unpublished data; Williams and Hynes (1976); Alkins—Koo (1989/90); Boulton and Lake (1992a), Heckman (1998), del Rosario and Resh (2000)

communities in the episodic streams were dominated by dipterans and beetles (e.g. *Hydroporus planiusculus* and *Hydraena vandykei*), and contained very few mayflies, stoneflies, and caddisflies.

4.4 The temporary water community—local scale comparisons

Despite the above similarities among intermittent pond and stream invertebrate communities from around the world, there are sufficiently large differences in local climate, topography, geology, riparian vegetation, etc. to create regional- and habitat-specific characteristics. Added to these differences are distributional properties of individual species, for example, endemism versus cosmopolitanism. The following sections attempt to address this variation by examining, on an individual case history basis, a cross-section of temporary waters for which detailed studies exist.

Case Histories—Intermittent Waters

4.4.1 Nearctic intermittent ponds

Sunfish Pond, Ontario, Canada, is a typical temperate, intermittent vernal pool. The *vernal* categorization, as noted earlier, was established by Wiggins (1973) and indicates a pool that derives its water primarily from rain and melting snow in early spring and which becomes dry in early summer, leaving a water-free basin for 8–9 consecutive months of the year, including winter. Ponds that retain water in the autumn, winter, and spring and only become dry for 3–4 months in the summer, are termed intermittent *autumnal* pools.

Community succession

Thus far, 98 taxa have been identified from Sunfish Pond. Figure 4.18 shows examples of these species arranged, not by taxonomic groups, but according to their seasonal occurrence in the pond. A succession is evident which, at first glance, seems fairly continuous but which can be divided into several distinct faunal groups:

Group 1 contains animals that could be found during virtually the entire aquatic phase of the pond. During the dry phase they could be dug up from the substrate as semi-torpid adults or immature stages. If placed in water they revived within minutes. Included in this group were all of the bivalves (*Sphaerium* and *Pisidium*), snails (*Lymnaea* and *Gyraulus*), and oligochaete worms (*Lumbriculus*, *Nais*, and Enchytraeidae), two species of the beetle genus *Hydroporus*, a copepod (*Acanthocyclops bicuspidatus thomasi*), and a very abundant chironomid (*Einfeldia*).

Group 2 comprises taxa present as active forms within a few days of the pond filling in the spring. These species mostly completed their life cycles within 4–6 weeks and disappeared (by entering a

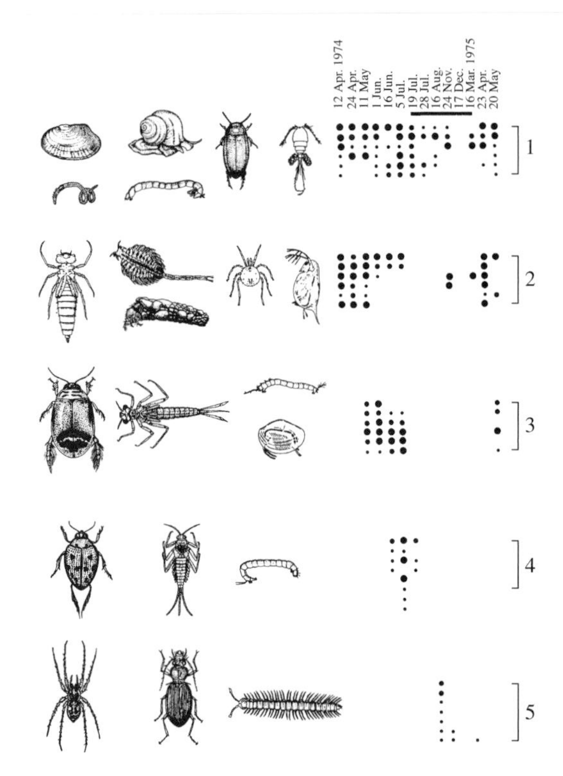

Figure 4.18 Summary of the fauna of a nearctic intermittent pond, Sunfish Pond, Ontario, arranged in successional groups (hydroperiod is indicated by the horizontal black bar; circles indicate relative abundance on a decreasing scale of 'abundant, common, and rare').

resting stage—usually as eggs or dormant/diapausing immatures—or by leaving the pond as emerged adults) well before (4–6 weeks) the pond dried up. This, the largest, group contained most of the microcrustaceans (ostracods, cladocerans, harpacticoids), a fairy shrimp (*Chirocephalopsis bundyi*), mites (*Thyas, Euthyas, Hydrachna, Hydryphantes,* and *Piona*), mosquitoes (*Aedes sticticus*), cased caddisflies (*Limnephilus* and *Ironoquia*), bugs (*Gerris, Sigara,* and *Notonecta*), some midges (*Trissocladius, Eukiefferiella. Phaenopsectra, Parachironomus,* and *Polypedilum*), beetles (*Agabus, Anacaena, Cyphon,* and *Neoscutopterus*), and a dragonfly (*Libellula*).

Group 3 contains taxa which appeared 2–5 weeks after pond formation in the spring. Taxa present included *Lynceus* (a conchostracan), a damselfly (*Lestes*), midges (*Micropsectra, Corynoneura, Ablabesmyia,* and *Psectrotanypus*), and beetles (*Dytiscus, Rhantus, Acilius, Berosus, Helophorus grandis, Hydrochara,* and Helodidae/Scirtidae), some of the latter appearing only as adults in search of food and not breeding in this particular pond. Life cycles of species in Group 3 were typically completed in 5 weeks.

Group 4 taxa appeared 2–3 weeks before the pond dried up (approximately 10 weeks after filling). The taxa included beetles (*Gyrinus, Laccophilus, Hydaticus, Hydrovatus, Helophorus orientalis, Haliplus,* and Georyssidae), mayflies (*Cloeon* and *Stenonema*), and midges (*Cricotopus*).

Group 5 taxa appear only in the dry phase. They were primarily terrestrial or riparian species and include isopods (*Oniscus*), millipedes, centipedes, spiders (e.g. Lycosidae), and beetles (Staphylinidae, Ptiliidae, Heteroceridae, Noteridae, and Curculionidae), together with slugs (*Limax*).

Figure 4.19 shows seasonal variation in the total number of taxa seen in the pond. Species richness was greatest during the hydroperiod, but the drying basin by no means lacked inhabitants.

4.4.2 Nearctic intermittent streams

The differences in community composition between permanent standing and running waters may be very considerable. However, because many intermittent streams may flow more slowly and form pools in their drying beds, significant overlap may occur between their faunas and those of intermittent ponds—although the lotic phase of the habitat adds to the diversity of niches available. Moser Creek, Ontario, the example we shall consider, was (it was subsequently destroyed by the installation of agricultural drainage tiles) a small 400 m long by 0.8 m wide, intermittent stream which flowed for about 7 months of the year (November–April). During May and early June it consisted of a series of unconnected pools which dried up completely by July.

Community succession

The invertebrates found in this stream can be divided into three successive groups based on the water conditions found in the habitat throughout

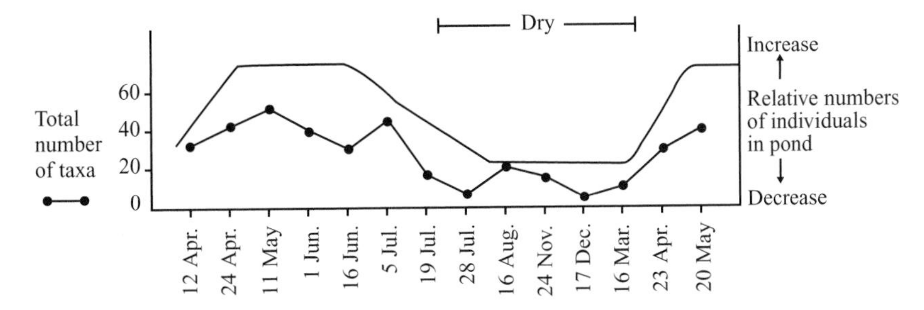

Figure 4.19 Seasonal variation in the total number of taxa (solid circles) present in Sunfish Pond, together with an estimate of the relative numbers of individuals present (solid line).

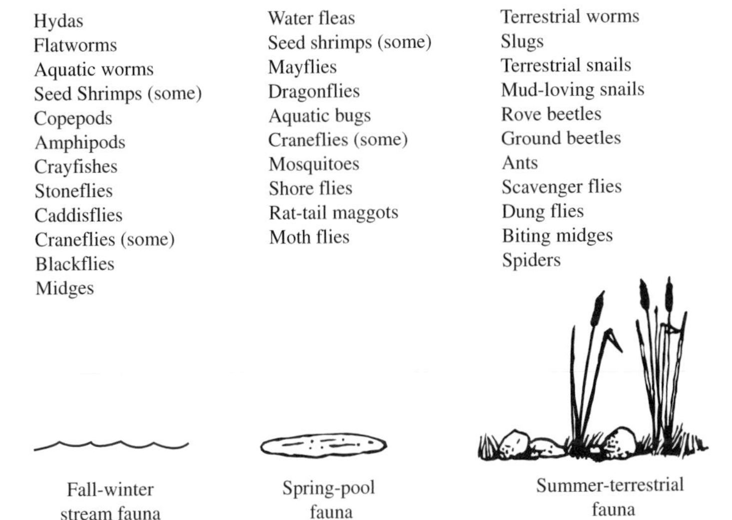

Figure 4.20 Summary of the fauna of a nearctic intermittent stream, Moser Creek, Ontario, arranged in successional groups according to major phases in the hydroperiod.

the year. The major components of these three groups are summarized in Figure 4.20. Inclusion in a particular group indicates that this is where the species passed the most active stages of its life; some overlap is inevitable.

The *fall–winter stream fauna* consisted of those species that appeared shortly after the stream started flowing in the autumn and most reproduced successfully before the flow stopped in the spring. Many of them, for example, some of the midges (e.g. *Trissocladius*) and the cased caddisflies (*Ironoquia punctatissima*), grew quickly before the water cooled down and then were ready, shortly after ice breakup, to pupate, emerge, mate, and oviposit. Other midges (e.g. *Diplocladius*) and the amphipods (*Crangonyx minor*) grew steadily throughout the winter and matured a little later. A few midge species (belonging to the genera *Chironomus* and *Micropsectra*) grew so slowly that their period of activity spanned two groups.

The *spring-pool fauna* comprised those species that dominated the system after flow had stopped and only shallow pools remained. These pools were excellent breeding environments due to the ease with which they warmed up and the abundant macrophyte and algal food resources that developed in them. There were two basic categories of species that used these pools. The first, which included mosquitoes (*Aedes vexans*) and some of the aquatic beetles (e.g. *Anacaena limbata* and *H. orientalis)*, had been present since the previous autumn (the mosquitoes as eggs and the beetles as adults), but they did not produce any larvae until flow had ceased. Once hatched, the larvae grew quickly and matured before the pools dried up. The second category consisted of other beetles (e.g. *Tropisternus*, *Hydroporus*, and *Rhantus*) and the water strider, *Gerris remigis*. These flew or walked in as adults just as the pools were forming, laid their eggs, and typically left. Their larvae quickly hatched and matured just as the pools dried up. Occasionally, an unusually dry spring sped up evaporation so that the life cycle was not complete when the water disappeared. However, some of the species in this group had pupae that were capable of resisting drought for a short while, enabling the adults to emerge successfully.

The *summer-terrestrial fauna* consisted mostly of riparian species that moved onto the streambed once it had dried. Some, such as the earthworms, were probably attracted by the dampness whereas others, like the slugs and snails, came to feed on the exposed algal mats. Beetles (e.g. Scarabeidae and Staphylinidae) and ants scavenged the bed for dead and dying individuals left over from the aquatic phase, and the spiders (largely Lycosidae) in turn followed these, their prey. A few specialized taxa belonged in this group also, for example, dipteran flies of the families Sepsidae and Sphaeroceridae (e.g. *Leptocera*), which are generally attracted to damp, decaying material on which they lay their eggs.

It is clear that each of the three hydroperiod phases was characterized by several major taxa. Further, there is obvious similarity between the fauna of the spring-pool phase and that seen in intermittent ponds. However, elements substantially under-represented in Moser Creek and other intermittent streams, are the microcrustaceans (ostracods, copepods, and especially cladocerans) and large branchiopod crustaceans (fairy shrimp, clam shrimp, and tadpole shrimp). Their absence may be due to unsuitable habitat conditions during the lotic phase, or, in the case of the larger, pelagic crustaceans, to the presence of predatory fishes which often migrate into intermittent streams from permanent channels in the catchment. Other, macrocrustaceans, particularly amphipods, are often common in intermittent streams, but exhibit a more cryptic, benthic behaviour. Notable additions to the generic intermittent stream fauna are stoneflies, blackflies, and a greater diversity of chironomids, most of which require a current. Greater details of the life cycles of some of these species will be given in Chapter 5.

4.4.3 Australian intermittent streams and billabongs

Australia is an arid continent exhibiting both low levels of precipitation and high rates of evaporation. Its drainage basins reflect these features in that many show wide variation in streamflow on both seasonal and annual timescales. Over half of the land mass is drained by intermittent streams and rivers (W.D. Williams 1981,1983). Intermittent flow and high evaporation cause water levels in entire river or pond systems to fluctuate sufficiently to have produced a biota well adapted to the cyclical loss of water. Available data on abiotic characteristics indicate the same, wide variation in factors such as pH, dissolved oxygen, temperature, and conductivity as have been noted in the Northern Hemisphere (Boulton and Suter 1986).

Despite current concerns over the management of low flow systems, only a few comprehensive studies of their faunas exist. However, these suggest that Australian temporary streams are perhaps richer in species than their counterparts elsewhere. Further, species richness appears to increase with increasing permanence of the water body. Most of the macroinvertebrate communities comprise insects (>75%) with the Diptera (chiefly tipulids, chironomids, simuliids, and ceratopogonids) being dominant. Other insect groups present are the Coleoptera (dytiscids and hydrophilids), Trichoptera (leptocerids and hydrobiosids in Victoria and South Australia), Plecoptera, Hemiptera, Odonata, and Ephemeroptera, in roughly decreasing order of proportional representation. Regional differences are apparent (e.g. the plecopteran fauna of South Australia is relatively poor) and this may explain the fact that the proportions of various insect groups represented are somewhat different from those in Northern Hemisphere streams of similar latitude (Boulton and Suter 1986). During their terrestrial phase, two intermittent streams in Victoria were observed to be colonized by a 'clean-up crew' not unlike that reported for Moser Creek, Canada (Section 4.4.2). The Victorian fauna consisted of carabid and hydraenid beetles, lycosid spiders, ants, and terrestrial amphipods (Boulton and Suter, 1986).

In Australia, as elsewhere, relatively little is known, quantitatively, of the fate of detritus and its derivatives, their partitioning in the foodweb, and their importance to the energy budget of the system compared with say autochthonous input (Boulton and Suter 1986). What is known is that intermittent streams running through *Eucalyptus* forests receive a distinct peak in input of leaf litter during summer and this coincides with the period of low or zero flow (Lake 1982). This is somewhat different from temperate, Northern Hemisphere streams where most of the input of deciduous leaves is in the late autumn, a time when many intermittent streams are flowing. When flow began in Brownhill Creek in South Australia, Towns (1985) recorded a pulse of organic matter, consisting of both coarse particles and dissolved materials, which was carried downstream. Boulton and Lake (1992b) confirmed that concentrations of benthic organic matter (BOM) in the Werribee and Lerderderg Rivers were highest immediately after eucalypt leaf fall, but low during high discharge in the winter and spring. However, the amount of BOM after October floods actually increased, from both upstream and riparian sources. Correlations between detritivore densities and BOM proved to be habitat-specific, for example, there was a strong and positive relationship on riffles of the Lerderderg River. A later study on this river determined the fundamental role of detritus to its food base, and also the dramatic changes that occurred in both community structure and feeding interactions over time. Despite large temporal variation, spatial variation in community structure was low (Closs and Lake 1994).

Another notable feature of Australian temporary waters are billabong systems, which are seasonally created backwaters (lagoons) unique to northern Australia. Magela Creek is a tributary of the East Alligator River in the Northern Territory. This region has a tropical monsoon climate and thus has distinct wet and dry seasons with 97% of the 153 cm annual rainfall occurring from October to May. During this time, the Magela floodplain is a continuous body of water covering some 190 km^2 to depths of between 2 and 5 m, and having water velocities that may exceed 1 $m\,s^{-1}$. During the dry season, it consists of a series of discrete billabongs of varying size (Morley *et al.* 1985). These billabongs undergo seasonal and diurnal fluctuations in both physical and chemical properties of their waters. For example, surface temperatures vary between 22°C in July and 39°C in November, and diurnal fluctuations are greatest in the dry season. Turbidity tends to increase throughout the dry season but is less in the wet season. Conductivity increases during the dry season and pH ranges between 6.0 and 7.0. Oxygen levels vary according to temperature and the amount of photosynthetic activity of macrophytes and phytoplankton, but at no time is there complete deoxygenation (Marchant 1982a). The billabongs in the Magela system can be subdivided into two basic types: (1) those on the floodplain which are separated from the creek by a levée but which have an intermittent connection with the creek thus enabling them to fill and drain—these have been termed *backflow billabongs*; and (2) *channel billabongs* which occur on the main channel of the creek and have separate inlets and outlets (Walker and Tyler 1979). The former are usually shallower than the latter and there are consequently some differences between the types of environment that they present for the biota. Marchant (1982a) recorded some differences in the composition of the littoral faunas between these two types, but chiefly among the less common taxa. The littoral fauna appears to be rich in species with particularly high densities of Ephemeroptera, Trichoptera, Mollusca, Hemiptera, and Chironomidae occurring in macrophyte beds during the wet season. Taxa predominant in the dry season include Coleoptera (especially adult Dytiscidae), tanypodine chironomids, Ceratopogonidae, some Hemiptera, some Gastropoda (e.g. *Ferrissia*), and the prawn *Macrobrachium*. Less common taxa found at different times of the year include Tricladida, Oligochaeta, Hirudinea, Porifera, Hydridae, Gordiidae, Hydracarina, Ostracoda, and Conchostraca.

In the shallow Magela billabongs the greatest densities and diversities of animals occur during the late wet season/early dry season with as much as a five-fold factor difference in the numbers of individuals compared with other times of the year. Seasonal fluctuations in density and diversity do not appear as marked in the channel billabongs. Temporal fluctuation is probably the result of changes in macrophyte abundance, with maximum animal biomass coinciding with maximum plant biomass (April–July). The plants provide both food for macroinvertebrates and protection from predators. It is likely, as in many other aquatic habitats, that the macrophytes are not eaten directly but after they have died and become part of the general pool of detritus on the billabong bottom. Epiphytic algae on the live macrophytes may be another important source of food.

The billabong fauna survives the dry season by a variety of methods including hibernation in the bottom mud—Gastropoda; resistant eggs—Ephemeroptera; and recolonization from other billabongs, particularly those in the main channel which are deeper and therefore virtually all permanent. In fact, during the time that the Magela system was studied, none of the main billabongs dried up totally. The beginning of the wet season is characterized by a rapid resurgence of the fauna, especially in the shallow billabongs, with many species having short life cycles (e.g. *circa* one month) with fast rates of larval growth (Marchant 1982b).

The general features of the fauna of the Magela billabongs parallel those known for similar habitats (e.g. floodplain river systems; see Welcomme 1979, and Chapter 8) in other regions of the tropics. Further, these habitats support at least as many species as much larger temperate Australian lakes. Outridge (1987) has argued that the high species richness found in the Magela communities is due to rarefaction and predictable environmental heterogeneity, related to the monsoonally influenced variations in flow and water quality.

4.4.4 Turloughs and water meadows

Turloughs are shallow seasonal lakes lying in depressions underlain by limestone karst, and were first recorded from Ireland where they are regionally common. Nineteenth-century geological surveys identified these waters, although placenames, such as 'Turloughmore' and 'Killaturly' clearly show an earlier familiarity (Coxon 1987). One of the first studies of these habitats was made by Praeger (1932) who examined the macroflora. He identified three characteristics of the basins: an absence of trees and shrubs; an upwards extension of plants, such as mosses; and the presence of an unusual mixture of both water-loving and drier soil species. He observed that many of the latter showed signs of 'dwarfing', likely the result of close grazing, but also were rare elsewhere (e.g. *Viola stagnina*). Turloughs were also the first recorded habitat of an anostracan (*Tanymastix stagnalis*) in Ireland (Young 1976).

Coxon (1987) consolidated earlier attempts to define turloughs to include the following hydrological criteria, based on 90 sites: (1) seasonal flooding, with a minimum depth of 0.5 m; and (2) evidence of emptying to groundwater, for example, via a sinkhole, or an intermittent spring aperture. Details of Coxon's analysis of the macroflora were given in Section 4.2.4.

Few studies have addressed, comprehensively, the faunal communities of these habitats, tending instead to focus on prominent elements, such as crustaceans, beetles, and molluscs. For example, Duigan and Frey (1987), examined the cladocerans in a series of turloughs in County Galway and discovered the presence of *Eurycercus glacialis*, a relatively large species thought to be excluded from permanent waters due to fish predation. Reynolds *et al.* (1998) deemed the inhabitants to comprise some species that are opportunistic and widespread, together with others well adapted to turlough conditions. The community is believed to be dominated by detritivores, with herbivores and predators occurring in more permanent places. Often, there is a zonation evident that appears to be related to species sensitivities to factors such as depth, temperature, and hydroperiod. In a summary of some of the invertebrates characteristic of turloughs, Reynolds (1996) lists one flatworm (*Polycelis nigra*), one snail (*Lymnaea palustris*), the mayfly *Cloeon simile*, and 15 species of crustacean. Bilton (1988) supplements this list with eight species of aquatic beetle (primarily dytiscids and hydrophilids), known to be rare or unknown elsewhere in the British Isles. In addition, the non-aquatic phase of Irish turloughs has been shown to be important habitat for terrestrial beetles, particularly staphylinids and carabids. For example, Good and Butler (2001) recorded five staphylinid species new to Ireland from four turloughs in south Galway. It is possible that risk of desiccation of some of the fauna may be lessened by the extensive algal paper mats that form over the basin surface as the water recedes Reynolds (1983).

In 1992, Campbell *et al.* documented the presence of a turlough (Pant-y-Llyn) in Wales. This shared many of the hydrological features of Coxon's Irish sites—of which, these authors reported, some 30% had become hydrologically compromised due to adjacent engineering and drainage projects in the intervening six years. Campbell *et al.* also compared turloughs to some of the Breckland meres in southeastern England, but concluded the latter to be different as they do not have an annual hydroperiod and are thus more episodic in nature.

Garcia-Gil *et al.* (1992) reported the existence of a turlough-like lake in the Lake Banyoles karstic area near Girona, Spain. Clot d'Espolla is a basin fed by a number of springs that flow only when heavy rains sufficiently charge the surrounding aquifer. The flora is dominated by fast-growing species, such as *Ranunculus aquatilis* and *Chara* sp., drought-resistant species, such as *Scirpus maritimus*, and species that can survive periodic inundation, such as *Agrostis stolonifera*. The fauna is dominated by amphibians and invertebrates. The former is represented by eight species, the highest diversity in the region: *Rana perezzi* and *Hyla meridionalis* (frogs); *B. bufo*, *A. obstetricans*, and *Pelobates cultripes* (toads); *T. helveticus* and *T. marmoratus* (newts); and *S. salamandra* (salamander). The most conspicuous invertebrate (30 individuals m^{-2}) is *T. cancriformis*. Several bird species frequent the lake to feed (*Egretta garzetta*, *Anas platyrrhynchos*, and *Gallinula chloropus*).

In a more detailed study of Espolla lake, Boix *et al.* (2001) collected 113 taxa, approximately one-third of which are known to be characteristic of temporary waters. Some 64 taxa were considered rare (they were found on fewer than 10% of sampling days). Insects dominate (82 taxa) the system: Ephemeroptera (1 species); Odonata (2 species); Heteroptera (17 species); Coleoptera (26 species); Trichoptera (6 species); and Diptera (34 species). Especially abundant were the beetles *Agabus nebulosus* (as larvae) and *Berosus signaticollis* (as adults), as well as the mayfly *Cloeon inscriptum*, and the chironomids *Psectrocladius* gr. *sordidellus* and *Cricotopus bicinctus*. Besides *T. cancriformis*, six other branchiopod crustaceans were present, including *C. diaphanus*. Other invertebrates included flatworms, nematodes, oligochaetes, ostracods, copepods, an isopod, and an amphipod.

Water meadows are much shallower, low-lying grasslands that are frequently semi-natural or man-made, and are strongly influenced by local water management and agricultural practices. Although shallow (<30 cm) they may contain water for much of the year, and often support rich plant communities comprising grasses, sedges, cattails, and flowering plants. Invertebrates present include microcrustaceans, dipterans (chironomids and mosquitoes), and odonates. Vertebrates include frogs, toads, ducks, and various wading birds.

4.4.5 Indian wetlands

India has a rich variety of wetlands. Excluding rivers, the total area of these habitats is estimated to be 58,286,000 ha, or around 18% of the country. However, approximately 70% is comprised of cultivated paddy fields and so exists in a much altered, or man-made, state. Many natural wetlands, perhaps as much as one-third, have been destroyed through drainage, landfill, pollution, growth of exotic weeds (especially *Salvinia molesta* and *E. crassipes*), and over-exploitation of fish resources (Abbasi *et al.* 1997). An initial inventory in 1989 identified 572 natural wetlands. Two, one natural—the more permanent, brackishwater Chilka Lake (in the eastern state of Orissa), and, one man-made—Keoladeo National Park (Bharatpur), were designated under the Convention of Wetlands of International Importance (Ramsar Convention) as being especially significant waterfowl habitats (Scott 1989). In total, 93 sites were identified as being of international importance. However, Keoladeo goes completely dry almost every year now, and another Ramsar site, the saline Lake Sambhar, hold water for only a short period each year (B. Gopal, Personal Communication).

A later survey listed 2,167 natural (3.2%) versus 65,253 man-made wetlands. These have been assigned to the following regions: the reservoirs of the Deccan Plateau in the south, together with the lagoons and the other wetlands of the southern west coast; the vast saline expanses of Rajasthan, Gujarat, and the gulf of Kachchh; freshwater lakes and reservoirs from Gujarat eastwards through Rajasthan (Kaeoladeo Ghana National park) and Madhya Pradesh; the delta wetlands and lagoons of India's east coast (Chilka Lake); the freshwater marshes of the Gangetic Plain; the floodplain of the Brahmaputra; the marshes and swamps in the hills of northeast India and the Himalayan foothills; the lakes and rivers of the montane region of Kashmir and Ladakh; and the mangroves and other wetlands of the island arcs of the Andamans and Nicobars (Puri 2003).

Efforts to study and conserve Indian wetlands began in earnest only in the 1980s and aim to provide descriptions of wetland sites in order to provide basic information on size and location, habitat types, principal vegetation, ownership, degree of protection, land use, fauna, threats, research, and conservation (Abbasi 1997). The focus is on applying biological methods of conservation rather than adopting engineering solutions. In addition, a national wetland mapping project has been initiated in an attempt to provide an integrated approach to conservation (Puri 2003).

Much of the survey information gathered to date has focused on macrophytes and vertebrates, especially birds (see Abbasi 1997) and mammals (see Table 4.7). Aquatic invertebrate data are scarce, but there are indications of considerable biological diversity—although the level of taxonomic resolution in studies is often not high (Abbasi and Mishra 1997). Loktak Lake, a permanent, shallow 286 km^2 wetland in Manipur State, has yielded 176 species of invertebrate (alongside 249 species of vertebrate, 32 species of phytoplankton, and 233 species of macrophyte). Some 34 species (including 2 molluscs and 1 annelid worm) are currently in decline (Wetlands International 2003). However, invertebrate inventories seem typically ancillary to studies of shallow-water fisheries (e.g. Jhingran 1991). Clearly then, the aquatic invertebrate faunas of Indian wetlands, which represent a major subset of global wetlands (Gopal 2003), are in urgent need of detailed study.

A third of neighbouring Bangladesh is flooded annually by the monsoon rains, producing forested wetlands, such as those of the Sundarbans, which occupy about 7,800 km^2. Climate and a coastal location, result in the Sundarbans environment oscillating between salt and fresh water, and between heavy flooding and drought during the long dry season. The fish fauna of these wetlands is very rich, with around 115 species, including both fresh and brackish water forms. The cyprinids *Catla catla* and *Labeo calbasu* are of high economic importance. So, too, are the highly diverse decapods (shrimp, prawns [28 species], crayfishes, lobsters, and crabs [26 species]). A rich molluscan fauna includes filter-feeding clams—which are important in the foodweb leading to birds, mammals, and reptiles, and gastropods—the shells of which are used extensively to make lime (Ismail 1990).

4.4.6 Tropical floodplains

In areas which are subject to large seasonal changes in rainfall (e.g. monsoon regions of the tropics), many rivers cyclically overflow their banks. These dramatic fluctuations in water level result in local periodic episodes of flood and drought that affect not only the river itself but also tributaries and ponds on its floodplain. Associated with these cycles are significant changes in water chemistry and primary and secondary production. The aquatic floras and faunas of these floodplain water bodies are thus subject to spatial and temporal fluctuations in their environment of the nature of those discussed in Chapter 3. Despite this, some floodplain waters have diverse and abundant biotas, especially fishes, although these have yet to be studied in depth (Welcomme 1979).

Because of the individual nature of the topography and flow characteristics of any one river, the water bodies on the floodplain vary both qualitatively and quantitatively among systems. In addition, current-induced shifts in sediment frequently change the nature of the streams and pools within single systems. Figure 4.21 illustrates the geomorphological features likely to be encountered on a tropical floodplain. They are:

A. the river channel;
B. oxbows or oxbow lakes, representing the cut-off portions of meander bends;
C. point bars, loci of deposition on the convex side of river curves;
D. meander scrolls, depressions, and rises on the convex side of bends formed as the channel migrated laterally downvalley and towards the concave bank;
E. sloughs, areas of dead water, formed both in meander-scroll depressions and along the valley

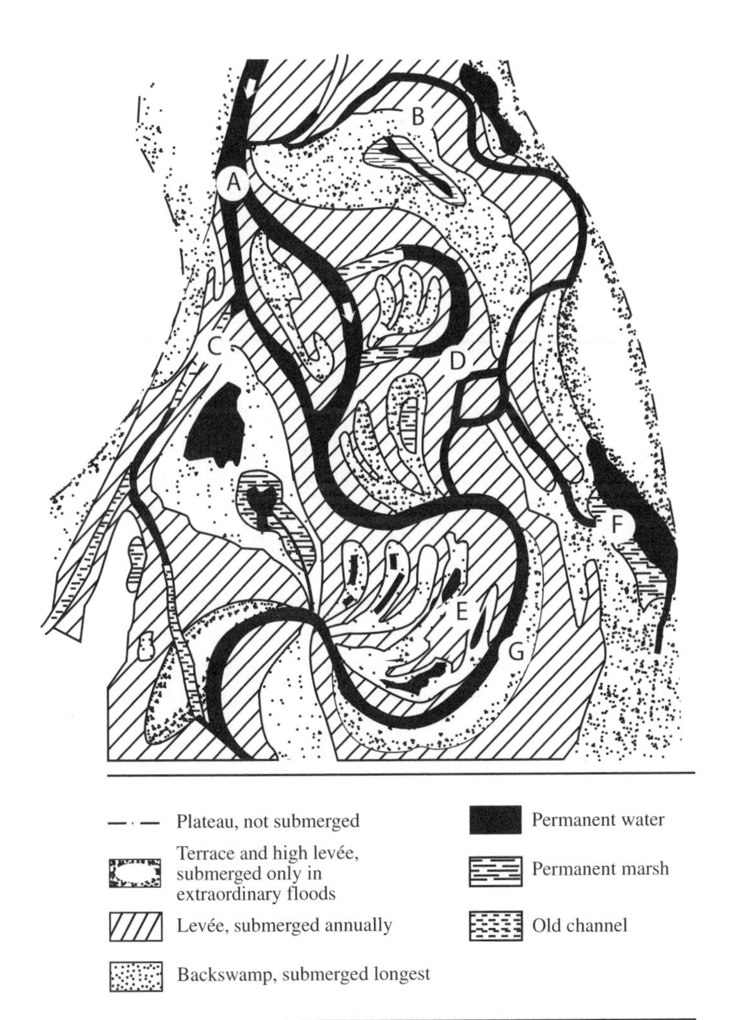

Figure 4.21 Diagram of the main geomorphological features of a tropical floodplain (letters refer to descriptions in the text; redrawn from Welcomme 1979).

walls as flood flows move directly downvalley, scouring adjacent to the valley walls;
F. natural levées, raised berms, or crests above the floodplain surface adjacent to the channel, usually containing coarser materials deposited as floodwater flows over the top of the channel banks. These are most frequently found at the concave banks. Where most of the load in transit is fine-grained, natural levées may be absent or nearly imperceptible;
G. backswamp deposits, overbank deposits of finer sediments deposited in slack water ponded between the natural levées and the valley wall or terrace riser;
H. sand splays, deposits of flood debris usually of coarser sand particles in the form of splays or scattered debris. (Leopold *et al.* 1964)

The main channel and its branches (A and B, respectively, on Figure 4.21) are usually permanently water-filled, although they do not always flow year round. Ancilliary to these are channels and streams that flow across the levées and connect the main channel with lentic water bodies on the floodplain. Many of these lesser channels are intermittent but they are important routes for seasonally migrating fishes. The lentic habitats are those left after the main channel has overflowed and returned to its normal level. Many are temporary in nature and they include sloughs in oxbows (C), meander scroll depressions (D), backswamps (E), or channels left by a previous course of the river (F). During the peak of the flood, these water bodies are frequently temporarily merged into one (Welcomme 1979).

Most of the major seasonal floodplains occur in the tropics and subtropics and, in fact, most tropical rivers have fringing, internal or coastal deltaic floodplains somewhere along their course, for example, the San Antonio River, the Orinoco River, and the Amazon River (Central and South America); the Senegal, Niger, Zambezi, Nile, and Volta rivers (Africa); and the Euphrates, Tigris, Indus, Ganges, and Mekong rivers (Asia). There are, however, floodplain rivers in temperate regions of the world, for example, the Danube (Europe) and Amur (Asia), although many of these have been altered by flood-control programmes. Perhaps the largest floodplain system in the world is the 'Pantanal' (area approximately 93,000 km^2) of the Rio Paraguay in the border area of Bolivia and Brazil. Here the annual flood cycle temporarily transforms savanna into a vast wetland.

In the tropics, basins on the floodplain lose water, through evaporation and seepage, throughout the dry season; thus they tend to diminish in size with increase in time since the last flood. This has the effect of concentrating the nutrients dissolved in the water, which is an important process in regions like the Amazon where the levels of nutrients in the main river channels are extremely low. Productivity in the floodplain lakes is thus much higher than in the river. In Amazonia, these Várzea lakes (Figure 4.22) are filled annually and are situated on the floodplain between higher ground (terra firma), which is never flooded, and the lower igapó forest which is an area subject to prolonged or even permanent submersion. In central Amazonia, the vertical height between the dry season level of the water in the river (the level of the igapó forest) and the wet season, high water mark (the level of the upper Várzea) is as much as 10 m.

Phytoplankton biomass increases in floodplain water bodies as they evaporate and reaches a peak in the dry season. This is quite simply due to the concentrating effect of the nutrients. However, once nutrients have been depleted, the populations crash and production is low until the next flood. In many regions, phytoplankton production is limited to the surface layers because of colouring of the water which limits light penetration. In western Amazonia, for example, waters derived from the Andes are loaded with suspended particles derived from basic rocks and have a cloudy-white appearance. In central Amazonia, waters are high in humic substances and are consequently dark-brown or even black, as in the Rio Negro watershed.

Epiphytic algae may be more important, in terms of production, than phytoplankton (Ducharme 1975). However, most of the primary production in

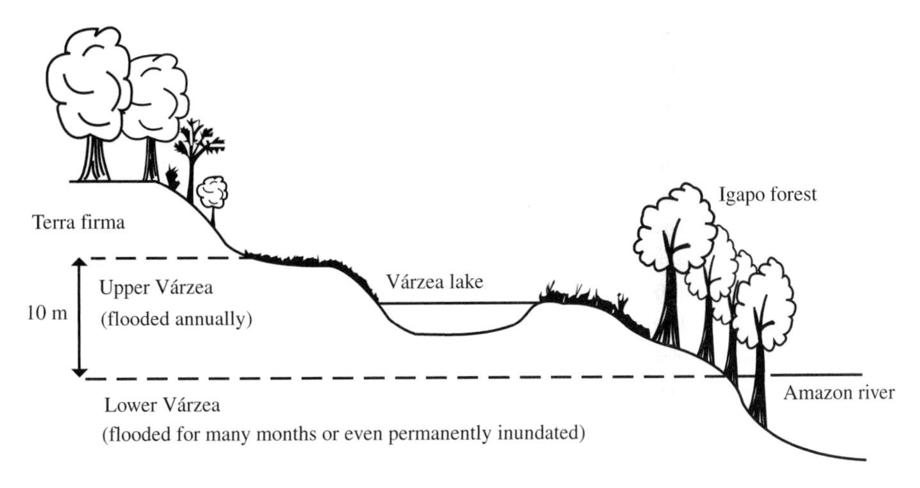

Figure 4.22 Position of a Várzea lake on a floodplain in central Amazonia.

the floodplain waters is derived from macrophytes. The latter's role is in:

- providing diverse habitats for plants and animals;
- acting as a filter and trap for allochthonous and autochthonous materials which are nutrients not only for the plants themselves but also for associated aufwuchs, invertebrates, and fishes;
- increasing the concentration of elements in littoral areas of lakes and in newly flooded waters of the floodplain, via a nutrient pump effect;
- contributing to autotrophic production, through decay, by forming a rich detritus which is used as food by a variety of organisms;
- functioning as a nutrient sink which, by locking up nutrients during the flood phase and releasing them to the soil during the dry season (through decay, burning and subsequent production of ash, and by being grazed and returned as dung), results in the conservation/retention of nutrients in the floodplain, rather than their being carried downstream (Howard-Williams and Lenton 1975; Welcomme 1979).

These macrophytes include a variety of grasses (primarily on savanna-type floodplains), which are mostly rhizomatous and flood-resistant, and dense forests of flood-resistant trees (in the rainforest regions of Africa, South America, and Asia, for example, the igapó forests of Amazonia). Other important elements in the flora are the floating forms, both free-floating species and those that form floating mats or meadows. Many of the free-floating species have cosmopolitan distributions, for example, water hyacinth (*E. crassipes*), water lettuce (*Pistia stratiotes*), water velvet (*Azolla* sp.), and salvinia (*Salvinia* sp.). Part of the success of these plants in these waters is due to their ability to rise up and down with the water level thereby remaining in optimal light conditions. Many submerged plants are forced to become dormant for that part of the year when they are covered by turbid waters. Growth of floating forms is so high that the water hyacinth, for example, can double in size every 8–10 days given warm, nutrient-rich water (Wolverton and McDonald 1976). The long, flowing, densely branched root clusters of many of these species (adapted to absorb the few nutrients from the water) provide sites of attachment for invertebrates—the 'perizoon' community. This community is believed to be the richest for aquatic species within Amazonia and contains representatives of most lotic taxa, except the Plecoptera; bivalves and tubificid worms are also rare. Although the level of taxonomic knowledge is still very low, mites and rotifers appear to be particularly species rich. Numerically dominant are microcrustaceans (copepods, cladocerans, and ostracods), dipterans, mayflies, and caddisflies. Dominant species are the conchostracan *C. hislopi* and snails of the genus *Biomphalaria* (Junk and Robertson 1997).

Zooplankton is more abundant in the waters of the floodplain than in the main channel, as flowing water does not suit pelagic species. Here, their numbers are controlled by factors, such as phytoplankton biomass (food), water temperature regime, length and severity of drought, availability of oxygen, turbidity, and amount of vegetation. Moghraby (1977) has shown that in many species that live in floodplain pools alongside the Blue Nile, adults or eggs enter a diapause, as water temperature drops and turbidity increases at the onset of a flood, and survive in the bottom sediments until temperature and turbidity return to dry season levels. In Amazonian floodplain lakes, the zooplankton is dominated by rotifers (300 species), cladocerans (20 species), and copepods (40 species, chiefly calanoids) (Robertson and Hardy 1984).

Little is known of the benthos of tropical meandering, slow-flowing, silty floodplain rivers, but diversity and abundance are thought to be low (Monakov 1969) although this may vary according to substrate composition (Blanc *et al.* 1955), as in other benthic habitats. The benthos of permanent lakes on the floodplain seems dominated by oligochaetes, molluscs, and chironomids, and biomass is extremely variable (e.g. low in profundal areas but often high in sheltered, shallow bays), due perhaps to lack of oxygen in deeper waters and periodic mass emergence of some of the insect species (Rzoska 1974). Temporary ponds and lakes on the floodplains, again, have been little studied but molluscs, particularly pulmonate snails, appear to be prominent in the benthos (Blanc *et al.* 1955; Welcomme 1979) although their diversity may be low (Bonetto *et al.* 1969). Ephemeroptera, Trichoptera, Chironomidae, Hemiptera, and Mollusca are abundant and widespread in ponds of the Kafue River, Africa (Carey 1967). Reiss (1977) showed two peak periods of abundance in the littoral benthos of the blackwater Lago Tupé system in the Amazon. On the rising flood, 2,559 animals m^{-2} were collected, compared with 1,248 animals m^{-2} at low water; interim density was 623 m^{-2}. Chironomids dominated the fauna at low water (but mites, corixid bugs, and oligochaete worms were also present) although phantom midges (Chaoborinae) and ostracods replaced them as the water level rose. In whitewater floodplain pools, similar double peaks have been recorded, one before flooding and a second after high water. Benthos in the igapó forests is sparse (Sioli 1975). Periodic anoxia near the bed, or desiccation, have been observed to result in forced invertebrate migrations, either vertically into more-oxygenated water, or horizontally to permanent waters (Reiss 1976; Junk and Robertson 1997).

Fittkau (1975) proposed three methods by which the benthos could survive between-flood droughts:

(1) migration, principally seen in the large decapod crustaceans;
(2) dormancy, principally the method seen in the molluscs;
(3) recolonization by aerial adults, as seen in the insects.

Further discussion of floodplain biotas, together with river–floodplain trophic dynamics is referred to in Chapter 8.

4.4.7 Intermittent saline ponds, lakes, and streams

In addition to having to contend with water loss from their habitat, the biota of temporary saline waters has to deal with change in ionic proportions. As the water evaporates, the ions become more concentrated thus adding an osmoregulatory stress-factor to those environmental forces already acting on the biota. In saline ponds, the ions most commonly present are sodium and chloride, but magnesium, calcium, and sulphate ions may also occur. Salinity may vary, over one year, from <50 to >300 $g\,l^{-1}$ (full-strength seawater is 35 $g\,l^{-1}$), and the pattern of variation may change between years (Figure 4.23). As well as acting through changes in osmoregulation, fluctuations in salinity may influence organismal respiration as less oxygen can be dissolved in saline water than in fresh water. Other factors, such as turbidity and high seasonal and diel fluctuations in water temperature, experienced by freshwater pond floras and faunas are experienced also by the inhabitants of saline ponds.

It was initially believed that the diversity of organisms in intermittent saline habitats was low and that those species present were more or less cosmopolitan in distribution. This is now thought not to be the case (W.D. Williams 1981), as the following groups are now known to be represented, although frequently not in the same habitat or geographical area: ciliates, foraminiferans, spirochaetes, gastropods, oligochaetes, anostracans, copepods, cladocerans, chironomids, ephydrids, ceratopogonids, culicids, stratiomyids, and birds. W.D. Williams (1985) pointed out the notable absence of some obligate temporary water forms, such as the notostracans and conchostracans, and also the general lack of frogs and fishes. Faunal diversity tends to be negatively correlated with increasing habitat salinity although the relationship is not clear cut. In fact, within wide salinity boundaries, salinity may not be a prime determinant of species presence (W.D. Williams 1984). Diversity also may be linked to habitat predictability, with lakes subject to predictable filling (e.g. Lake Eurack, Australia) having more species than unpredictably filled ones (e.g. Lake Eyre, Australia). Although some seasonal succession in the species is evident, it does not appear to be as well defined as in intermittent freshwater ponds (Figure 4.24).

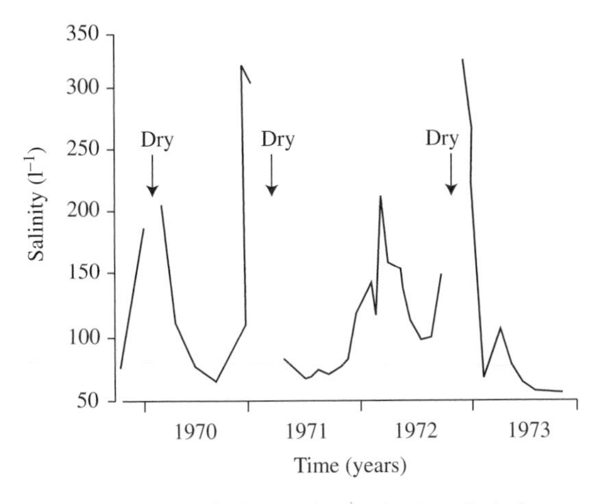

Figure 4.23 Salinity of Lake Eurack, Victoria, Australia (redrawn from W.D. Williams and Buckney 1976).

Little is known of the micro- and macroflora of intermittent saline ponds. The bacterial genera *Halobacterium* and *Halococcus* have been found in highly saline waters. Above 100 g l^{-1} the blue-green alga *Aphanothece halophytica* occurs and its distribution appears cosmopolitan. Below 100 g l^{-1}, other blue-greens appear, for example, species of *Oscillatoria*, *Phormidium*, *Microleus*, and *Spirulina*. The green alga *Dunaliella* occurs at high salinities but others occur only at moderate salinities. Few submergent macrophytes tolerate much salinity but *Ruppia* (Potamogetonaceae) tolerates salt well.

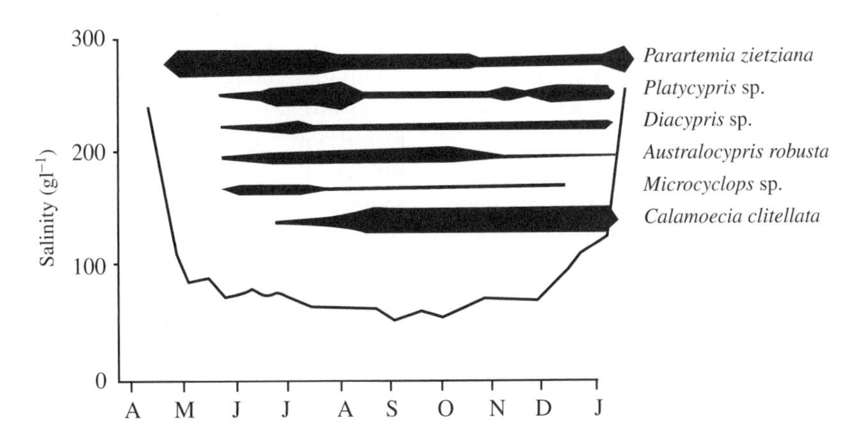

Figure 4.24 Selected examples of the seasonal distribution of species in Lake Eurack, Victoria, Australia (redrawn from Geddes 1976).

Table 4.10 Maximum salinities at which various species have been collected from intermittent salt lakes

Major group	Taxon	Salinity ($g\,l^{-1}$)
Bacteria	*Halococcus* sp.	350
	Halobacterium spp.	350
	Ectothiorhodospira spp.	340
Cyanobacteria	*Spirulina* sp.	100
	Aphanothece halophytica	350
	Oscillatoria limnetica	250
Chlorophyta	*Dunaliella salina*	350
Angiospermophyta	*Ruppia* sp.	230[a]
Mollusca	*Coxiella striata*	112[a]
Anostraca	*Artemia salina*	330
	Parartemia zeitziana	353[a]
Ostracoda	*Diacypris whitei*	180[a]
	Limnocythere staplini	205
Copepoda	*Calamoecia salina*	131[a]
	C. clitellata	113[a]
Isopoda	*Haloniscus searlei*	192[a]
Insecta	*Ephydra* sp.	222
	Culicoides variipennis	220
	Tanytarsus barbitarsis	95[a]
Pisces	*Taeniomembras microstomum*	70[a]

[a] Australian endemics.

Source: From W.D Williams (1985)

Scirpus maritimus (bullrush) is an important emergent halophyte in North America (W.D. Williams 1985). Salinity maxima at which various species have been recorded in the field are summarized in Table 4.10. It is clear from this that adaptation to high salt levels in combination with resistance to desiccation is something that quite a wide variety of organisms have achieved.

Problems stemming from living in water bodies subject to varying levels of salinity are chiefly ones of osmoregulation. W.D. Williams (1985) identified two mechanisms by which organisms cope with this problem: osmoregulation and osmoconformity, the former is typical of animals in saline habitats, the latter is found in all plants and in some animals. In osmoconformers, high internal pressures are maintained by the accumulation of organic or inorganic ion osmolytes. In the Cyanobacteria, sucrose, betaine, glutamate, and glucosylglycerol have been identified as osmolytes (Borowitzka 1981). In the Halobacteria, proteins within the cytoplasm undergo wide amino acid substitution and this enables cytosolic proteins to function under high concentrations of inorganic solutes—in this case potassium (Kushner 1978). In many halophytic macrophytes, the osmolyte is praline.

The majority of animals living in intermittent saline ponds tend to osmoregulate by taking in the water and then excreting unwanted ions. This is typically the method seen in *Artemia* (Anostraca), *Ephydra* (Diptera), and *Haloniscus* (Isopoda). However, some animals appear to osmoconform although really they may be osmoregulating at the cellular level, that is, osmoregulation may be a function of cellular activity, or cells must have a high tolerance of salt (W.D. Williams 1985). The copepod *C. salina* of Australia is an osmoconformer (Bayly 1969). Some animals seem to be able to switch so that, for example, the snail *Coxiella striata* osmoregulates weakly at low salinities but osmoconforms at high salinities (Mellor 1979).

In some osmoconformers (e.g. the alga *Dunaliella*, Halobacteria, and Cyanobacteria), the osmolytes have not only an osmoregulatory function but they appear also to reduce the effect of cold and heat on enzyme systems (Borowitzka 1981). The latter adaptation is particularly useful in temporary saline ponds in arid regions where water temperatures frequently exceed 30°C.

In Australian, arid-zone floodplain wetlands, Timms and Boulton (2001) found salinity to be a useful tool (alongside turbidity and water regime) in distinguishing among wetland types and in identifying lentic invertebrate communities. Within the Paroo River catchment (central-southeastern Australia), these authors were able to characterize both zooplankton and littoral species associated with both low and high salinity lakes. The latter were dominated by *Parartemia minuta* (Anostraca), *Apocyclops dengizicus* (Cyclopoida), and *Diacypris* spp. (Ostracoda), whereas the former were dominated by the ostracods *Trigonocypris globulosa* and *Mytilocypris splendida*, *Daphniopsis queenslandicus* (Cladocera), the corixid *Micronecta* sp., and the branchiopod *Eocyzicus* sp. The most saline lakes (70–162 $g\,l^{-1}$) contained the least number of species. Shallow, saline lakes of western Victoria were also dominated by crustaceans, although the chironomid *Tanytarsus barbitarsis* and larvae of a ceratopogonid were very common in salinities of between 20 and 100 $g\,l^{-1}$, whereas *Procladius* spp. and *Chironomus duplex* were more characteristic of salinities of between 3 and 30 $g\,l^{-1}$ (Timms 1983).

Alcocera-Durand and Escobar-Briones (1992) have described some of the characteristics of a now extinct lacustrine complex in the Mexico Basin. During Aztec times (~AD 1250) the complex comprised six interconnected water bodies within an endorheic basin. Today, there are only intermittent floodplain lakes, remnants of the huge saline Lake Texcoco; a tangle of channels fed by treated wastewater, representing the remains of Lake Xochimilco; and Mexico City has been built on the dried out bed of Lake Mexico. However, the fauna of present-day Lago Viejo, which is now largely fresh water and eutrophic, due to wastewater input and diversion of the Rio Hondo, retains some species known from the old complex, for example, the fishes *Chirostoma jordani* (Atherinidae) and *Girardinichthys, viviparus* (Goodeidae), an axolotl (*Ambystoma mexicanum*), and *Cambarellus montezumae* (a crayfish). Interestingly, odonates were very abundant in the lakes, but did not survive after aquatic vegetation was removed from Lago Viejo in 1965. The shore fly, *Ephydra hians*, still persists in the remnant saline pools of Lake Texcoco, and is documented as having been a significant food source (in all its life stages: larvae, pupae, and adults) for native Indians, since prehispanic times. Indigenous peoples also ate empty corixid eggshells, as a delicacy, and caught and fed adults to chickens. Another former perennial, large lake, Totolcingo, in the eastern portion of the Mexican Plateau, is now only rarely filled. When flooded in 1993, it persisted for only one month and exhibited a pH of around 10, with a conductivity of up to 30,000 $\mu S\,cm^{-1}$. The fauna was very limited during this time, comprising chiefly *E. hians*, *L. hoffmeisteri*, and a beetle, *Berosus* (Alcocer *et al.* 1997).

There are few studies of saline intermittent streams, despite the fact that salinization in arid and semi-arid regions is becoming an increasing global problem due to greater water abstraction. However, Bunn and Davies (1992) studied the macroinvertebrates fauna at two sites on Thirty-four Mile Brook, an intermittent stream in south-western Australia. In the 1980s, this stream exhibited up to 19 $g\,l^{-1}$ of total soluble salts (chiefly as sodium chloride) during intermittent flow conditions in March, dropping to 1 $g\,l^{-1}$ in July—a time of high rainfall. For comparison, the study also included samples from the nearby Hotham River (a permanent stream where the maximum total soluble salts measured was 6 $g\,l^{-1}$). Spring, late-summer, and mid-winter sampling produced a total of 68 benthic taxa from these two rivers. This fauna was dominated by crustaceans, especially ostracods and amphipods, with the paucity of insects (only 14.5%) being attributed to the high salinity. Some of the species present are also known from athalassic salt lakes across southern Australia, for example, *Austrochiltonia subtenuis* (amphipod), and the ostracods *Candonocypris* sp., *Sarscypridopsis aculeata*, and *Alboa warooa*. No mayflies were found, and only a single specimen of the stonefly *Leptoperla australica*. Ten caddisfly species were recorded, but at very low densities, and it was thought that they may have represented 'wash-ins' from less saline headwater tributaries. Chironomid larvae made up 10.6% of the total invertebrates collected, and were represented by 18 species, of which species of *Tanytarsus* were the most abundant. Other insects present included, in low numbers, dytiscid (7 species), gyrinid, and hydrophilid beetles, dragonflies (3 species) and damselflies (2 species), a single corixid (?*Sigara* sp.), and several dipteran families (Ceratopogonidae, Simuliidae, Thaumaleidae, Dolichopodidae, Empididae, Stratiomyidae, and Tabanidae). Other, non-insects were mites, a planorbid snail, a leech, oligochaetes, nematodes, a flatworm (Dugesiidae), and a hydra. Surprisingly, in an ordination analysis of their dataset, Bunn and Davies found few of the measured environmental parameters, including salinity, to be useful in explaining the patterns observed.

Case Histories—Episodic Waters

4.4.8 African desert and semi-arid rainpools

In the whole of northern Sudan and southern Egypt, there are no standing waters other than those created by rain that collects in depressions in the Nubian Desert (Rzoska 1984). These rainpools are several hundred square metres in area but are no more than 50 cm deep. The area is subject to high air temperatures and severe winds, and the rainfall itself is irregular. Around Khartoum, there are typically 8 dry months in a year with perhaps 4–10 episodes of rain during July, August, September, and October. Mean annual rainfall is between 160 and 180 mm, and only episodes of greater than 15 mm form rainpools. As soon as they are formed, the hot dry air (28–41°C) and winds cause rapid evaporation such that the typical duration of these pools is only 7 to 15 days. The main elements in the fauna of these pools, recorded by Rzoska, are crustaceans, particularly, the Branchiopoda (Table 4.11). In addition there are nematode worms, protozoans, chironomid larvae and other adult insects, and two rotifers, *Asplanchna* and *Pedalion*. With such a short-lived habitat, the fauna is highly specialized and shows rapid development. The Conchostraca, for example, appeared in the Khartoum pools within three days of pool formation, and were mature and bearing eggs by day five. The cladoceran *Moina dubia* was observed to be fully grown within 72 h and bred on the third and fourth days, first parthenogenetically and then by the production of ephyppia (see Chapter 5). The copepod *Metacyclops minutus* was fully grown within 48 h and then bred immediately. High population densities were achieved rapidly, reaching maxima of 460 *M. dubia* per litre of water, and 800 *M. minutus* per litre, on top of which there were many more immature individuals (up to 1700 l^{-1}). The larger species, such as the notostracans, appeared in the pools on day seven.

Table 4.11 The crustacean fauna of temporary rainpools near Khartoum

Major group	Taxon
Anostraca	*Streptocephalus proboscideus*
	Streptocephalus vitreus
	Branchipus stagnalis
Conchostraca	*Eocyzicus klunzingeri*
	Eocyzicus irritans
	Leptestheria aegyptiaca
	Limnadia sp.
Notostraca	*Triops granarius*
	Triops cancriformis
Cladocera	*Moina dubia*
Copepoda	*Metacyclops minutus*
	Metadiaptomus mauretanicus

Source: After Rzoska (1984).

Rzoska speculated on the nature of the food chains in these pools. Based on other studies it is known that the Anostraca and Conchostraca are filter feeders, that *Triops* is a scavenger and a carnivore, and that the microcrustaceans are fine-particle feeders. Rzoska suggested that protozoans and bacteria might form the basis of the trophic pyramid, although he did, once, observe a bloom of the blue-green alga *Oscillatoria*.

In the arid western and southern regions of Namibia, extensive shallow depressions (pans) fill occasionally after rainfall, but typically are dry for many years. As in the Nubian Desert, the biota comprises species specially adapted to withstand such extended droughts and to respond rapidly to rainfall (Curtis *et al.* 1998). Again, crustaceans predominate due to their ability to hatch, mature, and reproduce within 24 h (Allan *et al.* 1995). The crustacean fauna is surprisingly rich, and there is a high degree of endemism, particularly among the ostracods (Martens 1988) and the anostracans (Hamer and Brendonck 1997). Other elements in the fauna include the small, drought-resistant snail, *Bulinus reticulatus*, an endemic species of fish belonging to the genus *Nothobranchius*, which has desiccation-resistant eggs with arrested development, and a few species of insect, chiefly beetles. The pans are also used, opportunistically, by several species of sand frog (*Tomopterna* spp.) and *Pyxicephalus adsperus*, the African bullfrog. All of these amphibian species are arid-adapted and have broad sub-regional distributions (Curtis *et al.* 1998).

In neighbouring South Africa, shallow pans at Bain's Vlei are similarly characterized by crustaceans, but insects have been observed to colonize where the hydroperiod is extended. Large pans tend to support richer faunas than smaller ones (Meintjes 1996).

4.4.9 Rain-filled rockpools of African 'kopjes'

Shallow, very small depressions in bedrock which periodically collect rainwater are common features in many parts of the world. However, perhaps only in tropical and subtropical regions does the water in these habitats attain a sufficiently high temperature to enable organisms to rapidly and successfully complete their life cycles. On isolated hillsides in tropical Africa, McLachlan and Cantrell (1980) identified three types of such pool. These differ mainly in depth, which causes a gradation in habitat life span following rain; none, however, contains water during the six-month dry season. The relative life spans of these pool types are: (*a*) approximately 24 h; (*b*) several days; and (*c*) several weeks. The faunas of these pools are dominated by dipteran larvae which can achieve high densities ($50–300 \times 10^3$ larvae per pool) but which appear highly specific to a particular pool type. Pools of Type *a* are populated by larvae of the chironomid *Polypedilum vanderplanki* which are physiologically very tolerant of drought conditions; Type *b* pools are inhabited by larvae of the biting midge *Dasyhela thompsoni* which bury themselves in the mud of the pond bottom when the water evaporates; whereas Type *c* pools, the longest-lived, are populated by larvae of the chironomids *Chironomus imicola* and *C. pulcher*, which have little tolerance of desiccation and must disperse before drying.

Food in these pools is generally abundant and is typically allochthonous in origin, consisting of fruit, pollen, flowers, and the dung of small carnivores (McLachlan 1981). There are few predators during the aquatic phase but dormant stages are susceptible to scavenging by terrestrial taxa, such as pheidolid ants, particularly when the water has just evaporated but the mud is still damp. Once the mud dries totally, predation ceases.

Many of these rockpools support virtual monospecific populations. This might be explained in terms of the rigours of the environment or perhaps some other factor. McLachlan and Ladle (2001) concluded that the main selective pressure shaping species' adaptation in these pools was their spatial consistency together with their unpredictable duration.

4.4.10 Rain-filled rockpools of southwestern Australia

A series of shallow, water-filled depressions in the granite outcrops of southwestern Australia was studied by Bayly (1982). This region is characterized by dry summers and wet winters, and pools form in mid-May. The pools are generally shallow (most are less than 10 cm deep), have a pH of 6.0 or less, and have a conductivity of less than $180\ \mu S\,cm^{-1}$. Some of the pools have a pH of less than 4.0, probably the result of plant growth on the pool bottoms. Daytime water temperatures range between 10.7 and 17.2°C. The fauna is dominated by microcrustaceans, particularly cladocerans (including an endemic species, *Neothrix armata*), ostracods, and copepods, together with mites (especially oribatids) and chironomids. Some succession of species is evident. Other groups represented are turbellarian flatworms, nematodes, anostracans, conchostracans, hemipterans, beetles, and ceratopogonids. The latter group is represented by *Dasyhelea*, the same genus found in rockpools in Africa. However, apart from this the composition of the faunas is not very similar between the two continents, even though the hydroperiod of the Australian pools appears comparable to the Type *c* pools in Africa.

Food input in some pools is derived from the breakdown of lichens and mosses growing on the granite surfaces whereas, in others, marsupial dung is the major source of energy.

4.4.11 Rain-filled pools in Malaysia

Laird (1988) provided an example of the biota of another type of episodic pool, in Singapore. Here, the climate, although tropical is much more humid (on average 80–85%), and rainfall more frequent throughout the November to January rainy season. The 'puddle' was approximately 1 m across, with a maximum depth of 4 cm, and never persisted for more than 9 days after a rainfall. pH ranged from 6.6 to 7.2, alongside water temperatures of 25 to 37°C. Thirty-two collections were made from this puddle between August 23rd, 1955 and December 20th, 1956, some when the puddle was dry, when soil samples were taken. The prokaryotes found in this puddle were listed in Table 4.1, in addition there were several species belonging to the following protoctist groups: Euglenophyta, Chlorophyta, Zoomastigina, Rhizopoda, and both stalked and motile Ciliophora. The invertebrate component comprised two rotifers, nematodes, microcrustaceans, mites, springtails, several hemipteran species, a dytiscid beetle, a chironomid (*Chironomus*), and two species of *Culex* (Table 4.12). In addition, myrmicine ants were frequently seen scavenging at the edges of the puddle and over the moist bed during drying.

In terms of frequency of occurrence, the bacteria, protoctists, bdelloid rotifers, nematodes, crustaceans, and surface-dwelling hemipterans were always present. The mosquitoes, *Cx. annulus* and *Cx. fuscocephalus*, were present only on four sampling occasions, but pupal exuviae were found only on one, suggesting that this puddle was a marginal habitat for the two species—this despite the fact that they are considered to be 'transient pool' specialists (Colless 1957). These same two species were also present, sometimes singly, sometimes together, in 10 other, similar, puddles in the vicinity. Occasionally, three other species were collected from these latter puddles: *Anopheles kochi*, *Culex tritaeniorhynchus summorosus*, and *Cx. nigropunctatus*. All five mosquito species frequently failed to complete their life cycles, which Laird (1988) attributed to: (a) unpredictability of hydroperiod (only once in 16 months of observation did his primary puddle retain water for as long as 9 days); (b) relatively low abundance of suitable larval food (e.g. phytoflagellates); (c) lack of shading, which allowed mid-day water temperatures to climb to near-lethal levels (high 30s°C); and (d) possibly the abundance of several species of epibiont, such as *Sphaerotilus natans* and *Vorticella microstoma*.

Table 4.12 The fauna of an episodic rainpool in Singapore

Major group	Taxon
Rotifera	*Philodina* sp. [Bdelloida]
	Asplanchna brightwelli [Ploima]
Nematoda	indet.
Cladocera	*Moina* sp.
Cyclopoida	*Mesocyclops thermocyclopoides*
	Microcyclops varicans
Acari	*Hydrachna tenuissima*
Collembola	*Sminthurides aquaticus*
Hemiptera	*Gerris adelaides*
	Limnogonus fossarum
	Microvelia sp.
	Micronecta quadristrigata
	Plea liturata
Coleoptera	*Canthydrus flammulatus*
Hymenoptera	*Pheidole megacephala*
Diptera	*Chironomus* sp.
	Culex annulus
	Cx. fuscocephalus

Source: From Laird (1988)

4.4.12 Mediterranean ramblas

The mediterranean climate is characterized by wet autumns and winters and very dry summers, the latter lasting for up to five months. Globally, this biome comprises coastal zones bordering the Mediterranean Sea; central and southern California; coastal Western Australia and South Australia; the Chilean coast; and the Cape Town region of southern Africa. Rainfall is typically around 40 cm, annually, and results in a vegetation dominated by shrubs and low-growing species, with areas of larger trees, all of which possess some xerophytic properties. These subtropical regions, called 'maquis' in Europe, and 'chaparral' in North America, support distinctive episodic waters.

Although sparsely studied, a survey of the ramblas of southeastern Spain by Moreno *et al.* (1997) provides some information on their biotas. These authors defined ramblas as wide, usually dry, channels that flow significantly only during

Figure 4.25 Example of a rambla in southeastern Spain: the Rambla del Orón, Murcia (courtesy of M.R. Vidal-Abarca).

flash-flood events (Figure 4.25)—although in some reaches, groundwater seepage into the bed, or wastewater from human activities, can extend the hydroperiod. The Rambla de la Viuda, in the Valencia Region of Spain, is one of the most hydrologically studied such systems. Between 1959 and 1984, an average of around three rainfall events, per year, were recorded in this rambla. These totalled around 31 days each year (8%) during which some runoff was detected; the average length of a runoff episode was 8.2 days (Segura 1990). Detailed recording of a single hydrograph in the Rambla de Poyo basin, showed that in the first 15 mins, discharge rose from 0 to 193 $m^3 s^{-1}$. Within 3.5 h, the flow had stopped. Surface discharge was calculated to have represented 8% of the rainfall over the basin (Camarasa Belmonte and Segura Beltrán 2001). Flash floods in the Rambla del Judio, Murcia (Figure 4.26), occur at a rate of approximately 3 to 4 in a 10-year period (M.R. Vidal-Abarca, personal communication).

When water flows down these channels, it is frequently saline (e.g. up to 9 $g l^{-1}$, in the Rambla del Moro, and between 5.5 and 27.5 $g l^{-1}$ in the Rambla del Judio; Vidal-Abarca *et al.* 1996), due to solution of evaporites accumulated among the bed materials. Sometimes, such water conditions, especially if supplemented by groundwater flow, support helophytes, such as the reed *Phragmites australis, Arthocnemum macrostachyum*, the saltwort *Sarcocornia fruticosa*, and the seablite *Suaeda vera* (Gomez 1995). In a series of ramblas in

Figure 4.26 Photograph of the Rambla del Judio, Murcia, just after a flash flood (courtesy of M.L. Suárez).

southeastern Spain, Moreno *et al.* (2001) collected 47 species of macroalgae and submerged macrophytes, some of which showed affinities with either saline or freshwater conditions. Most showed physiological, regenerative, or dispersal adaptations for dealing with water loss.

The fauna of ramblas is dominated by aquatic Coleoptera and Heteroptera, typically species with good colonization powers and adaptations to water level fluctuation. From 25 sites, Moreno *et al.* (1997) collected 63 species: 45 beetles (representing the families Haliplidae, Dytiscidae, Hydraenidae, Hydrophilidae, and Dryopidae) and 18 bugs (Gerridae, Hydrometridae, Hebridae, Veliidae, Nepidae, Naucoridae, Pleidae, and Corixidae). Highest diversity (38 species) occurred in groundwater-fed reaches, whereas episodic pools yielded only from two to 11 species, the most frequent of which were *Laccobius hispanicus* (Hydrophilidae), *Meladema coriacea* (Dytiscidae), and *Ochthebius quadrifoveolatus* (Hydraenidae). Most of these species were collected from static pools after flood events, and the predominance of Coleoptera and Heteroptera confirms this lentic state. What is unknown from Moreno *et al.*'s study was the presence of any species during the brief lotic phase of these habitats. However, Ortega *et al.* (1991) carried out a detailed inventory of the invertebrates recolonizing the Rambla del Moro after an autumn flood event. After the flash flood, many small pools formed on the floodplain, and these became repositories for some of the taxa washed out of the main channel. On the first day of sampling, 25 taxa were found in a single pool, compared with 20 in the stream channel. Thereafter many taxa (especially molluscs, mayflies, dragonflies, hemipterans, beetles, and dipterans) began to move back into the channel; the pools dried up within 15 days.

Gasith and Resh (1990) reviewed aspects of streams from mediterranean-type climates from around the globe, and emphasized that the latter strongly influenced their ecology. Although these authors considered seasonal events to be predictable, perhaps questioning their designation, here, as episodic waters, they concluded that flooding and drying 'vary markedly in intensity on a multi-annual scale'. The biota was seen as being under considerable *abiotic* pressure from floods, yet there is a period (spring-early summer) when more moderate conditions and improved resource availability permit the biota to recover. As populations recover, *biotic* factors, such as predation and competition, become more influential. However, with the onset of drought, abiotic regulation returns. Gasith and Resh underlined the paucity of knowledge of these stream types by proposing 25 hypotheses that need testing in these systems. These included relating the influence of stream hydrographs to such properties as faunal richness, abundance, and diversity; species coexistence; the consequences of protracted riparian leaf-litter input; and the contribution of autotrophy.

Streams in the rain shadow of the Sierra Nevada mountains, southwestern USA, frequently flow only from November to March, in association with cyclone fronts, and occasionally during the summer after heavy thunderstorms. Most exist only as headwater springs that flow permanently or periodically for perhaps only 200 m before disappearing underground. The biota has in some cases been found to be species rich, and to include endemics. Further, populations have been found to exhibit properties, such as reduced flight ability (brachyptery or aptery), that prevent futile dispersal across the extremely harsh intervening landscape that lies between these isolated habitats (Myers and Resh 1999).

4.4.13 Polar meltwater streams

Meltwater streams are common around the margins of Antarctica, in summer, and may vary in length from a few to tens of kilometres; some flow from ice sheets to the sea whereas others flow inland to lakes. In the McMurdo Sound region of southern Victoria Island (78° S; 165° E) the streams are characterized by highly variable flow patterns based on diel, seasonal and annual periodicities. Typically, streamflow begins in mid-November or early December and continues into January, but there is variation among years. By late January/early February, the streams freeze solid. Smaller streams are particularly subject to diel changes in flow, with discharge changing as much as two hundred-fold according to whether the source-glacier face is exposed to direct sunlight or is in shadow (e.g. range 0.006 $m^3 s^{-1}$, in shadow, to 0.1 $m^3 s^{-1}$, in direct sunlight; as measured in the Whangamata Stream originating at the Commonwealth Glacier; Vincent and Howard-Williams 1986).

Associated with these dramatic changes in discharge are wide fluctuations in water temperature and dissolved nutrients. Water temperatures are invariably low but vary both from day to day (e.g. from 3.0 to 5.5°C) and diurnally (e.g. from 0°C at 2:00 a.m. to 6.0°C at 6:00 p.m.). Suspended solids tend to be positively correlated with discharge, whereas dissolved nutrients (e.g. nitrates and soluble reactive phosphorus) are highest when the streams start to flow, but their concentrations decline thereafter (Vincent and Howard-Williams 1986). In the lakes of South Georgia, higher phosphorus concentrations occur in those closer to the sea, and are attributed to waste input from marine vertebrates, such as penguins, petrels, and seals, along with a slightly longer hydroperiod. Higher nutrient levels also correlated with higher invertebrate abundance (Hansson *et al.* 1996). Nutrient enrichment from birds (Adélie penguins and skuas) has also been recorded in a stream at Hope Bay, on the Antarctic Peninsula (Pizarro *et al.* 1996).

The epilithic community of the rock surfaces in the McMurdo Sound streams is simple, compared with those of temperate streams, and is dominated by filamentous blue-green algae (especially *Nostoc*, *Phormidium*, and *Oscillatoria*) but also contains bacteria, fungi, and microherbivores such as protozoans, rotifers, tardigrades, and nematode worms. Larger herbivores, such as aquatic insects and crustaceans are completely absent from this region of Antarctica, although three species of collembolan and three species of mite are known to occur in riparian mosses (Vincent and Howard-Williams 1986). The epilithic algal communities at the Hope Bay stream were dominated by mats or filaments of the green alga, *Prasiola crispa*, but other forms were also common, particularly a blue-green, *Leptolyngbya fragilis*, and several chrysophytes: *Hydrurus foetidus*, *Chrysococcus* cf. *rufescens*, and *Phaeogloea mucosa*. The chrysophytes were more prominent near the mouth of the stream, whereas *P. crispa* was more characteristic of the source. In contrast to the McMurdo Sound streams, 57 algal taxa were found in this stream (Pizarro *et al.* 1996).

In the whole of Antarctica, there appear to be only two insect species that occur naturally, and both are chironomids. Both are restricted to the maritime region (which includes the west coast of the Antarctic Peninsula and its offshore islands, the South Orkney, South Shetland, and South Sandwich Islands) (Convey and Block 1996). *Belgica antarctica* (Orthocladiinae) is endemic to this region, and its larvae are to be found year-round associated with moss, grass, algae, animal detritus, and nutrient-enriched soils around seal wallows and bird nesting sites (Peckham 1971). Larvae are also known from mats of *Phormidium* in summer meltwater streams (Richard *et al.* 1994). The adults have reduced wings. *Parochlus steinenii* (Podonominae), in contrast, is fully winged, and its larvae are more typically aquatic, living in the sediments of streams and lakes—although it has also been collected from wet moss (Wirth and Gressitt 1967). This species has a wider distribution, occurring also in the high Andes and Tierra del Fuego. There is a third species, *Eretmoptera murphyi* (Orthocladiinae), which is believed to be endemic to South Georgia (sub-Antarctic) and was accidentally transplanted to South Orkney, with soil, in 1967 (Edwards 1980). *E. murphyi* is also a brachypterous terrestrial species whose larvae live in disturbed peat and soil associated with coastal lichens and mosses (Convey and Block 1996).

Meltwater streams also occur in the Arctic, where the overall richness of invertebrate species is much higher (e.g. 1,650 species of insect, both terrestrial and aquatic, in the North American sector; Danks 1990) than in Antarctica. This is probably related to the comparatively warmer climate; much of Arctic North America corresponds to the Southern Ocean islands, the Falkland Islands, and Tierra del Fuego (Convey and Block 1996). Stocker and Hynes (1976) studied three inflow streams into Char Lake on Cornwallis Island (74°43′N, 94°59′W). These can be considered episodic because although they are generated each year, around June, the associated avalanche of snow, stones, and slush causes their location to be ill-defined and shifting within the local gravel basin. Stream temperatures largely remained around 1°C, reaching a peak of only 4°C where water ran in very thin sheets close to the lake. The fauna was dominated by chironomids, *Diamesa* (near the headwaters), and the orthoclads *Diplocladius*, *Orthocladius*, *Limnophyes*, and *Metriocnemus* (downstream). Also present were mites, a tipulid, three species of enchytraeid worm, three species of ostracod, nematodes, collembolans, and two copepods (*Cyclops magnus* and *Attheyella nordenskioldii*). Some sections of the channels supported growths of diatoms (12 species), clumps of the cyanophyte *Gloeotrichia*, and filaments of the green alga *Zygnema* and *Lyngbya* (cyanophyte).

4.5 Global commonality among communities

As previously noted, bacteria, protozoans, algae, and fungi have been accounted for, in detail, in only a small number of studies of temporary waters; comparison is therefore difficult. However, where given due treatment, some of these groups appear to be quite species rich and numerically abundant, especially in intermittent water bodies (e.g. see Tables 4.1 and 4.2, and Figures 4.1, 4.3; and 4.4). Episodic waters may contain a partial subset of those species found in intermittent waters (Table 4.1), but insufficient studies have been made to be confident of this. However, the presence of 57 algal taxa in an episodic stream on the Antarctic Peninsula (Section 4.4.13) is noteworthy. The high levels of detritus typically present at the beginning of the hydroperiod, together with increasing water temperatures, are likely to promote high abundances of bacteria early on in most temporary waters (Figure 4.1), which are soon supplemented by protozoans and algae, especially in sunlit locations. Often, species richness increases within each of these groups as the hydroperiod progresses (e.g. Figure 4.4).

Macrophytes are better known, and common elements are identifiable within major climatic zones. Most occur along a shoreline gradient related to the degree of inundation. For example, in temperate regions, cat-tails (*Typha*), rushes (e.g. *Eleocharis* and *Juncus*), and grasses (e.g. *Glycera* and *Agrostis*) characterize the margins of intermittent water bodies, especially where there is slow or no flow. Other plants, such as *Callitriche* (water-starwort), tolerate submergence better and are found in shallow water. Forms such as *Nuphar* and *Nymphaea* (water lilies) grow in deeper water. A number of terrestrial plants, such as *Rumex*, invade the bed as the water dries up. In the tropics, there are similar plant morphotypes that respond to water level, and also large floating forms, such as water hyacinth (*Eichhornia*) and water lettuce (*Pistia*). Episodic habitats seem generally to support fewer higher plants species, apart from some mosses and lichens, perhaps a consequence of the unpredictability of both the appearance and retention time of the water, together with an oftentimes bedrock basin.

Among the invertebrates, clearly, in most temporary fresh waters, it is the insects and crustaceans that dominate the fauna, although part of this may be an artifact of bias against other groups in terms of the level of taxonomic treatment applied, or available. For example, nematode worms are invariably present, often in large numbers, but are seldom identified other than by their phylum. Further, both oligochaetes and mites, if identified beyond family or genus levels, may prove to be species rich.

The degree of similarity in invertebrates varies according to habitat type. For example, Section 4.3.1 identified the following common elements among intermittent ponds: snails (typically Planorbidae); oligochaetes (especially *L. hoffmeisteri*); usually one species of Anostraca, Notostraca, and/or Conchostraca (often *Triops*); microcrustaceans (especially chydorid cladocerans [*Moina*], ostracods and copepods [*Diaptomus* and *Canthocamptus*]); aquatic mites; springtails; odonates (Zygoptera and Anisoptera [*Anax*]); chironomid [*Chironomus*], culicid, and ceratopogonid dipterans; and a high diversity of Hemiptera (*Gerris, Hydrometra* and *Microvelia*) and aquatic Coleoptera (typically *Agabus, Berosus, Haliplus, Cyphon*, and *Helophorus*). In addition, there are likely to be representatives of overlooked meiofaunal taxa, such as microturbellarians, gastrotrichs, rotifers, and nematodes.

Among intermittent streams, and again keeping in mind geographical separation, differences in climate, endemism, and local hydrological characteristics, the following taxa are likely to be present (Section 4.3.2): macroturbellarians; nematodes; gastropods (including planorbids); sphaeriid clams; naidid, tubificid, and enchytraeid worms; cyclopoid and harpacticoid copepods; amphipods; decapods; water mites; leptophlebiid mayflies; stoneflies (although the families represented vary considerably); corixid, gerrid, and veliid hemipterans; elmid, dytiscid, hydrophilid, and psephenid beetles; hydropsychid caddisflies; tipulid, ceratopogonid, simuliid, and chironomid dipterans. Compared with intermittent ponds, these streams tend to share fewer genera, have fewer microcrustaceans, and no large branchiopods, but have greater representation of rheophilic taxa.

From limited studies, the invertebrate faunas of episodic streams appear quite variable, both qualitatively and quantitatively. For example, streams in Oregon were reported dominated by dipterans and beetles, with poor representation of mayflies, stoneflies, and caddisflies (Dieterich and Anderson 2000). Beetles and hemipterans dominated Spanish ramblas (in their pool phase) (Section 4.4.12), whereas Antarctic streams were populated by three species of chironomid (Podonominae and Orthocladiinae), together with rotifers, tardigrades, and nematodes. Chironomids (Diamesinae and Orthocladiinae) also dominated streams in the

Arctic, along with nematodes, enchytraeid worms, ostracods, and two species of copepod (section 4.4.13).

Episodic pools show some similarity in their invertebrate communities, although dominance differs. For example, although crustaceans and dipterans are typically present, crustaceans (especially branchiopods) have been shown to dominate African desert rainpools and rain-filled rockpools in southwestern Australia, whereas dipterans (chironomids and ceratopogonids) dominate African rockpools. Other taxa typically found in the first two, and in rainpools in Malaysia, include nematodes, rotifers, mites, and often collembolans, hemipterans, beetles, and mosquitoes (Sections 4.4.8–11).

Saline temporary water communities contain representatives of a number of the organismal groups found in temporary fresh waters. However, two main features characterize them: (1) both the abundance and richness of crustacean species typically far exceed that seen in the insects, which are mostly limited to dipterans; and (2) the biota contains many species that have become virtually restricted to these habitats as a result of specialized adaptations to their osmoregulation (e.g. *Halococcus* sp., *Halobacterium* spp., *Dunaliella salina, Artemia salina, C. salina*; Table 4.10).

Many temporary waters support populations of vertebrates. Fishes mostly occur in intermittent tributaries or floodplain pools where they may feed or reproduce, but where periodic and predictable reconnection to permanent waters is virtually assured. Species are typically those with a tolerance of high water temperatures and degrading water quality, and fast egg development. Tropical forms, such as lungfishes (e.g. *Lepidosiren*), may survive drought, *in situ*, by virtue of specialized adaptations, in this case dormancy, buried in mud, in the adult stage. A small number of fishes have drought resistant eggs, for example, killifishes (Cyprinodontidae) and *Cynolebias*, a tooth carp (Rivulidae).

Frogs, toads, and salamanders are common inhabitants of intermittent lentic waters, where they feed and breed. They are less commonly found in episodic and lotic waters. Many reptiles frequent temporary water, mostly to feed, but also to moisten their skins prior to shedding, and to thermoregulate. A few, such as the anaconda, breed in wetlands.

Mammals and, especially, birds use temporary waters as a source of food and drinking water, and often raise their young nearby. Species composition is highly dependent on locality.

4.6 Permanent versus temporary water faunas

Wiggins *et al.* (1980) maintain that temporary waters represent discrete types of freshwater habitat in which the dry phase imposes such rigorous environmental conditions that only a limited number of species can survive in them. Nevertheless, as we have seen, species from most of the major groups occurring in fresh water have succeeded in doing so. This reflects the richness of resources in many of these waters and the powerful selection forces at work in all these organismal groups. Continuity of water in an aquatic environment is surely a fundamentally necessary factor. However, a large proportion of temporary water species are insects or mites, groups which evolved on land, and the problems of adapting to water loss may, therefore, not be that difficult; taxonomic diversity and species richness in temporary waters is frequently high. There are, perhaps, no less rigorous factors controlling the community composition of permanent waterbodies.

Comparison of community composition between permanent and temporary waterbodies reveals relatively little overlap. Two examples from Canada are given in Table 4.13. In example A, the intermittent stream is actually a tributary of the permanent stream, potentially allowing easy access between the two. However, only 11.3% of the taxa overlap. In example B, the two streams, although not connected, are only 200 m apart yet the similarity index is only 13.3%. In contrast, in some parts of Australia, much of the 'typical' stream fauna present in permanent streams can be found successfully reproducing in nearby intermittent streams (Boulton and Suter 1986). Dance and Hynes (1979) found that two tributaries of the

Table 4.13 Similarity in taxa between intermittent and permanent nearctic streams

	Intermittent stream		Permanent stream
Example *A*			
Number of taxa	33		46
No. of taxa in common		8	
Similarity index		11.3%	
Example *B*			
Number of taxa	23		28
No. of taxa in common		6	
Similarity index		13.3%	

same Ontario river, one permanent, the other intermittent, both had rich midge faunas. At the species level, however, some of the Chironominae and Tanypodinae occurred only in the intermittent branch, while the Diamesinae and some Orthocladiinae only occurred in the permanent one.

In a study of calanoid copepods from permanent and intermittent ponds in Arizona, Cole (1966) collected six species in abundance. *Diaptomus albuquerquensis, D. siciloides,* and *D. clavipes* were common to both types of pond. *D. nudus* occurred only in one permanent pond and *D. novamexicanus* and *D. sanguineus* were restricted to intermittent ponds. Comparative measurements of each of the three species common to the two pond types revealed that cephalothorax lengths and clutch sizes were greater in the intermittent pond populations. Congeneric occurrence of adjacent-sized and similar-sized species was common in the intermittent ponds. Cole cited reasons for these observations as being either genetic or the result of superior food supply in the intermittent ponds. Predation on large individuals might select for smaller individuals in permanent ponds. Kohler *et al.* (1999) tested the related hypothesis that the presence of fish predators in temporary waters should exclude large, active invertebrate predator species. The study site comprised floodplain ponds, where, during flooding, fishes had access to both permanent and temporary ponds, and remained in these habitats after the floodwater receded. In both the Mississippi and Sangamon Rivers, removal of large species occurred across the pond permanence gradient.

Otto (1976) compared the biology of three species of cased caddisfly from southern Sweden. *Limnephilus rhombucus* occurred in a pond, *Glyphotaelius pellucidus* occurred in an intermittent pool, and *Potomophylax cingulatus* was collected from a small stream. *L. rhombicus* was found to be poor at coping with water level fluctuations and with fast water. *G. pellucidus* was inferior to the other species in interspecific interactions but endured low oxygen concentrations best. *P. cingulatus* was confined to areas of high oxygen but was superior in resisting the force of water currents.

Delucchi and Peckarsky (1989) compared the life history patterns of several insect species in an intermittent and a permanent stream in New York State over a three-year period, with a view to testing the hypothesis that intermittent stream populations grow and develop faster, and emerge earlier. They found that species living primarily in intermittent streams ('specialists') emerged by mid-June, compared with late summer for permanent stream species. However, only one of two facultative species (those found in both permanent and intermittent streams) and one specialist species showed greater size and/or advanced development at sites that dried up compared with populations at permanent sites—however, these differences were apparent only in the early spring, some time before the stream actually dried. The authors thus concluded that there was little evidence that drying of the stream exerted significant selection pressure on any of the *facultative* species to produce earlier emergence at the intermittent site. They also maintained that the intermittent stream did not demonstrate a unique insect fauna—as seven of the eight species studied were common to both stream types.

In a different study, Way *et al.* (1980) were able to detect intraspecific differences between the two habitat types. They found that a population of the clam *Musculium partumeium* in an intermittent pond had a univoltine life cycle spanning 13 months. In contrast, a population from a permanent pond passed through two generations in the same period of time. Springtime recruits to the intermittent pond population had high C:N ratios, suggesting that they have a high content of fats and/or carbohydrates in their tissues which sustains them during the dormant phase that begins at birth. The life history characteristics of *M. partumeium* do not appear to fit exactly any of the popular theories on life history tactics—for example, that temporary pond populations tend to show more features of K-selected species rather than *r*-selected species (see Chapter 5) (Burky *et al.* 1985). Clams in the permanent pond appeared to mimic the life cycle pattern of the intermittent pond population in that they went through a period of 'dormancy' during the summer that coincided with the dry period of the intermittent pond. It has been suggested that this species actually evolved in temporary habitats and thus the summer dormancy noted in permanent pond populations may be a genetic legacy.

The brine shrimp *A. salina* represents a complex of sibling and semi-species that inhabit permanent lakes and temporary ponds which are salty. In Mono Lake, a permanent habitat in California, the *Artemia* population is highly specialized to live in a permanent lake. The environmental cues for hatching and cyst formation seen in temporary ponds in Nevada (desiccation, increased salinity, high temperature, rehydration) are not present in Mono Lake. Instead, the Mono Lake shrimps have evolved a cold, pre-incubation hatching requirement, one more suited to a stable temperate lake. In addition, the cysts of the Mono Lake population sink to the lake bottom and have poor resistance to desiccation, properties related to a less crucial need for dispersal in a permanent environment. Again, the ability to survive passage through the digestive system of birds (thought to be an important mechanism of dispersal in *Artemia*) is not evident in the Mono Lake population. *Artemia* in Mono Lake have the fewest number of generations per year (two) known, and this may support predictions from life history theory that species in stable environments tend to have fewer generations per year, together with slower development of individuals (Dana 1981).

So far, we have treated permanent and temporary freshwaters as separate entities. It is important to remember, though, that a continuum often exists between the two. Caspers and Heckman (1981) in their study of orchard drainage ditches adjacent to the Elbe estuary found a definite seral progression towards a terrestrial climax. Distinct sets of physical and chemical conditions were evident along this gradient, together with several typical species aggregations appropriate to these conditions. In central Saskatchewan, Driver (1977) noted that midge diversity in small prairie ponds was dependent on the stage of development of the plant community along a moisture gradient. Generally, chironomid diversity increased gradually from intermittent to permanent ponds. Based on this, he suggested that the relative permanency of these ponds could be assessed by examining the chironomid faunas. This may be useful in situations where more obvious temporary water indicators, such as the Branchiopoda, are, perhaps for reasons of zoogeography, absent.

Figure 4.27 summarizes the changes in community composition that might be expected along such a continuum. In ponds and streams that are dry for only a few weeks each year (left-hand side of the graph) the community will consist largely of facultative species, that is, those found primarily in permanent waterbodies but which, because of wide physiological tolerance or flexibility of life cycles, are able to survive for short periods out of water. At the other end of the scale, in temporary waters that dry up for many months in a year, or

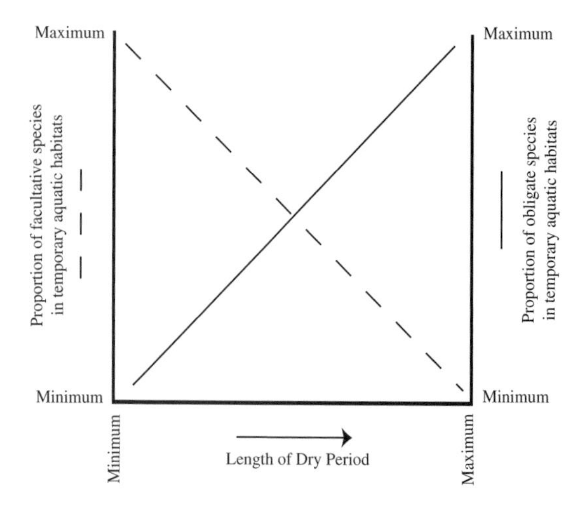

Figure 4.27 Predicted changes in community composition along the temporary–permanent aquatic habitat continuum.

even for periods of several years (as, for example, do some of the endorheic lakes in Africa and Australia) the community will be comprised almost entirely of obligate species that are highly adapted and largely restricted to temporary waters. Details of the adaptive traits that enable species to be distributed at different points along the continuum will be revisited in Chapter 5.

4.7 Insects versus crustaceans versus 'the rest'

Figure 4.28 summarizes much of the distributional data on invertebrates referred to in the preceeding sections, and underlines the predominance of insects and crustaceans. Insects are well represented in most water bodies, although much less so in saline waters (Williams 1999), however, species richness decreases from permanent to episodic habitats. In particular, chironomids span the permanent–temporary water continuuum, but only a few species are present in episodic waters. In highly saline waters, chironomids tend to be replaced by ceratopogonid and ephydrid dipterans. Crustaceans, too, are well represented but there are several notable discontinuities: large branchiopods, while prominent in intermittent and episodic standing fresh waters, are typically absent from permanent and running water bodies, probably in reponse to predation and inability to deal with currents, respectively. Copepods and ostracods span the continuum, probably a result of their superior diapausing abilities (Fryer 1996); they are also unaffected by salinity gradients.

The insect and crustacean fractions of temporary water communities both contain species with broad (e.g. holarctic) distributions, some of which are known from permanent waters (e.g. the cladoceran *D. pulex*—but see Korinek and Hebert, 1996; and the chironomid *Tanytarsus gracilentus*), and species that occur only, or predominantly, in temporary waters (e.g. the tadpole shrimp *T. cancriformis*; and the hydrophilid beetle *A. limbata*). Limited dispersal powers in some species of crustacean may lead to a greater incidence of endemism than in insects, although transport of eggs via wind, bird's feet, and human activities has produced the cosmopolitan distributions seen, today, in many species.

Although insects and crustaceans coexist, and have come to interact, in temporary waters, their evolutionary histories are quite different, arising as they do from terrestrial and aquatic ancestors, respectively. It is to be expected, therefore, that each may demonstrate different adaptations for surviving the dry phase. In addition, the fact that crustaceans complete their entire life cycles in temporary waters whereas many insects have to recolonize each year may lead to some differences in the timing of occurrence of these two groups. It has been hypothesized that these, in turn, may contribute to the strong seasonal succession so characteristic of the fauna. Further, in a study of a West African (Senegal) intermittent pond, Lahr *et al.* (1999) proposed that because of phylogenetic restrictions in life histories, successional stages were characterized by major taxonomic groups. They showed that crustaceans (all arising from dormant stages) dominated the first two of four phases of the hydroperiod, while insects, that were largely present through aerial recolonization, dominated later. Some crustaceans (e.g. anostracans) were also present in the later phases, probably due to slower development rates. Lahr *et al.* (1999) assumed that the greater diversity in development rates among temporary water crustaceans was an evolutionary consequence of their *permanent* association with the habitat, resulting in a greater variety of life history strategies than in aquatic insects. A tendancy towards dominance of early pond stages by crustaceans has been noted in other localities, for example, the American Southwest (Sublette and Sublette 1967; Moorhead *et al.* 1998), Louisiana (Moore 1970), and South Africa (Meintjes 1996).

In contrast, in intermittent ponds in northeastern North America (e.g. Michigan: Kenk 1944; Ontario: Wiggins *et al.* 1980; Williams 1983), early phases in the hydroperiod are characterized by both crustaceans (e.g. branchiopods, copepods) and insects (e.g. coleopterans, hemipterans, midges, mosquitoes, caddisflies), with insects becoming even more abundant in the later phases. Perhaps the differences in crustacean dominance between the northeastern North American and the more

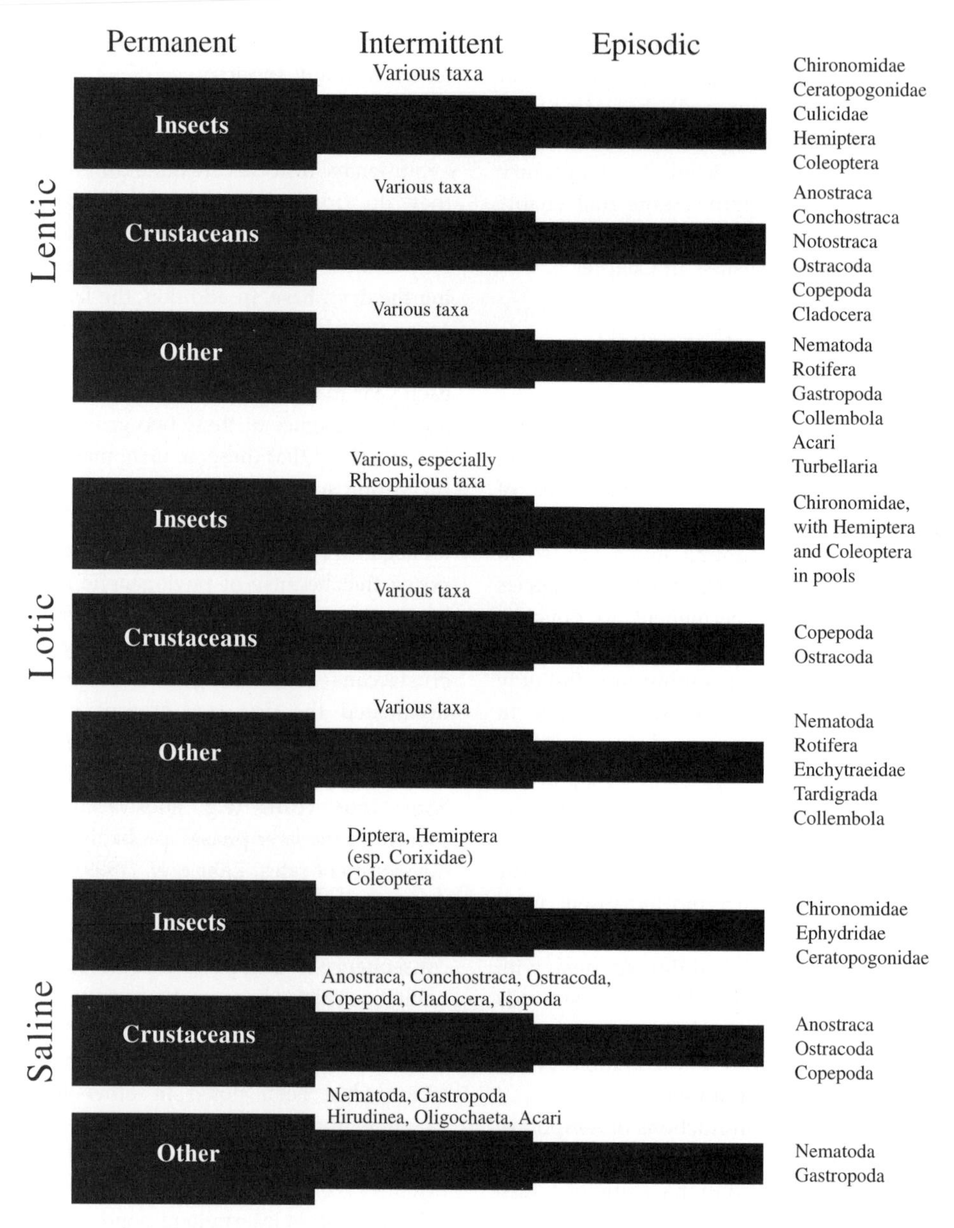

Figure 4.28 Summary of the relative proportions of insects, crustaceans, and other invertebrate taxa along a habitat continuum from permanent to episodic waterbodies. Habitats are broadly grouped into lentic, lotic, and saline waters, and predominant taxa are named.

southerly located ponds are related to the more rapid recolonization of aquatic habitats by insects after Pleistocene glaciation. Insects predominate in other freshwater habitats in previously glaciated regions (Williams and Williams 1998). None of the other locations cited above was impacted as much by ice (Matthews 1979).

Other invertebrate taxa capable of living throughout the continuum include nematode worms, rotifers, some annelids, snails, collembolans, mites, tardigrades, and some turbellarian flatworms (Figure 4.28). Among non-arthropods, only nematodes and some highly specialized snails have been recorded from saline episodic pools.

CHAPTER 5

Population dynamics

5.1 Introduction

As previously noted, in general, temporary waters exhibit amplitudes in both physical and chemical parameters that are much greater than those found in most permanent ponds and streams. The organisms that live in these types of habitat have therefore to be very well adapted to these conditions if they are to persist. Survival depends largely on exceptional physiological tolerance, effective migration abilities, and/or life history[1] modifications. Each of these will be considered in more detail, later.

First, however, it will be useful to recap some of the environmental features of temporary waters that are most likely to affect the biota. In temporary streams, the period and range of water flow are important factors affecting the flora and fauna, as the current may range from torrential, during spring floods, to zero, in summer pools. The degree and rate of descent of the groundwater table together with the permeability of the bed substrate are important also in so far as they control the formation and duration of the pool stage that invariably follows the cessation of flow. These pools allow the survival of certain species that would not normally be able to live in a lotic environment and, at the same time, invite colonization from many purely lentic forms. In tropical regions, the water temperature of temporary pools approaches the upper limit for biological processes, whereas in polar regions it seldom gets more than a few degrees above freezing. In temperate regions, water temperatures range from near zero to the low 30s centigrade thus subjecting species to temperatures near both the high and low thresholds for aquatic life. High temperatures in such pools encourage the rapid growth of algae that may supplement the food supply of some of the fauna, but another consequence is depletion of oxygen dissolved in the water—especially at night. As noted, along with increased levels of photosynthesis dramatic changes in pH can be expected, and these may affect the transport of materials across cell membranes.

By far the most influential environmental parameter affecting species, and especially their population dynamics, is the loss of water during the dry period. For an aquatic organism, the imposition of what is in effect a terrestrial phase in its habitat must be a considerable obstacle to completing its life cycle. McLachlan and his co-workers (McLachlan and Cantrell 1980; McLachlan and Ladle 2001) studied the survival 'strategies' of fly larvae in tropical rain pools, and found that the duration of the pool was important in determining which species were present. Pools lasting a few weeks after each rain, favoured the chironomids *Chironomus imicola* and *C. pulcher*, which have larval lifespans of only around 12 days. These species rely on egg-laying females to reinvade newly filled pools. In situations where the hydroperiod is shorter than these species' minimal life span, some physiological mechanism is required in order to survive the dry period *in situ*. In practice though, these shorter-lived pools are inhabited by larvae of another chironomid, *Polypedilum vanderplanki*, and larvae of the biting midge, *Dasyhelea thompsoni*. Larvae of *P. vanderplanki* are poor at invading newly flooded pools, but are able to tolerate

[1] 'Life history' is herein defined as the life cycle together with qualitative and quantitative information on associated factors, which may differ among individuals or populations (Butler 1984).

virtually a complete loss of body water, and are therefore able to survive drought in the dry mud. They are, consequently, always the first species to appear after refilling, a factor that gives them an advantage in very small pools. Larvae of *D. thompsoni* are not quite as good at surviving drought but are better at invading; they therefore occupy pools of intermediate size and probably inhabit a larger number of pools than any of the other three species.

The preceding discussion emphasizes the fact that adaptation is generally a multidimensional phenomenon, thus adaptation to a single environmental factor is not likely to ensure survival (Alderdice 1972). The organisms living, successfully, in temporary aquatic habitats are more likely therefore to be adapted to deal with a number of habitat factor combinations. However, studies of adaptation to such multivariable phenomena, particularly in temporary waters, are very rare (W.D. Williams 1985). Significantly, in a study of the population ecology of larvae of the shore fly, *Ephydra hians* (Ephydridae), in two saline lakes, biotic interactions were found to limit larval abundance at low salinity levels, whereas physiological stress limits abundance at high salinity (Herbst 1988).

Although adaptations shown by individual species are thus varied, members of the same major taxonomic groups tend to demonstrate them at similar stages in their life cycles. For example, cladocerans, mayflies, lestid odonates, and mosquitoes largely survive drought as eggs (Lehmkuhl 1973; Ingram 1976; Wood *et al.* 1979); other microcrustaceans, together with amphipods survive as immatures; snails, some dragonflies, many beetles, and hemipterans survive as adults (Macan 1939; Nilsson and Svensson 1994); and stoneflies survive as diapausing early instar nymphs (Harper and Hynes 1970). Chironomids may diapause as larvae or eggs (Thienemann 1954; Williams and Hynes 1977a). This topic will be revisited, in more detail, later. Figure 5.1 represents a summary of where these taxa survive the dry phase in an idealized intermittent stream or pond basin. It shows eight basic means by which the fauna can cope with a summer drought, ranging from survival as adults on the wing to those species that burrow into the substrate or make use of the remaining pools. It is important to realize, however, that not all these methods will be open to the inhabitants of both streams and ponds as, for example, in some streams the residual pools may dry up completely, whereas in some regions burrowing crayfishes or equivalent tunnelers may not be present.

A major problem in studying the dynamics of populations, or communities, living in temporary waters is that of obtaining sufficient numbers of organisms, in replicated samples, to allow rigorous statistical analysis of the data, without destructive sampling of the habitat. This is especially cause for concern in small water bodies (e.g. small woodland ponds, rainpools, container habitats), but also in larger, episodic, habitats, such as glacial meltwater streams, or steep channels in arid mountainous regions—where the biota may occur at very low densities, and any significant removal may affect future generations. Location of sampling sites is also important. While obtaining good replicate samples is central to accurate and precise measures of populations, concomitant sampling of less common microhabitats is essential for determining species richness, so as to include rare species. If both measures are required, then the convenience of using a single protocol may have to be abandoned. Further, if high habitat disturbance is likely to result, then some form of finite correction factor may have to be used in calculating variances. In addition, the practice of live replacement after faunal assessment may help to minimize sampling impact.

Sampling for rare species can be of utmost importance, in many ecosystems, for the following reasons: (1) although most rare species may not be very important in terms of the functional properties of communities, they frequently make up the largest component of a habitat's species richness. Further, many of these species may be endemic to these habitats, or rare individuals may enhance the gene pools of regional metapopulations; (2) in contrast to the above, under certain circumstances, rare species can be responsible for significant ecological changes, as in the case of large predators (e.g. amphibians and fishes), or where a few

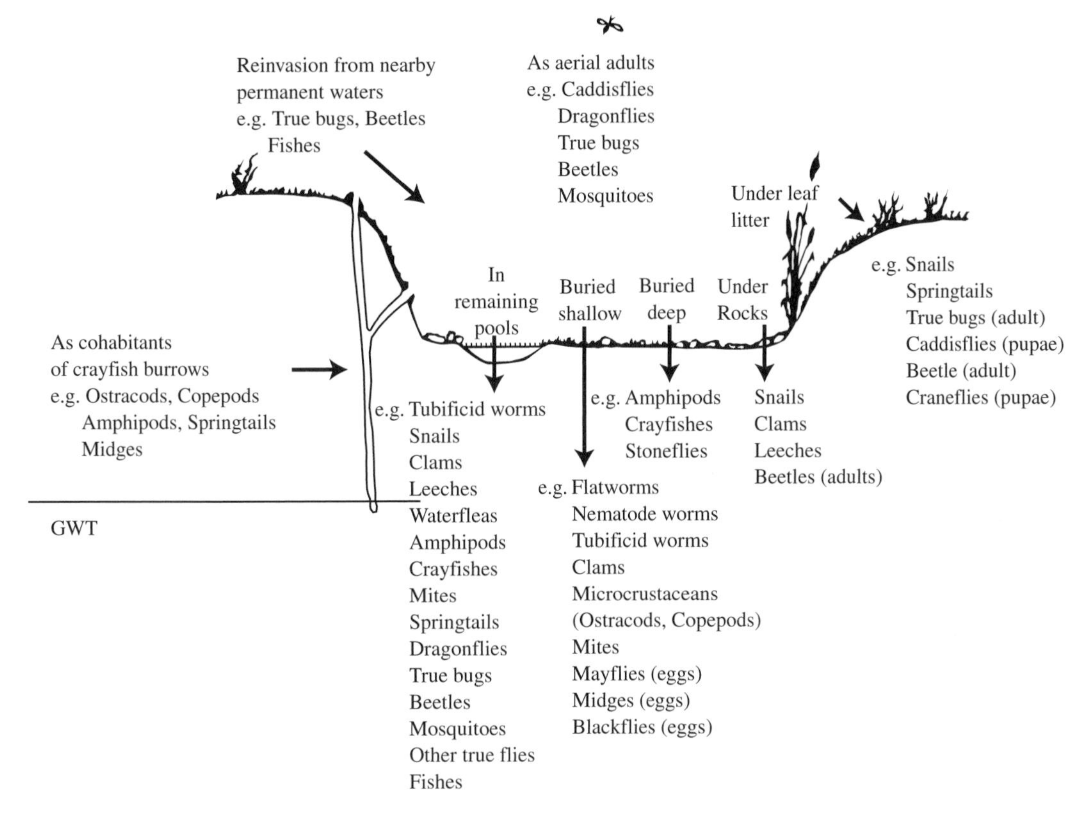

Figure 5.1 Pictorial summary of the drought-survival methods commonly seen in various aquatic taxa in an idealized intermittent stream or pond basin.

invading species bear diseases, or parasites (e.g. bacterial or fungal pathogens, or mites). Whereas there are few recorded cases of the latter in temporary waters, they are well documented for terrestrial insects (Venette *et al.* 2002); and (3) despite the current penchant for high accuracy in formal biodiversity estimations, many aquatic bioassessment studies fall short of this target. More significantly, inadequate sampling may lead to false conclusions being drawn. For example, in a modelling exercise based on a large dataset of river invertebrates, Cao *et al.* (1998) showed that species abundance patterns were significantly different among sites of varying water quality—with more rare species occurring at the least impacted site. Deletion of rare species at the same level of rarity, in the model, yielded unequal effects at different sites, leading to an underestimation of differences in species richness among sites. Small sample size, which implicitly excludes rare species, also can influence the comparison of species richness—especially when combined with a uniform sampling protocol. The reliability and sensitivity of many commonly used community analysis and bioassessment methods can therefore be compromised. Rare species are critical for accurate community studies and bioassessment.

5.2 Seasonality and variability in life cycles

It was proposed, above, that species living successfully in temporary aquatic habitats are likely to exhibit adaptations to a number of different habitat factors. This will often result in life history trade-offs that are observable as negative associations between traits. An example would be wing polymorphism in crickets, where early fecundity is negatively associated with flight capability (Roff 1986). Such trade-offs form a key assumption in optimality models of how life histories evolved, and are seen, by some, as providing an explanation for the widespread occurrence of variable life history traits observed in natural populations (Stearns 1992; Zera and Harshman 2001).

Life history modification in aquatic invertebrates, for example, is influenced by both internal factors (such as physiology, behaviour, and morphology), which tend to restrict life history traits within certain genetically pre-determined ranges, and factors in the environment (such as water loss, temperature, food, photoperiod, and other biota) some of which also can impose range limits (reviewed by Williams 1991). With regard to the latter, using worldwide data on the reproductive behaviour of 131 species of aquatic insect, Statzner *et al.* (1997) attempted to test the prediction that habitat provides the template upon which evolution forges characteristic life history traits (the 'habitat template concept' of Southwood 1977). They found that patterns in reproductive and habitat-use traits corresponded to predictions only on the small scale, which they believed underlined problems such as the effect of the scale of perception of the environment by organisms. Statzner *et al.* also believed that their findings supported the view of Hildrew (1986) that a mixture of life history strategies in a single water body contributes diversity to natural communities, but hampers generalization of the processes structuring the community.

For the most part, the faunas of temporary waters seem to exhibit traits of *r*-selected species (*sensu* MacArthur and Wilson 1967), especially high powers of dispersal, rapid growth, short lifespan, small size, and opportunistic/generalistic feeding, although comprehensive dietary studies are rare (see Table 5.1 for a comparison of the main attributes of *r*- [exploitative] and K- [interactive] selected species types). On the negative side, they may suffer from poor competitive abilities (e.g. McLachlan 1993). Much work needs to be done to test if temporary water biotas do in fact conform to these traits, as some studies would suggest otherwise. For example, the position on the *r*- and K-selection continuum of the pitcher plant mosquito *Wyeomyia smithii* is believed to shift with time, perhaps associated with the highly specialized resource available in these very small temporary waters (Istock *et al.* 1976). Colonization of such a monophagous resource may result in adaptive strategies following this resource more closely than other species in larger temporary waters where a variety of food resources are available, and where *r*-selection and rapid growth are the norm. Further, Bradshaw and Holzapfel (1983) were able to demonstrate that, in terms of its response to photoperiod, this species shows a unique degree of geographical tracking. At the southern end of its range (Florida) *W. smithii* undergoes severe density-dependent selection resulting in variable larval mortality during summer, autumn, winter, and spring. However, the severity of this density-dependent selection

Table 5.1 Comparison of features of *r*- and K-selected species

r-selected	K-selected
Typically found in unstable habitats	Typically found in stable habitats
High powers of dispersal	Low powers of dispersal
Poor competitive ability	Good competitive ability
High intrinsic rate of natural increase	Lower intrinsic rate of natural increase
Large number of eggs produced	Fewer eggs produced
Early reproduction in life cycle	Later reproduction in life cycle
Small body size	Large body size
Short lifespan (e.g. annual)	Longer life span (several years)
Opportunistic/generalistic feeders	Specialized feeders

diminishes with increasing latitude, and also altitude. Accompanying this, populations also show increased growth, shorter generation time, higher fecundity, and a reallocation of reproductive effort to earlier adult ages. As a result, *W. smithii* shifts from being bi- to univoltine.

In a comparison of wing length in 10 species of *Chironomus*, McLachlan (1985) found an unexpected increase in mean length with increasing habitat duration. According to *r*-selection predictions, species from more ephemeral habitats should have greater flight capacity by virtue of long wings. Interestingly, McLachlan (1983) has shown that adult females of the African chironomid *Chironomus imicola* emerging from remote episodic rockpools are uniformly large compared with females emerging from pools that occur in clusters. One explanation is that females from remote pools have greater distances to fly and are more fecund. In comparison with males from the same remote pools, the females have longer and broader wings that beat more slowly and have a longer stroke; wings that beat faster with a shorter stroke in males benefit competitive aerobatic manoeuvring in mating swarms (McLachlan 1986). Broader wings, combined with possibly greater energy reserves by virtue of a larger body, may effectively increase dispersal range in the females.

Variations in aquatic insect life history traits among three basic types of temporary water are modelled (based on Stearns 1976) in Figure 5.2. This interpretation is preliminary in light of the limited empirical data available, but it is presented both as a platform for critical re-examination of the application of *r*–K selection theory to the inhabitants of temporary waters and in the hope that it will stimulate further research. The loss of water in all three waterbody types is seen as providing sufficient selective pressure for commonality of traits such as egg diapause, however, while some *r*-selected traits (e.g. good powers of dispersal, medium to high fecundity) benefit inhabitants in all three types of water, there is discrepancy in the perceived selective advantage of others (e.g. longevity). Indeed, in some cases (e.g. growth rate and pattern), opposing traits seem to apply, but then Pianka (1970) has argued that no organism is entirely *r*- or K-selected and that, generally, a balance is struck that maximizes the adaptive value of features drawn from either type of selection. Models, other than those based on *r*–K continuum theory, that also underline the importance of habitat in determining life history strategies may prove to be more applicable to temporary freshwaters. These models have been reviewed by Southwood (1988) and Williams (1991).

Life history adaptation appears to become more complex in temporary waters that are highly saline. W.D. Williams (1985) produced a habitat template in which the two axes are the extent of predictability of the habitat and salinity levels (Figure 5.3). The figure shows the presumptive distribution of *r*- and K-selection types together with A-, or adversity, selection. The figure re-emphasizes the importance of *r*-selection in temporary waters. However, again, no organism is likely to be entirely *r*-, K-, or A-selected.

Many of the orders and families of insects that are most successful in temporary waters (summarized in Table 5.2) are widely separated phylogenetically, representing both exo- and endopterygotes. Some have an ancient lineage (the Ephemeroptera and Odonata arose in the Upper Carboniferous), whereas others are more recent (e.g. the earliest known chironomids are from the Lower Cretaceous). However, most of the traits listed are common to all the groups, and several of the others are possessed by at least some members of each group. Particularly pertinent traits are a highly flexible life cycle, temperature-linked development, possession of a diapausing or otherwise protected egg, and high powers of dispersal. Also noteworthy is the presence of a trend towards terrestrialization of immature stages, seen to some degree even in the strongly aquatic mayflies, odonates and caddisflies (Anderson 1967; Watson and Theischinger 1980; Erman 1981; Williams and Feltmate 1992; McCafferty and Lugo-Ortiz 1995; Nolte *et al.* 1996). Some, but not all, of these traits agree with *r*-selected species predictions. For further discussion of the evolutionary history of aquatic insects, crustaceans, and mites in temporary pools see Wiggins *et al.* (1980).

- long-lived, iteroparous adults with high powers of dispersal
- high fecundity; eggs are widely distributed within and among habitats
- staggered hatching of long-diapause eggs
- long development time of immature stages
- diapause capability in several life cycle stages
- opportunistic growth rate
- generalized feeding by immatures; adults can/may feed
- optional parthenogenicity
- developmental plasticity in life cycle

Seasonal, with long periodicity (P > 1 yr)

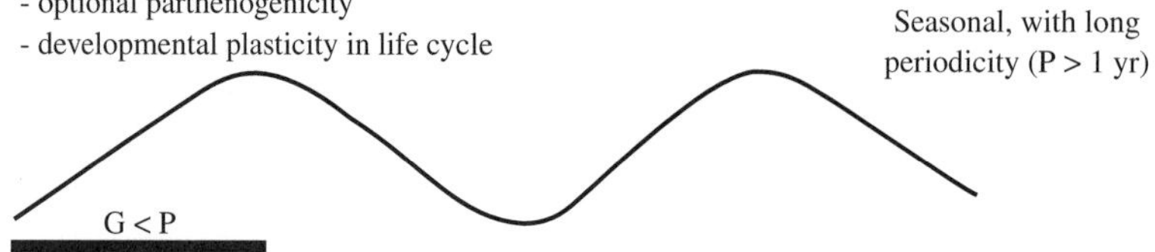

G < P

- shorter-lived, semelparous adults with good powers of dispersal
- intermediate fecundity
- shorter-term diapausing eggs in which hatching coincides with the historically optimum time that the water reappears (with some within clutch polymorphism in case of error)
- predictable, seasonally adjusted growth rate with well defined cohorts
- generalist feeding by immatures with some specialization on predictable food types
- penultimate instar with desiccation resistance (physiological or behavioural)
- high correlation between life history stages and environmental factors

Highly seasonal (P = 1 yr)

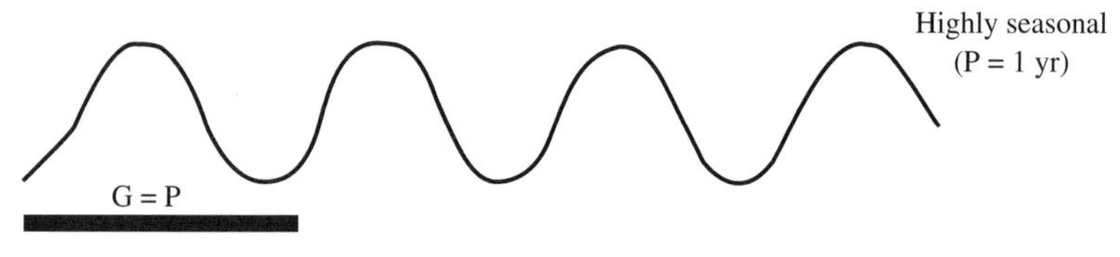

G = P

Environmental periodicity (hydrograph)

- short-lived, highly vagile adults with high fecundity
- potentially long-diapause eggs that have staggered hatching and rapid breaking of diapause upon stimulation by 1 or 2 key environmental cues
- poorly defined cohorts with differential development rate for subgroups of siblings leading to dwarf and giant individuals
- opportunistic feeding by immatures, including cannibalism of dwarfs
- extraordinary resistance of immature stages to dehydration (e.g. *Polypedilum vanderplanki*)

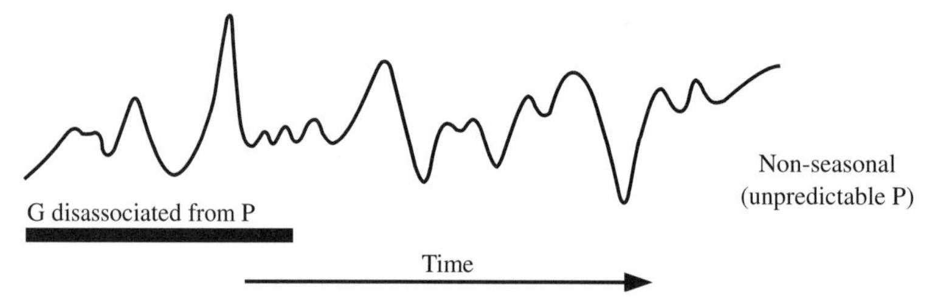

Non-seasonal (unpredictable P)

G disassociated from P

Time

Figure 5.2 A predictive model of the desirable adaptive traits of aquatic insects in three basic types of temporary water that differ in their seasonal periodicity and predictability. The dark line indicates the length of generation time (G) of a hypothetical insect species relative to cycle periodicity (P) of the environment—the latter could, in this context, be equated to the hydrographs of three waterbody types. The concept is based on one proposed by Stearns (1976).

The drought survival method used reflects the type of life cycle exhibited by a particular species. Figure 5.4 gives examples of the different types of life cycle shown by some of the more abundant invertebrates in Sunfish Pond, an intermittent pond in southern Ontario, and illustrates relative growth rates. The species are categorized into five groups according to the seasonal succession outlined in Chapter 4 (Figure 4.18) for the fauna of this pond.

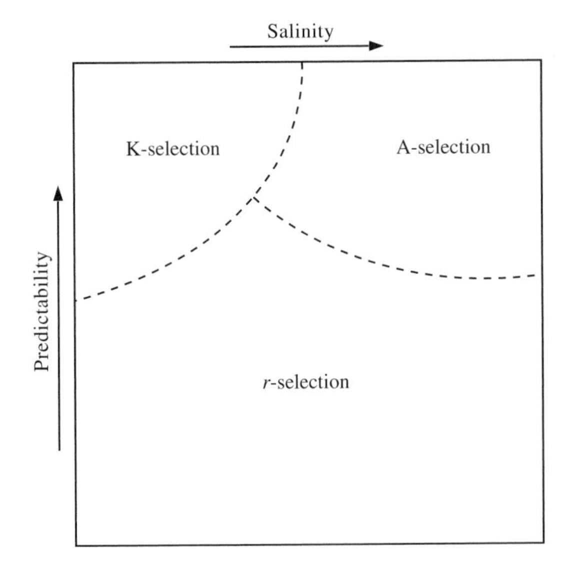

Figure 5.3 Habitat template showing the distribution of *r*-, K-, and A-selection according to habitat salinity and predictability (redrawn from W.D. Williams 1985).

In *Group 1*, the chironomid *Einfeldia* appears to have two periods of growth. Larvae in all four instars are present in mid-April and the earlier ones grow rapidly until only third and fourth instars and pupae are evident in mid-May. Adults are abundant by the end of May. The resulting eggs hatch quickly and the larvae grow rapidly in early June, passing into their second instar. Thereafter growth slows and the summer dry period is passed as early instars that can be recovered from the pond substrate. As soon as water reappears in the spring, the larvae continue their growth. Two of the common molluscs, *Sphaerium* sp. and *Lymnaea humilis* show gradual increases in growth throughout the pond phase with peak growth in the warm water of mid-July just before dry-up. Virtually no growth occurs in the dry phase. Species in Group 1 are typically subjected to the temperature range 0–27°C during the aquatic phase of their habitat.

Table 5.2 Summary of some of the main adaptive characteristics of the aquatic insect orders and dipteran families most commonly encountered in temporary waters

Trait	Ephem	Odonat	Hemip	Coleop	Tipul	Culic	Cerat	Chiron	Ephyd	Trichop
Global distribution	√	√	√	√	√	√	√	√	√	√
Occur in lotic and lentic waters	√	√	√	√	√	√	√	√	√	√
Highly adaptable life cycle	√	√	√	√	√	√	√	√	√	√
Highly diverse immature forms	√	√	√	√	√	√	√	√	√	√
Development strongly linked to environmental temperature	√	√	√	√	√	√	√	√	√	√
Egg diapause	√	√	√	√	√	√	√	√	√	√
Desiccation-protected eggs, or staggered egg hatching	√	√	?	√	?	√	?	√	?	√
Parthenogenesis known	√	No	√	√	√	√	√	√	√	√
High powers of dispersal	No	√	√	√	√	√	√	√	√	√
Long-lived adults	No	√	√	√	√	√	√	Some	Some	√
Generalist feeders	√	No	Some	√	√	√	√	√	√	Many
Terrestrial ancestors	No	No	√	√	?	?	?	?	?	?
Trend towards terrestrialization known for some immature stages	√	√	√	√	√	No	√	√	√	√
Water surface/air breathers	No	No	√	Most	Most	√	Some	No	√	No

Source: Information derived from Williams and Feltmate (1992).

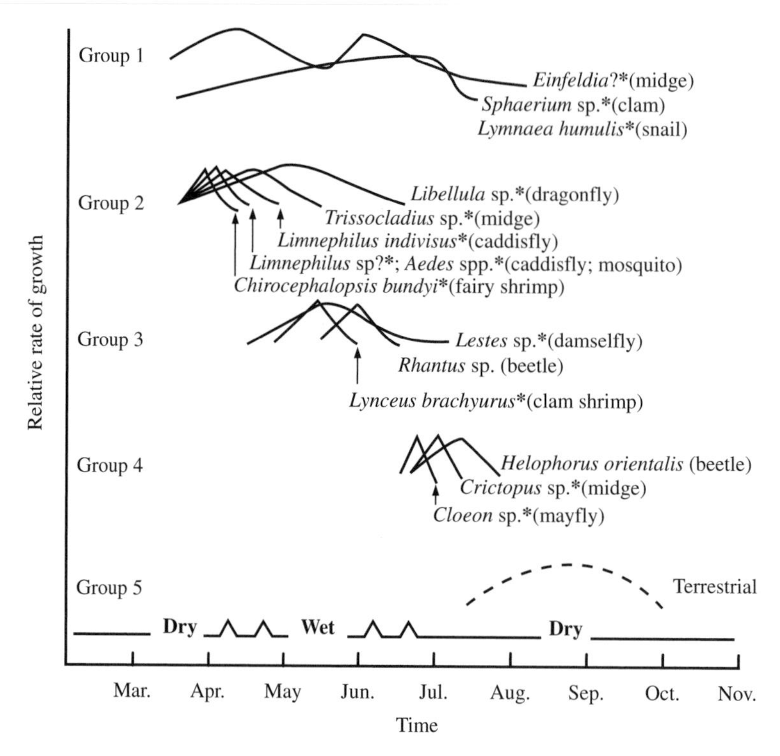

Figure 5.4 Diagram illustrating the different life cycle patterns exhibited by the invertebrate fauna of an intermittent pond in Ontario. The curves approximate life cycle duration and relative growth rates. Species marked with an asterisk are those known to employ dormancy in their life cycles, whereas those with a question-mark/asterisk are suspected of using dormancy (based on data in Williams 1983).

The species assigned to *Group 2* show a variety of life cycles. Many, such as the fairy shrimp *Chirocephalopsis*, the caddisflies *Limnephilus indivisus* and *Limnephilus* sp., and three species of *Aedes* mosquito grow very rapidly at the start of the pool phase. Although the life cycles of the *Aedes* species are very similar (all being univoltine and present as early instars, concurrently), some staggering of development through late instars and emergence is evident. *Aedes trichurus* is the first to emerge, followed a week or so later by *Ae. fitchii*, and finally *Ae. sticticus*. The chironomid *Trissocladius* shows a more gradual growth rate spanning a longer period—though it still emerges well before the end of the pool stage. Nymphal development of the dragonfly *Libellula* takes virtually the entire aquatic phase of the pond. Species in Group 2 are generally subjected to a less wide temperature range (0–17°C) during development.

Group 3 species again show variable growth rates. The dragonfly *Lestes* grows gradually throughout much of the pond phase, while the clam shrimp *Lynceus brachyurus* and the beetle *Rhantus* grow more quickly, but at different times. These species develop in the temperature range 12–22°C.

Among the *Group 4* taxa, the mayfly *Cloeon* shows extremely rapid growth, completing its life cycle in 2–3 weeks. The chironomid *Cricotopus* grows quickly also, taking about 4 weeks, and completes its larval development in the moist pond substrate in mid-July. The hydrophilid beetle *Helophorus orientalis* shows slower growth and passes much of its larval development in the moist bottom material. Water temperatures prevailing at the time of Group 4 development are in the range 17–27°C.

Suggested life cycles and growth patterns of the species of *Group 5* are largely speculative as these primarily terrestrial animals were not sampled at other times of the year. It is likely that they grow at a reasonable pace given the warm temperature and plentiful food supply in the moist pond basin.

In temporary streams, there is the added dimension of moving water for part of the habitat cycle. This affects the life cycles of the fauna. Figure 5.5 shows examples of the life cycles of invertebrates in several intermittent streams in Ontario. There are three types of cycle evident among the fall-winter stream fauna (see Chapter 4, Figure 4.20, for description of habitat phases):

(1:A) in which the eggs hatch immediately on the resumption of flow and the larvae develop quickly. Maximum size is reached just before ice cover brings cooler temperatures, and the larvae remain at a low level of activity throughout the winter to emerge early in the spring, for example, the caddisfly *Ironoquia punctatissima*, the chironomids *Trissocladius* and *Micropsectra*, and the stonefly *Allocapnia vivipara*;

(2:C) in which the opposite occurs, that is, slow hatching and larval development at first, followed by rapid growth as the water begins to warm up in the early spring. Most are mature before the stream stops flowing, for example, the microcrustaceans *Acanthocyclops vernalis* and *Attheyella nordenskioldii*, the cranefly *Tipula cunctans*, the ostracod *Cypridopsis vidua*, the amphipod *Hyalella azteca*, and the beetle *Agabus semivittatus*; and

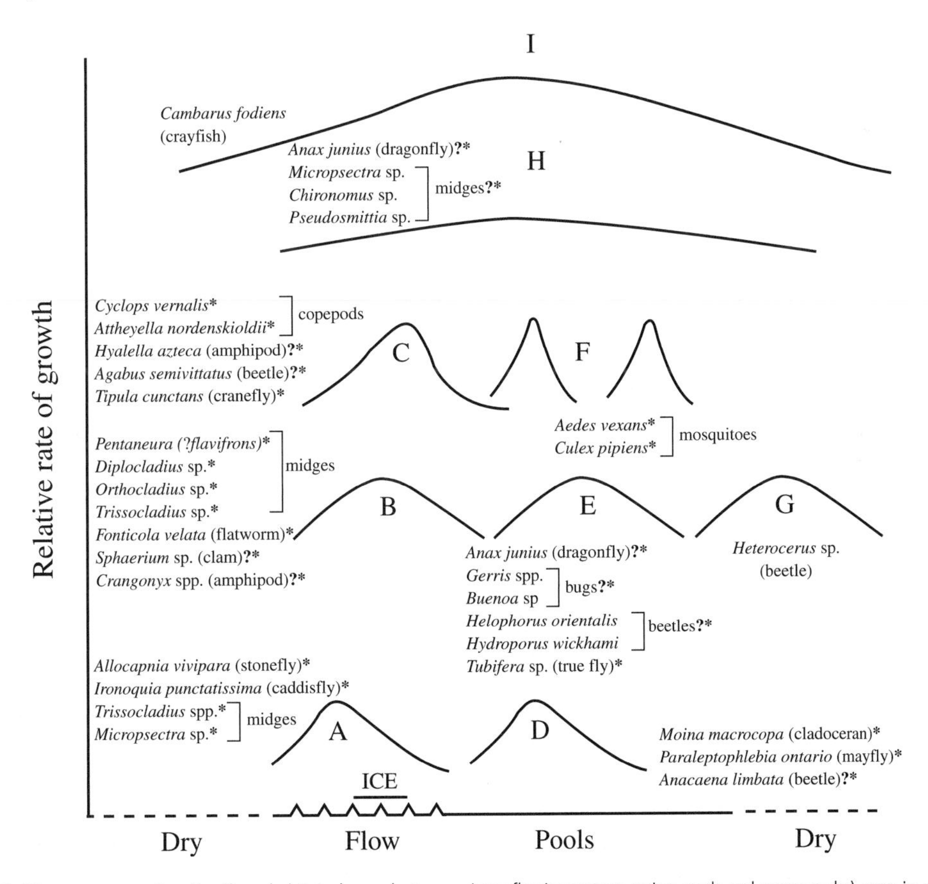

Figure 5.5 Diagram representing the three habitat phases (autumn-winter flowing water; spring pool; and summer dry) seen in small intermittent streams in Ontario, together with the different life cycle patterns exhibited by the invertebrate fauna. The curves approximate life cycle duration and relative growth rates. Species marked with an asterisk are those known to employ dormancy in their life cycles, whereas those with a question-mark/asterisk are suspected of using dormancy (based on data in Williams and Hynes 1977).

(3:B) in which there is a more even growth rate spread over the whole period of stream flow, for example, the flatworm *Fonticola velata*, the amphipods *Crangonyx minor* and *C. setodactylus*, the chironomids *Diplocladius*, *Orthocladius*, *Trissocladius* sp., and *Arctopelopia (Pentaneura) ?flavifrons*, and the clam *Sphaerium*.

Similarly, three types of life cycle are evident among the spring-pool fauna:

(1:D) in which rapid growth occurs only at the beginning of the pool stage, for example, the mayfly *Paraleptophlebia ontario*, the beetle *Anacaena limbata*, and the cladoceran *Moina macrocopa*. This growth pattern lessens the chances of these species getting caught by an early drought;
(2:E) in which there is an even growth rate spread over the entire pool stage, for example, the waterstrider *Gerris*, the beetles *H. orientalis* and *Hydroporus wickhami*, the dragonfly *Anax junius*, the backswimmer *Buenoa*, and the dipteran *Tubifera*; and
(3:F) in which multiple cycles occur in those species capable of very fast development, for example, the mosquitoes *Aedes vexans* and *Culex pipiens*. In some drier than normal years, the last generations of these species may not be completed. However, mosquitoes, in particular, have various adaptations, such as staggered egg hatching, (a process examined in more detail later) which ensure the survival of at least part of the population.

Growth rates for the summer-terrestrial fauna have not been studied, but it is likely that they follow the pattern of the mud-loving beetle, *Heterocerus*, which seems to have an even growth rate throughout the summer (**G**).

Two other life cycle patterns are evident:

(1:H) shown by the chironomids *Pseudosmittia*, *Chironomus* and *Micropsectra*, and part of the *A. junius* population, in which growth proceeds at a steady rate throughout the year, with emergence taking place near the end of the pool stage; and
(2:I) shown by the crayfish, *Fallicambarus fodiens*, in which growth spans 2 years, with peaks in development during the aquatic phases.

These variations in life cycle create a faunal succession from the stream stage through the pool stage to the terrestrial.

5.3 Phenotypic and genotypic variation

Roff (1992) has pointed out that whereas most life history traits show considerable phenotypic variation, most empirical studies focus on trying to predict some *optimum* value. Phenotypic variation may result from plasticity of a single genotype or of different genotypes within a population. Roff argues that selection should favour the genotype that produces the most fit offspring in all environments encountered. Phenotypic variation may thus be expected to be largely a result of a reaction norm. However, the significant amounts of genetic variation observed in natural populations suggest that either no single genotype is most fit across all environments, or that mutations are generating variation at a pace close to its erosion by natural selection. Roff draws attention to the continuing debate on the role that mutation plays in the maintenance of genetic variance, and the fact that in virtually all studies genotype-by-environment interactions have been found. An obvious conclusion is that functional constraints that prevent any single genotype from achieving maximum fitness are commonplace. The extent to which adaptations result from phenotypic plasticity versus genetic differentiation requires a great deal more study, and temporary water populations would seem to be well suited to this end.

Single species in temporary waters have been observed to be represented by several distinct ecotypes due to the uniqueness (chiefly in terms of physical/chemical properties) of each habitat (Bowen *et al.* 1981). Variation in a life cycle, or part thereof, appears to be an important adaptation to local environments, and must have a role in diversification and speciation of populations in such habitats (Tauber and Tauber 1981). Adaptation to extreme environmental factors in individual water bodies, together with inbreeding, may eventually lead to reproductive isolation (Templeton 1980). How does the evidence stand for ascribing trait expression to genetic influence or phenotypic plasticity? Several studies point to trait expression being largely the result of phenotypic plasticity. For example, *Artemia salina*, the brine shrimp, is one of the most studied organisms from temporary waters, especially in terms of its genetic variability and speciation which result primarily from adaptations to temperature and ionic composition (Collins and Stirling 1980). *A. salina* is a complex of many sibling species and semi-species and its taxonomic status is still in a state of flux. Populations are particularly adapted to chloride waters, and the phenotypic plasticity that allows them to survive and breed in a wide range of anionic ratios represents the greatest known anionic tolerance of any metazoan species.

When populations of *A. salina* belonging to different sibling species are inter-mated there are no offspring. This complete reproductive isolation has been attributed either to a change in chromosome number or to parthenogenesis (Barigozzi 1980). However, in a careful study of the *A. franciscana* cluster of non-parthenogenetic populations, Bowen *et al.* (1981) found incomplete reproductive isolation due to adaptations to habitat—each population being adapted to waters where the predominant anion might be chloride, carbonate, or sulphate. They obtained viable F_1 and F_2 offspring from matings between these populations, but pointed out that two populations which prove to be cross-fertile in artificial media in the laboratory may well be reproductively isolated in their native ponds—because of the inability of one or both to complete their life cycles in the ionic composition or temperature regime of the alien habitat. These authors concluded, tentatively, that the phenotypic plasticity of a population of *A. fransiscana* from Fallon Pond, Nevada was due to one genotype with a wide norm of reaction rather than to the presence of many genotypes. In contrast, populations of the anostracan, *Branchinecta sandiegoensis*, in 'ephemeral' pools in coastal California showed a high degree of genetic differentiation—which was thought to result from low gene flow and founder effects resulting from habitat fragmentation and a lack of potential vectors for cyst dispersal (Davies *et al.* 1997).

Wissinger *et al.* (2003) showed that three species of caddisfly in high-altitude wetlands in Colorado possess life history properties (rapid larval growth, ovarian diapause, and terrestrial deposition of desiccation-resistant eggs) that allow them the flexibility to survive in both permanent and temporary habitats (Figure 5.6). In reality, however, their survey showed that *Asynarchus nigriculus*, tended to be rare or absent in permanent ponds, whereas *Limnephilus externus* and *L. picturatus* were largely rare or absent from most temporary ponds. Experiments indicated that biotic interactions (e.g. intraguild predation) act to limit each species to a subset of potentially exploitable habitats.

Life history plasticity has also been demonstrated for the stonefly *Nemoura trispinosa* along a permanent–temporary water gradient in cold-water springs in Ontario (Williams *et al.* 1995). Nymphs generally were able to tolerate a wide range of environmental conditions and were found in 78% of the springs sampled, however, population densities differed markedly. Maximum annual water temperature proved to be the factor that most influenced nymphal growth rate (although this was a non-linear relationship), whereas degree of habitat permanence was related to generation time (Figure 5.7). In the laboratory, Johansson and Rowe (1999) manipulated the environmental cues of populations of the damselfly, *Lestes congener*, to see if this induced life cycle changes. In a group where nymphs were hatched under a late-season light regime, development accelerated, producing early maturation, but at a smaller size. However, there was a cost, as increased activity rates, in order to find food, in these 'time-constrained' nymphs resulted in a higher incidence of cannibalism.

Some very elegant evidence in favour of phenotypic plasticity comes from a transplant experiment. Hornbach *et al.* (1991) compared the life cycles of two populations of the sphaeriid clam *Musculium partumeium* in an intermittent and a permanent pond in Minnesota. Habitat permanence strongly influenced the life cycle. When clams from the two populations were transferred between ponds, the transferees always displayed the life cycle appropriate to the new pond, rather

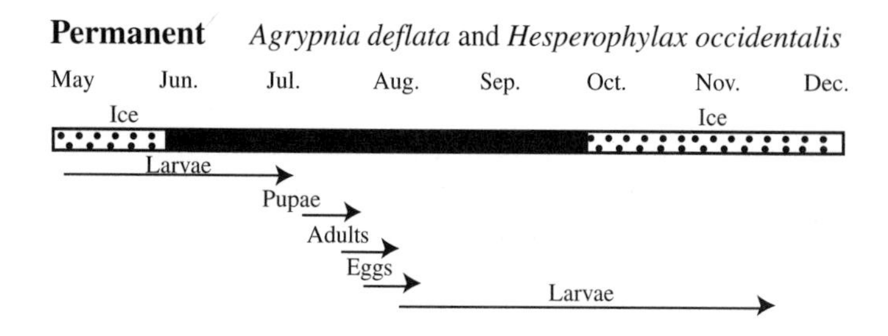

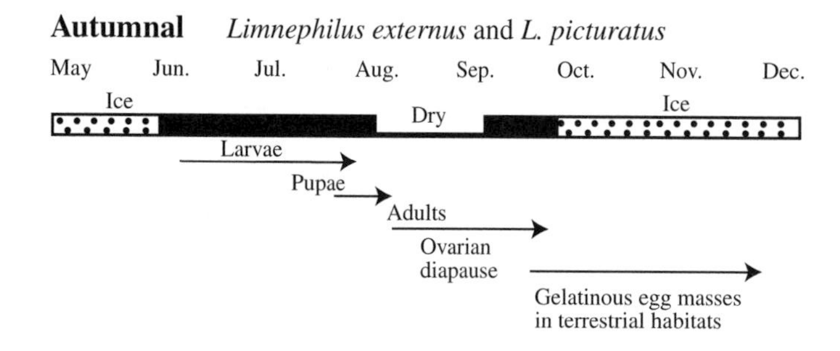

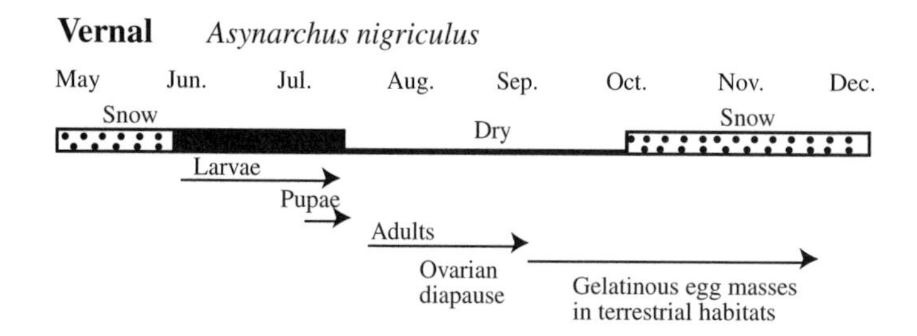

Figure 5.6 Summary diagram of the life history patterns that allow caddisfly species to exploit permanent and temporary habitats. *Asynarchus nigriculus* can complete its life cycle in all three wetland pool types. *Limnephilus externus* and *L. picturatus* can complete development in autumnal and permanent pools, whereas *Agrypnia deflata* and *Hesperophylax occidentalis* are restricted to permanent habitats (redrawn after Wissinger *et al.* 2003).

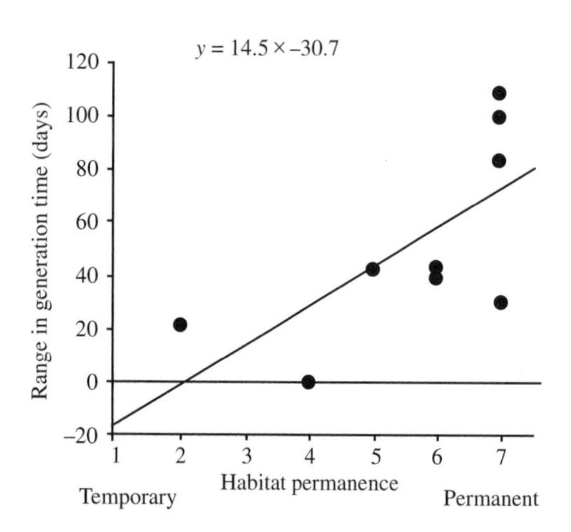

Figure 5.7 Plot of range in generation time of the stonefly, *Nemoura trispinosa*, against degree of habitat permanence.

than that of their old pond. In contrast, transplanted females of the copepod *Diaptomus sanguineus* were unable to sense a change in pond type, or to adjust egg production accordingly, leading Hairston and Olds (1984) to conclude that populations of this species appear to be adapted to the specific conditions of isolated ponds rather than to a broader geographical region containing a variety of pond types. Interestingly, Innes (1997) has shown that populations of *Daphnia pulex* in temporary ponds demonstrate increased sexual reproduction, and a decreased influence of the environment on sex determination, compared with populations in permanent ponds. Armbruster *et al.* (1999) have shown that intraspecific hybridization in the pitcher-plant mosquito, *W. smithii*, such as might result from outcrossing with neighbouring populations, did not increase either the likelihood of population persistence under, or adaptation to, a chronically stressful environment. This was despite initially beneficial effects of hybridization to fitness, which later proved to be transient.

Some of the best demonstrated phenotypic plasticity has come from studies of amphibians. For example, Denver *et al.* (1998) showed that tadpoles of the spadefoot toad, *Scaphious hammondii*, responded to reduction in habitat water volume by accelerated metamorphosis, and that the response was reversible if water was added. The environmental cues promoting these changes did not appear to involve concentration of chemicals in the water, but rather were related to a reduced swimming volume and perhaps proximity to the water surface. In an unusual study of two colour morphs of the green toad, *Bufo viridis*, in pools in the Negev Desert, Blaustein and Kotler (1997) were able to offer an example of a more mechanical explanation for differential development. They found that charcoal-coloured tadpoles grew much more quickly than sand-coloured conspecifics. They concluded that, by absorbing more solar radiation, the dark morph was better adapted for life in such pools—although they did admit that an alternative explanation might link the light morphs to a genetic defect that retards development.

Quite apart from the plasticity seen in larval development, there is also flexibility in habitat choice by adults, although this has been seldom studied in amphibians. Murphy (2003) examined the egg-deposition-site choice in a neotropical frog, *Edalorhina perezi*, which reproduces in isolated forest pools in the Amazon Basin. Murphy detected a predictable pattern in combined pool risks that was driven by a trade-off between drying and the risk of tadpole predation by insects. Persistent pools presented more favourable conditions early and late during the 150 days rainy season, whereas 'ephemeral' pools were more favourable during mid-season. During the last 50 days, however, the majority of adults laid their eggs in the ephemeral pools, even though neighbouring, more persistent pools were more benign. Murphy concluded that *E. perezi* responds to variation in juvenile risk at several scales when it selects egg-laying sites, but that adults are imperfect at tracking risk. Some adult amphibians seem to be extremely bad at assessing such risk. For example, Williamson and Bull (1999) found that larvae of the frog, *Crinia signifera*, failed to achieve metamorphosis in all of the years that a group of temporary pools in South Australia was studied.

Sexual dimorphism is a phenotypic/genotypic trait common to many species. In *C. imicola*, flight is sexually dimorphic and has been related to selective pressures of its habitat (rain-pools on isolated hills in tropical Africa), and to the disparate roles of flight between males and females. The episodic nature and isolation of these habitats require a mandatory dispersal of gravid females during which they may have to fly for many kilometres. Consequently, females of this species show characteristics necessary for sustained flight, namely relatively long, broad wings that beat slowly and have a large amplitude of beat. Males, on the other hand have shorter, narrower wings that beat quickly with a short stroke, and their flight periods are typically shorter. These characteristics are particularly suited to aerobatic flight that is necessary for the successful acquisition of females in a competing swarm of males (McLachlan 1986). Bradshaw *et al.* (1997), working on *W. smithii*, found that females not only show a higher growth rate than males, but also harbour a greater genetic variation for development time.

Another form of reproductive polymorphism has been observed in the mayfly *Stenonema femoratum*. This species exhibits tychoparthenogenesis, which is characterized by hatching of a small proportion of unfertilized eggs (usually $<10\%$) from females that normally reproduce sexually. Ball (2002) estimated the tychoparthenogenetic capacity of females in populations spanning a temporary–permanent water gradient. Although there was no difference in capacity for one year, in another intermittent streams showed a hatching success of unfertilized eggs that was twice that of permanent waters; capacity seemed to be related to water temperature.

In a study of the life cycles of mayflies and stoneflies in temporary streams in western Oregon, Dieterich and Anderson (1995) underlined differences in a species' ability to adjust its development to drought occurring in different seasons. In this region, summer droughts are predictable and annual, however, winter and spring droughts result from climatic stochasticity and are not predictable. Development of life cycle specialization to these latter, episodic, events was deemed unlikely. Many of the species in these streams exhibit prolonged egg hatching and adult emergence periods, and have a wide range of nymphal stages present at any given time. Such asynchronous development is seen as reducing risk by spreading life stages over time, so that eradication of entire populations by catastrophic events is unlikely. These authors concluded that lack of predictability, together with the severity of events, provides a powerful force towards maintenance of life cycle plasticity.

The question of how many different phenotypes can be maintained by environmental variance, and whether a strategy of producing phenotypes having separate genotypes would win out against one that produces them all from a single genotype has been posed by Sasaki and Ellner (1995). They discovered that as environmental variance increases, and all types are in play, the number of phenotypes maintained increases. However, these phenotypes are distributed in discrete blocks, and not continuously. Further, genotypes that derived several phenotypes always invaded a polymorphic population of similar, but genetically fixed, phenotypes. They were able to do so because of a bet-hedging advantage (see Section 5.7.1) that, through production of a variety of offspring, reduced the temporal variance in per-generation fitness (Stearns 2000).

5.4 Physiology of desiccation

Water is crucial to the inhabitants of temporary waters, and, as we have seen, at the scale of the hydroperiod it is an important regulator of population and community structure and dynamics. However, the role of water must also be considered at the cellular level, as desiccation tolerance is a key factor to survival of the individual. Water has a unique set of properties that suit it to being a universal solvent for life processes (e.g. high melting and boiling points, a wide range of temperature in which it exists in a liquid state, a high dielectric constant which is important to its action as a solvent, and it readily forms hydrogen bonds). Water limitation is thus a major stress factor for organisms (Rothschild and Mancinelli 2001). Fortunately, desiccation tolerance (or anhydrobiosis) is quite a widespread phenomenon, being found in many microorganisms, in plants, and in animals such as rotifers, nematodes, tardigrades, and in the larval cysts of crustaceans that live in temporary waters (Proctor and Pence 2002).

Anhydrobiosis involves the ability of the organism to survive loss of all cellular water (perhaps with the exception of that bound tightly to macromolecules) without sustaining irreversible damage (Hochachka and Somero 1984). Anhydrobiosis differs from diapause (see Section 5.6.1), where metabolism is not brought to a complete standstill. Crowe (1971) has subdivided desiccation-tolerant organisms into two groups: (1) species capable of anhydrobiosis only in their early developmental stages—including the spores of bacteria and fungi, plant seeds, eggs and early embryos of some crustaceans (e.g. *Artemia*), and the larvae of certain insects; and (2) species capable of anhydrobiosis during any stage in their life cycles—including certain protozoans, rotifers, nematodes, and tardigrades. These groups share many similar biochemical mechanisms associated with the initiation and termination of this state.

For plants, desiccation tolerance has been defined as 'the ability to dry to equilibrium in moderately dry air (50–70% relative humidity, at 20–30°C), and then resume normal function when rehydrated' (Alpert and Oliver 2002). Terrestrial plants, for which there is a better study base, have evolved two main strategies for dealing with water loss. Most have adaptations that lessen their chances of drying out, such as external wax layers, stomata that can regulate water loss, and reduced leaf area. A much smaller number of species have evolved a strategy of 'drying without dying' by which they are able to desiccate during drought, but later rehydrate and resume growth. There are records of plants losing all of the free water in their cells, yet being able to recover after many years (Gaff 1977).

Up until the 1970s, tolerance of desiccation in plants was believed to be largely via mechanical adaptations, such as having flexible cell walls or small vacuoles, or lacking plasmodesmata (narrow channels that act as cytoplasmic bridges allowing transport of materials, including water, between plant cells). However, Bewley (1979) proposed an alternative view, currently widely accepted, that much desiccation tolerance results from inherent properties of protoplasm. Bewley defined three aspects of desiccation tolerance critical to plant survival: (1) limitation of damage to a repairable level; (2) maintenance of physiological integrity in the dry state, for extended periods; and (3) upon rehydration, mobilization of mechanisms that repair damage suffered during both desiccation and rehydration. Evidence suggests that, in desiccation-tolerant plant tissues, relatively little damage actually occurs during drying. Instead, there takes place an ordered 'collapse' of the cellular environment that results in little ultrastructural damage. When rehydrated, all plant tissues have been observed to leak solutes, but in desiccation-tolerant species this is a transient event resulting from lipid-phase transitions that occur in the plasma membrane. As drying proceeds in plant tissues, membranes pass from a liquid crystalline phase to a gel phase, a process that is reversed during rehydration. However, in seeds it is believed that these membrane-phase changes fail to occur because the presence of the hard seed coat impedes the passage of water to the dried cells. Such slowed penetration may induce a protective 'pre-hydration' state in which the membranes are in a liquid crystalline state before liquid water envelops the rehydrating cells. Apart from the cell membrane, rehydration damage is frequently visible in some of the cell organelles (e.g. swollen chloroplasts and mitochondria), however, these regain their normal structure within 24 h. Some monocots reversibly lose their chlorophyll and dismantle their chloroplasts during desiccation. Cellular protection is also enhanced by the synthesis of proteins and sugars just prior to, or during, drying, and by the production of antioxidants and enzymes involved in oxidative metabolism (Alpert and Oliver 2002).

In preparation for anhydrobiosis, invertebrates frequently follow a sequence of adjustments to their morphology (e.g. as in bdelloid rotifers and tardigrades), and also synthesize protective chemicals. In tardigrades, species that demonstrate the greatest infolding of cuticle (maximum reduction of body surface area), during the tun stage, prove to be the most desiccation resistant (Wright 1989). Details for *Artemia* and nematodes, in which desiccation resistance has been well studied, follow.

The *Artemia* cyst consists of a non-cellular shell surrounding an embryo in the early gastrula stage. This embryo consists of approximately 4,000 undifferentiated cells which have no obvious morphological features that indicate them to be desiccation–adapted. Despite this, the embryo can tolerate virtual complete loss of water yet remain viable for many years. Table 5.3 shows the relationship between the water content of *Artemia* cysts and metabolic activity, as outlined by Clegg (1981). Below 0.3 g of water present per gram of initially dehydrated cyst material ($g\,g^{-1}$) no metabolic reactions take place, and any chemical transformations that do occur may do so without the contributions of enzyme catalysis—this is the *ametabolic domain*. Some metabolic activity may begin at hydration levels near 0.3 $g\,g^{-1}$ but this is not reflected by a measurable increase in oxygen uptake and consists of only a few types of metabolic reaction—this is the *domain of localized or restricted metabolism* and spans the hydration range of 0.3–0.65 $g\,g^{-1}$. Above 0.65 $g\,g^{-1}$, oxygen consumption rates become detectable and rise with increasing water content. Metabolic activities here are qualitatively the same as those found in fully hydrated cysts—this is the *domain of conventional metabolism*. Restoration of cellular water results in rapid reactivation of metabolic and developmental processes.

Clegg proposed that the initial addition of water to the cysts is needed to form hydration layers around macromolecules. Some of this 'bound' water may have remained in the cyst even under extreme desiccation, but hydration levels of 0.15 $g\,g^{-1}$ allow full repletion of this store. Further

Table 5.3 Hydration-dependence of cellular metabolism in cysts of the brine shrimp, *Artemia*

Cyst hydration ($g\ H_2O$ per g cysts)	Metabolic events initiated	Domain
0 to 0.1	None observed	Ametabolic
0.1	Decrease in ATP concentration	
0.1 to 0.3 ± 0.05	No additional events observed	
0.3 ± 0.05	Metabolism involving several amino acids, Krebs Cycle and related intermediates, short-chain aliphatic acids, pyrimidine nucleotides, slight decrease in glycogen concentration	Restricted metabolism
0.3 to 0.6 ± 0.07	No additional events observed	
0.6 ± 0.07	Cellular respiration, carbohydrate synthesis mobilization of trehalose, net increase in ATP, major changes in the free amino acid pool, hydrolysis of yolk protein, RNA and protein synthesis, resumption of embryonic development	Conventional metabolism
0.6 to 1.4	No additional metabolic events observed	

Source: Information derived from Clegg 1981, and Hochachka and Somero (1984).

hydration (up to 0.6 $g\,g^{-1}$) may replenish the 'vicinal' water component of the cyst—this is water which is relatively 'organized' within the cells and is associated with longer range interactions with membrane surfaces and macromolecular complexes such as cytoskeletal elements. Yet more hydration (>0.6 $g\,g^{-1}$) is thought to contribute to the 'bulk' water, that is, the water associated with other, general, cellular components. The bulk aqueous phase appears to be necessary to allow effective transport of metabolites, fuel sources, and the like, between cellular compartments.

There is a strong correlation between survival in the desiccated state and accumulation of polyols (polyhydroxy alcohols) in the cells. Polyols seem to be important in osmoregulation and resistance to freezing, and may be necessary in anhydrobiosis as well. Two polyols are evident in cysts of *Artemia*: glycerol comprises about 4% of cyst dry weight, whereas trehalose accounts for up to 14%. In anhydrobiosis, polyols may act as water substitutes, creating hydrogen-bonded interactions with polar or charged entities of the cell, and, in addition, they may stabilize protein structure at low water activities. The latter occurs because glycerol is excluded from the highly structured water surrounding proteins. The addition of glycerol therefore promotes the compact, folded structure of proteins thus preventing them from unfolding and becoming denatured. When water returns to the cell, and there is a need for rapid reactivation of metabolism, the necessary enzymes are at hand in a functional state (Gekko and Timasheff 1981).

In nematodes, anhydrobiosis is not confined to any particular stage in the life cycle. As in *Artemia* cysts, the stimulus for anhydrobiosis appears to be reduced availability of water. This begins a series of progressive morphological changes that take place between 24 and 72 h after initiation. The entire animal contracts longitudinally and becomes coiled, intracellular organelles such as muscle filaments undergo ordered packing, and membrane systems show ordered change. The sequence of events is initiated by water loss, but the morphological changes themselves are under the endogenous control of the animal itself. The main metabolic adjustment that occurs during dehydration is one of redistribution of carbon from storage products, such as glycogen and lipid, into large intracellular pools of trehalose and glycerol. In nematodes, these two polyols reach levels of 10 and 6%, respectively, of the animals' total dry weight. Similar changes in polyhydroxy alcohols are known to occur in other anhydrobionts. Glycerol may be important in stabilizing membranes (Crowe 1971). There are few data available on the metabolic events at different stages of hydration in nematodes but it is likely that, as in the cysts of *Artemia*, metabolism comes to a halt during anhydrobiosis.

During rehydration of desiccated nematodes, metabolic events appear to be the reverse of those leading to entry into anhydrobiosis, although the rate at which they happen is very different: up to 72 h for dehydration but only a few hours for rehydration. This is presumably linked to the relative times the water body takes to dry up (long) and refill (short), and thus is significant in terms of the effective timing of occurrence of the species in the habitat.

Anhydrobiosis bestows many advantages upon those species that have managed to incorporate it into their biology. These include: (1) enhancement of dispersal, particularly by wind; (2) allowing, or improving chances of, survival in habitats subject to severe drought; (3) synchronization of biological processes (e.g. feeding, growth, reproduction, etc.) with favourable episodes in the environment; (4) increased longevity; and (5) pre-adaptation for survival of other adverse environmental factors (e.g. temperature extremes and anoxia).

Are there limits to anhydrobiosis? Rothschild and Mancinelli (2001) cite the following mechanisms of death due to extended anhydrobiosis: irreversible phase changes to proteins, lipids and nucleic acids (such as denaturation and structural breakage), and accumulation of reactive oxygen species during drying, especially under exposure to solar radiation.

5.5 Adaptations

As previously noted, the suitability of temporary water bodies as habitats for aquatic organisms is highly varied. Consequently, many different types of life cycle 'strategies' have evolved to enable species to exploit the resources of these environments. Much of the success of individual species in a particular type of temporary pond or stream is due to special adaptations of some aspect, or aspects, of the organism's physiology and behaviour. The array of such adaptations that has arisen is almost as varied as that of the habitats themselves (Table 5.4). Examples of these adaptations will be considered under three broad groupings: behavioural avoidance, timing of growth and emergence, and dormancy and hatching response.

5.5.1 Behavioural avoidance

Several methods of avoiding drought were indicated, previously (Figure 5.1 and Table 5.4). Some species, notably adult insects and especially the Coleoptera and Hemiptera, fly away from drying water bodies and spend the dry phase at nearby permanent waters, flying back upon refilling of the basin. Other adult insects may spend the dry period resting in sheltered places. In France, for example, adults of the caddisfly, *Stenophylax*, which emerge from intermittent streams, fly to nearby caves where they hibernate under stable environmental conditions (Bouvet 1977). Fishes also may migrate to and from temporary waters, such as floodplain pools and tributaries that dry up, where the necessary connection exists with a permanent water body, and the adaptations seen in some species were discussed in Section 4.2.6.

Many species resort to burrowing to avoid drought. For some, this consists merely of crawling under debris (leaves, dry algal mats, etc.), or rocks on the bed. Others burrow to varying degrees into the substratum of the bed and most enter some form of resting state. Many are invertebrate species, but some vertebrates survive this way also, for example, the tropical lungfishes and the spadefoot toad. Specialized freshwater crayfishes, such as *Fallicambarus fodiens*, a 'primary burrower' (*sensu* Hobbs 1981), can burrow to depths of a metre or more, the extent of the shaft depending on the depth of the GWT from the bed. As the groundwater table rises and falls so does the resting station of the crayfish (Figure 5.8). Pellets of mud from the excavation are brought to the surface and deposited as a chimney around the entrance to the burrow. In dry air conditions, these chimneys are plugged probably as a means of preventing moisture loss from the burrow. Most of the above activity takes place during darkness,

Table 5.4 Summary of the physiological and behavioural mechanisms by which various aquatic organisms survive desiccation

Organism	Mechanism/stage in life cycle	Reference
Algae/flagellated protozoans	Modified vegetative cells with thickened walls; mucilaginous sheaths; accumulation of oil in cells; heat-resistant asexual cysts	Evans (1958) Belcher (1970)
Sponges	Reduction bodies; gemmules	Frost (1991)
Flatworms	Dormant eggs; resistant cysts enclosing young, adults, or fragments of animals; cocoons	Castle (1928) Ball *et al.* (1981)
Rotifers	Survive as dehydrated individuals; some bdelloids secrete protective cysts	Pennak (1953)
Nematodes	Eggs; larvae; adults	Poinar (1991)
Bivalves	Young and adult stages	Thomas (1963)
Gastropods	Adults form a protective epiphragm of dried mucus across shell opening; adults and young survive in moist air and soil under dried algal mats on habitat bed	Strandine (1941) Eckblad (1973)
Oligochaetes	Dormant eggs; resistant cysts enclosing young, adults, or fragments of animals	Kenk (1949)
Leeches	Survive as dehydrated individuals; some species construct small mucus-lined 'cells'	Hall (1922) Davies (1991)
Tardigrades	Resistant 'tun' stage	Everitt (1981)
Mites	?resistant larvae; in most species, larvae attach to migrating insect hosts and leave the habitat, returning with the water. Larvae remain attached to host throughout its stay in permanent water body	Wiggins *et al.* (1980)
Anostracans	Resistant eggs	Hinton (1954)
Notostracans	Resistant eggs	Fox (1949)
Conchostracans	Resistant eggs	Bishop (1967)
Cladocerans	Ephippial eggs; adults survive in moist soil	Chirkova (1973)
Copepods	Diapausing eggs, late copepodites, adults (in cysts)	Yaron (1964)
Ostracods	Resistant eggs; as near adults in moist substrate	Delorme (1991)
Amphipods/ Isopods	Immatures near the groundwater table	Clifford (1966), Williams/Hynes (1976)
Decapods	Juveniles and adults in burrows at groundwater table	Crocker/Barr (1968)
Collembolans	Resistant eggs	Davidson (1932)
Stoneflies	Diapausing early instar	Harper/Hynes (1970)
Mayflies	Resistant eggs	Lehmkuhl (1973)
Odonates	Resistant nymphs; recolonizing adults	Daborn (1971)
Hemipterans	Recolonizing adults	Macan (1939)
Beetles	Semi-terrestrial pupae; burrowing adults; recolonizing adults; resistant eggs	Jackson (1956), Young (1960), James (1969)
Caddisflies	Diapausing eggs; resistant gelatinous egg mass; terrestrial pupae; recolonizing adults; larvae deep in substrate	Wiggins (1973), Williams/Williams (1975)
Chironomids	Resistant late instar larvae, sometimes in cocoons of silk and/or mucus; recolonizing adults; ?resistant eggs	Hinton (1952), Danks (1971), Jones (1975)
Mosquitoes	Resistant eggs; resistant late instar larvae and pupae	Horsfall (1955)
Other dipterans	Resistant eggs, larvae, and pupae	Hinton (1953), Bishop (1974)
Fishes	Recolonization by adults; dormant adults in substrate; resistant eggs	Kushlan (1973), Williams and Coad (1979)
Amphibians	Recolonization by adults; dormant adults in substrate	Wilbur and Collins (1973)

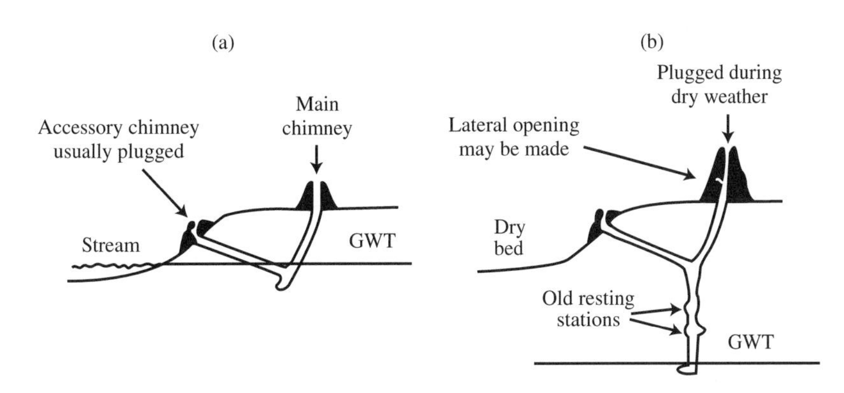

Figure 5.8 Diagrammatic sections through burrows of the crayfish, *Fallicambarus fodiens*: (a) shows the typical Y-shaped burrow in early spring; (b) shows modifications to the burrow during periods of receding groundwater table (after Williams *et al.* 1974).

especially on warm humid nights. Activity increases after heavy rainfall as this causes collapse of the chimneys and perhaps also of the tunnels. During the autumn, when the GWT rises, the crayfishes start to move their resting stations towards the soil surface, and if the vertical shaft of the burrow is not too steep this merely entails climbing up as the water rises and establishing new, or reoccupying old, resting stations. However, if the shaft is perpendicular or almost so, the crayfish often fills in the lower parts of the shaft beneath it.

The water in the bottom of crayfish burrows serves as a refuge for smaller species and the term 'pholeteros' has been coined to describe these assemblages (Lake 1977). They have been recorded from burrows of crayfishes in North America and Australia, and consist primarily of microcrustaceans, oligochaete worms, amphipods, isopods, collembolans, and chironomid larvae. The water in the burrow is generally of low pH and contains very little oxygen (e.g. 8–12% saturation), the pholeteros must therefore be physiologically adapted to these conditions. The crayfishes themselves are usually located at the water–air interface, and it has been suggested (Grow and Merchant 1980) that they may be extracting oxygen from the moist air in the burrow and not from the water.

In intermittent streams and rivers, the hyporheic zone (the zone of saturated sediment beneath and lateral to the channel) has long been thought to act as a refuge for benthic organisms during drought (Williams and Hynes 1974). Boulton *et al.* (1992) found support for this hypothesis on two continents. In a Sonoran (USA) desert stream, 69% of the taxa present in the dry season were found in the hyporheic zone. In two Australian intermittent streams, 35% of the fauna sought refuge in this zone, with 48% retreating underneath dried litter and rocks; 26% were cohabitants of crayfish burrows. Nematodes and oligochaetes were especially common in these hyporheic zones, and also present were turbellarians, gastropods, isopods, mites, mayflies, stoneflies, tipulids, ceratopogonids, and chironomids. In addition, amphipod crustaceans, particularly the genus *Crangonyx*, also are known to burrow in bottom sediments to avoid adverse conditions (Williams and Hynes 1977a). In a desert stream, Clinton *et al.* (1996) recorded a GWT recession rate of around 11 cm per week, to greater than 1 m below the bed surface by August. Shallow hyporheic areas supported mainly insect larvae and cyclopoid copepods, whereas bathynellaceans, isopods, and harpacticoids tended to occur deeper.

5.5.2 Timing of growth and emergence

Figures 5.4 and 5.5 showed the duration of species' life cycles in temperate region intermittent ponds and streams, respectively, but they also indicate relative rates of growth throughout these cycles. The growth curves are of three basic types:

(1) a symmetrical (bell) curve, indicating that during the animal's life cycle, the rate of growth

first increases gradually and then decreases gradually as the animal approaches adulthood;
(2) an asymmetrical curve skewed to the left, indicating a rapid rate of growth in the early instars, perhaps to take advantage of warmer water temperatures before the onset of winter, or a temporary abundance of food, followed by a very slow growth rate. Such species will be in a near adult stage well in advance of any significant change in environment. This is a common type of growth curve in temporary waters. Some animals, for example, mosquitoes, show such rapid growth that they are capable of completing two or more generations within one environmental cycle (Figure 5.5); and
(3) an asymmetrical curve skewed to the right, indicating a slow rate of growth for much of the life cycle, followed by a rapid burst of growth near the end—perhaps to take advantage of specific springtime environment conditions. Such cycles apparently are not flexible prior to significant change in the environment, and this type of growth curve seems to be characteristic of a limited number of species.

Following this broad picture of the differences in growth rates among species, it is appropriate to consider some of the factors that control, or are correlated with, growth.

Water level. Chodorowski (1969) found that dilution of pond water by rain retarded the development rate of the mosquito, *Aedes communis*. Both drying up of the pool and overcrowding in the population accelerated growth, the former effect being slight but the latter being great. McLachlan (1983) working with laboratory populations of *C. imicola* similarly suggested that patterns of adult emergence were associated with larval density, adaptively appropriate to the corresponding stage of evaporation of pools in nature. At initial low densities, safe, early emergence produced small females; increased risk for larvae emerging late was compensated for by the production of larger and therefore fitter adults. At high density, when the habitat was about to dry up, females emerged earlier but were also larger. Again, Fischer (1967) found that the development rate of the damselfly *Lestes sponsa* accelerated in response to rapid water loss from an intermittent pond; not that all populations of temporary water species have generation times well synchronized with the end of the aquatic phase. Hildrew (1985) found that a rainpool in Kenya dried up on more than one occasion stranding many *Streptocephalus vitreus* (fairy shrimp)—many of which were still producing eggs. The strategy of this species seems to include extreme spreading of risk among the progeny, with rapid growth, early reproduction at a modest size and repeated mating to produce further clutches of eggs—but with considerable wastage of reproductive effort.

Broch (1965) studied the embryology of eggs of the fairy shrimp *Chirocephalopsis bundyi*, together with the ecological and physiological factors controlling development and hatching. He found that embryological development was divisible into three blocks. Each was dependent on a set of ecological parameters associated with a woodland temporary pool, and the development sequence was synchronized with the seasonal changes in these parameters. The first developmental block took place under high temperature and aerobic conditions during summer, at the pond margin. This was termed the summer phase of synchronization, and unless this stage was completed by autumn, there was no further development. The second block was a response to low temperature. This occurred in early autumn and under aerobic conditions resulting from the recession of water from the margin. The third block (prehatching to the metanauplius stage) took place in early winter at the dry pond margin. In the pond, prehatching began in a band around the periphery of the margin, and this prehatching band moved to higher levels of the margin as winter progressed. Freezing conditions removed water, as a liquid, from around the eggs and inhibited prehatching. The unhatched metanauplii overwintered in a dormant state, until hatching was triggered by low-oxygen tension in the water at the soil–water interface, when the pond filled in the spring. Thus it was the completion of embryonic development in a terrestrial environment, from autumn to early winter, that made synchronization of hatching with pond formation possible. Seasonal fluctuation of the water level in the pond was therefore the most important physical parameter in the early development of this, and probably other, species of Anostraca, although other parameters associated with the rise and fall of the water level were actually the controlling factors. In his rainpool in Kenya, Hildrew (1985) found that more eggs of *S. vitreus* hatched from cores taken near the margins than at the centre. This ensured that the pool must be nearly full before the eggs were inundated, thus increasing the chances of survival.

Temperature. Aquatic invertebrates are all poikilotherms, that is, their body temperatures reflect those of their surroundings. As a rise in water temperature generally increases their metabolic rate, invertebrates in temporary waters should grow faster during those phases when the water temperature is highest—provided that growth is not limited by some other factor such as food availability, or low oxygen levels. However, metabolism and growth rates will fall as upper (and lower) lethal temperatures are approached. If the relationship between metabolism and temperature was to be always as above, the life cycles and growth curves seen in any single water body would likely be very similar for all inhabitants. Clearly, this is not the case. Different species may have specific temperature ranges at which growth is maximal. Thus, for example, whereas some intermittent pond species grow faster in the low water temperatures that follow spring snowmelt, others grow faster in the high temperature water immediately prior to the summer drought. Such differences contribute to the characteristic succession of species in the community.

Food. Growth of an organism is obviously affected by the quality and quantity of food available to it. Food webs in temporary waters include, primary producers, herbivores, detritivores, and carnivores, in fact all of the components of normal, efficient communities in other biotopes (see Chapter 6). Although seasonal availability of certain food types contributes, along with temperature, to temporal succession of species, many temporary waters appear to have a plentiful and continuous supply of food, particularly detritus. However, Barlocher *et al.* (1978) have suggested that, in ponds where decaying vegetation is exposed to air (in many temperate, woodland pools this would be during autumn and winter), the protein content, and thus the nutritional value of the detritus, in spring, will be greater than in permanent ponds. Enhancement of protein levels is the result of colonization of the detritus by fungi and bacteria, which use the material both as a substrate for attachment and as a source of nutrients. These authors hypothesized that this protein-rich detritus in temporary pools might enable animals to grow faster, thus completing their life cycles safely within the hydroperiod.

Two other very important aspects of the timing of growth and emergence are delayed and staggered egg hatching. Service (1977) studied the pattern of egg hatching in a population of the mosquito *Aedes cantans* from southern England. Figure 5.9 shows the weekly percentages of eggs that hatched from batches soaked continually for 32 weeks from October to May. Each year, eggs started to hatch in week 14 (January). By week 21, 50% had hatched and by week 28, 95%. The longest spread of hatching of a single batch of eggs was 6 days. Embryonic development required only 15 days after which, if relative humidity was maintained at 85% or higher, the eggs remained viable for many months—to a maximum of 3.5 years, if not flooded. Ponds in this locality were usually dry from May to September, however, if they became flooded during this period some eggs would hatch. From September to January no eggs hatched, as they were in an obligatory diapause initiated by reduced temperature and/or day length. The factors terminating diapause were not determined, but Horsfall (1956), working on *Aedes vexans*, found that once eggs of that species had been exposed to adverse environmental conditions they had to be conditioned before they would hatch. The conditioning process varied according to the eggs' prior environment (see Table 5.5). After conditioning, the basic hatching stimulus was a decrease in the dissolved oxygen content of the pond water caused, for example, by intense microbial growth.

Staggered egg hatching is not confined to mosquitoes. *Limnadia stanleyana*, an Australian

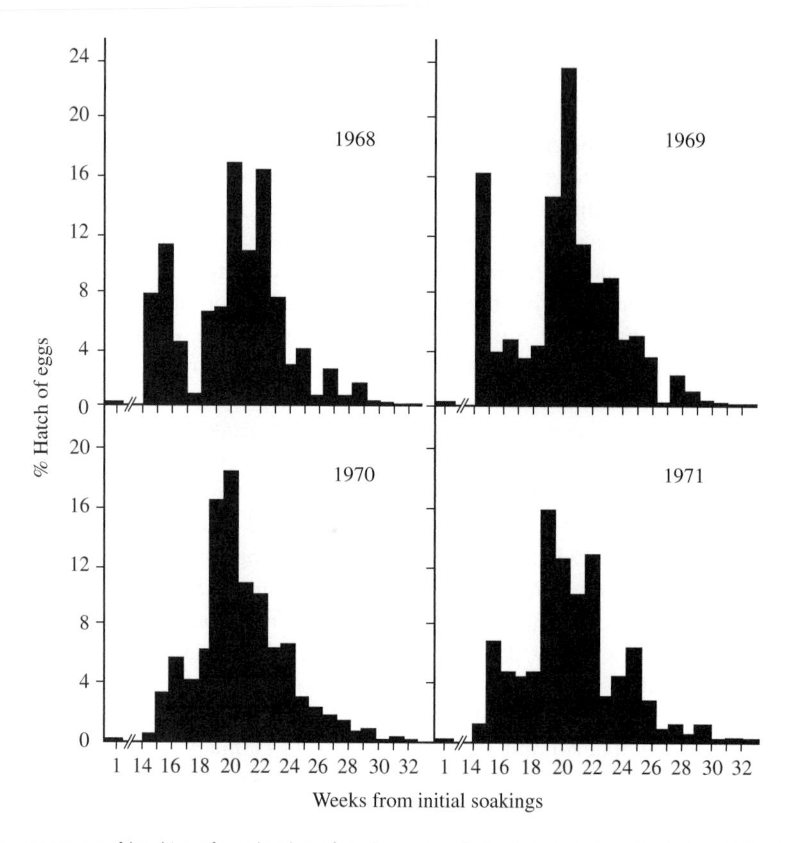

Figure 5.9 Weekly percentages of hatching of egg batches of *Aedes cantans* first rehydrated at the beginning of October, and then kept immersed for 32 weeks. Each year, eggs started to hatch in week 14 (beginning of January) (redrawn from Service 1977).

Table 5.5 Examples of the conditioning processes for diapausing eggs of the mosquito *Aedes vexans*

Prior environment/embryonic state	Conditioning sequence
Embryonated, in moist pond debris	Submergence, deoxygenation of the water
Embryonated, in air-dried debris	Submergence, de-watering, aeration of moist eggs, submergence, deoxygenation
Pre-embryonated, flooded	De-watering, aeration of moist eggs, submergence deoxygenation

Source: Information derived from Horsfall (1956).

conchostracan, lives in episodic rainwater pools in eastern New South Wales. Bishop (1967) reported that generations of this species are frequently destroyed, before maturing, because the rainfall in this area is unpredictable in amount and irregular in occurrence. Although the eggs of this species are drought resistant, the larvae are killed by drying. Thus, not all eggs hatch when the pools fill, and many remain in the mud, as a reserve, to hatch at a later time. Darkness and low oxygen concentration in the mud inhibit hatching, but each time the pools fill, some eggs are brought to the surface and hatch.

The fairy shrimp *Branchinecta mackini* occurs in temporary ponds in the Mohave Desert, California, and populations may last from as little as 3 days to as long as 4 months, depending on pond conditions. The populations survive drought as dehydrated eggs and hatch (within 24–36 h) as water begins to fill the pond basin, but only if the salinity of the water remains low (generally <1,000 μmhos). At low salinity, hatching is virtually continuous but, as salinity in these ponds generally increases rapidly after filling, the initial hatch is usually terminated quite quickly. Further episodes of hatching follow periods of influx of non-saline water as, for example, from rain or melting ice. For this species in these particular ponds, the duration of hatching is inversely proportional to the rate of increase in salinity until, above 1,000 μmhos per day, hatching is completely inhibited. In the laboratory, it has been shown that egg hatching is actually controlled by a combination of salinity and dissolved oxygen, operating in various proportions, to either initiate or inhibit hatching. Geddes (1976) has similarly shown that the resting eggs of *Parartemia zietziana*, which occur in 'ephemeral' saline ponds in Victoria, Australia, hatch in response to a salinity drop over a range of 50–200 ppt. From the results of their study, Brown and Carpelan (1971) proposed that the existence of a salinity-oxygen controlling mechanism might explain a difference between branchiopods found in humid and arid regions. Stimulation of egg hatching must be due to some factor that changes at the time of origin of the habitat. In humid regions, the factors seem to be temperature and oxygen. In arid regions, where ponds undergo significant changes in salinity, control of egg hatching may be by both salinity and oxygen, in which case, temperature simply controls the rate of development of the embryo.

As a final example of timing of growth and emergence, Morse and Neboiss (1982) recorded the unusual phenomenon of viviparity in all five species comprising the *Triplectides australis* group (Trichoptera), in which females bear living first-instar larvae through ruptured ventral membranes between segment VIII and the gonopod plate. These species are highly successful inhabitants of temporary waters in the drier parts of Australia, likely attributable to a reduction in the amount of time required for larval development in an external liquid medium.

5.5.3 Dormancy and hatching response—mechanics and evolution

In many environments, regulation of an organism's development occurs via the *direct* action of local factors on biological processes. Appropriate growth rates and limits are thereby adjusted so as to assure the successful completion of the life cycle. Single species frequently show different growth characteristics among populations occurring under different habitat conditions (Danks 1987). Such direct control, as, for example, is seen in the effect of environmental temperature on the growth rate of poikilotherms, appears to apply even where there is seasonal variability. Clearly, however, there must be limitations in habitats that are highly seasonal (e.g. in continental as opposed to maritime climates) as in these, to use the temperature example again, environmental extremes often exceed the tolerable threshold for a species. In such highly fluctuating environments although external factors exert control they are more likely to be used, *indirectly*, as cues that initiate or terminate developmental stages.

Dormancy is an adaptation that allows developmental hiatuses to occur, enabling organisms to deal with unfavourable changes in their environments. Dormancy is herein defined as being 'a state of suppressed development' which may take either the form of *quiescence*, an immediate direct response to a limiting factor, or *diapause*, a more profound interruption of development that diverts the metabolic programme away from direct developmental pathways and results in a hiatus that is not controlled directly by environmental factors and precedes the onset of adverse conditions (Danks 1987).

There are several hypotheses as to the origin and subsequent evolution of dormancy, ranging from those which lean towards a phylogenetic explanation of commonality (see Thienemann 1954; Vepsalainen 1978; Corbet 1980; Hairston and Cáceres 1996), to those which imply that habitat pressures are the primary driving forces (Fryer 1996), having produced multiple, independent origins (see Svensson 1972; Grodhaus 1980). In the broad geographical context, dormancy seems common in temperate and polar regions, where it is employed typically to survive cold temperatures, and several studies have shown that, within single species, its occurrence increases as a species spreads away from the tropics (e.g. Nielsen 1950; Readshaw and Bedford 1971). This does not dictate, however, that dormancy originated in cooler climes, as seasonality is also prevalent in the tropics although it takes less familiar forms (e.g. monsoonal rainfall, food-plant growing seasons). Indeed, a tropical origin has been proposed (see review by Tauber and Tauber 1981). Interestingly, tropical species often maintain some expression of dormancy within their populations, despite generally stable habitat conditions, presumably as the ultimate fail-safe device (Denlinger 1978). There are also suggestions that dormancy, specifically as diapause, has evolved as an adaptation to drought (Ushatinskaya 1959) and/or dealing with unfavourably high environmental temperatures (Pritchard *et al.* 1996). In the latter contexts, dormancy is a particularly favoured mechanism for survival in aquatic habitats in arid and semi-arid regions (e.g. in episodic rainpools) (W.D. Williams 1985; Brendonck and Persoone 1996). Among hypotheses as to the origin of dormancy that are largely independent of geography are those that cite freedom from competition (Alekseev and Starobogatov 1996), avoidance of food limitation (Hansen 1996), and escape from predation (Fryer 1996). Here, evidence is sought of dormancy as an adaptation that allows invertebrates to achieve some synchrony between their life cycles and specific phases of temporary freshwater habitats.

To illustrate the importance of dormancy as a strategy for survival/persistence in temporary and unpredictable habitats, consider faunal diversity as one moves from aquatic habitats that are permanent to those that are dry for short, predictable periods (intermittent), to those that are short-lived and episodic (Table 5.6). The frequency of occurrence of dormancy is very low in stable water bodies, and typically seems restricted to cold-adapted species that cannot tolerate summer warm-water temperatures, such as nymphs of the stonefly *Allocapnia vivipara*. Such habitats in eastern North America have communities dominated by insects, crustaceans, and non-arthropod taxa. As water bodies become subject to periods of drought, the incidence of dormancy increases dramatically, although frequently communities comprise the same major taxa, but different species. At the extreme end of the scale, in short-lived, episodic rainpools, the fauna is truncated, typically consisting of crustaceans and dipterans exhibiting highly specialized powers of growth and developmental arrest (which in the case of *P. vanderplanki* is more likely to be quiescence rather than true diapause).

There seem to be relatively few insect species capable of living in highly episodic waters. Whether, in these extreme habitats, this is due to limited individual habitat size, limited food resources, and therefore severe competition, insufficiently rapid colonization abilities, or lack of comprehensive study is largely unknown. Many of the insect taxa most commonly encountered in less extreme temporary freshwaters possess suitable traits (as was demonstrated in Table 5.2), but perhaps the latter are insufficiently advanced for episodic waters.

Despite the evolution of various other (behavioural and physiological) adaptive traits, dormancy, particularly as diapause, seems to be a common solution to invertebrate survival and persistence in most temporary waterbody types (as was seen in Table 5.4). Whether this is in reponse to actual water loss or associated, unfavourable temperature change will be discussed later. Clearly though, diapause permits some synchrony with habitat phase (Harper and Hynes 1970; Elgmork 1980), but not just with the wet–dry components (as was illustrated in Figures 5.4 and 5.5). These two figures are now revisited with respect to the incidence of dormancy (marked with an asterisk) in the life cycles depicted.

Recall that Figure 5.5 showed the three habitat phases seen in small intermittent streams in Ontario: an autumn-winter flowing water phase; a spring pool phase; and a summer terrestrial phase. In these streams, approximately 50% of the taxa have at least one dormant stage in their life cycles, and a further 36% are suspected of dormancy.

Table 5.6 Comparison of the frequency of occurrence of dormancy in the faunas of freshwater habitats that differ in degree of permanence/stability

Habitat Description:	Cold-water spring[a]	Permanent river[b]	Rain pools in N.S.W.[c]	Intermittent stream[d]	Intermittent pond[e]	Desert rain pool[f]	Rain pools in Africa[g]
Duration of aquatic phase:	12 months	12 months	11 months	8 months	3 months	1–2weeks	~24 h
Total no. of taxa:	66	73	18	63	97	16	1
No. taxa known or suspected to exhibit dormancy in life cycle:	2[h]	4	9	54	72	14	1
% of total fauna exhibiting dormancy:	3.0	5.5	50.0	85.7	74.2	87.5	100.0
Dominant organisms:	Insects Mites Crustaceans Molluscs	Insects Crustaceans Molluscs Oligochaetes Turbellarians	Insects Crustaceans Nematodes	Insects Crustaceans Molluscs Oligochaetes Turbellarians	Insects Crustaceans Mites Molluscs Oligochaetes	Crustaceans	*Polypedilum vanderplanki* [chironomid]

[a] Williams and Hogg (1988).
[b] Bishop and Hynes (1969).
[c] Bishop (1974).
[d] Williams and Hynes (1976).
[e] Williams (1983).
[f] Rzoska (1984).
[g] McLachlan and Cantrell (1980).
[h] Due to insufficiently detailed study of individual taxa, these values are estimations only.

Potentially, therefore, 86% of the fauna depend on arrested development at some stage in their life cycle in order to survive in such streams. In permanent streams in Ontario, this figure is likely to be less than 10%.

The *flowing water fauna* comprises species that appear shortly after the streams have started to flow again in October/November. Most complete their life cycles before flow stops in the spring and many then enter some form of dormancy. Cessation of flow brings with it warmer water temperatures and lower oxygen levels, both unsuitable for cool, rheophylic species (Hynes 1970). Algal blooms in the resulting pools significantly increase pH (e.g. from 8.0 to 9.4; Williams and Hynes 1976a) which may be another factor promoting dormancy in certain species. In Figure 5.5, the life cycle patterns have been assigned subjectively, based on field-collected data and life cycle determination, to a series (A to H) of groups representing taxa with similar life cycle, growth rates, and period of occurrence.

Most of Group A taxa hatch from eggs immediately upon flow resumption. The nymphs/larvae grow quickly and are in penultimate instars prior to ice cover. During mid-winter they remain at a low level of activity, and most emerge as adults in early spring. Subsequent occurrence varies according to species, but each employs dormancy. For example, the winter stonefly *A. vivipara* emerges first, through cracks in the ice in March, mates, and females crawl back under the ice and oviposit in the water. Early instar nymphs are to be found diapausing deep in the moist bed substrate in summer where the ambient temperature suits their cool-adapted physiology. Typically, the body of an inactive nymph becomes filled with fat globules, which give it a characteristic pale colour, the head and antennae become reflexed under the thorax, and the end segments of the cerci become thin and bald (Figure 5.10). Harper and Hynes (1970) have argued that, in contrast to an egg diapause, which must occur where the eggs fall after oviposition, a nymphal diapause allows the nymph to actively seek out a suitable site in which to become dormant. In similar fashion, final instar larvae of the caddisfly *Ironoquia punctatissima* crawl out of the water and spend several months as diapausing larvae/pupae, shaded among the roots of riparian vegetation, and sealed in their larval cases. Adults emerge in late summer and oviposit on the dry stream bed (Williams and Williams 1975), and the, presumably diapausing, eggs hatch when flow resumes. The chironomids *Trissocladius* and *Micropsectra* emerge at the end of the stream phase and lay eggs which remain dormant through the, for them, unfavourable environmental conditions present during both the pool and summer terrestrial phases. The rapid initial growth of Group A taxa perhaps is associated with the plentiful food available and also the moderate late-autumn water temperatures. This food is in the form of left over vegetation which decomposes faster after having been exposed to air and has an enhanced protein content upon flooding—compared with vegetation that has been continuously submerged in permanent waters (Barlocher *et al.* 1978).

Other chironomid species show the more gradual growth pattern characteristic of Group B. All emerge around the time of flow cessation, and all appear to pass the pool and terrestrial phases as diapausing eggs. This same growth pattern is seen in two species of the amphipod genus *Crangonyx*, the clam *Sphaerium*, and the flatworm *F. velata*. The amphipods survive the drought as small, inactive (?diapausing) individuals near the groundwater table, the flatworm as diapausing cysts in the moist bed, and the clams as either small individuals in the bed, or a few large individuals in the pools (Williams and Hynes 1976).

Group C taxa grow more slowly at first, but faster later in the running water phase. Both of the copepod species survive the drought in two states: *Cyclops vernalis* chiefly as near adults found diapausing in the moist substrate, but also as active individuals in the water of crayfish burrows; and *A. nordenskioldii* either as non-diapausing, near adults in the vicinity of the GWT, or as encysted, near adults nearer the streambed surface. The cyst consists of a single layer of a transparent, gelatinous substance to which sand grains become attached (Figure 5.11). Near adults of benthic ostracod species and small *H. azteca* (Amphipoda) can be found in the capillary fringe but do not appear to be in diapause. Pupae of the cranefly *T. cunctans* are found among grass roots on the stream bed in summer. This diapausing stage lasts until adults emerge in October. Eggs are found on the damp bed shortly thereafter, but do not hatch until they have been flooded with water for several

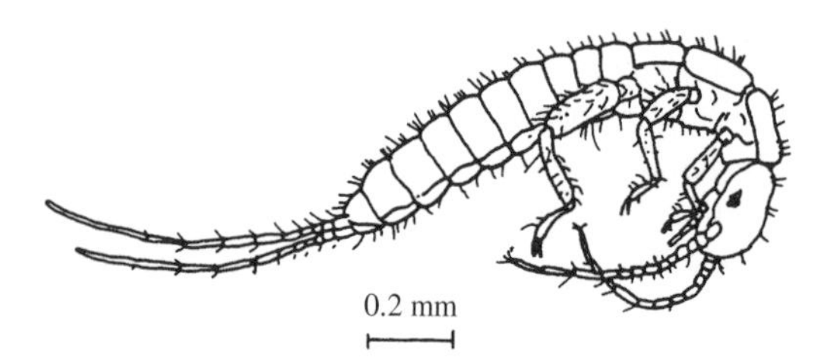

Figure 5.10 A diapausing nymph of the stonefly *A. vivipara* (redrawn from Harper and Hynes 1970).

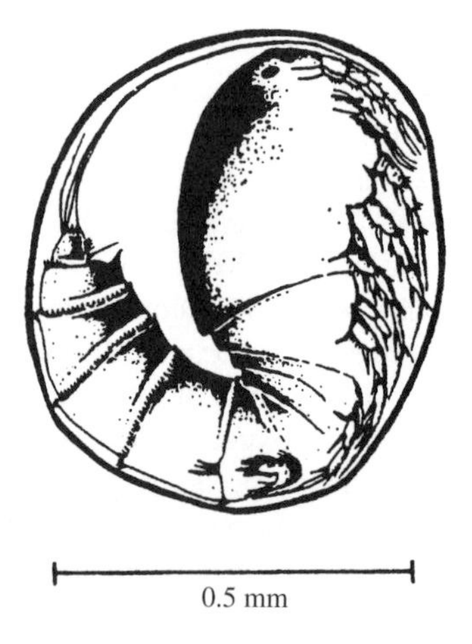

Figure 5.11 Summer drought-resistant cyst of the harpacticoid copepod *A. nordenskioldii*.

weeks. Adult *A. semivittatus* emerge at the end of the stream phase and fly away. Presumably they overwintered as diapausing adults.

The *pool fauna* comprises species that appear shortly after stream flow has ceased in the early spring. Most complete their life cycles before the pools evaporate in June. Advantages to be gained, by these lentic species, might include warmer water temperatures, together with plentiful food in the form of the algae that develop once flow ceases and in the form of prey for incoming predators (e.g. hemipterans and dytiscid beetles; see Arts *et al.* 1981; Williams 1983). Again, three basic life cycle patterns are proposed.

Group D taxa show rapid growth as the pools form and include the mayfly *P. ontario*, the cladoceran *M. macrocopa*, and the hydrophilid beetle *A. limbata*. The first two survive the drought as diapausing eggs *in situ* whereas the beetle likely leaves to overwinter as an adult in nearby permanent ponds from where it disperses, in true pioneer species fashion, to locate new temporary ponds the following spring (Fernando and Galbraith 1973).

Group E includes beetles and hemipterans some of which breed in the pools, others of which use the pools as feeding grounds. Some of the beetles are still in the pupal stage when the pools evaporate, however, development is completed among drying algal mats and other bed materials. Most of the Group E taxa likely overwinter as adults. Both the resident and non-resident populations of the dragonfly *A. junius* probably survive the drought as diapausing eggs.

Mosquitoes (Group F) are able to complete at least two generations during the pool stage. They deposit diapausing eggs on the stream bed prior to the pools evaporating. The occurrence of more than one generation in such pools (and in temporary ponds *per se*) raises interesting questions concerning the mechanism involved in such facultative diapause, where a species is capable of suppression of diapause in the first generation and of its expression in the last. In many insect populations, shift of a whole generation from a non-diapausing to a diapausing course of development is initiated by a critical daylength, typically with short days triggering diapause (Danilevskii 1965; Cohen 1966), or by a combination of photoperiod and temperature (Solbreck 1978). Because of inter-year variability in both the start and longevity of the pool phase, in temporary waters photoperiod would seem a less reliable cue for diapause initiation—unless each generation produces a diapausing fraction with the latter increasing as the season advances (Vinogradova 1960). More effective, might be sensitivity of a particular life cycle stage in the development of a subsequent generation to cues associated with demise of the habitat (e.g. a critically high water temperature, or an increase in conductivity or decrease in dissolved oxygen, all of which signal decreasing water volume). In comparison with the amount of work done on photoperiod and temperature, that involving diapause induction via chemical cues is very sparse in insects. Although diapause in nymphs of the mayfly *Leptophlebia cupida*, known to live in temporary pools (Burks 1953), has been correlated with low oxygen concentrations (at high water temperatures), whether temperature was the real controlling factor is unknown (Kjellberg 1973). For large, branchiopod crustaceans, Brendonck (1996) provides evidence that emergence from cysts can be stimulated by low conductivity, given suitable temperature and light conditions, and dissolved oxygen levels.

The mud-loving beetle *Heterocerus* (Group G) bred in the moist stream bed, after the pools had disappeared. Whether any form of diapause is employed by such members of the *terrestrial fauna* is unknown, however, it is likely that adults are forced into hibernation during the cold Canadian winter.

With the possible exception of the crayfish *Fallicambarus fodiens*, most of the species in Group H (i.e. those having life cycles that spanned the entire aquatic period of the streams, or had semivoltine cycles) are believed to encorporate a dormant stage into their life cycles.

Figure 5.4 showed the two habitat phases seen in a small intermittent pond in Ontario: a 3–4 month (spring-midsummer) aquatic phase; and an 8–9 month (late summer-early spring) terrestrial phase (Williams 1983). Again, much of the fauna (74% of species) is either known or suspected to employ dormancy—the lower percentage compared with the intermittent streams example, above, being due to the high number of aerially colonizing Coleoptera and Hemiptera, more typical of temporary lentic habitats. Noteworthy examples of species employing dormancy (marked with an asterisk) are: larvae of the chironomid, *Einfeldia*, which pass the drought as second instars buried in the pond mud; and the damselfly, *Lestes* sp., the branchiopods *Lynceus branchyurus* and *C. bundyi*, and the three species of *Aedes*, all of which survive the drought as diapausing eggs.

Dormancy in temporary waters is not just seen in invertebrates, however. In Africa, for example, as the water level in its habitat drops, the lungfish, *Protopterus*, burrows into the bottom mud and secretes a cocoon of mucus around itself. This cocoon is made waterproof by a layer of lipoprotein, and, inside, the fish lies folded upon itself with its head adjacent to a small opening at the top of the cocoon. Inside the cocoon of *Protopterus aethiopicus*, the fish's rate of oxygen uptake is gradually reduced to approximately 10% that of an active fish. Heartbeat drops to about three beats per minute, and the fish loses some water from its tissues. Part of the cocoon extends into the fish's pharynx and this acts as a respiratory tube. Urine production stops and nitrogen excretion changes from producing toxic ammonia to urea, which, being less toxic, can then accumulate in the blood and tissues with no adverse effects. In this state, *Protopterus* can survive for several years. Termination of dormancy is rapid, within minutes, and is stimulated by immersion in water, which begins to asphyxiate the fish as it cannot survive in water solely by gill-breathing—it has to have access to atmospheric air. It is believed that such a rapid 're-awakening' must involve release of a special hormone into the bloodstream (Beadle 1981). Another lungfish, *Lepidosiren para-doxa*, which lives in the temporary swamplands of central South America, survives drought in a similar manner. In Australia, several species of gudgeon (Gobiomoridae) and native minnow (Galaxidae) are known to bury themselves in mud or moist soil as their habitats dry up, but little is known of their physiology or adaptations. Examples of dormancy in temporary water amphibians (e.g. in the spade-foot toad, *Scaphiopus couchi*) were given in Chapter 4.

Dormancy as well as allowing species survival in temporary waters, is also a factor in producing the highly visible seasonal succession of species in these habitats. However, it has to be admitted that much of the evidence for the link between dormancy and habitat phase synchronization is correlative, and that manipulative and experimental approaches need to be applied for further corroboration and refinement.

It is probably appropriate, here, to make some remarks about the resistance to environmental adversity that dormancy appears to bring about. In general, dormancy allows individual organisms to suspend active developmental and maintenance processes during unsuitable environmental conditions. Typically, this involves a lowered metabolism (Davey 1956) which reduces energy expenditure and often provides protection from prevailing environmental adversities, such as drought, extreme low or high temperatures, and lack of oxygen (Hairston and Cáceres 1996). It has been argued, however, that cold-hardiness and drought resistance (which involve changes in the solute composition of the organisms' cells) require separate biochemical changes from those that induce diapause (e.g. Bohle 1969), with the former simply coinciding with the latter (Salt 1961). Indeed, examples exist of extreme cold-hardiness occurring in the absence of diapause (Hoy and Knop 1978). Cold-hardiness is experienced by the invertebrates of aestival streams and ponds that 'dry out' in winter due to complete freezing of the water column (Daborn 1974; Johansson and Nilsson 1994). The stresses produced by freezing (withdrawal of cellular water to the cell exterior to freeze as extracellular crystals) have been compared with those of dehydration due to excessively high temperatures (Meryman 1974; Crowe and Crowe 1982), although relationships between the latter and diapause have not been well studied (Danks 1987). However, in the case of *P. vanderplanki*, when dehydrated to 3% of their normal tissue water level in the laboratory, larvae have tolerated temperatures as low as −270°C and as high as +102°C (Hinton 1951). In the Crustacea, Khlebovich (1996) considers the embryonic diapause that ensures persistence through drought to be the same adaptation allowing survival during complete freezing of water basins.

Control of the onset and termination of developmental arrest may result from simple water loss from the organism's tissues, as appears to be the case for *P. vanderplanki*, or may require a more formal sequence of internal changes in response to external cues (Morris 1967; Evans and Brust 1972; Zaslavski 1996). Interestingly, in temporary waters, the cues for termination are not always the return of water to the habitat. For example, Grodhaus (1980) has shown that diapause termination in certain temporary pool chironomids is cued to falling temperatures, which forestalls reactivation occurring during false starts to the wet season in the autumn. Similar safeguards apply in other taxa. For example, in large branchiopods in highly episodic ponds, the egg hatching pattern is erratic. Although the bulk of the eggs hatch together upon return of the water, hatching is typically spread out over several days or sometimes weeks. In subtropical/desert species, low hatching percentage occurs, a reflection of the likely low chances of successful life cycle completion (Brendonck 1996).

There are also instances where, within the same species, dormancy may occur in some populations but not in others. For example, in Europe the adults of caddisfly species that live in intermittent ponds undergo an adult diapause while the ponds are dry. In Iceland, however, where these same species live in permanent ponds, there is no diapause as it would prevent completion of life cycles within a single year (Gislason 1978). Khlebovich (1996) cites several examples of populations from temporary waters in which invertebrates (crustaceans, rotifers, and ciliates) have been observed to lose their capacity for diapause when reared under fully aquatic conditions in the laboratory. Such loss is hypothesized to result from either direct mutation causing 'breakage' in the sequence of biosynthesis controlling diapause, or loss of phenotypically controlled, inactive genes in populations no longer subject to the selection pressures operating in temporary waters.

An interesting question is whether the dormancy associated with protection from drought, and more specifically its expression as diapause (including the physically protective layers surrounding the organism, as well as the biochemical processes that arrest development), has its lineage in terrestrial or aquatic organisms. We have seen that it is the insects and crustaceans that dominate the extreme end of the temporary water spectrum, that is, episodic freshwater pools. These two major groups of arthropods are believed to have had separate evolutionary lines. Although primitive arthropods likely arose from primitive segmented annelids (polychaetes) or from some common ancestor, unlike the crustaceans, insect evolution and diversification has primarily taken place on land (Williams and Feltmate 1992). One could argue, therefore, that adaptation to drought would be more difficult in crustaceans than in aquatic insects which are descended from terrestrial stock and most of which maintain some trend towards terrestrialization (see Table 5.2). However, this is not borne out by the respective occurrences of these two groups in temporary waters. Further, there is wide belief that diapause has evolved independently in arthropods and non-arthropods several times (Lees 1955), even within closely related taxa, including genera (Farner 1961). This suggests that it may be a relatively simple adaptation to acquire—perhaps because, physiologically, normal development is controlled by only a few strong hormones (Danks 1987). Interestingly, diapause may be primed by the same hormones that promote migration, the other main strategy employed by temporary water invertebrates to avoid unfavourable habitat conditions (Rankin 1974). Another point worth considering is the fact that virtually all of the crustaceans found in temporary freshwaters (isopods that colonize at the end of the pool phase are an exception) are aquatic during their entire life cycle whereas most of the insects (some aquatic beetles and hemipterans are exceptions) have an adult terrestrial stage. What consequences this may have had for the origins of dormancy in these two major invertebrate groups are largely unknown.

Fryer (1996) has put forward the notion that temporary ponds probably represent the most 'permanent' (in terms of persistence of a habitat type throughout geological time) of all freshwater habitats. Aquatic invertebrate life cycles thus have had a very long time to become adapted to even the most extreme of temporary waters. In the cases of three major invertebrate groups (Branchiopoda, Copepoda, and Ostracoda), diapause is believed to have been the key factor in their extensive adaptive radiation and ecological success—and this may be true for some non-crustacean invertebrates, too. Indeed, in the Branchiopoda, a diapausing egg stage has been universally adopted (Fryer 1996). Dormancy frequency, through many taxonomic groups, is clearly very high in the invertebrate communities of all the temporary freshwater examples examined here, but is much lower in permanent waters.

Among the chief forces driving dormancy selection are the advantages (increased food resources, greater habitat area, reduced competition, lack of predators, etc.) bestowed upon species that can survive in seasonal environments. Seasonality thus is characterized by many different factors any one or combination of which may select for dormancy. In temporary freshwaters, dormancy also may have arisen in response to achieving synchrony between an organism's life cycle and a specific habitat phase or food resource opportunity. There is strong evidence, for example, that dormancy is a key factor in not only enabling aquatic invertebrates to exploit the resource opportunities available in temporary streams and ponds (e.g. decaying plant material left over from the terrestrial phase), but also in allowing species to re-establish during the habitat phase most favourable to them (Elgmork 1980). An example of the latter might be the need to be exposed to, or avoid, a suite of interdependent factors characteristic of a particular phase (e.g. the cool, highly oxygenated, flowing water of the stream phase) or to just one key factor (e.g. loss of water, or unacceptably high pH). There remains also the tenet that dormancy represents an adaptation that has permitted the survival of slow-moving, largely defenceless, invertebrate species (such as the Anostraca and Notostraca) in temporary freshwaters by virtue of the latter being generally free of carnivorous vertebrates. In light of the evidence reviewed here, it is perhaps unreasonable to look for a single evolutionary origin of the dormancy phenomenon.

5.6 Adaptive 'strategies' of colonizing organisms

The remaining sections of this chapter deal primarily with those animals that avoid the drought phases of temporary waters by leaving the habitat. Later, when the water returns, they either recolonize the same habitat or colonize a similar one within their dispersal range. A number of such species use temporary waters exclusively as feeding and breeding grounds and spend the rest of their time in permanent water bodies.

Often, there is some confusion among the terminology associated with organism movement. Here, as the above indicates, 'dispersal' is used to denote the outward spreading of organisms, or propagules, from some point of origin or release (essentially one-way movement from one home site to another; Lincoln *et al*. 1998); 'colonization' indicates successful establishment at another site; and 'recolonization' indicates re-establishment at a site formerly occupied prior to drought.

What factors make good colonizers? According to Lewontin (1965), the species should be capable of effective dispersal, have high somatic plasticity (i.e. respond readily, in body form at least, to changes in environmental conditions), and have high interspecific competitive ability. Few species possess all these attributes to a high degree and often it is a case of finding and occupying habitats where one may be exploited. For example, many species which are poor interspecific competitors have high powers of dispersal. In this way they are able to colonize newly flooded habitats first—either annually or seasonally—rapidly produce a new generation, and then emigrate before the arrival of species that may be competitively superior but which colonize at a slower rate, or before the habitat dries up again. However, in many temporary waters the brevity of the hydroperiod may prevent much competition from occurring, and success therefore depends more on power of dispersal and plasticity of the life cycle. Once the former has enabled the species to reach the habitat, the latter may help to maintain it there.

Species that devote much of their energy to dispersal are often referred to as fugitive, or pioneer, species and animals that show this trait can be likened to the 'weed' species of plants. MacArthur and Wilson (1967) put forward the idea that natural selection takes different directions depending on whether a species is living under crowded or uncrowded conditions. Species living in uncrowded conditions, as noted earlier, are termed *r*-selected species whereas those living near the carrying capacity of their population are termed K-selected species.

All animals have the capacity for dispersal and it may take one of two forms, active or passive. The former involves the animal reaching a destination 'under its own steam', so to speak—usually by flight or crawling, whereas the latter relies upon the animal being transported by some external agent such as the wind or a larger animal. Although a greater element of chance in reaching a suitable new habitat occurs in passive dispersal/colonization, as we shall see, it is not entirely out of the control of the transportee.

5.7 Active colonization

Active seasonal migration typically necessitates substantial powers of movement, coupled with mechanisms for locating and evaluating new bodies of water. Flight is the most conspicuous medium, and this has been studied most in the Hemiptera and Coleoptera, but also in the Odonata and some Diptera. However, non-insects such as crayfishes, leeches, amphipods, amphibians, and even molluscs (Kerney 1999) have been observed to move overland between adjacent ponds. Such dispersal may be restricted to periods when the local climate is favourable (e.g. moderate to warm temperatures and high humidity in the crayfish *Orconectes rusticus*; Claussen *et al.* 2000). In general, dispersal in freshwater taxa is often difficult to study, directly, and so some episodes, even biologically significant ones, may go undetected (Bilton *et al.* 2001).

Historically, some of the earliest observations on insect colonization were made in the United Kingdom Grensted (1939), for example, erected a small (1.5 m diameter) open air canvas tank in his garden 'with no higher scientific purpose than the amusement of small children in hot weather'. Within 24 h of having been filled with tap water, it had been colonized by over one hundred water beetles representing eight different species, although *Helophorus brevipalpis* (Hydrophilidae) was the most common. Since his garden was a 'long way' from water, Grensted concluded that these species flew freely in suitable weather and then mostly in the afternoon and evening. This phenomenon was confirmed by Pearce (1939) who was 'shewn a small birdbath, in Sussex, that had been repaired and filled only a day or two previously and which could not hold more than a couple of gallons when full. It was literally swarming with *H. brevipalpis*. There must have been several hundred specimens present'. The nearest permanent water was a lake about 0.5 km away.

Timing is very important in the active colonization process. Typically, in temperate regions, adults having overwintered in some permanent body of water, disperse in early spring in search of newly formed ponds. Here, eggs are laid and the young grow quickly under conditions of plentiful food and reduced competition. Adults of this new generation mature shortly before the dry phase and fly to new overwintering habitats in a second dispersal in summer (Fernando and Galbraith 1973). Adult caddisflies of the genus *Stenophylax* are known to fly away from drying streams to hibernate in nearby caves. When flow resumes, they return to lay their eggs (Bouvet 1977). Other limnephilid adults have been observed to return, under high relative humidity, to dry pond basins where they deposit their eggs under logs or in other protected locations. Embryonic development proceeds, in the absence of water, within a protective gelatinous matrix and the larvae emerge when the pond refills (Wiggins *et al.* 1980). Batzer and Resh (1992) were able to show that all of the dominant insects (beetles, corixids, and several dipterans) living in a seasonally flooded marsh persisted only through repeated colonization from nearby permanent habitats, rather than by physiological tolerance. Boulton (1989) similarly found that most of the insects living in two temporary streams in southern Australia over-summered in nearby permanent streams. Timing is perhaps less crucial in tropical regions, where temperature changes are small, however, there may be other environmental events (e.g. availability of a particular food resource) with which a species' colonization pattern needs to become coordinated.

Although there seems to be evidence for return to the same habitat, some individuals appear to disperse more widely to lay their eggs at other locations, and, by the same token, others may fly in from distant sites. Rigorous direct evidence is, however, scarce due to lack of practicable tracking technology, but studies using mark-recapture methods, morphological variation (e.g. Fluctuating Asymetry), or DNA comparisons (e.g. Colbourne *et al.* 1998; Turner and Williams 2000), indicate some continuous interchange within at least regional metapopulations. However, using a combination of traditional Malaise traps and novel harmonic radar techniques, Briers *et al.* (2002) were able to show that dispersal distances of individual adult caddisflies (Rhyacophilidae) from small, upland streams in Wales were quite short; 90% of individuals flew less than 11 m. There was no evidence that surrounding vegetation impeded dispersal, but local air temperature and wind speed were influential. The study showed that much of the flight activity occurred during the day, and that males were the more active sex. In an extension of this study, but this time examining adult dispersal of the stonefly *Leuctra inermis* labelled with ^{15}N as nymphs, Briers *et al.* (2004) found a small number of isotopically enriched individuals at adjacent streams between 0.8 and 1.1 km away from the source population, including a headwater stream in a different river system.

In temporary lotic waters, recolonization of dry beds may be by both oviposition by returning adults, or by immature stages derived from upstream movement or downstream drift from pools and other refugia, or from upward migration from the hyporheic zone (e.g. Williams and Hynes 1976b; Delucchi 1989). Miller and Golladay (1996) found that periodic loss of water in an intermittent stream in Oklahoma favoured the establishment of spate-tolerant *Caenis* mayflies over other taxa. Moreover, summer drying affected riffle community structure during low-flow periods via: (1) limiting the establishment of flow-dependent taxa; (2) truncating the recovery of invertebrate densities after the spring spate; and (3) favouring drought-tolerant species. Recolonization in this stream was rapid, with 21 taxa present only 4 days after the return of water. Other studies have reported times in the region of 4 weeks or more (e.g. Williams 1977), and, in the case of newly created streams, in excess of 16 weeks (Williams and Hynes 1977b).

5.7.1 Bet hedging—survival and distributional legacies for the next generation

'Bet hedging' is a term first introduced by Bernoulli, a mathematical physicist, in 1738. It advocates minimizing risk by spreading it across a set of independent events, and has been variously adopted by other disciplines, ranging from economics to biology. Within the latter, it has been applied by evolutionary and behavioural ecologists to a variety of observational, experiment, and conceptual studies. Although bet hedging can occur in environments that are neither variable nor heterogeneous (there simply needs to be risk at all places and times), it tends to be more studied in variable environments (Stearns 2000). Consider a population of organisms living in a pond where juvenile survival is variable and occasionally high. Under such circumstances, neither K- nor *r*-strategies would seem optimal: *r*-selection traits may result in reproduction coinciding with poor environmental conditions and thus loss of the next generation, whereas K-selected traits would not maximize offspring numbers. We noted, earlier, in this chapter that in reality many species may not be entirely *r*- or K-selected, and in this case, the species' strategy might best be one of an iteroparous K-type.

Bet-hedging is a strategy employed by many inhabitants of temporary waters, and again, as noted earlier, is typically seen in hatching responses that ensure that not all juveniles emerge before the hydroperiod has become properly established. For example, in a study of bed materials collected from a dry pond in Israel, Schwartz *et al.* (2002) found a variety of rehydration responses by the crustaceans present. Conchostracans hatched most readily and showed the earliest peak hatch. In contrast, cladoceran hatching was highly variable, with some propagules hatching more than 4 weeks after wetting. Copepods employed an intermediate strategy. These authors concluded that bet-hedging strategies might have evolved, (limited perhaps by particular physiological plasticity of the species) in response to selection pressure from intra- and inter-specific competition and predation. In a study of a temporary pool in Morocco, Thiéry (1997) observed that females of *Triops numidicus* (Notostraca) laid their, slightly sticky, eggs along the periphery of the pool. Here, the eggs remained out of water when limited rainfall only partially filled the pool. This was deemed a strategy to ensure hatching only when there was sufficient water present, and hence length of hydroperiod, to allow completion of the life cycle. In this same study, Thiéry provided examples of the maximum densities of branchiopod eggs that can occur in temporary water sediments (per 100 cm^2): *Chirocephalus diaphanus*: 1,700 eggs; *Lepidurus apus*: 1,150; *T. numidicus*: 285; *Branchipus schaefferi*: 238; *Leptestheria mayeti*: 39; and *Tanymastigites perrieri*: 28.

Brendonck and Riddoch (1999) demonstrated a novel example of bet-hedging in the anostracan *Branchipodopsis wolfi* in South Africa. This species lives in short-lived rock pools, exhibits rapid maturation, is highly fecund, and practices staggered egg hatching. However, it also produces two types of resting egg: both have a shell with polygonal sculpturing, but some are sticky and collect debris as soon as they are deposited, whereas others remain clean (Figure 5.12). The sticky egg type is believed to be less liable to wind dispersal than the smooth, and together, they appear to represent a dimorphic dispersal strategy. The authors concluded that production of egg types with different dispersal potentials alongside the generation of multi-year egg banks make *B. wolfi* an extreme bet-hedger with the means to respond to unpredictable habitat availability/suitability in both time and space.

Kam *et al.* (1998) showed that, in response to intraspecific competition, females of the arboreal frog, *Chirixalus eiffingeri*, are capable of strongly influencing the growth, development, and survivorship of their tadpoles. The latter, which

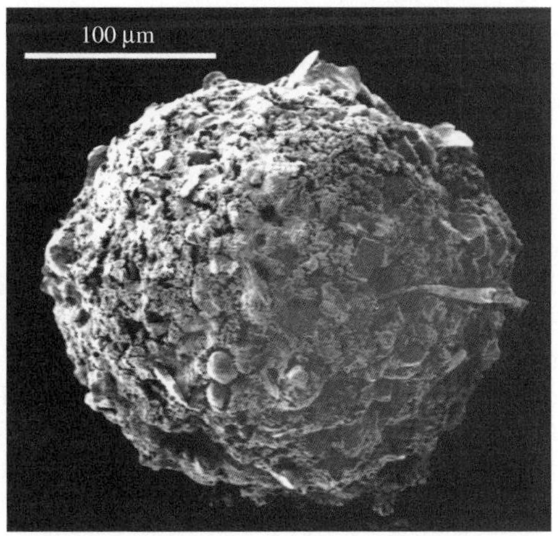

Figure 5.12 The two egg types in *B. wolfi* (Anostraca): smooth (left) and sticky (right) (reproduced with permission of L. Brendonck).

develop in water retained by bamboo stumps, are sustained by trophic eggs provided by the female. As tadpole densities increased, females increased the rate of food provision according to their own energy reserves. Body size at metamorphosis was density-dependent, but the time to reach metamorphosis was not—a property different from that observed in pond-breeding species. Another density-dependent, bet-hedging response has been observed in ovipositing females of the mosquito *Aedes aegypti*, in the laboratory. Females consistently chose to lay their eggs in water in which fed larvae had been reared. Water from larvae starved for from 3 to 5 days proved to be progressively less attractive, and even repellent, suggesting that females were capable of assessing the suitability of larval habitat and distributing their eggs accordingly (Zahiri *et al*. 1997).

In a study of the oviposition flight patterns of four species of dipteran in tropical rainpools, McLachlan and Ladle (2001) noted that the larvae of each occurred at high densities, on their own, and were highly predictably in particular pools. A species' loyalty to a specific pool proved to be absolutely consistent from year to year, with ovipositing females always laying their eggs in certain pools but not others. The authors concluded that pool choice was based on a cue linked to water depth, made by females, rather than by interspecific competition among larvae.

Other examples of bet-hedging were given in Sections 5.5.2 and 5.5.3. However, despite such a suite of safeguards, there is increasing evidence that populations in individual, small freshwater habitats frequently fail, to perhaps become re-established in a subsequent year by new colonists from a regional metapopulation. Notably, in highly detailed, long-term emergence records from a small stream in Germany, only the dominant species were reliably detected every year (Wagner and Gathmann 1997). A similar conclusion was reached during a taxonomically rigorous, 3-year-study of small, coldwater springs in Ontario (Gathmann and Williams, unpublished data), but with an important additional implication: springs may be regarded as habitat islands in a hostile environment for all crenobiont (obligate) and, to a lesser degree, crenophilous (facultative) species, and because of the rather limited size of most springs, the absence of a particular species in the emergence record from a given year may well reflect a preceding extinction of the local population. Significantly, Stagliano *et al*. (1998) recorded an average of only 25% of emerging insects returning to the water surface in a small wetland pond in the southeastern United States, the rest were presumed either to have been consumed by terrestrial predators or to have emigrated to other systems. Importantly, the percentage return varied dramatically among species (e.g. close to 100% in the chironomids *Dicrotendipes* and *Tanypus*, but only around 6% in the chironomid *Ablabesmyia*), and so a scenario in which an emerging species fails to re-establish itself in a small pond or spring in any given year seems very plausible. A further example comes from an intriguing 2-year-study by Nurnberger (1996) which examined the local dynamics and dispersal of the gyrinid beetle *Dineutus assimilis* in 51 ponds located within a 60 km^2 area. Populations occurred in 31 of these ponds but varied dramatically between years, and included nine extinction events. This species proved to be a highly successful colonizer of newly available sites (nine colonization events also were observed), with dispersal distances (determined from marked individuals) ranging from 100 m to at least 20 km. Adults typically dispersed after diapause, and there was considerable variation in reproductive success—likely the result of local pond conditions. Immigration rates varied widely within a season.

5.7.2 Flight periodicity and dynamics

Aerial colonization of temporary waters is largely limited to the spring and summer months in temperate latitudes, although it may occur all year round in the tropics if not influenced by other climatic factors, such as monsoons. The best studied groups are the Hemiptera and Coleoptera. Macan (1939) found that six species of Corixidae (waterboatmen) were commonly on the wing in Cambridgeshire and that colonization of water bodies was frequent. He showed that a succession of species occurred depending on the percentage of organic matter present in the ponds, and he hypothesized that high mobility was necessary to maximize niche separation. Knowlton (1951) noted a flight of corixids from one pond at 8:40 p.m. and their landing at another pond at 10:00 p.m. Temporary pond species migrated more frequently than those living in permanent waters, and immature specimens seemed more inclined to fly during dispersal times.

Fernando (1958) working on ponds in Oxfordshire showed that emigration of aquatic beetles from one pond and their subsequent colonization of others occurred seasonally. He found that *H. brevipalpis* had two distinct dispersals in Britain, one in the spring (April–May) and the other from the middle of June to the end of August—the size of the second being much greater than that of the first (Figure 5.13). Those females that took part in the spring dispersal carried mature eggs which were laid in the newly colonized habitats, whereas the late summer flight was one of adult dispersal. Similar flight patterns were seen in two (*Dytiscus marginalis* and *Acilius sulcatus*) of the six dytiscid species studied in temporary ponds in Cheshire, England by Davy-Bowker (2002). Working in Canada, Fernando again found similar dispersals for *H. orientalis* and another beetle, *A. limbata*, but their maximum seasonal abundances never coincided—when *Helophorus* ceased its colonization flight in early July, *Anacaena* began its summer dispersal. With the decline in flight of *Anacaena* at the end of August, there was another flight by *Helophorus*. These staggered flights were probably the result of different responses to climatic factors and may have resulted in reduced competition between the two species for the same habitat. In addition, a bimodal, daily flight pattern with approximately equal morning and evening peaks was evident for these two beetle species. Landin (1968) observed a similar pattern for *H. brevipalpis* in Sweden, but, as Figure 5.14 shows, the number of beetles captured during the evening was 3–4 times that captured during the morning. Temperature may determine when an insect species flies through fixed values constituting thresholds for this activity. *H. brevipalpis* has a low temperature threshold of between 11 and 15°C, and there is some evidence for an upper threshold of around 25°C (for the Canadian species the range is

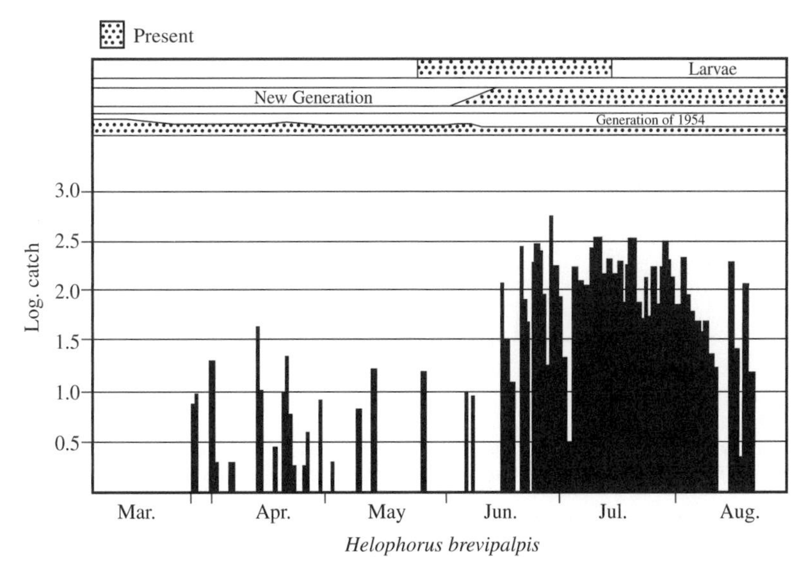

Figure 5.13 Logarithm of the catches of *H. brevipalpis* in a field experiment in Oxfordshire during 1955. Catches were made using six artificial habitats (4 × 4 ft), three glass traps (5 × 3 ft), and three glass traps (2 × 3 ft). A daily census was made from March to September (redrawn from Fernando 1958).

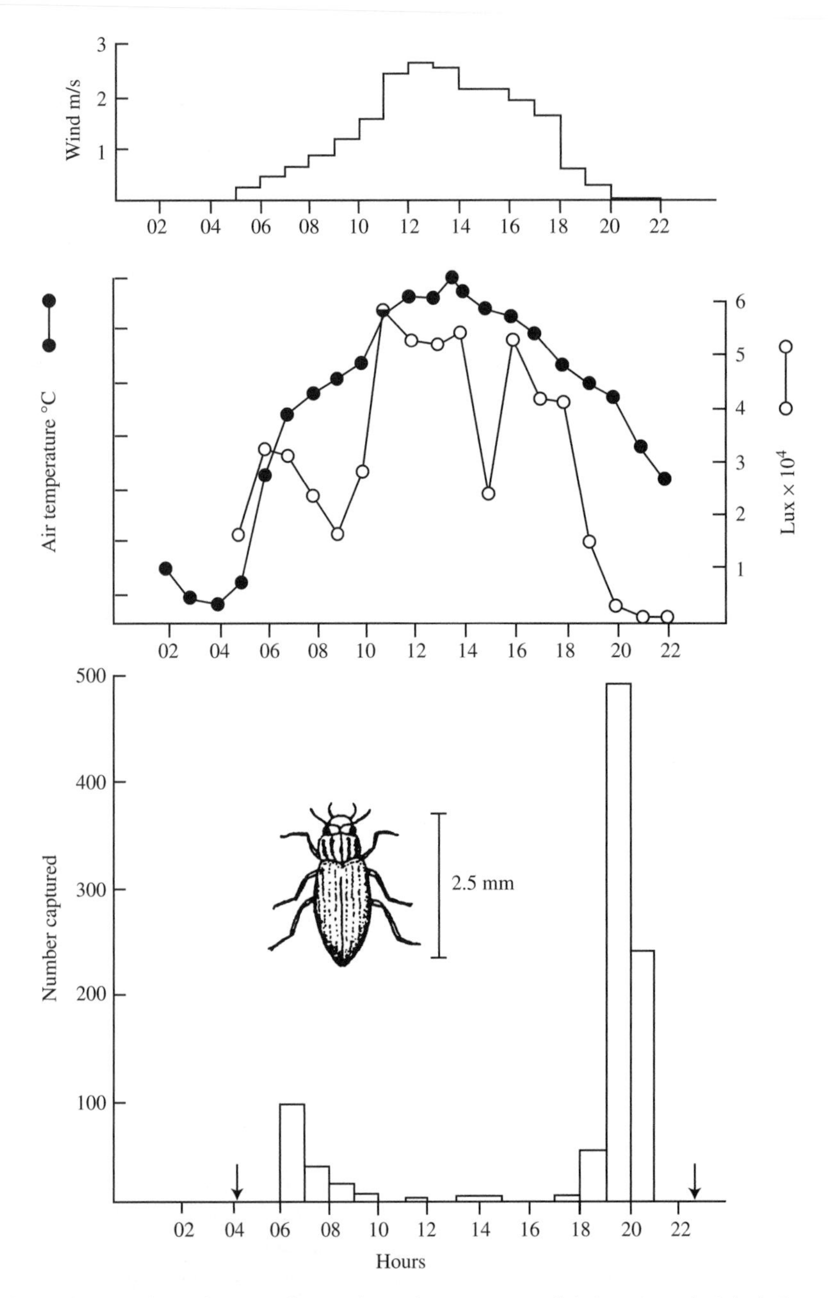

Figure 5.14 Flight periodicity in *H. brevipalpis*, in Sweden, together with air temperature, light intensity, and wind velocity on the 22 July, 1965. Wind velocity was measured from 5 a.m. to 10 p.m.; sunrise was at 3:12 a.m., and sunset at 8:34 p.m. (redrawn from Landin 1968).

15–27°C). Beament (1961) has shown that the permeability to water of the cuticles of the water beetles *Dytiscus* and *Gyrinus* increases six-fold above a cuticle surface temperature of 23°C. If similar values apply to *Helophorus* and *Anacaena* (and a great many aquatic insects have cuticles which change permeability rapidly at low temperatures) then it is likely that the reason for diurnal periodicity in flight is that summer midday temperatures exceed that which allows normal cuticle permeability, thus subjecting the beetle to dramatic loss of water from its tissues.

Nilsson and Svensson (1995) found that the immigration rate of dytiscid adults in the Swedish Arctic was significantly higher in ponds located in clearings than those in shaded, spruce-swamp forest. Newly established, artificial ponds, were colonized by dytiscids only if they were large (1.6 m^2), and the 10 species included the flightless *Hydroporus melanarius*.

5.7.3 Flight initiation and termination

Many factors control the pattern of insect colonization of temporary waters. Fernando (1958) believed colonization in aquatic beetles to be the end product of a series of steps in each individual's behaviour. These steps are: (1) dispersal, which provides the basis for colonization; (2) location of habitat; and (3) selection of a habitat. The steps are products of evolution caused by two types of factor, *proximate* factors, like temperature and light, and *ultimate* factors, like food. Beetles respond directly to proximate factors by adaptations of physiology and behaviour, thus water- or air-temperature change, change in light intensity and duration, rain, or a combination of these factors may cause them to leave the water and take to the air from where they may locate a suitable new water body. Davy-Bowker (2002) observed that *A. bipustulatus*, unlike some other beetle species, did not leave ponds in England as they dried up, but remained in terrestrial vegetation in the damp pond basins for several months. They flew away only when the basins dried out completely. It is possible that certain internal physiological states, such as ovarian development in the female, or hunger may initiate flight. For example, Landin (1980) found that, in the spring, flying *H. brevipalpis* females have larger oocytes than non-flying ones, and that the guts of fliers are invariably empty. Ultimate factors, such as food and substrate, may not be directly connected with the proximate factors but the insect may often anticipate their being favourable for itself or its offspring because of the evolution of seasonal dispersal.

Flight terminates with pond/stream selection and either colonization by the adult insect itself or egg laying by the female to establish a new generation. The actual process of habitat selection may involve one or more of the insect's senses. Popham (1953), for example, showed that sight was important in corixids, as flying adults orient themselves at a set angle to the incident light and thus the reflection of light from a water surface is sufficient for them to home-in on. This characteristic explains why many species of aquatic bugs, beetles, and dragonflies swoop down and land on the shimmering hot surfaces of asphalt roads, or the surfaces of freshly polished automobiles. Kriska *et al.* (1998) found that swarming mayflies are highly attracted to the strongly horizontally polarized light reflected from the dry surface of asphalt roads, which mimic a highly polarized water surface—the darker and smoother the asphalt, the higher the attraction. Additional features of roads, such as their conspicuous and elongated form, especially where there is no canopy, together with heat emanation, may enhance the attraction. In a detailed study of the polarized pattern of freshwater habitats via video polarimetry, Horvath and Varju (1997) showed that the patterns of small water bodies are highly variable in different spectral ranges, according to ambient illumination. For example, under a clear sky, and in the visible range of the spectrum, calm water surfaces reflecting light from the sky are most strongly polarized in the blue range. However, under an overcast sky radiating diffuse white light, small pools are characterized by a high level of horizontal polarization in most spectral ranges. These differences have consequences for water-seeking insects which rely on highly horizontally polarized light during habitat selection, and thus are not attracted by water bodies that reflect vertically polarized light, or by horizontally polarized light with a low degree of polarization. That some aquatic insects operate in the short-wavelength (ultraviolet or blue) region of the spectrum (an adaptation to the high degree of polarization of reflected light in this spectral range) has been demonstrated by Schwind (1995). In contrast, some other species appear to operate in long-wavelength regions that seem better adapted to the spectral characteristics of the underwater visual environment. Significantly, eutrophic ponds exhibit different polarization signatures, from, for example, dark-water ponds.

Pajunen and Jansson (1969) observed that adult corixids could discriminate pool size, moving from shallow pools in which they lived in the spring, to pools deep enough for over-wintering in the late autumn. Water movement, shade, colour, alkalinity, salinity, and emergent vegetation all have been implicated in the habitat selection/oviposition stimulation processes of temporary water mosquitoes. An ability to detect hydroperiod characteristics has been demonstrated in the giant water bug, *Abedus herbeti*, although its response is to too much water rather than too little (Lytle 1999). This species abandons its desert stream habitats prior to flash floods by detecting increased rainfall prior to flood events. Adults leave after perceiving a threshold of 8 min of torrential rainfall (juveniles respond after 29 min), and move uphill away from the water, stopping in sheltered areas away from the active channel. Distances travelled are as much as 23 m, and return may be within 24 h. The perception cue seems to be impact of rain on the stream surface, rather than chemical or physical cues associated with rainwater.

One highly understudied aspect of the dispersal-colonization process concerns the fate of males. However, Benjamin and Bradshaw (1994) examined the effects of body size and flight activity on male reproductive success in *W. smithii*, the pitcherplant mosquito. They found that small males were more likely to experience reproductive failure than large males—although among males that did reproduce, there was no difference in lifespan, lifetime offspring sired, or cost of reproduction with size. Small size was associated with early emergence to take advantage of a more plentiful supply of unmated females. Cage experiments showed that males, regardless of size, allowed a larger flying space in which to feed and mate experienced decreased longevity, sired fewer offspring, and showed an increase in the cost of reproduction compared with males in small cages, suggesting an energetic trade-off between flight and reproductive success. In males of a small, isolated population of the damselfly *Calopteryx haemorrhoidalis*, Beukema (2002) noted that site fidelity increased with increasing sexual maturity, and culminated in the establishment of distinct territories. So strong was this trait that 50% of artificially displaced individuals returned to their original territories, typically on the same day. Clearly, such fidelity limits disperal opportunities to pre-reproductive males. Site fidelity in females was similar to that of immature males.

Most remarkable is the process by which many temporary water insect species and amphibians select and lay eggs in ponds and streams when they are dry. How does the female predict that the dry bed upon which she lays her eggs will later develop into an aquatic habitat suitable for sustaining aquatic larvae? The limited evidence available suggests that colonizing females may be attracted by some special feature of the bed. For example, a moisture gradient emanating from the soil-water, or the scent of rotting aquatic vegetation have been suggested as attracting gravid mosquitoes. The distribution of adult tipulids has been shown to be influenced by factors affecting local evaporation rates (relative humidity, air movement, and insolation) from woodland floodplain soils (Merritt and Lawson 1981). Corbet (1963) thought that female dragonflies might be attracted by a specific preference for a plant such as *Typha* which only grows where the ground is susceptible to seasonal flooding. The aphid *Hyalopterus pruni* has been shown to use colour in selecting the reed species upon which it feeds (Moericke 1969). Aphids, in general, have been prime targets for research on flight termination mechanisms in insects. Findings indicate that 'green leaf volatiles'—the alcohols and aldehydes released by all plants as part of their general metabolism, together with other volatiles characteristic of a particular species—are used as ultimate factors in habitat/resource selection. Although many such compounds contribute to the identification signature of each plant, small molecular weight compounds, being the most volatile, are more likely to be involved in long-distance attraction (Compton 2002). Anderson and Bromley (1987) concluded that the terpenes characteristic of trees and shrubs may be the primary cues used by aphids when returning to wooded areas in autumn. Detection of these chemical is via antennal sensilla (Keil 1999), however, their precise use in homing-in on resources is likely to vary with species. For example, for some, visual cues may be important in guiding an insect to an area, but once there, volatiles may take over to provide more precise landing.

Several researchers have examined the effects of parasites on dispersing adult insects. For example, in each of the following cases, the flight capabilities (e.g. distance flown, total flight time) were compromised: *Aedes togoi* infected with the filarial worm *Brugia malaya*; *Ae. aegypti* infected with *B. pahangi*, and *Anopheles stephensi* infected with the malarial parasite *Plasmodium cynomolgi* (Husain and Kershaw 1971). As well, the presence of parasites frequently shortens the adult life span, as has been shown for *Ae. aegypti*, *Ae. sollicitans*, and *Ae. trivittatus* infected with the nematode *Dirofilaria immitis* (Christensen 1978). Compromise is believed to be due largely to depletion of host nutrients, and/or damage to body parts, such as flight muscles or the gut. In a study of *Aedes sierrensis* infected with the ciliate *Lambornella clarki*, Yee and Anderson (1995) concluded that the parasite primarily impacted the lipid resources required for adult survival rather than the glycogen resources needed for flight—which may be adaptive for the parasite if host flight is more important than host longevity. Dwyer and Hails (2002) have used mathematical models to examine how insect dispersal affects the population dynamics of insect-baculovirus interactions, including how dispersal affects the possibility that genetically engineered virus strains (such as might be used to control mosquito populations) might outcompete wild-type strains. A possible mechanisms might be through the novel pathogen reducing its host's dispersal distance (through gradual paralysis of its muscles) thereby allowing it to overlap the region in which the wild type exists, and thus invade natural populations.

5.8 Passive colonization

Unlike species that are capable of controlled, active colonization, those that rely on passive means are subject to a much greater degree of chance in populating new habitats. Although adult and larval stages are sometimes involved, it is more common that a resting, reproductive body acts as the transportee. A large proportion of freshwater species has such a stage in their life cycles, which is not the case for marine species—probably a consequence of the much greater persistence of marine basins. Talling (1951) cites examples of these stages: algal spores, sponge gemmules, polyzoan statoblasts, and the resting eggs of rotifers, fairy shrimp, cladocerans, copepods, and flatworms. Such propagules possess both extraordinary resilience to desiccation and temperature extremes, and mechanisms that increase the chances of them attaching to animal vectors. Talling also outlined the factors governing the chance element in colonizing ponds (Table 5.7).

Agents of passive dispersal can be assigned to two broad groups: the first is *abiotic* in nature, the second, *biotic*: Very light organisms can be transported by air currents, for example the parachute seeds of *Typha*, and strong winds acting on the drying beds of temporary waters can carry away viable algal cells and cladoceran ephippia. Rzoska (1984) proposed that the hot winds of the Sahara Desert sweeping over and picking up the dry debris in shallow pond basins, are responsible for spreading the resistant stages of many taxa, particularly those of the Notostraca, Conchostraca, Anostraca, Cladocera, and Copepoda (see also Brendonck and Riddoch 1999). Maguire (1963) examined the colonization of small, water-filled jars by these aerial forms and concluded that: (1) the number of different organisms per jar decreased with increased height above the ground; (2) they similarly decreased with increased distance from the source pond; and (3) levelling off of the numbers of different organisms was approached after 6 weeks near the ground but after only 2 weeks in the jars kept just over 1 m above the ground. Disseminules frequently were washed into the jars from nearby vegetation and soil

Table 5.7 Factors governing the chance element in the colonization of ponds and streams (based on Talling 1951)

A. Factors governing the dispersal of reproductive bodies
   1. Intrinsic properties of the organism
      (a) Formation of drought-resistant reproductive bodies (e.g. cysts, ephippial eggs, reduction bodies)
      (b) Capacity for active overland dispersal (minimal in passive forms)
      (c) Frequency of occurrence in nature (density and patchiness)
   2. External agencies of dispersal
      (a) Wind
      (b) Inflowing stream and water drainage (includes downstream 'drift')
      (c) Animal vectors (e.g. waterfowl, adult insects, aquatic mammals)
   3. Local factors
      (a) Age of waterbody
      (b) Area of waterbody
      (c) Distance from similar habitats

B. Factors governing successful colonization by reproductive bodies
   1. Intrinsic powers of multiplication and competitive ability
   2. 'Openness' of the habitat (accessibility in terms of the degree of cover)
   3. Destruction by other organisms (e.g. predation, fungal/microbial attack)

surfaces by rain. Temporary waters, such as those on floodplains, that are periodically connected to streams may be colonized by invertebrates such as midge larvae, snails, and oligochaetes that are rafted in on algal mats which have been sloughed off the bed. Few data are available on the distances travelled by passive propagules. In some cases, prevailing winds may disperse them considerable distances, on the other hand, Pajunen (1986) has estimated the mean dispersal distances of daphniid ephippia to reach unoccupied rockpools, from the closest source populations, to be between 5 and 9 m. This suggests that colonization may, in some species, be a slow process, with dispersal only proceeding in short jumps.

Charles Darwin was among the first to observe passive colonization through biotic agents (sometimes termed 'phoresy') in the form of organisms attached to ducks and water beetles. Since then there have been numerous records of a variety of taxa found attached to flying waterfowl. For example, Proctor (1964) recovered viable eggs of fairy shrimp, tadpole shrimp, clam shrimp, cladocerans, and ostracods, together with algae, from the lower digestive tracts of wild ducks; and amphipods have been seen clinging to the feathers of dead mallards that were some distance from, and had been out of water for several hours. An interesting correlation has been derived between the distribution of sibling species of the brine shrimp *Artemia* and bird migration routes in North America (Bowen *et al.* 1981). In Europe, the distribution of populations of the bryozoan *Cristatella mucedo* has been similarly linked to a commonly used waterfowl migration route, leading Freeland *et al.* (2000) to hypothesize that waterfowl may well function as vectors linking sub-populations of this species within a northwestern Europe metapopulation. Further, again in North America, the distributions of separate genetic lineages of *Daphnia laevis* are known to be associated with three well-defined waterfowl flyways (Taylor *et al.* 1998). Bilton *et al.* (2001) have drawn attention to the range of adaptations that increase the chances of propagule attachment to animal vectors. These include sticky surfaces, hooks, and spines (Figure 5.15(a,b)), together with buoyancy to avoid deposition in bottom sediments, and also accumulation of large numbers at times that coincide with peak waterfowl migration.

Smaller hosts also may contribute to the passive dispersal/colonization of certain groups: algae, protozoans, and aquatic fungi have been collected from flying adult hemipterans, caddisflies,

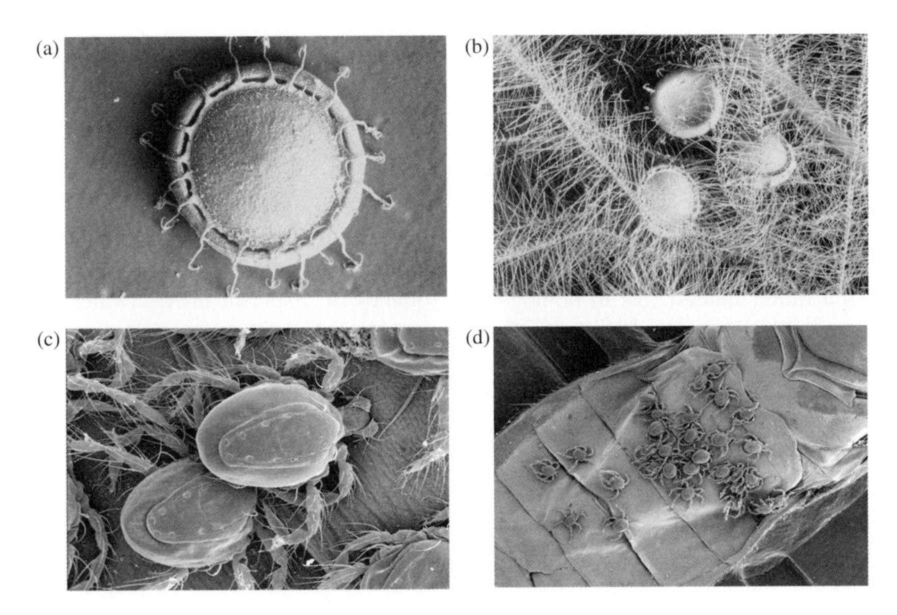

Figure 5.15 Examples of passive dispersal in freshwater invertebrates: (a) statoblast of *Cristatella mucedo* (Bryozoa), which contains gas-filled cells and has hooks for attachment to feathers (b) and fur; (c) and (d) attachment of the water mite *Eylais* sp. (Acari) to the body of the water boatman *Sigara falleni* (reproduced with permission of D.T. Bilton).

craneflies, chironomids, mosquitoes, aquatic beetles, and odonates. Fryer (1974) found bivalve molluscs attached to the leg setae of corixids, and ostracods often attach to notonectids. Viable eggs of *A. salina* have been recovered from the faeces of crayfish species with which they coexist, and clams may sometimes be found clamped tightly on the tips of crayfish walking legs. Schneider and McDevit (2002) found that propagules of four species of the algal genus *Vaucheria* are able to survive passage through earthworm guts, and thus may subsequently be wafted into the air as dust as burrow castings dry. From data derived from transplant experiments and gut analyses, Bohonak and Whiteman (1999) estimated that the metamorphic stage of the salamander *Ambystoma tigrinum nebulosum* is capable of dispersing thousands of eggs of the fairy shrimp *Branchinecta coloradensis* among ponds, annually. Maguire and Belk (1967) have observed the transfer of *Paramecium* and other small aquatic organisms between the water-filled bracts of *Heliconia* flowers by terrestrial snails of the genus *Caracolus*.

Among the best examples of habitual passive colonizers are water mites. The larvae are parasitic on insects and have good organs of attachment in the form of specially modified claws and mouthparts. Favoured hosts are the Hemiptera, especially water striders and water boatmen (Figure 5.15(c,d)), but mosquitoes and midges are parasitized too, frequently as pupae. As the pupa rises to the water surface to emerge into the adult, the mites very quickly transfer their hold from the sloughed pupal skin to that of the imago and are thus carried away from the pond by the insect. By selecting host species specific to temporary waters, the mites are virtually assured of successful transference to suitable new habitat. Lanciani (1970) studied 20 species of *Eylais* (Eylaidae) from small, shallow ponds in New York State and found that extensive resource partitioning occurred. Coexistence of sympatric species was found to be possible as a result of these mites exploiting host resources in different ways—typically, they parasitize a range of beetle and hemipteran species, but when the same host species is shared the mites remain segregated through differences in attachment site, in habitat, and season of parasitism. The study found that 11 species of hemipteran and 30 species of beetle were parasitized by these mites. In an extensive study of the effects of insect-mediated dispersal on the genetic structure of postglacial water mite populations in North America, Bohonak (1999) concluded that dispersal probably plays a rather limited role in homogenizing populations; allozyme population structure seems more strongly dependent on historical patterns. However, in a subsequent study, Bohonak *et al.* (2004) found that *Arrenurus angustilimbatus* (in which the larvae parasitize and disperse on adult mosquitoes) exhibits a broader and more continuous geographic distribution than the related *A. rufopyriformis* (which does not make use of dispersal hosts). Further, populations of the latter were more morphologically divergent than the former, and allozyme heterogeneity was higher and population differentiation lower in *A. angustilimbatus*.

In their review of dispersal in freshwater invertebrates, Bilton *et al.* (2001) highlight two important aspects of the populations of species that employ passive mechanisms. First, they examined the role of propagule banks in sediments—such as have been found for rotifers, cladocerans, and copepods, and argue that these may be essential to allowing species temporal escape from unfavourable habitat conditions (for up to 332 years in the case of *Diaptomus sanguineus*; Hairston *et al.* 1995). In other words, dispersal through time may be just as crucial as dispersal through space to the long-term survival of metapopulations. Okamura and Freeland (2002) have further suggested that temporal gene flow from hatching propagules retained in pond sediments may often be of greater significance to population genetic structure than spatial gene flow. Second, Bilton *et al.* point to the utility of modern molecular markers in studying the frequency of dispersal in freshwater invertebrates—for example, in the case of distinguishing between the extent to which population structure may have resulted from ongoing gene flow versus historical events, such as fragmentation and range expansion (e.g. after glacial retreat).

In a comparison of taxa that disperse by active (Coleoptera) versus passive (microcrustaceans) means, Rundle *et al.* (2002) found that beetle species richness was positively correlated with pond permanence and maximum area, whereas crustacean richness was relatively equal across sites, and did not correlate with the environmental variables measured—although a high proportion (25%) of species were more numerous in smaller, more temporary habitats. This was despite the fact that both groups appeared to exhibit the same degree of dispersal. The authors suggested that, at small spatial scales, beetles, which are capable of multiple dispersal events, are more likely to colonize larger and more permanent ponds. In the case of microcrustaceans, the influence of dispersal in driving between-site variation in species composition may well be over-ridden by other traits, especially those associated with survival in temporary waters.

CHAPTER 6

Community dynamics

6.1 Introduction

Biological communities, in general, are sometimes difficult to define, as: (1) several different parameters may be used (e.g. taxonomic, spatial, or functional—especially trophic); and (2) their boundaries, especially in a spatial context, are often difficult to delineate. Species interaction, however, is a requirement of the definition. Fortunately, temporary water communities frequently have quite clear spatial boundaries marked by the transition between water and land (e.g. the water contained within a pitcher plant, or watering trough), although the edges of, for example, episodic ponds and intermittent streams may represent ecotones where two communities (aquatic and terrestrial) merge. There is also the problem of within habitat changes over time. Should the community of an intermittent pond, for example, be defined only by those species present during the hydroperiod, or should it include semi-terrestrial, and even terrestrial species of plants and animals that inhabit the drying bed? Definition using strictly taxonomic criteria might suggest that two or more distinct communities occupy these two habitat phases. However, trophic criteria would point to a mutual nutrient supply and its processing by two groups of species sharing the same habitat but at different times. Such groups could just as well be considered to be part of a single community.

Community ecologists sometimes compartmentalize local associations of interacting species into smaller ecological units termed 'guilds'. Within a guild, interactions are strong, but between guilds the interactions are weaker (Pimm 1982; Cohen 1989). Schluter and Ricklefs (1993) advocate that more detailed appraisals of species niche relationships and food webs will enable ecologists to relate local interactions to the regulation of diversity and to the organization of community structure. Unfortunately, species interactions within many temporary waters are still insufficiently known to be able to apply both community subdivision and niche definition with confidence. What does seem intuitive, however, is that many temporary water communities have both *spatial* and *temporal* dimensions—the latter shown not only by the wet and dry guilds/assemblages discussed above, but also by the strong succession of species that occurs throughout both phases.

The currency of community ecology is largely a measure of how many species are present and in what abundances. Several descriptors have been proposed, but the most commonly adopted are *species richness* (the total number of species present) and *species diversity* (which combines the species total with a measure of relative abundance, and for which there are a number of widely used indices, for example Shannon-Weaver). In addition, *species abundance relationships* are graphical means of expressing community characteristics, as are classification and ordination techniques that attempt to describe *species composition*. Much of contemporary community ecology centres around the sources of species diversity within and among natural communites, and many explanations and models vie for supremacy. Broken down into equilibrium and non-equilibrium camps, these include, for the former: niche diversification, compensatory mortality, and intransitive competitive solutions; and, for the latter: explanations that invoke gradually changing environments, lottery models, and the intermediate disturbance hypothesis (Morin 1999). Although the datasets are somewhat limited for temporary water communities, there seems to be some evidence, for example, to support higher species numbers in waters subject to intermediate disturbances: crustacean richness was highest in waters with a hydroperiod length of between 150 and 250 days, compared with shorter and longer times (Table 3.2); similarly invertebrate richness proved to be higher in an intermittent pond than in either a stable coldwater spring, or episodic rainwater pools (Table 5.6). Further, in a study of chironomid assemblages in relation to water level fluctuations in the Danube, Reckendorfer *et al.* (1996) found that species richness and diversity were highest at intermediate levels. While such temporary water findings lend provisional support for some models of species diversity (in this case the intermediate disturbance hypothesis), more studies in replicate habitats, greater taxonomic resolution, and manipulative experiments are required before generalizations can be made.

Part of the controversy around equilibrium and non-equilibrium explanations for diversity is due to uncertainty as to whether many natural communities are close to or far from equilibrium, and this is particularly true for temporary waters. Again, the available information is sparse, however, Williams and Hynes (1977) provided some data from an examination of the colonization and extinction rates of invertebrates entering a newly created stream channel in Ontario. Stability appeared to be approaching after 109 days of flow (colonization rate = 0.18 species per day; extinction rate = 0.1 species per day), but was disrupted by a pollution episode. At the end of 373 days the colonization rate exceeded the extinction rate by five times (0.15 and 0.03 species per day, respectively). Two explanations were offered: (1) that the MacArthur–Wilson equilibrium model (1963) might not apply as well to habitats that are seasonally invaded by recurring temporary inhabitants (see Schoener 1974); or (2) that the stream community was still not a stable one. In this particular stream, water depth was shallow, and flow was minimal in summer, when water temperatures rose. These conditions attracted many species, especially aquatic beetles, dipterans, and hemipterans, known to show dispersal flights to shallow, warm waters for feeding and breeding—this produced a rise in the colonization rate. After maturing, the new generation would have left for overwintering habitats, producing an apparent increase in the exinction rate for the community. Such intermittent use of temporary aquatic habitats may negate the application of general models to the structure of their communities.

To a large extent, many of the *structural* aspects of temporary water communities have already been covered in Chapter 4—The Biota. Conclusions from that review indicate that many physicochemical factors strongly influence community structure, often directly, but also indirectly through trophic and successional processes. Among these factors, habitat duration (length of the hydroperiod) emerges as being perhaps the most important. For example, in a study of the relative influences on the invertebrate community of a Sonoran Desert stream, the dry phase was deemed to be more important than spates (Boulton *et al.* 1992), although floods may increase in influence as intermittency becomes less prolonged (Poff and Ward 1989). Erman and Erman (1995) showed a clear relationship between increasing variation in spring discharge in the Sierra Nevada, between pre-drought and drought conditions, and decreasing caddisfly richness (Figure 6.1). Schneider and Frost (1996) concluded that the processes structuring temporary pond communities in Wisconsin were hierarchically organized by habitat duration. Schneider (1999) produced a graphical model to illustrate the effects of hydroperiod on community organization in these ponds (Figure 6.2). It proposes that whereas predation plays an important role in structuring the communities of long hydroperiod ponds, this is not the case in ponds with short hydroperiods.

From a study of the changes in diversity of marine, soft-sediment infaunal communities ranging across a gradient of decreasing environmental 'stress' (shallow to deep water), Sanders (1968) proposed that where environmental conditions are severe and unpredictable, adaptations are primarily

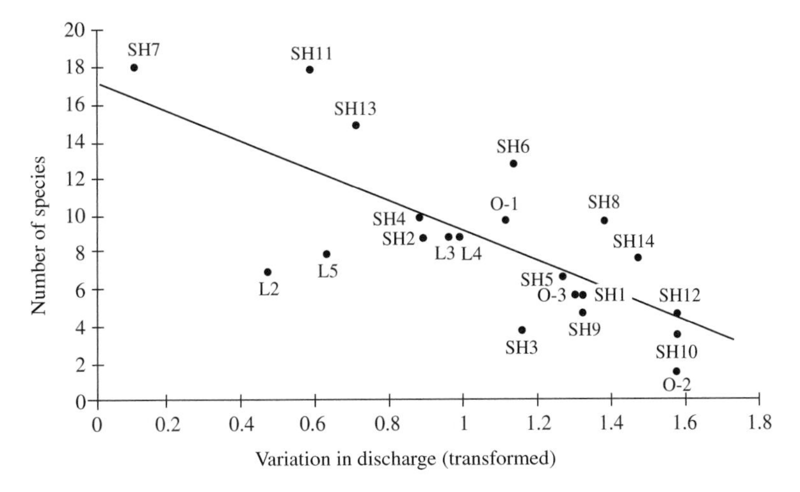

Figure 6.1 Relationship between increasing variation in spring discharge in the Sierra Nevada, between pre-drought and drought conditions, and decreasing caddisfly richness (redrawn from Erman and Erman 1995).

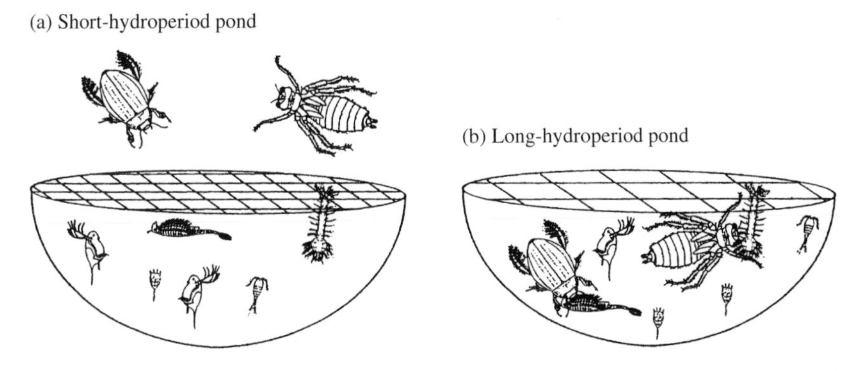

Figure 6.2 A model of community organization in Wisconsin snowmelt ponds. The grid over the pond symbolizes hydroperiod length: (a) has a fine mesh (short hydroperiod) and only allows establishment of species with rapid growth/development, or desiccation-resistant stages in their life cycles. Taxa present include rotifers, fairy shrimp, cladocerans, copepods, and mosquitoes. Excluded taxa, such as predaceous beetles and odonates, have life cycles that are too long to be completed in the time available. Although the cladocerans may prey on the rotifers, in general, predation plays a minor role in structuring the community; (b) has a coarser mesh (longer hydroperiod) which allows colonization of larger predators that feed on many of the other residents, thus reducing their populations, or even excluding them altogether. Predation in this second type of pond plays a major role in structuring the community (redrawn from Schneider 1999).

to the physical environment and such communities are physically controlled. Conversely, where environmental conditions are benign and predictable, adaptations are primarily directed at other species so as to optimize biological interactions. Menge and Sutherland (1976) criticized this view because of its failure to recognize: (1) that animals are usually physiologically well-adapted to their environments and thus are rarely stressed by it; and (2) that all communities have interactions that occur between (predation) as well as within (competition) trophic levels. They argued that it is more appropriate to label a community as being 'physically controlled' only if environmental episodes (catastrophies) are a primary and direct cause of distribution and abundance patterns; in such habitats, biological interactions are likely to be of lower intensity.

In intermittent waters, where the dry period is cyclical and predictable, the communities should be species rich and consist almost exclusively of obligate temporary water forms, well adapted to environmental stress. Empirical evidence (Section 4.6) suggests that this is often the case, although there are clear exceptions (e.g. Boulton and Suter 1986). Such communities are ultimately controlled by physical constraints, but they will be subject also to proximal control by biotic factors. In episodic waters, subject to unpredictable loss of water, the proximal factors controlling community structure are likely to be physical ones, and species richness should be predictably low. Rain-filled rockpool communities in Africa, as an example, show such low diversity and are controlled primarily by physical factors (McLachlan and Cantrell 1980). Temporary streams in Hong Kong also seem to be good examples of these types of waters, as a combination of variable precipitation pattern, steep local relief, and thin soils create unpredictable habitats that support only one or two species of mosquito (D. Dudgeon, Personal Communication). Communities in permanent waters that may experience isolated episodes of drought will also be subject, at those times, to control by physical factors although, normally, they are more likely to be controlled by biological ones.

The remainder of this chapter is devoted largely to more *temporal* and *functional* aspects of temporary waters communities.

6.2 Community succession

In several of the preceding chapters, a theme of succession of species throughout different phases of temporary water habitats has become strongly apparent. Morin (1999) defined succession as being a community-level phenomenon that results from the full array of interspecific interactions, historical effects, and spatial dynamics that operate in developing communities. Further, he identified some of the potential causes of seasonal patterns of abundance or activity (phenology)—some having a basis in adaptive responses to interactions with other species, others perhaps merely reflecting physiological constraints or stochastic events: (1) temporal resource partitioning; (2) tracking of a seasonal resource—such as a food type or microhabitat; (3) avoidance of a predator; (4) facilitation—where one established species may facilitate the establishment of a new arrival; (5) differences in physiological tolerance—such as to the various phases of the wet/dry cycle; and (6) chance—such as in the arrival of aerial colonizers.

Some, such as Hanski (1987) have argued that, by the very nature of the process, succession will not lead to a stable community structure in individual habitat patches but, rather, cause and/or reflect their absorption by the larger, surrounding environment. Hanski pointed out that systems may be unstable at the small scale, but stable at another. Even though a particular species may exist as temporary local populations, it persistence on a larger scale (metapopulation) may be assured. Further, local instability may actually promote the exceptionally high species richness frequently observed in ephemeral habitats (e.g. Kenk 1949; Paine and Levin 1981). However, if, at the local scale, the population of a species may fail in any given year (as was documented for wetland pools and small springs in Section 5.7.1) does this necessarily mean that community stability of an individual habitat is lost? Perhaps the available niches, guilds, or trophic compartments are filled by equivalent species. Due to lack of long-term studies, there seems relatively little empirical evidence available to show how consistent are the species present in individual temporary waters—although the larval insect communities of *Heliconia* bracts in Puerto Rico are known to be remarkably consistent from year to year in both species composition and relative abundance (Richardson and Hull 2000). Despite this general lack of evidence, several authors have attempted to classify temporary water communities into subgroups (response guilds) (see Chapter 4). A few of these studies have been based on data collected over several years/hydroperiods and thus may serve as an indication of community stability over time.

Wiggins *et al.* (1980), for example, produced a classic study of the communities of invertebrates in intermittent woodland ponds in southern Ontario over the period 1960–77 (with intensive sampling between 1972 and 1977). Although the study did not present details of variation in species presence between individual years, it subdivided the community into four different components based on adaptations for surviving the dry period and seasonal patterns of recruitment. These patterns (Figure 6.3) identified four groups: Groups 1–3 are particularly well adapted to these ponds and spend the dry phase in some drought-resistant stage (typically as eggs); Group 4 is characterized by species that are not well adapted, physiologically, and must disperse to permanent waters prior to the dry phase, to recolonize, later, when the new hydroperiod has become established. For Sunfish Pond, Williams (1983) divided the community into five distinct faunal groups (see Section 4.4.1). For intermittent streams, again in southern Ontario, the communities fell quite naturally into three successive species groups: a fall-winter running water fauna; a spring pool fauna; and a summer terrestrial fauna (see Section 4.4.2).

Higgins and Merritt (1999) observed that in woodland ponds in Michigan heavy mid- to late-summer rains can result in dry (or nearly dry) basins flooding again. This renewed hydroperiod is generally short (a few weeks), and has been termed the *aestival* phase, to distinguish it from the preceding *vernal* phase. In the Michigan ponds, several taxa appeared to be specifically adapted to the aestival phase, for example: mosquitoes (in particular *Aedes vexans, Ae. trivittatus*, and species of *Psorophora*), microcrustaceans, planarian worms, gastropods, and opportunistic migrants (such as *Anopheles* mosquitoes, beetles, and hemipterans). The strategy seen in *Ae. vexans* of ovipositing near the centre of these basins, ensures hatching during the shallower aestival phase, and is in contrast to the practice of ovipositing near the margins of the vernal extent of flooding seen in spring species of *Aedes* (Enfield and Pritchard 1977). First instar larvae of *Ae. vexans* appear within a few hours of flooding, and adults begin emerging within several days.

Over six hydroperiods in Espolla Pond, a 3 ha, intermittent, karstic water body in northeastern

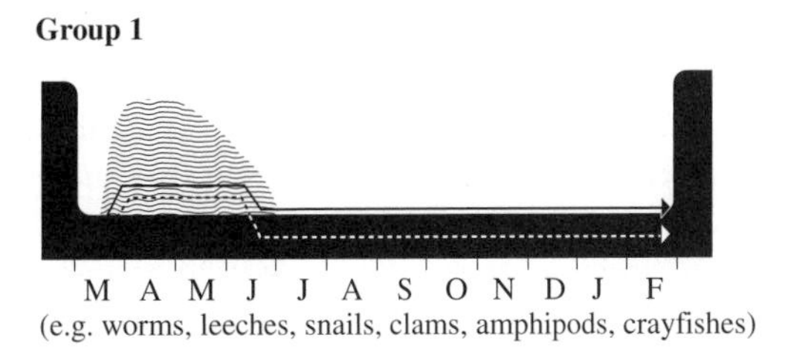

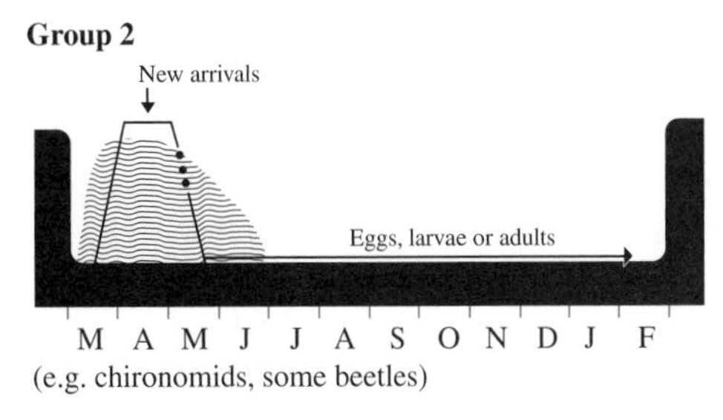

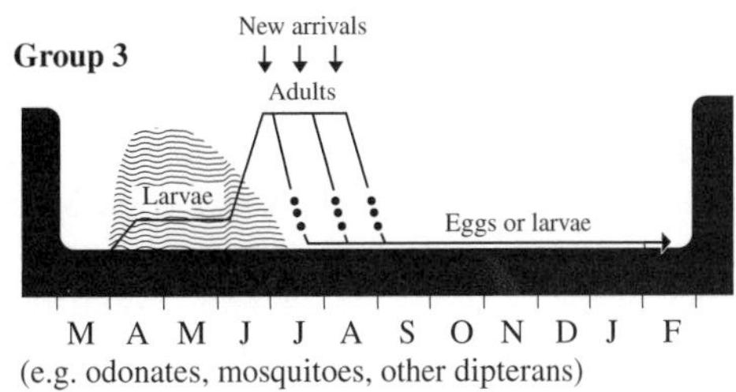

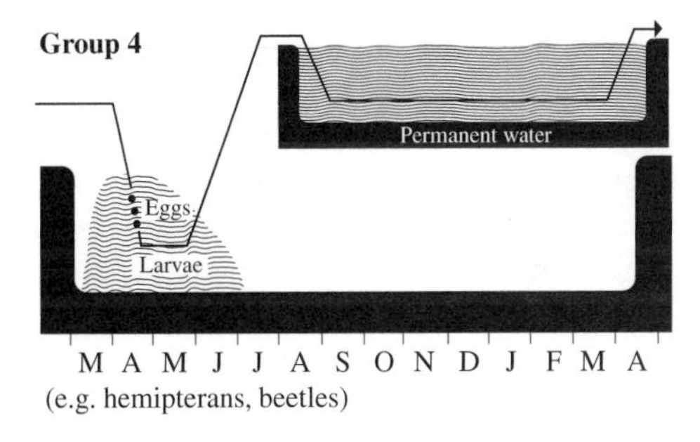

Figure 6.3 Subdivision of the communities of invertebrates in intermittent woodland ponds in southern Ontario based on seasonal patterns of recruitment and adaptations for surviving the dry period (redrawn from Wiggins *et al.* 1980). Group 1: *overwintering residents*, solid line represents exposed resistant stages; dashed line represents burrowers. Group 2: *overwintering spring recruits*, dots indicate oviposition. Group 3: *overwintering summer recruits*, dots indicate oviposition. Group 4: non-wintering spring migrants.

Spain, Boix *et al.* (2004) detected three successional phases, based on a multivariate analysis of changes in taxon abundance. The first and third phases were related to changes in hydrological variables (pond drying and water turnover), and to seasonality, whereas the second phase was more associated with stable hydrological conditions. In addition, these authors were able to delineate five successional phases, based on biomass size-spectra with similar structure, as revealed by a fit to a function derived from a Pareto Model I (Vidondo *et al.* 1997). The first and fifth phases corresponded to the flooding and drying phases identified using the taxon-based model. However, the second model produced a greater degree of resolution for the middle phase: In Phase 2, the shape of the biomass-size spectrum depended on season, with microcrustacean size classes (copepods and ostracods) dominating community biomass during winter hydroperiods, but macrofauna (*Triops cancriformis* and anuran larvae) dominating during spring/summer hydroperiods. Phases 3 and 4 were dominated by macrofauna size classes, and featured *Triops cancriformis, Fossaria truncatula* (a snail), and anuran larvae. Cladocerans contributed to microcrustacean biomass in Phase 4.

In the bracts of *Heliconia caribaea*, population succession seems related to bract development stage. Specifically, ceratopogonid larvae appear first, followed by psychodids, syrphids, and culicids, with tipulids occurring last. Richardson and Hull (2000) concluded that such clear succession provides strong support for temporal niche partitioning—the mechanisms for which include variation in oviposition and development time.

6.3 Colonization and competition

Colonization—defined as the establishment of new populations—is essential to species inhabiting temporally and spatially limited habitats, such as temporary waters. Characteristically, it requires high mobility in order to establish more local populations than are likely to be lost through inevitable local extinctions. Many communities are subject to disturbances that maintain a spatial mosaic of habitat patches, and local extinction–immigration dynamics are key to this variation (Hanski 1987). In temporary waters, however, colonization of existing basins or channels seems to be largely a *re*-establishment process that takes several forms including: (1) recolonization via mature individuals that have spent the drought in nearby, alternative habitats (e.g. in permanent waters, buried, or in protective vegetation); (2) hatching of dormant eggs or emergence from resting stages already present in the basin sediments; and (3) novel colonization by species new to an individual habitat (e.g. by flight or passive transport), in all likelihood from regional metapopulations.

As we have noted, long-term studies of individual temporary waters are rare, and thus it is difficult to determine how similar are the communities that become re-established each year. On the other hand, the similarity in, for example, the invertebrates found in the same habitat types, both on a regional basis (e.g. intermittent ponds: Wiggins *et al.* 1980; Williams 1983; Higgins and Merritt 1999; and tank bromeliads: Armbruster *et al.* 2002), and on a global scale (see Chapter 4), suggest that some communities are quite repeatable. Indeed, Therriault and Kolasa (2001) observed that temporary coastal rockpool communities in Jamaica exhibited lower variability in community structure than permanent pools. Many ecologists believe that repeatable patterns of community development imply the existence of what have been termed 'assembly rules' (Diamond 1975; Morin 1999). Evidence for such control includes detection of various non-random patterns, for example, influence of early colonizing species on the success of later arriving species, the habitual co-occurrence of certain species, or the habitual exclusion of others. Several such observations have been made in temporary waters, suggesting the existence of some deterministic pathways of community development. Morin cautioned, however, that if, for a given community and a given number of species, there exist many possible pathways or sequences of species invasion, establishment, and extinction, all of which ultimately result in a particular community, then assembly rules may be more academic in interest.

While temporary water communities would seem to be ideal for the study of assembly rules, there have been few such attempts—most studies having been done on plants, birds, and mammals. However, McLachlan's long-term study of dipterans in rain-filled rock pools in tropical Africa seems to provide some evidence. Recall that these habitats support very simple communities, comprising primarily the larvae of three species: *Chironomus imicola, Polypedilum vanderplanki*, and *Dasyhelea thompsoni* (McLachlan and Cantrell, 1980). In the case of the two chironomid species, pools either contain *P. vanderplanki* or *C. imicola*, rarely both, and they do not replace each other in a seasonal succession within the same habitat. McLachlan (1985) showed that 'possession' of a particular pool is not determined by the first species to arrive, nor is it the result of non-equilibrium conditions. The outcome is determined by interacting variables, some of which are physical in nature, the others biological (Figure 6.4). McLachlan ranked the physical variables of pool location, duration of the water-phase, and the amount of bottom sediment as primary factors. Biotic variables, including competition (based on body size), population density, accumulation of 'metabolites' and stirring up of the sediment by tadpoles, were ranked as secondary but still important. Because experimental introductions of *P. vanderplanki* larvae into pools normally inhabited by *C. imicola* failed, McLachlan ruled out chance as a causal factor. *P. vanderplanki* seemed to be excluded by scarcity of sediment (in which they normally shelter), the presence of tadpoles (which disturb the sediment), and the presence of 'metabolites' (essence) of *C. imicola*, which condition the water in some way that discourages colonization or establishment of *P. vanderplanki*. *P. vanderplanki* persists only in pools where there is enough sediment, where there are no tadpoles, and where the pool basin is shallow enough to be flushed by rain, thus removing the 'essence' of *C. imicola*. The two species only really compete in short-lived pools. There, because *P. vanderplanki* survives dry periods as partially grown, diapausing larvae as soon as the pools are refilled it reappears and, being larger than the first instar larvae of *C. imicola* (which cannot tolerate drought and have to recolonize via eggs laid by migrating females), it can out-compete the latter.

Study of progressively more complex assembly rules for temporary water communities, beginning with the hypothetical example shown in Figure 6.5, would seem to be long overdue. The example, for an intermittent pond in southern Ontario, illustrates possible pathways of species colonization in a five-species model that leads to three alternative, much simplified, community outcomes. Species 1 and 2 are primary producers (the cyanobacterium *Synechococcus* sp., and the filamentous green alga *Oedogonium* sp., respectively). Larvae of species 3, the mosquito *Aedes sticticus*, consume both species 1 and 2, and are, in turn, eaten by both species 4 (nymphs of the dragonfly *Sympetrum obtrusum*) and species 5 (larvae of the dytiscid beetle *Rhantus* sp.). Species 4 consumes species 1, 2, and 3, and is itself eaten by species 5. The latter is capable of consuming all of the other species, to varying degrees. Based on such a model, Law and Morton (1993) proposed that one possible assembly rule would be that where multiple, alternate permanent sets of species exist, no alternate permanent set will be a subset of any other permanent set. Another rule proposes that species have higher probabilities of entry into an assembling

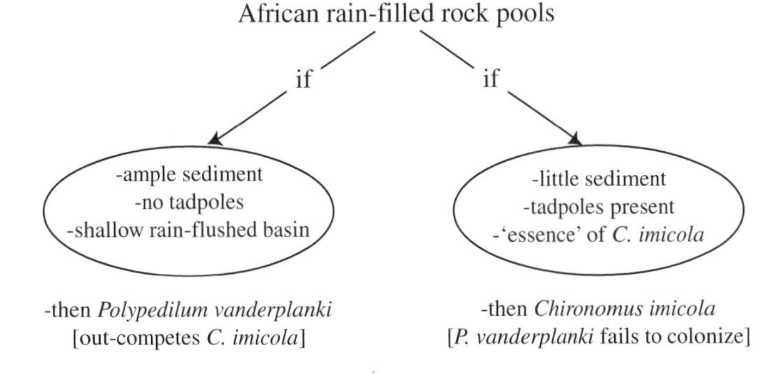

Figure 6.4 The outcome of dipteran species occupancy in African rain-filled rock pools is determined by interacting variables, some of which are physical in nature, the others biological (from the work of McLachlan and co-workers).

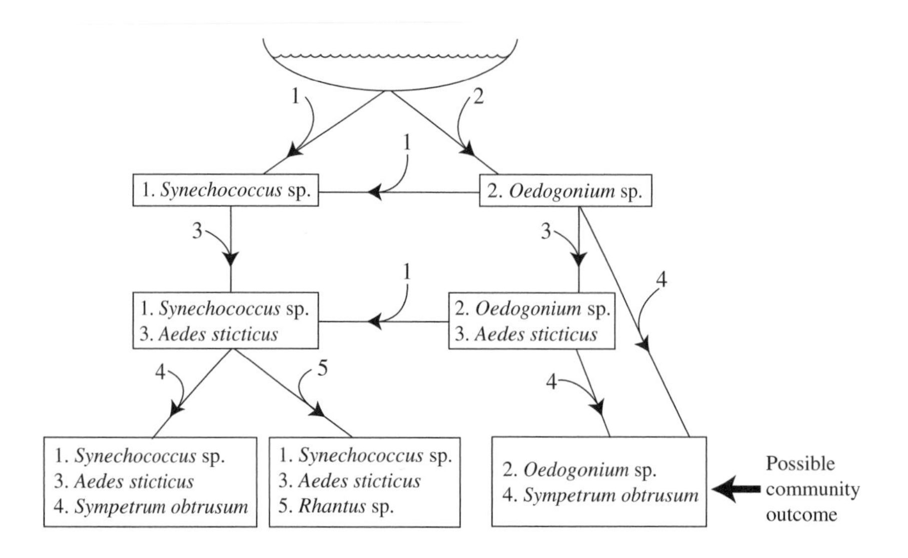

Figure 6.5 A model for an intermittent pond in southern Ontario, illustrating possible pathways of species colonization in a five-species model that leads to three alternative community structure outcomes. Species 1 and 2 are primary producers. Species 3 consumes both species 1 and 2, and is, in turn, eaten by both species 4 and species 5. Species 4 consumes species 1, 2, and 3, and is itself eaten by species 5 (based on a model proposed by Law and Morton, 1993).

community if their guild is under-represented (Fox 2000). Such propositions could fruitfully be addressed in temporary waters. For example, in a 2-year-study of three sympatric species of the dragonfly genus *Sympetrum* in a southern Ontario pond, Febria and Williams (unpublished data) observed that, in the first year, nymphal populations occurred in the sequence *S. obtrusum*, *S. internum*, and *S. costiferum*. In the second year, however, only *S. obtrusum* occurred. While it was thought that unusually cold water temperatures at the beginning of the second year's hydroperiod may have been a factor in the non-appearance of *S. internum*, and *S. costiferum*, there was also a high indication of interspecific competition for food. Several studies have shown that the effects of both competition and predation increase with length of the hydroperiod (e.g. Schneider and Frost 1996; Higgins and Merritt 1999).

Part of the process of colonization involves habitat selection, another key element in determining the makeup of a community. Habitat selection has many facets, and studies have documented species exhibiting settlement behaviours based on ultimate factors (see Section 5.7.3) that are often anticipated as being favourable or unfavourable. These factors include the availability of prey species, the presence of potentially competing species, and predation avoidance. Skelly (2001), for example, has gathered evidence that amphibians breeding in temporary waters are capable of sensing and altering their oviposition patterns in response to a variety of cues, foremost among which is the threat of predation (Table 6.1). Walton (2001) made the interesting observation that tadpole shrimp (*Triops newberryi*) swimming at the water surface in ponds in California, particularly during crepuscular and nocturnal periods of low dissolved oxygen concentration, discouraged oviposition by insects. Stav *et al.* (2000) similarly observed that the presence of nymphs of the dragonfly *Anax imperator* repelled oviposition by the mosquito *Culiseta longiareolata* in a temporary pool. By this means, and by direct consumption of *Culiseta* larvae, *Anax* virtually eliminated the dominant grazer in the pool community, causing significant changes in community structure.

Table 6.1 Summary of the influence of different factors on habitat selection (oviposition site) by anurans

Species	Factor	Interpretation of causal mechanism
Hyla chrysoscelis	Conspecific larvae	Competition avoidance
	Fishes	Predation avoidance
	Salamander larvae	Predation avoidance
	Heterospecific larvae	No effect
	Adult salamanders	No effect
	Insect larvae	No effect
Hyla pseudopuma	Conspecific larvae	Predation avoidance
	Depth	Desiccation avoidance
Hyla versicolor	Snails	Infection avoidance
Rana sylvatica	Fishes	Predation avoidance
Rana temporaria	Fishes	No effect
Bufo americanus	*R. sylvatica* larvae	Predation avoidance
Hoplobatrachus occipitalis	Conspecific larvae	Predation avoidance

Source: From Skelly (2001).

As noted above, competition is another powerful force acting to structure communities. Classically (Park 1962), it has been divided into *exploitative* competition (where species interact indirectly through some shared, finite resource) and *interference* competition (where there is direct interaction between species, such as for territory). More recently, Schoener (1983) has proposed six different mechanisms that encompass most known types of competitive interaction. These are: (1) *Consumption*—this is close to exploitative competition; (2) *Pre-emption*—which is primarily competition for attachment space, and is seen typically in sessile species; (3) *Overgrowth*—another form of competition for space but involving interference between individuals; (4) *Allelopathy*—chemical competition in which an individual secretes a toxin to inhibit or kill individuals of another species; (5) *Territoriality*—which is manifest by aggressive behaviour resulting in the exclusion of another organism from a defended space in which there is typically some resource; and (6) *Encounter*—this is a non-territorial contest for some resource, typically food. Relatively few studies have produced detailed and tested accounts of such competition types in temporary water communities, although many studies have inferred them.

Consumptive competition is likely to be quite common in many temporary waters, particularly among predators such as beetle larvae and dragonfly nymphs (e.g. Moore 1964; Anholt 1990; Fairchild *et al.* 2003), and perhaps also among the many detritivorous species—although whether detritus is a limited resource in, for example, intermittent ponds, is uncertain. In response to intraspecific competition for food, some temporary water amphibian species are known to show trophic polymorphism, in which changes in head and tooth structure occur, enabling the morphs to select different prey types (Blaustein *et al.* 2001). Kiesecker and Blaustein (1999) made the interesting observation that the presence of the pathogenic water mold, *Saprolegnia ferax*, had positive indirect effects on both *Hyla regilla* (the Pacific treefrog) and *Rana cascadae* (the Cascades frog) by regulating both intra- and interspecific competition. The presence of *Saprolegnia* differentially affected larval recruitment of the two species (larval recruitment of *R. cascadae* was reduced by 46.2% in the presence of *Saprolegnia*, whereas the survival of *H. regilla* was not affected). However, *R. cascadae* larvae that survived infection developed faster and were larger at metamorphosis. In the absence of the pathogen, *R. cascadae* exerted strong negative effects on the growth, development, and survival of *H. regilla*, but in the presence of *Saprolegnia*, the outcome of competition between the two species was reversed.

Bengtsson (1989) showed that interspecific competition acted to determine the coexistence of three species of *Daphnia* in Swedish rock pools. Typically, only one or two species occur in any one pool, but all three (*Daphnia magna, D. pulex*, and *D. longispina*) occur in the same locality. Supported by species-introduction experiments, Bengtsson found that extinctions were most often observed in pools containing all three species, but were less common in two-species pools, and absent in single-species pools. Persistence of all three species in the locality was thought to be the result of a competitive hide-and-seek mechanism among pools. Working with container-dwelling mosquitoes in southern Florida, Juliano (1998) studied the mechanisms by which *Aedes albopictus*, an introduced species, appears to be causing a decline in populations of *Ae. aegypti*, a resident. In water-filled tyres, *Ae. aegypti* did well only when it was alone, when it was present at low densities, and when there was abundant food. *Ae. albopictus*, on the other hand, proved to be a superior competitor in this environment where it was able to do well at high combined density and with lower per capita food resources. As in the case of *Daphnia* in Sweden, persistence of *Ae. aegypti* occurred because of variation in the intensity of interspecific competition among sites.

Using species from temporary pools in the Negev Desert, Blaustein and Margalit (1996) demonstrated, experimentally, that the outcome of competition and intraguild predation between larvae of *C. longiareolata* and *Bufo viridis* depended on the order in which they appeared in the pools—with the second species to arrive always being disadvantaged. This illustrates that, in many instances, it is not competition alone that determines community structure, but a combination of biological factors, frequently with environmental factors. For example, in a series of experimental ponds, Wilbur (1987) was able to demonstrate that the relative abundance of four anurans (*Rana utricularia, Scaphiopus holbrooki, Bufo americanus*, and *Hyla chrysoscelis*) was the outcome of interactions between competition and predation against a backdrop of length of the hydroperiod.

Pre-emptive competition is perhaps less common in temporary waters, although it might apply among sponge species requiring sturdy attachment sites, or, at a much smaller scale, to sessile ciliates. Overgrowth also may be of less significance in temporary waters, but again may occur in sponges or microorganisms. There are a few records of both inter- and intraspecific, allelopathic-like competition. For example, fewer eggs of the conchostracan *Cyzicus* were observed to hatch in rock pools in northern Israel that contained salamanders than in those that did not (Spencer and Blaustein 2001). Zahiri and Rau (1998) showed that as the biomass of larvae of *Ae. aegypti* increased in relation to the volume of rearing water, attraction to gravid females first rose to a peak and then declined. Any further increase in biomass rendered the water highly repellent, and dilution of the repellent did not restore attraction. These authors concluded that the density-dependent action of oviposition attractants and repellents may serve to maintain larval populations near some optimal level.

Territoriality is known to occur in burrowing crayfishes (Williams *et al.* 1974) and odonate nymphs (Baker 1981), and, in the latter, it has been shown to enhance the mating success of adult males (Harvey and Corbet 1985). Encounter competition likely occurs among individuals of temporary water predator species when contesting prey in non-territorial situations.

A final point worth considering, is that competitive interactions and outcomes within a given community may not be immutable. From an analysis of a variety of case-histories, Schoener (1983) concluded that about 50% showed annual variation in competition. However, this variation took the form of changes in competition intensity rather than its presence or absence.

6.4 Temporary waters as islands in time and space

'Populations that cannot persist in a single habitat patch can persist in a collection of different patches' is an axiom central to both the theories of metapopulation dynamics (Levins 1969) and island biogeography (MacArthur and Wilson 1967) (Mouquet *et al.* 2001). The latter states that isolated oceanic islands are continuously being colonized by new species while some previously arrived species are lost or become extinct from the biota. The fauna and flora *per se* may never be static, in terms of species complement (i.e. they have 'open' communities), but, at sometime, the immigration rate will equal the extinction rate and a state of dynamic equilibrium will be attained. This theory has been applied not only to oceanic islands but to other habitats isolated in space, such as coral heads, baskets of sterile rocks on the beds of streams, and the faunas of mountain peaks. It may also be applicable to 'habitat islands' that are isolated in time, such as the aquatic stages of temporary waters. Ward and Blaustein (1994) tested the five predictions of island biogeography theory in Negev Desert pools, and found three (including the species–area relationship) to be upheld. Significantly, however, the study showed that although there was a turnover of species, species richness did not remain constant, but increased steadily throughout the study. Further, there was no correlation between extinction rate and pool size. Flash floods emerged as being of overriding importance in these particular habitats, as they flushed all of the species out of the pools, thus causing the colonization process to be restarted.

Ebert and Balko (1982) produced a standard graphic model of colonization and extinction rates, showing the equilibrium points for species' numbers, modified for time rather than space (Figure 6.6). They equated frequency of occurrence of a habitat in time with distance from a source in space, arguing that patches of habitat, such as isolated ponds, are a long distance in time from a source of potential colonizing species, and that resting stages (spores, cysts, eggs, seeds, etc.) may have to travel for considerable periods of time between favourable habitat conditions. As in the original model for islands in space, 'islands' distant in time have lower rates of colonization than do 'islands' that are closer (i.e. temporary waters that fill more frequently). In the original space model, rates of extinction were shown to be related to the physical size of the island whereas, in the time model, the analogue is duration of the habitat. Pools that dry up frequently may be considered to be islands of short duration, having a small size in time, whereas permanent waterbodies are islands of long duration and are large with respect to time.

The time model (Figure 6.6) predicts that infrequent, short-lived (episodic) waterbodies will have fewer species at equilibrium (S_1) than frequently occurring (intermittent) waterbodies of longer duration (S_2). However, many species living in temporary waters are affected by both time and space factors, and thus a more complete model

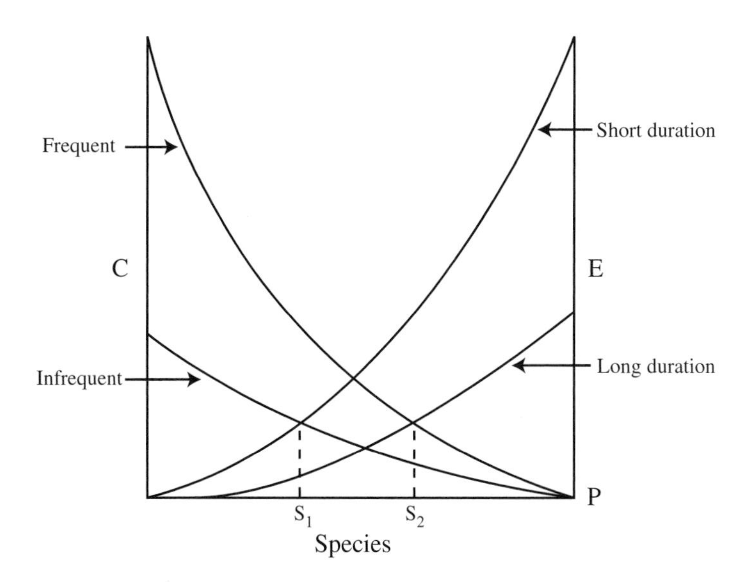

Figure 6.6 Diagram of colonization (C) and extinction (E) curves for 'islands' in time, indicating the analogs with 'islands' in space. P = the number of species in the species pool; S_1 represents the number of species at equilibrium for an island which occurs infrequently and has a short duration (e.g. an episodic pond); S_2 represents the equilibrium number of species on an island habitat which occurs frequently and has a longer life (e.g. an intermittent pond), (redrawn from Ebert and Balko, 1982).

should accomodate both. Ebert and Balko therefore proposed another model (Figure 6.7). This depicts both components of time and space, with axes for colonization and extinction rates drawn perpendicularly to the X-Y species plane. Two extreme temporary water bodies are shown on the X- and Y-axes as P_1 and P_2. The X-Y plane represents the species pool of all possible colonists, within an individual taxonomic group, onto all possible types of 'islands'; there would be different X-Y planes for each taxonomic group being considered. Because of temporal instability, the water body may be suitable for some species but not for others. Some species will be part of species pools over much of P_1 to P_2, indicating that they have very broad tolerances/adaptations and normally occur in many habitats from those that are highly episodic to those that are permanent. Others have very narrow tolerances and occur over a very restricted region of the species plane.

Rotation from P_1 to P_2 on the graph represents movement along a gradient of temporal habitat instability. The most episodic habitats are nearest the X-axis. As one approaches the more permanent waters, nearest the Y-axis, species that are highly adapted to temporary waters tend to drop from the species pool but species that can only survive in permanent waters are added to the potential species pool. Rotation from P_1 to P_2 not only describes the change in potentially colonizing species but represents a gradient of the influences of temporal versus spatial factors of the habitat. On the X-axis, the rates of colonization and extinction are largely determined by time factors, such as the frequency and duration of the habitat. On the Y-axis, colonization and extinction rates are largely determined by spatial factors such as the size of the water body and the distance from a source of potential colonizers. The rates of colonization and extinction between P_1 and P_2 are functions of frequency, duration, size, and distance. In a study of the impact of drought on ponds in Scotland, Jeffries (1994) found that extinction rates were high for taxa typical of permanent or temporary ponds. Notably, colonization rates were high for taxa from temporary ponds, but low for those from permanent ponds.

As in the original concept of the equilibrium theory, as proposed by MacArthur and Wilson, the

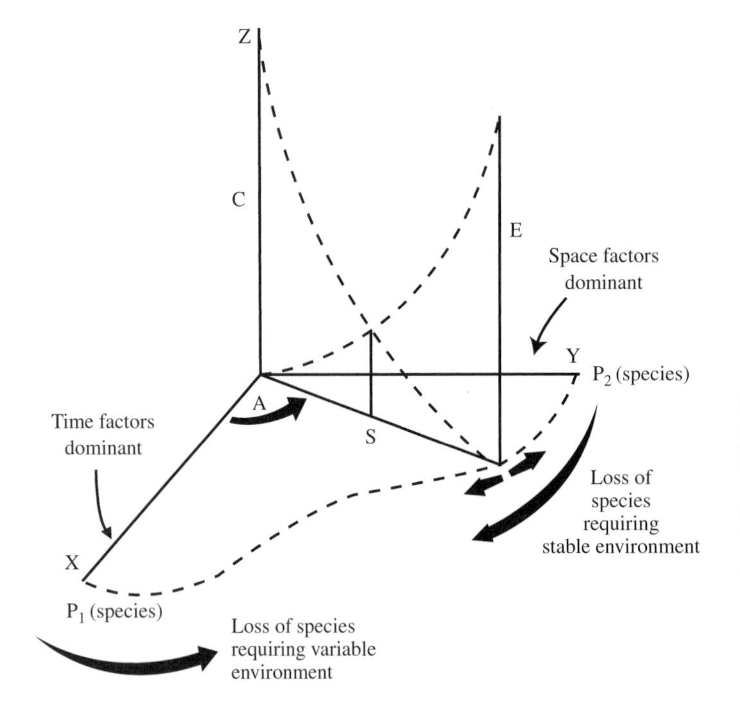

Figure 6.7 A three-dimensional representation of colonization (C) and extinction (E) rates (the Z axis) on a species plane. Points on the line from P_1 to P_2 represent the realized species pools that can supply colonists to 'islands'. A realized pool near the X-axis would be for a species group that perceives habitats more like islands in time than in space; a realized species pool nearer the Y-axis would be for species groups that perceive habitats more as islands in space (redrawn from Ebert and Balko, 1982).

equilibrium number of species (S) is, on Ebert and Balko's 'species plane', the point where the rates of colonization (C) and extinction (E) are equal. The angle, A, is a function of the duration and frequency of the water body, such that: $A = j(f, u)$.

'j' will have a maximum value of $\pi/2$, when 'f' (the frequency of habitat occurrence) and 'u' (habitat duration) are both equal to 1.0. That is, the species pool is P_2 in a permanent water body—which, by definition, occurs each year and persists for the entire year, with a probability of 1.0. By including A in the determination of S, it is possible to account for changes in species composition as well as changes in species number.

There are a few cautionary notes to be added to the application of this innovative, multidimensional model, however. There are other factors besides space and time that influence the rates of colonization and extinction, for example, weather (Donald, 1983), food quality and availability, and interactions between species (see Sections 6.3 and 6.5). Further, as the pool of colonists may vary seasonally this may prevent some habitats from ever attaining a true equilibrium (e.g. see Williams and Hynes 1977; Williams 1980). Hanski (2001) has pointed out that, in the case of small populations, such as in many temporary waters, emigration may substantially increase the risk of local extinction, whereas immigration tends to reduce it. Further, dispersal in metapopulations often allows species to exist in habitats where conditions are tenuous (including sink populations), and is also a major factor in allowing the persistence of both prey species and inferior competitors in multispecies communities. Hanski believes that spatio-temporal variation in the availability of suitable habitat is a central determinant of dispersal rate, the evolution of which, in changing environments, is an area of metapopulation dynamics ripe for important theoretical and empirical discoveries.

6.5 Trophic relationships

Modern food-web theory is complex, rife with terminology, and sadly lacking in quantitative, experimental validation. Unfortunately, the latter is especially true for temporary water communities. Some light has been shed through meta-analyses of studies on a range of food-web types, from which some common properties have been identified (Cohen 1978; Pimm 1982; Lawton and Warren 1988; Martinez 1992; Kitching 2000). Those most relevant to temporary waters include (based on Morin 1999): (1) many food webs seem to have constant ratios of predator to prey species, or ratios of basal to intermediate to top predator species—although the influx of predatory beetles and hemipterans in the later stages of the hydroperiod of intermittent streams would suggest otherwise; (2) frequently, food webs are 'interval' in nature—indicating that overlaps between predators occur, which would agree with observations in intermittent ponds, for example; (3) cycles and loops may be present—these are most often seen in communities that contain species present in a range of size classes, and where larger individuals of one species are able to consume smaller individuals of another and vice versa, as is the case with, for example, amphibians and dytiscid beetle larvae; (4) food chains are relatively short—this would seem to agree with what is known in temporary waters, although length may vary with habitat type (e.g. very short in some pitcher plant communities, but longer in floodplain lakes); (5) omnivory, initially thought to be relatively rare, appears now to be more commonplace, especially in webs where insects and parasitoids are plentiful (insects and mites are certainly abundant in intermittent ponds and streams); (6) connectance (a measure of the number of possible links in a food web) is thought to decrease in more variable environments, whereas species interaction strength is believed to increase—intuitively, both might seem supported (e.g. the strong, and highly focused, predation effects of dytiscid larvae on fairy shrimp). Interestingly, Beaver (1985) has observed that among different tropical pitcher plant species, those with less complex food webs showed greater connectance; (7) webs do not tend to be compartmentalized, or subdivided—this would appear not to be true for some temporary waterbodies (see Section 6.5.1). Kitching (2000), in a thoughtful reassessment of the quantification of foodweb statistics as applied to phytotelmata, advocates setting out clear baseline rules before web construction. He points out that errors may be made, for example, by not considering whether, when calculating predator : prey ratios, predator species should be included within the 'prey' count when some species are used as food by top predators—leading to a single species being counted on both sides of the ratio. Kitching further illustrates that connectance can be strongly influenced by the number of trophic levels present, and the extent to which the species are specialists or generalists—he gives the example of two webs: a complex tree-hole web which exhibits potentially many possible trophic links, but in which each predator uses only a subset of the prey types present; and a much simpler plant-axil web which contains only a single, generalist predator species and no top predators; the latter web, like Beaver's *Nepenthes* example, displays the higher connectance value.

Many temporary water habitats have food webs that are detritus-based, whether the latter is derived from leaf-fall from surrounding vegetation, the decay of plants that grow in the basin sediments during the dry phase, fine-particle wind-blown debris, or the breakdown of animal bodies trapped within pitcher plants. Other, more open, habitats may support a higher diversity of photosynthetic species that live on the sediment surface, in the water column, or as emergent or semi-emergent growth forms (see Sections 4.2.2 and 4.2.4). Even woodland temporary streams and ponds may have a significant autotrophic phase early in their hydroperiod, before trees leaf-out. At such time, benthic and planktonic algae provide a vital food source for early filter-feeders, such as fairy shrimp and microcrustaceans. Some streams, such as those in southwestern Spain, have a distinct autotrophic lentic stage that *follows* the heterotrophic lotic phase (Molla *et al.* 1996).

Barlocher *et al.* (1978) have described the process by which leaf litter is incorporated into the food web of temporary ponds. Leaves falling into dry basins in autumn are initially colonized by microbes of terrestrial origin, especially microfungi. These organisms begin the process of decomposition and increase the protein component of the leaves (via an increase in fungal biomass) above that seen in comparable leaves kept under water. Once the hydroperiod begins, however, protein levels decrease rapidly, due to leaching, releasing nutrients into the water column. These nutrients are not only proteins, but the degradation products of other microbial processes, for example, the breakdown of lignin, and the depolymerization of celluloses and hemicelluloses, so released, into simpler carbohydrates (Chamier 1985; Ljungdahl and Eriksson 1985). The build-up of biofilms on the remaining leaf substrates populates the pond basin with a rich microbiota, many of which secrete extracellular compounds of use to others (Nalewajko 1977). Gradually, the leaf material is converted into fine particulate material and microbial biomass. This reduction process may be the reason why, compared with permanent streams, for example, the number of leaf-shredding detrivore species in intermittent woodland ponds is low (Figure 6.8; Higgins and Merritt 1999). The paucity of shredders has also been noted in intermittent streams (Figure 6.9).

Different types of temporary waters have different degrees of trophic complexity (see below), and the ways in which temporal heterogeneity of the habitat can affect community organization may

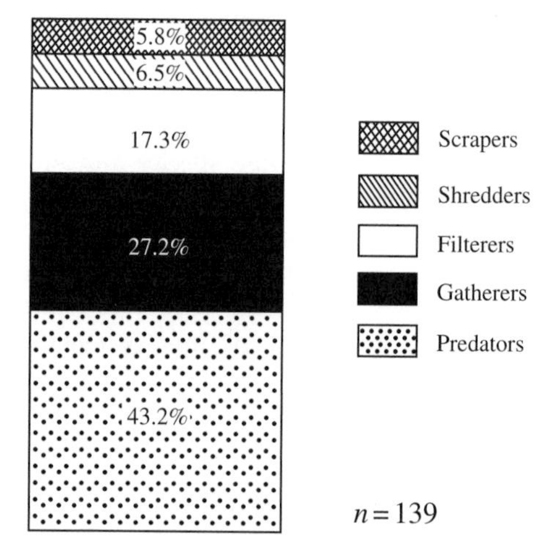

Figure 6.8 Relative proportions of invertebrate functional feeding groups (by genera) in intermittent woodland ponds in the temperate zone (redrawn after Higgins and Merritt 1999).

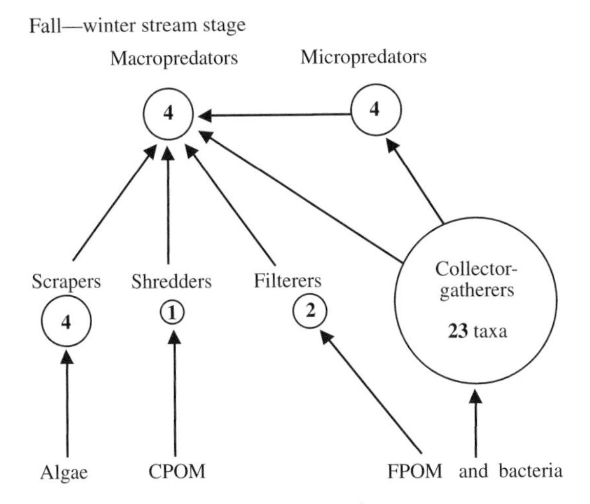

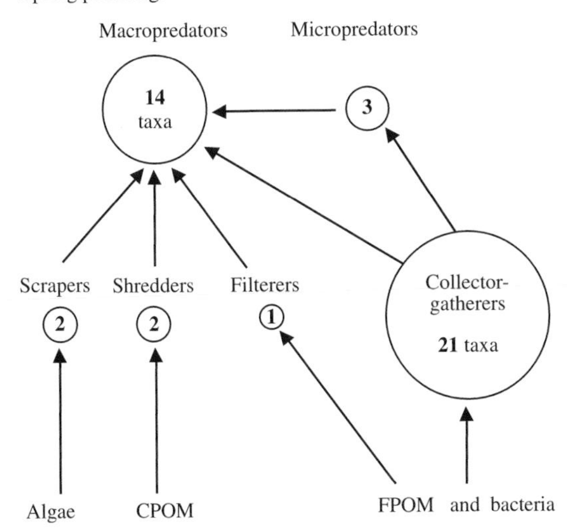

Figure 6.9 Comparison of the trophic structures of the invertebrate communities of Kirkland Creek, an intermittent stream in southern Ontario, during the fall-winter stream stage (upper) and the spring pool stage (lower). Taxa present (see Williams and Hynes 1976) were assigned to functional feeding groups based on unpublished data and the dietary information provided in Merritt and Cummins (1984). No quantities have been assigned to the various links, and the size of the circles simply reflects the number of taxa present in a particular feeding group.

increase with this factor. Menge and Sutherland (1976) identified some important relationships for animal communities, in general: trophic complexity is a function of temporal heterogeneity; and the number of trophic levels in a community seems dependent, at least to some extent, on the rate and predictability of primary and secondary production, which is likely influenced by habitat stability and predictability. The, somewhat limited, data on the trophic structures of temporary water communities suggest that there are fewer trophic levels as temporal heterogeneity increases. In other communities, this may be due to the elimination of more specialized consumers of high trophic status because the variety of resources becomes less predictably available. Another potentially important effect of temporal habitat heterogeneity on higher trophic species stems from the environment being unfavourable for certain periods (e.g. during drought). This reduces the time during which consumers can forage and may enable some prey species to avoid predation.

In communities in general, the often observed reductions in within-habitat species richness, along gradients of temporal heterogeneity, may be largely due to the increased incidence of competitive exclusion as trophic levels become lost or ineffective (Menge and Sutherland 1976). However, the importance of interspecific competition for food resources in small seasonal ponds may be dependent on whether the resources are limited as, for example, may be the case in dytiscid beetles (Nilsson 1986). Also, where food resources are ephemeral, the probability of interspecific competition occurring is thought to be low, thus allowing the coexistence of very similar beetle species (Price 1984; Larson 1985). Dytiscid diversity is less in more permanent waterbodies which may be a reflection of the more intense competition found there (Nilsson 1986). The beetles, themselves, are known to sometimes exert considerable influences on some prey populations (e.g. *Acilius* larvae on *Daphnia*; Arts *et al.* 1981), as has also been shown for the effects of the predatory flatworm *Mesostoma ehrenbergi* on *Daphnia* in small ponds (Maly *et al.* 1980), and for the influence of introduced fishes on the entire macroinvertebrate community of the pool stage of an intermittent stream (Smith 1983). Holyoak (2000) created a laboratory experiment to examine how spatial dynamics (habitat patchiness) might affect the structure of a simple food web (four species).

Increasing habitat fragmentation lengthened the persistence of some species, altering the extent of food-web collapse through extinction. Extended persistence was associated with immigration reprieving a basal prey species from local extinction. Increasing fragmentation reduced the population density of the top predator, but allowed increases in the abundance of two other predator species. Clearly, species interactions affect the structure of temporary water communities.

Many communities contain keystone species, that is, species whose roles are so critical in supporting the food web, that their elimination causes the food web to collapse and also results in changes in the local ecosystem. Unfortunately, there have been few quantitative analyses of such changes in temporary waters. However, in a study of temporary ponds in the Carolinas, Fauth (1999) found that the eastern newt, *Notopthalmus viridescens*, and the lesser siren, *Siren intermedia*, were keystone species in ponds at northern sites, but not in southern ponds. In the latter, another salamander, *Ambystoma talpoideum*, not present in the north, was a keystone. This species functioned independently of the densities of other predators, and also of environmental factors. In the southern ponds, *Tramea carolina* (Odonata) also proved to be a keystone species, although the influence of nymphs was not as strong, and context-dependent.

6.5.1 Trophic continuum or disjunct phases?

An ecological feature, identified earlier, and characteristic of most temporary waters is species succession through the different phases of the habitat. For example, intermittent streams in temperate regions go through three stages: (1) a fall-winter running water stage; (2) a spring pool phase, as flow stops and a series of disconnected pools remain; and (3) a summer terrestrial stage (Williams and Hynes 1976). There is sufficient replacement of species throughout these stages that the structure of the faunal community differs substantially over time. Similar successions of species have been recorded in semi-arid foothill streams in central California (Abell 1984), in a desert stream in central Arizona (Boulton *et al.* 1992), and in two streams in southern Australia (Boulton and Lake 1992). Despite the obvious conclusion that such structural changes must be accompanied by functional changes in the community, very few quantitative attempts have been made to address the latter. Assignment of species in Kirkland Creek, Ontario, to major functional feeding groups illustrates these changes (Figure 6.9; further details of this study site are given in Williams and Hynes 1976). Compared with the fall-winter running water stage, which is dominated by collector-gatherers, the spring pool stage shows a marked increase in macropredators—primarily due to aerial colonization by adult beetles and hemipterans. Both adults and their offspring prey on members of the aquatic community. In two intermittent streams in Australia, predator numbers also peaked before the habitat dried; scraper densities rose in the late spring/early summer coincident with increased periphyton growth; and collector-gatherers and collector-scrapers were well-represented throughout the year (Boulton and Lake 1992).

Unfortunately, trophic details of the drought phase of temporary waters are rare. Existing information is restricted to ancillary observations, and tends to be non-quantitative. However, the study by Cameron and LaPoint (1978) is an exception, as it reported on the decomposition of Chinese tallow leaves by terrestrial invertebrates in Texas pond basins. Using litter bags with either large or small pores, they found that the leaves lost weight significantly faster in the former. After 10 months, there was no detritus left in the large-pore bags (representing a loss of 0.33% per day), but more than 20% left in the small-pore bags (0.21% loss per day) after 12 months. Although initial weight loss from both bag types was due to leaching of soluble materials, much of the loss from the large-pore bags was due to ingestion and reduction by the terrestrial isopod *Armadillidium vulgare*. It was apparent, however, that the leaves were initially unpalatable to this isopod, and thus mechanical processing did not begin (February–March) until some time after leaf fall (August to December), when wetting and some microbial colonization and breakdown had taken place.

In a remarkable, taxonomically detailed study, Closs and Lake (1994) compared the food-web structure over time, and at three sites, on the Lerderderg River, an intermittent stream in southern Australia (Figure 6.10). Despite being around 1.5 km apart, the sites showed considerable similarity in community structure. However, temporal variability (over four phases corresponding to early flow, main flow, diminishing flow, and post-flow) was high, with species richness increasing markedly as the period of constant streamflow lengthened. The community was detritivore-dominated, in terms of both benthic densities and species present. Notably, the proportion of predators increased slightly towards the end of the year, suggesting that their recolonization lagged behind that of detritivores. Increased numbers of predators also resulted in an increase in the mean length of food chains as the year progressed. Despite patterns in the Lerderderg River food webs agreeing with those in other web-based studies (e.g. Boulton 1988), which might point to some defining control mechanisms, Closs and Lake cautioned about drawing such conclusions without considering methodological equivalency.

An interesting comparison to make would be how food-web structure might change across a gradient of water body persistence, that is, from episodic to intermittent to pemanent. Unfortunately, such full spectrum studies are rare. Two exceptions are the studies by Schneider (1997) and Wissinger *et al.* (1999). In the latter the authors constructed energy pathways for communities in permanent, autumnal, and vernal subalpine wetlands in Colorado (Figure 6.11). These authors found that permanent basins contained the most diverse communities, and included multiple year-classes of larval salamanders. The community was dominated by small-bodied zooplankters, chironomids, small dytiscid beetles, and cased caddisfly larvae. In comparison, semipermanent, autumnal basins had less diverse communities dominated by species largely absent or rare in permanent basins (e.g. fairy shrimp, large-bodies cladocerans and copepods, and large dytiscid beetles, with the latter replacing salamander larvae as the top predators on benthic species). Small vernal basins (the most ephemeral basin type) contained no salamanders and were dominated by what was seen as a subset of the autumnal basin community: several zooplankters, mosquito larvae, and one species each of corixid, beetle, and caddisfly. Wissinger *et al.* concluded that biotic interactions were largely responsible for the community differences observed. Schneider (1997) examined predation and food-web structure along a habitat duration gradient in temporary ponds in northern Wisconsin. He found that food-web statistics exhibited a complex relationship with measures of habitat variability. For example, connectance was highest in short-duration, highly variable habitats, and lowest in habitats that were intermediate in duration and variability. In addition, the number of links and links per taxon increased with increasing duration. The relationship between food-web statistics and habitat duration agreed with experimental evaluations of predation in these habitats that indicated increasing importance of predation with increasing hydroperiod. However, the statistics failed to detect threshold effects in this relationship. Connectance emerged as being the most reliable statistics with regard to taxon number, but was the least sensitive to changes in habitat characteristics.

Other information relevant to a habitat persistence gradient is potentially available from comparative food-web studies in which ecosystem (as opposed to habitat *per se*) stability has been assessed. For example, Montoya and Solé (2003) examined patterns of trophic connections in 12 of the most detailed food webs published, combining freshwater, marine, and terrestrial habitats. They concluded that the shape of the network of trophic interactions was highly dependent on the number of species present, as has been found by others (e.g. Martinez *et al.* 1999). These properties might be a consequence of assembly processes, as they can be partially reproduced via simple, multi-trophic models where web dynamics are characteristically dominated by weak interaction strengths between species—a finding that is supported by empirical data. As species richness is known to vary across the habitat-persistence gradient (see Chapter 4), changes in the structure of

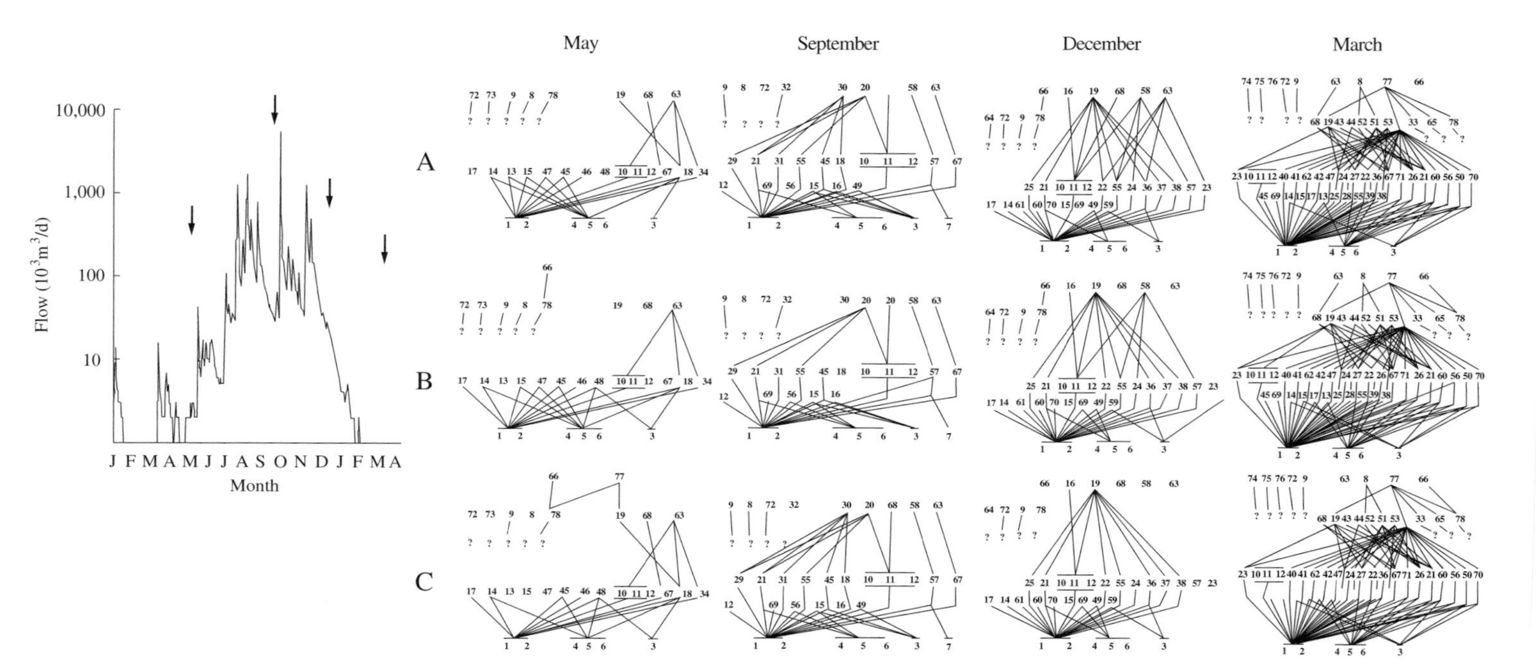

Figure 6.10 Food webs over spatial (three sites: A, B, and C) and temporal axes in the Lerderderg River, Australia. The sampling times, May, September, December, and March (indicated by the vertical arrows on the left-hand graph) correspond to streamflow phases of early flow, main flow, dimishing flow, and post-flow, respectively. Species for which prey are not indicated represent suctorial feeders. Each species (see key below) is represented by a letter, following its name, that indicates the major taxonomic group to which it belongs. These letters are: Tu = Turbellaria, O = Oligochaeta, M = Mollusca, G = Gastropoda, B = Bivalvia, D = Diptera, C = Chironomidae, S = Simuliidae, T = Trichoptera, Le = Leptoceridae, Gs = Glossosomatidae, Ca = Calocidae, E = Ecnomidae, Hb = Hydrobiosidae, Ep = Ephemeroptera, Lp = Leptophlebiidae, P = Plecoptera, Gr = Gripopterygidae, Au = Austroperlidae, Cl = Coleoptera, Dy = Dytiscidae, Gy = Gyrinidae, Cr = Crustacea, Co = Copepoda, Os = Ostracoda, Am = Amphipoda, Cd = Cladocera, Hy = Hydracarina, P = Pisces, Ga = Galaxidae, (A) = adult insect. *Key*: (1) Amorphous detritus, (2) Dead cellular plant material, (3) Algae spp., (4) *Tetraedron* sp., (5) *Tabellaria* sp., (6) *Navicula* sp., (7) Fungi, (8) *Hydra* sp., (9) *Cura pinguis* Tu, (10) Tubificidae sp. O, (11) Naididae sp. O, (12) Lumbricidae sp. O, (13) *Ferrissia* sp. MG, (14) *Angrobia* sp. MG, (15) *Physastra gibbosa* MG, (16) *Glacidorbis hedleyi* MG, (17) *Sphaerium tasmanicum* MB, (18) *Chironomus* sp. DC, (19) *Paramerina* sp. DC, (20) *Coelopynia* sp. DC, (21) *Riethia* sp. DC, (22) *Cricotopus annuliventris* DC, (23) *Rheotanytarsus* sp. DC, (24) *Parakiefferiella* sp. DC, (25) *Thienemaniella trivittata* DC, (26) *Tanytarsus inextensus* DC, (27) nr. *Halocladius* sp. DC, (28) *Stempellina* sp. DC, (29) *Austrosimulium* sp. DS, (30) Tipulidae sp. 1 D, (31) Tipulidae sp. 2 D, (32) Ceratopogonidae sp. D, (33) Empididae sp. 1 D, (34) Orthocladiinae sp. 1 CD, (36) Chironomidae sp. CD, (37) Orthocladiinae sp. 8 CD, (38) Tanypodinae sp. CD, (39) *Eukiefferiella* sp. CD, (40) *Brillia* sp. CD, (41) Tipulidae sp. D, (42) *Polypedilum* sp. CD, (43) *Larsia* sp. CD, (44) *Ablabesmyia* sp. CD, (45) *Triplectides truncatus* TLe, (46) *Triplectides ciuskeus* TLe, (47) Helicopsychidae sp. T, (48) *Leptorussa darlingtoni* TLe, (49) *Agapetus kimmins* TGs, (50) *Tasmasia acuta* TCa, (51) *Ecnomus russellius* TE, (52) *Oecetis* sp. TLe, (53) *Apsilochorema* sp. THb, (54) *Nousia fuscula* EpLp, (55) *Koorononga pilosa* EpLp, (56) *Leptoperla bifida* PGr, (57) *Riekoperla rugosa* PGs, (58) *Illiesoperla australis* PGs, (59) *Austrocerca tillyardi* PGr, (60) *Acruroperla atra* PAu, (61) *Dinotoperla serricauda* PGr, (62) *Dinotoperla* sp. PGr, (63) *Necterosoma penicillatus* (A)CDy, (64) *Necterosoma penicillatus* CDy, (65) *Macrogyrus oblongus* CGy, (66) *Macrogyrus oblongus* (A)CGy, (67) *Eucyclops* sp. Co, (68) *Macrocyclops albidus* Co, (69) *Ilyocypris australiensis* CrOs, (70) *Afrochiltonia australis* CrAm, (71) *Pleuroxus* sp. CrCd, (72) *Frontiopodopsis* sp. Hy, (73) *Limnesia* sp. 2 Hy, (74) *Aspidiobates* sp. Hy, (75) Hydracarina sp. Hy, (76) *Limnesia* sp. Hy, (77) *G. olidus* PGa, (78) Terrestrial insects. (Redrawn from Closs and Lake 1994).

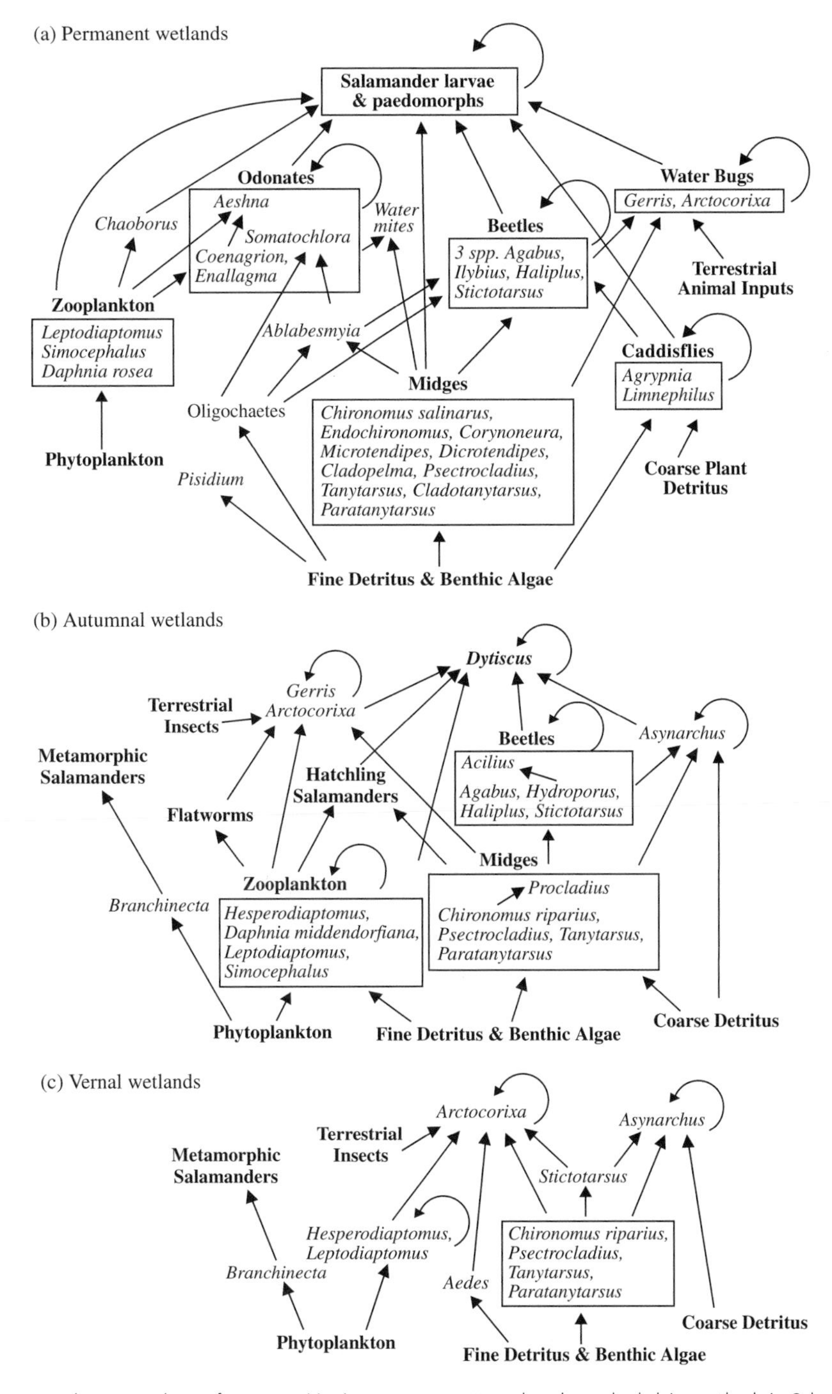

Figure 6.11 Constructed energy pathways for communities in permanent, autumnal, and vernal subalpine wetlands in Colorado (redrawn after Wissinger *et al.* 1999). The links are largely inferred from dietary data.

food webs may perhaps be implied. As noted previously, connectance is a measure fundamental to food-web characterization, and has been used to evaluate ecosystem stability. Several modelling studies have related lower values of connectance to higher system stability (e.g. Chen and Cohen 2001), enabling faster system recovery after disturbance. However, Law and Blackford's (1992) model suggests that more-connected systems exhibit a greater number of reassembly pathways, allowing them faster recovery. Montoya and Solé used such contrasting findings to point out that connectance, alone, is an inadequate measure to assess whether a species-rich community is more or less stable than a species-poor one. In its stead, they introduced a new measure, the link distribution frequency (defined as the frequency of species S_L with L links), and pointed to a need to develop new theoretical approximations to food-web assembly. In attempting to apply such models to temporary water communities, it should be remembered that a community's response and resilience to disturbance may differ depending on whether the latter is a random, 'one-off' event, or, as in the case of intermittent waters, a cyclically predictable one.

Some idea of how food-web characteristics may differ along the habitat-persistence gradient can perhaps also be gleaned from comparison of the few studies that have examined one end of the gradient. For example, Peterson and Boulton (1999) exposed natural epilithon from two permanent and two recently re-wetted sections of an intermittent stream to grazing by newly hatched tadpoles of the spotted marsh frog, *Limnodynastes tasmaniensis*. Their findings suggested that algal biomass in the intermittent stream was more accessible to ingestion and more digestible than epilithon from the permanent streams. Although there were clear differences in algal assemblages from the two stream types (diatom-dominated in the intermittent stream vs. a greater proportion of cyanobacteria in the permanent streams), Peterson and Bolton concluded that ingestion efficiency was more influenced by among-site differences in mat structure (cohesive vs. more flocculent) and differences in the degree of algal species stratification within mat assemblages of different age. Although, in this case the major trophic links (primary producer-herbivore) remained the same, there were changes in the primary producer assemblages (and thus possibly in connectance) in the intermittent stream that may allow the more rapid development of anurans that is known to occur in such waters (Newman 1989).

In a comparable study to that of Closs and Lake on the intermittent Lerderderg River in Australia, Tavares-Cromar and Williams (1996) constructed detailed food webs for Duffin Creek, a permanent stream in Ontario. Like the Lerderderg River, the food web of Duffin Creek was also detritus-based, but there was much less variability in total species richness over time in the latter (31–39, compared with 24–53 for the intermittent stream). However, in the Duffin Creek community, although the proportions of basal species were relatively constant, the proportions of intermediate and top species did change with time (e.g. from proportions of 0.36–0.78; October to February, respectively for top species), likely related to life history strategy events. The increase in the number of predator species in the intermittent stream resulted in an increase in mean food chain length. Weak links made up the largest proportion of total links in the Duffin Creek web, which agrees with the findings of other studies, but omnivory was more common than has been indicated in other webs and was likely due to ontogenic diet switching, which did not appear to be linked to changes in resource abundance. Tavares-Cromar and Williams surmised that some ecological systems may be more affected by temporal processes than others and that some food webs will be more variable. For example, a system in which generation times are fast, where diet switching is common, and where some members are present at certain times of the year but not others, will show a greater amount of variation than a system in which generation times are long, little dietary switching occurs, and the same members are always present at the same time.

Beaver (1985) produced an interesting analysis of food-web structure in *Nepenthes* pitcher-plant reservoirs and demonstrated substantial differences that appeared to be related to the size of the country of origin together with its degree of spatial and temporal isolation, the size of the local species pool capable of colonizing the pitchers, and to the number of *Nepenthes* species present. He found that relative to *Nepenthes* species near the evolutionary centre of the genus (Indo-malaysia), outlying species in the Seychelles, Sri Lanka, and Madagascar have: (1) fewer species living in them (predators and prey); (2) fewer and smaller guilds of species; (3) more apparently unfilled niche space; (4) less complex food webs; and (5) greater connectance (Figure 6.12). Despite these differences, however, the ratios of prey to predators, and of connectance to the total number of trophic types present remain almost constant.

6.5.2 Prey responses to predation risk

Although the populations of many species in temporary waters are strongly influenced by physicochemical conditions, the influence of predators, especially invertebrate taxa (Blaustein 1998), is also significant. Bilton *et al.* (2001) have proposed that the relative importance of predation in structuring communities increases with hydroperiod length, and also pond size. However, Schneider and Frost (1996) have suggested that the effect is largely time dependent, showing that, in systems with hydroperiods of between 50 and 100 days, predation has no substantial influence.

Early in the hydroperiod, prey species capable of rapid hatching and growth are able to take advantage of the microbially enriched water and bottom detritus, and of early phytoplankton blooms, in relative safety. Typically, these taxa, such as anostracans and mosquitoes, have few defences other than to attain a sufficient body size that protects them against the small instars of early colonizing predators. Interestingly, only a few species of predatory beetle (e.g. belonging to the dytiscid genus *Agabus*) overwinter as eggs and

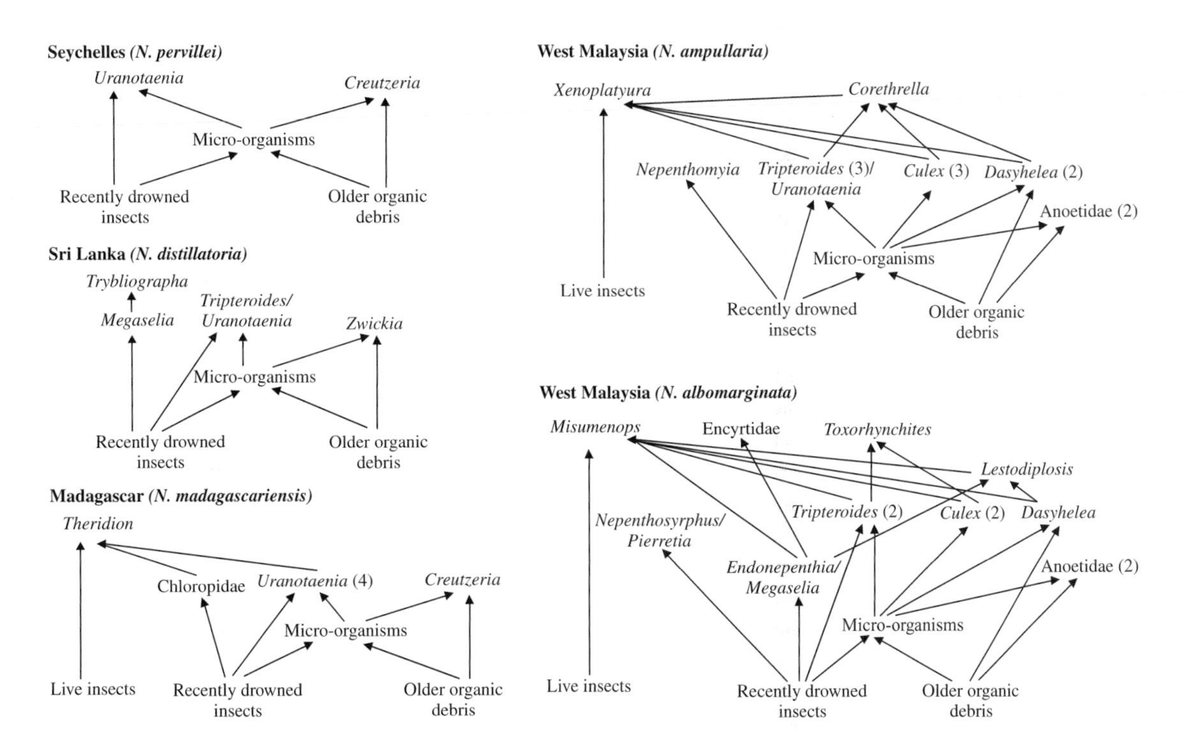

Figure 6.12 Food webs constructed for five species of pitcher plant (*Nepenthes*) in four countries. The numbers in brackets represent the number of species present where greater than one (redrawn after Beaver 1985).

are able to capitalize on these prey populations, and the eggs of the other major invertebrate predator group, the odonates, tend not to hatch immediately (Higgins and Merritt 1999). Predation pressure grows as the hydroperiod progresses and may be especially high as the water volume decreases towards the end of the hydroperiod when, despite a concentrating effect, the abundance of smaller species may drop dramatically, due to increased predation. Despite later arrival in the habitat, the protein-rich diet and high assimilation efficiencies of predators allows them to complete their life cycles.

Hanski (1987) has pointed out that heterotrophic succession in ephemeral habitats frequently involves generalist predators. Some of these exhibit non-random spatial distributions (as evident in the temporally coordinated, mass migration events of dytiscid beetles; e.g. Kingsley 1985) and can have a significant and direct impact on regional dynamics—as the mortality they cause affects the numbers of colonizers of new patches. Wellborn *et al.* (1996) have suggested that among species that inhabit temporary ponds, for example, distributions may often be constrained due to the fact that traits that enhance developmental rate and competitive ability also tend to increase susceptibility to predators. These authors drew up a schematic model of mechanisms generating community structure along a temporary-permanent habitat gradient in which predation vectors figure prominently (Figure 6.13). It is clear, then, that for certain species, temporary waters represent refuge habitats from predation. For example, larvae of the beetle genus *Cyphon* (Scirtidae) are characteristic of small seasonal ponds in Scotland, but not of nearby permanent ponds where *Notonecta glauca* is abundant. The later is an erratic colonizer of temporary waters, but when experimentally introduced into ponds containing *Cyphon* it reduced the latter's numbers by more than 50% (Jeffries 1996). In pools in the Negev Desert, Israel, another notonectid, *Notonecta maculata*, has been shown to act as an important organizer of invertebrate community structure in that it significantly reduces, and sometimes eliminates, larger pelagic or neustonic species, but does not affect the densities of smaller or benthic species (Blaustein 1998).

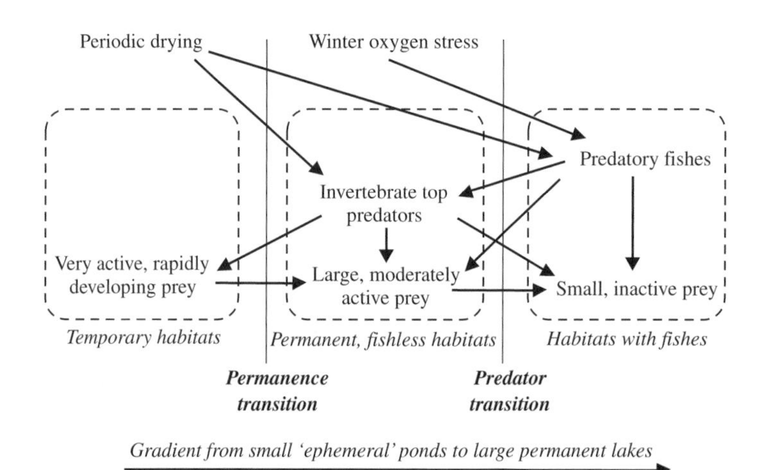

Figure 6.13 A schematic model of mechanisms generating community structure along a temporary–permanent habitat gradient in which predation vectors figure prominently. The arrows indicate directions of negative effects. Thick arrows signify very strong effects, such as constrain the distribution of affected species. Thinner arrows signify weaker trophic interactions that likely do not prevent the coexistence of interacting species. Strong interactions, resulting chiefly from predation, result in distinct transitions in community structure along the gradient. One major transition, the permanence transition, occurs between temporary habitats that contain few predators, and permanent habitats that contain many invertebrate predators. A second major transition in community structure, the predation transition, occurs between permanent fishless and fish-containing habitats (redrawn after Wellborn *et al.* 1996).

The mechanisms by which these reduction are achieved range from direct consumption (e.g. of ephydrids) to alteration of the oviposition behaviour by female mosquitoes (*Cs. longiareolata*) which deposit fewer egg rafts in pools containing *N. maculata*. Wissinger *et al.* (1999) studied foraging trade-offs along a habitat permanence gradient in the subalpine wetlands of Colorado and concluded that the tiger salamander *Ambystoma tigrinum* represented a keystone predator that strongly influenced the composition of both benthic and planktonic communities. When the two caddisfly species *Asynarchus nigriculus* and *Limnephilus externus* occurred together in temporary habitats, the high activity level and aggressiveness of the former resulted in strong asymmetrical competition (as a result of intraguild predation) with the latter. However, in permanent habitats, these behaviours increased the vulnerability of *A. nigriculus* to predation by salamanders.

On a much broader scale, Kerfoot and Lynch (1987) examined the role of predation refugia in terms of the historical interaction between predators and prey across the temporary–permanent water gradient, using planktivorous fishes and branchiopods as their model organisms. They argued that, initially, branchiopods were globally distributed in marine, estuarine, and freshwater environments. However, during the mid-Mesozoic marine 'revolution' (Vermeij 1977), there was a gradual restriction of conspicuous, large-bodied forms to fish-free environments, accompanied by the extinction of many taxa. These changes were thought to have been a direct result of the evolutionary increase in the intensity of fish predation during this period. Kerfoot and Lynch cited three pieces of evidence to support their claim: (1) the modern-day restriction of formerly widely distributed, large-bodied groups (Anostraca, Notostraca, and Conchostraca) to temporary or fish-free habitats; (2) latitudinal gradients observed in smaller-bodied taxa (Cladocera) that suggest exclusion of more visible species from regions of year-round fish activity (tropical lowlands); and (3) a historical correspondence between the evolution of fish feeding modes, the extinction of large-bodied species, and the appearance of modern microcrustacean assemblages. Where fishes do have access to temporary waters, they have been shown to influence community structure. For example, the presence of *Galaxias olidus* in the summer pools of an intermittent stream in southeastern Australia significantly reduced both the distribution and abundance of air-breathing nektonic insects. In contrast, non-surfacing taxa increased in abundance in the presence of these fish, possibly as a consequence of the removal of dytiscid beetles (Closs 1996). Rosenberger and Chapman (1999) demonstrated that the introduction of the Nile perch (*Lates niloticus*) into the Lake Victoria basin of East Africa coincided with the disappearance or decline of hundreds of indigenous fish species. However, this predator was not able to extend into adjacent wetland habitats where several of its previously open-water prey species achieved protection. Chan (1997) observed a similar shift in species abundance within the communities of *Nepenthes* pitcher plants in Singapore, where unique prey community structure in clumps of *N. ampullaria* containing large predatory species was attributed, in part, to selective removal of superior competitors and the resultant colonization of inferior competitor species that were more resistant to predation.

Sometimes, predators alter prey population structure rather than eliminating them altogether. For example, in Sycamore Creek, an intermittent stream in the Sonoran Desert, Arizona, the belostomatid *Abedus herberti* preferred live prey of small and medium size. However, it would also take large individuals if they were dead (Velasco and Millan 1998). In semi-permanent prairie wetlands in Minnesota, Zimmer *et al.* (2000) found that the presence of fathead minnows (*Pimephales promelas*) significantly lowered the abundance of most invertebrate taxa, although the populations of corixids responded positively to these fish. Blaustein (1990) found, similarly, that the presence of a predatory flatworm (*Mesostoma* nr. *lingua*) reduced the densities of two mosquito species (*Culex tarsalis* and *Anopheles freeborni*), as well as cladocerans and ostracods, in Californian rice fields.

Predator pressure also may induce phenotypic plasticity in prey populations. For example, in bog-ponds in Sweden to which perch (*Perca fluviatilis*) have access, nymphs of the dragonfly *Leucorrhinia dubia* develop significantly longer spines on abdominal segments 4, 6, and 9 than do nymphs in fishless ponds (Johansson and Samuelsson 1994). From observations that the densities of crustacean species living in Israeli temporary pools were dramatically reduced by the non-physical presence of larvae of the fire salamander (*Salamandra infraimmaculata*), Blaustein (1997) proposed that the eggs of *Arctodiaptomus similis*, *Ceriodaphnia quadrangula*, and *Cyzicus* sp. may have the ability to detect this predator via chemical cues and delay hatching.

In a more complex setting, Black (1993) reported that populations of *Daphnia pulex* from a small pond in Wisconsin were subject to predation both by larvae of the phantom midge *Chaoborus americanus* and by *Notonecta undulata*. In response to *Chaoborus*, *D. pulex* developed protective neck teeth and delayed maturity, but in response to *Notonecta* the cladoceran displayed rapid juvenile growth to a large size at first reproduction, little growth beyond maturity, and high reproductive output. Faced with these two predators simultaneously, *Daphnia* appeared to be able to interpret cues from both and responded with a unique combination of life history and morphological traits. Peacor and Werner (2001) recently examined the consequences of prey trait modification in response to predation risk in experiments using larval dragonflies (*Anax* sp.) as predators on bullfrog tadpoles (*Rana catesbeiana*). Based on their findings, they argued that trait modification can influence the rate at which prey can acquire a resource, and thus a predator can have a non-lethal impact on prey that can result in indirect effects on other members of the community. They termed these interactions 'trait-mediated indirect interactions' and suggested that such non-lethal effects may be as large as those resulting from actually killing prey. Indeed, they speculated that, across a broad range of ecological communities, many effects traditionally attributed to predators consuming prey, might actually result from predators inducing changes in the traits of their prey—such as behaviour (Stoks *et al*. 2003).

One aspect of predation that is exhibited within the oftentimes close confines of small temporary waterbodies is cannibalism. Sherratt *et al*. (1999) examined populations of two mosquito species that frequent tree holes for evidence of kin discrimination; they found none. Sympatric populations of the predatory *Toxorhynchites moctezuma* and the largely detrivorous *Trichoprosopon digitatum* were just as likely to consume members of their own species as of the other. The size-dependent nature of cannibalism in *Tr. digitatum* is thought to preclude preferential consumption of non-relatives, and the direct nutritional and indirect competitive benefits of indiscriminate cannibalism in *Tx. moctezuma* may outweigh the costs of consuming kin. Individuals of *Tr. digitatum* that, under experimental low food conditions, were given the opportunity to cannibalize survived better than those that were not (Church and Sherratt 1996). Fincke (1994) observed an interesting case of 'obligate' killing among nymphs of the giant damselfly *Megaloprepus coerulatus* in water-filled tree holes in Panama. In experiments using 0.4 litre holes, and despite the presence of abundant tadpole prey, the larger of two nymphs invariably killed the smaller one, but seldom ate it. This obligate killing was density dependent as, for example, in 3 litre holes with plentiful prey, conspecific killing stabilized at one nymph per 1–1.5 litres, which was similar to field-recorded densities. Fincke interpreted this cannibalism as a means of reducing the number of potential competitors for food, as well as potentially providing additional food.

CHAPTER 7

Other temporary water habitats

7.1 Introduction

Leeuwenhoek (1701) studied the faunas of roof gutters and concluded that the animals he found, predominantly rotifers and tardigrades, could be completely dried out without losing vitality—an obvious advantage to living in such a precarious environment. The purpose of this chapter is to briefly consider examples of such temporary freshwater habitats. The range in habitat type is vast, from the rainwater retained in footprints to reservoir shores, and most have been little studied even though many are commonplace. A few, such as tree holes and the chambers of insectivorous plants, have received more attention and will be dealt with in more detail.

The plants and animals living on the shores of permanent lakes and ponds in which water levels fluctuate widely are subject to many of the same adversities as temporary water inhabitants. There are many examples of wide fluctuation of water level in natural lakes, particularly in the tropics. Lake Chilwa in Malawi, Africa, is a prime example. It lies in a tectonic depression and covers an area of approximately 700 km^2. At high water, it spreads outwards to form flooded marshes that are dominated by *Typha domingensis* (bulrush) and *Aeschonomyne pfundii* (ambatch). In recent times, the change in level is of the order of 1–3 m, vertical height, but the locations of ancient beaches indicate that, in prehistoric times, fluctuations may have been as great as 30 m. The peripheral water is less saline than that of the main basin and supports a benthic fauna consisting primarily of chironomid larvae, dragonfly, and mayfly nymphs, leeches, and pulmonate snails. Of the 30 fish species recorded from the lake, only 3 are common in the open lake. Many fishes take refuge in swamps and scattered pools around the basin in times of drought (Beadle 1981). The importance of swamps and floodplains will be examined in more detail in Chapter 8.

Hynes (1961) observed the effects of such fluctuations on the littoral fauna of Llyn Tegid in Wales. This is a lake used for flood control and the study was made soon after this was implemented, a time when the water level range was about 5 m. The study showed that many of the original littoral species were wiped out, but that new and dominant species were arriving that could tolerate these conditions. Upon later reduction in the degree of waterlevel fluctuation, some of the temporary water forms were replaced by species from the original permanent water community (Hunt and Jones 1972). As global waterbodies become more and more subject to waterlevel control, the preponderance of this type of temporary aquatic habitat steadily increases. Ward and Stanford (1979) make reference to more than 12,000 dams, greater than 15 m high, that have been constructed on the world's major rivers. The total surface area of reservoirs thus created is in excess of 300,000 km^2. In these situations, temporary water habitats are created both on the shores of the reservoirs and on the margins of the receiving rivers.

On a much smaller scale, many temporary waterbodies are grouped under the term 'container' habitats. These include natural forms such as: the leaf axils of plants (especially of tropical bromeliads); teasels; bamboo habitats—some of which are formed when chrysomelid beetle larvae bore into large bamboo stems; the decaying pulp of split cocoa pods; rat-gnawed coconuts; split paw-paw stems; empty snail shells; and cup fungi (Mattingly 1969). Some of these habitats are extremely small. The total volume of water contained in the bract of a pineapple plant, for example, may be as little as $10\,cm^3$, yet it may support populations of two species of mosquito (Barton and Smith 1984). Such water-retaining structures on terrestrial plants were termed 'phytotelmata' by Varga (1928) and are found on at least 1,500 species of plant, belonging to 29 families. Although phytotelm habitats contain species representing many of the major aquatic insect orders, the Diptera are the most common, with more than 20 families reported. It is estimated that some 15 genera, containing 400 species, of mosquito inhabit these waterbodies, and although they show varying degrees of host–plant specificity, few species also occur in non-phytotelm habitats—this may be due to chemical oviposition cues of plant origin (Fish 1983).

In coconut habitats, there is a succession of mosquito species as the organic content of the water decreases and is diluted by rainwater. Mattingly (1969) determined that these rat-gnawed coconuts are important breeding sites for mosquitoes of the *Aedes scutellaris* complex, which are vectors of the debilitating filarial worm, *Wucheraria bancrofti*, that causes human elephantiasis. Such is the considered importance of these habitats to these mosquito vectors, that control of rats is a recognized ancilliary procedure for control of filariasis in the Pacific.

Man-made microhabitats also harbour important tropical vectors. For example, the mosquito *Aedes aegypti* breeds in water that collects in rain-barrels, cisterns, tin cans, jars, iron cooking pots, and old motorcar tyres; it transmits yellow fever.

Many species that are to be found breeding in container habitats specifically seek out these small habitats. As early as 1927, Buxton and Hopkins illustrated this in their study of the factors influencing oviposition of *Aedes polynesiensis* and *Ae. aegypti* on Pacific islands. Apart from finding that female egg laying was influenced by such factors as light and darkness, wind, rain, landing places, temperature, humidity, and salinity, they also observed a preference for small containers. More eggs were collected from containers 15 cm in diameter than from a series of smaller and larger containers. Drought resistance in such container-inhabiting species is provided by the egg stage (Lounibos 1980).

Another early study, Müller (1880), showed that container habitats are not the exclusive domain of insects, recording microcrustaceans in the axils of Brazilian bromeliads. Since then, a variety of crustacean taxa, particularly ostracods, has been recorded (Victor and Fernando 1978).

For insects, at least, container habitats may represent ancient environments. Barton and Smith (1984) summarized the evidence for this along the following lines: (1) much of the fauna is derived from the more primitive lines within aquatic insect taxa, or the fauna is of tropical origin. This is especially true of the Diptera—most of the families that are successful in container habitats belong to the suborder Nematocera, and many temperate zone species are derivatives of large, tropical groups. The pitcher plant mosquito of the nearctic (*Wyeomyia smithii*), for example, belongs to the tribe Sabethini, which is a successful and widely distributed group in the tropics; (2) breeding in container habitats is cosmopolitan in occurrence, having evolved independently in many taxa and in many regions; (3) ecological studies (although limited in number) indicate, particularly for the Diptera, a number of morphologically and behaviourally similar trends which suggest a long association between species and habitat; and (4) high specificity for certain types of container habitat by some species, for example *W. smithii*, is found only in pitcher plants.

7.2 Phytotelmata

A broad variety of plants retain water within cupped structures on their external surfaces, the latter including leaf axils, bracts, pitchers, and internodes. Istock *et al.* (1975, 1976) list the following examples: Gramineae (bamboos), Pandanaceae (screwpines), Palmae (palms), Agavaceae (*Dracaena*, etc.), Araceae (taros), Amaryllidaceae (*Crinum*, etc.), Musaceae (bananas and relatives), Bromeliaceae (pineapples, bromeliads), Cytinaceae (rafflesias), Nepenthaceae (climbing pitcher plants), and Sarraceniceae (terrestrial pitcher plants)—however, many of these have not had their aquatic communities examined in detail, although their mosquito components are better known. To these, largely tropical and subtropical phytotelmata, must be added a variety of tree hole habitats that are to be found throughout forested regions of the globe (Laird 1988). Kitching (2000) has recently reviewed the natural history and ecology of phytotelmata, with particular emphasis on food-web dynamics. That review also provides a very comprehensive record of the metazoan taxa known from phytotelmata, citing such unlikely groups as nereidid polychaete worms (*Namanereis catarractarum*, from tree holes in Indonesia), and yponomeutid moths (the giant larvae of *Proditrix nielseni* live in the water-filled axils of *Richea pandaniifolius*, an alpine epacrid from Tasmania).

7.2.1 Tree holes

A tree hole may be defined as any cavity or depression existing in or on a tree. Kitching (1971) designated two types, 'pans' which have an unbroken bark lining, and 'rot-holes' which penetrate through to the wood tissue of the tree (Figure 7.1). Both types collect leaf litter but rot-holes contain, in addition, a layer of fungus-rotted wood. Tree holes occur in a wide variety of tree species, especially deciduous ones. Relatively few animal groups are represented in tree-hole communities and these include only a limited number of insect orders, typically dipterans and beetles. Nematocerans such as mosquitoes, chironomids, ceratopogonids, crane-flies, moth flies (Psychodidae), and wood gnats (Anisopodidae) are the most frequently encountered dipterans, whereas marsh beetles (Scirtidae/Helodidae), and pselaphids (mold beetles) are the most common beetles. Occasionally, other taxa are present, for example, mites, some of which may be predators or parasites of the dipteran larvae. Fashing (1975) found *Naiadacarus arboricola* exclusively in tree holes of a number of tree species in Kansas. Small crustaceans, rotifers, and protozoans have been recorded also (Rohnert 1950).

Rohnert (1950) proposed a classification scheme for the faunas of tree holes based primarily on

Figure 7.1 An example of a tree-hole habitat at the base of a spruce tree (note the heavy accumulation of both coniferous and deciduous leaf litter).

their degree of specificity towards these habitats: *Dendrolimnetoxene*—which includes small taxa such as protozoans and rotifers that are readily dispersed by passive means, their occurrence in tree holes being largely accidental; *Dendrolimnetophile*—which includes aquatic and semi-aquatic species found in other habitats, such as forest floor litter, pools, and marshes (e.g. various families of Diptera, woodlice, and air-breathing snails); and *Dendrolimnetobionten*—which includes species of Diptera and Coleoptera that are obligate inhabitants of tree holes. This classification can also be applied to other container habitats.

Kitching (1971) found that in Wytham Wood, England, the inhabitants of tree holes fed saprophagously and had a variety of life cycles. Some, such as the beetle *Prionocyphon*, took two or more years to complete a generation, while the midge *Metriocnemus* completed up to three generations each year. Most species spent the winter as quiescent larvae or adults, and the densities of all species were higher in holes in the forest canopy layer than closer to the ground; the latter was attributed to better quality detritus in higher holes.

The probable mechanism for partitioning the detritus among the fauna of the Wytham Woods tree holes is illustrated in Figure 7.2. The two species feeding on large particle detritus have different habits—the scirtid (helodid) beetle breaks down leaf particles within the detritus layer, while the syrphid fly grazes on particles on the surface of the detritus. Potentially, the two midges would seem to compete for small-particle detritus, but the larvae are temporally segregated—the chironomid is most common in the summer, while the ceratopogonid is most common in the winter. Competition between the two mosquito species is avoided because *Aedes geniculatus* can exploit small particle detritus as well as organic matter in suspension, and it therefore has a food refuge. In addition, *Anopheles plumbeus* is comparatively rare, and in practice, the two species seldom coincide (Kitching 1983).

In a study of *Aedes triseriatus* in laboratory microcosms, Carpenter (1982) showed that the presence and type of leaf detritus were more critical for larval growth and development than the chemical composition of stemflow (the water that runs down tree trunks during rain)—although the latter also affected survival and adult emergence. In particular, stemflow pH and concentrations of ammonia, nitrate, and sulphate were identified as being influential.

Small habitats that are temporally and spatially variable, and thus unpredictable, should select for species that are phenotypically plastic, genetically polymorphic, and that exhibit frequent reproduction (iteroparity) and dispersed oviposition (Giesel 1976). Watts and Smith (1978) showed much of this to be true in a study of *Toxorhynchites rutilus*, a predatory, tree-hole mosquito. This species shows precocious oogenesis in that females emerging

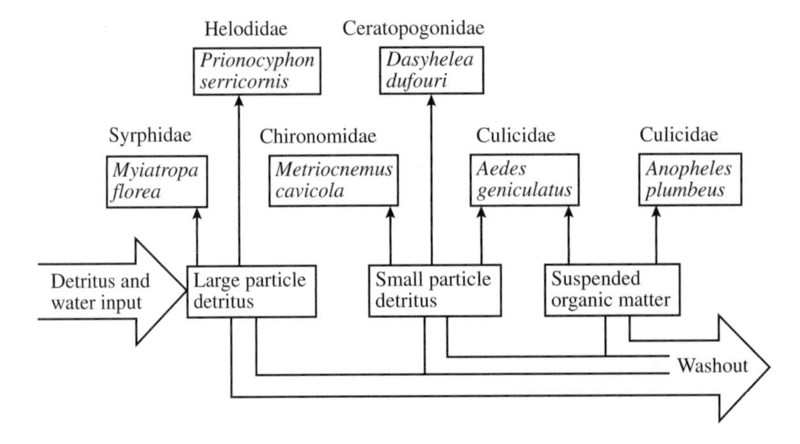

Figure 7.2 The food web for the tree-hole community of Wytham Woods, England (redrawn from Kitching 1983).

from the pupal skin contain nearly mature eggs. This is a feature common to many short-lived insects, but *Tx. rutilus* has a long lifespan compared with other mosquitoes. Ovarian precocity must ensure that some eggs will be deposited in the tree hole in which the female was raised. Coupled with this, *Tx. rutulis* exhibits oviposition behaviour that can be interrupted and which enables the female to lay between 1 and 7 eggs in a number of different tree holes. In addition, reproductive effort by females is continuous and asynchronous, rather than the periodic and synchronous pattern found in most other nematoceran Diptera. Presumably, this allows female *Tx. rutilus* a reserve of eggs that may be deposited whenever these small aquatic habitats become available. Lounibos *et al.* (1997) studied the long-term dynamics of *Tx. rutilus* with respect to its common prey species, *Aedes triseriatus*, and found that although populations of the latter frequently became locally extinct in tree holes, these events were not driven by the presence of the predator.

Some interesting information on the similarity of tree-hole habitats and their faunas between different parts of the world has been provided by Kitching (1983) who compared the microclimates of tree holes in Wytham Woods with those in Lamington National Park, Queensland, Australia. In Wytham Woods, a mixed, deciduous forest, the tree holes were situated most commonly among buttress roots and branches. The average surface area of each reservoir was 423 cm^2, and they were on average 25.2 cm deep. pH was around 6.4, and the mean conductivity was 339 $\mu S\,cm^{-1}$. The tree holes received an annual leaf litter input equivalent to approximately 934 $g\,m^{-2}$. Lamington Park is primarily subtropical rainforest and most tree holes occur among buttress roots. Mean reservoir surface area was 422 cm^2 and they were 15.8 cm deep on average. pH was around 5.9, and the mean conductivity was 226.5 $\mu S\,cm^{-1}$. Leaf litter input was estimated to be 903 $g\,m^{-2}$. The Wytham Woods tree holes supported a fauna consisting of one chironomid species, one ceratopogonid, two mosquitoes, a syrphid fly, and a helodid beetle. The Lamington Park tree-hole fauna comprised one chironomid species, one ceratopogonid, two mosquitoes, a helodid, three mites, and a frog; in more than one group the genera present were the same as in Wytham Woods (Table 7.1). The two systems were thus very similar, both faunistically and in terms of their physicochemical environments. Both had comparable trophic bases (largely saprophagic) but the Australian system had a substantial component of predators.

In topical regions, tree holes may support somewhat different taxa. For example, in the Kakamega Forest, Kenya, Copeland *et al.* (1996) found nymphs of the dragonfly *Hadrothemis camarensis* in 46% of available tree holes. Nymphs were more typically found in holes with larger surface areas and in those higher above the forest floor. Smaller nymphs fed predominantly on chironomid larvae, whereas larger nymphs ate culicid larvae. In a survey of tree holes in Brunei, Borneo, Kitching and Orr (1996) found a similar dominance of odonates. The damselfy *Pericnemis triangularis* occurred in 40% of the holes examined, with a mean density of 3.4 nymphs per hole, but a maximum of 40. Two other odonates were present at lesser frequencies and densities: *Lyriothemis cleis* (12%; mean = 0.28; max = 7) and *Indaeshna grubaueri* (2%). In total, the 52 sites sampled yielded 20 species, although the average number per hole was four (range 2–8). The fauna included one or more species of anuran, dytiscid and scirtid beetle, culicid, chironomid, ceratopogonid, phorid, syrphid, oligochaete, and copepod. The most frequently encountered were larvae of a single scirtid species (75% of holes), culicids (62%; *Toxorhynchites, Uranotaenia, Tripteroides*), chironominine chironomids (56%; *Polypedilum* sp. and *Paramerina ignoblis*), and ceratopogonids (35%; likely *Dasyhelea* or *Culicoides*). Kitching and Orr noted a similarity in the Brunei tree-hole community with those from other Old World tropical locations (northern Australia, New Guinea, and Indonesia), but pointed to the much enhanced macrosaprophage component (six species) in Borneo. They concluded that the latter was probably the result of this region's permanently humid, aseasonal, equatorial climate—the others having more well-defined wet and dry seasons. The Borneo climate likely: (1) permits a high and

Table 7.1 The faunas of water-filled tree holes in Wytham Woods, England and Lamington National Park, Queensland

Taxa	Wytham Woods	Lamington National Park
Insecta		
Diptera		
Chironomidae	*Metriocnemus cavicola*	*Anatopynia pennipes*
Ceratopogonidae	*Dasyhelea dufouri*	*Culicoides angularis*
Culicidae	*Aedes geniculatus*	*Aedes candidoscutellum*
	Anopheles plumbeus	*Aedes* sp.
Syrphidae	*Myiatropa florea*	—
Coleoptera		
Scirtidae/Helodidae	*Prionocyphon serricornis*	*Prionocyphon* sp.
Arachnida		
Acari		
Astigmata		
Hyadesiidae	—	New genus
Prostigmata		
Arrenuridae	—	*Arrenurus* sp.
Mesostigmata		
Ascidae	—	*Cheiroseius* sp.
Amphibia		
Anura		
Leptodactylidae	—	*Lechriodus fletcheri*

Source: From Kitching 1983

sustained rate of breakdown of the allochthonous detritus which provides the energy base for the tree-hole food web; and (2) maintains water in the tree holes throughout the year, providing a more predictable environment in which long-lived macrosaprophage species can exist.

Despite an overall richer fauna in the tropics, the above studies indicate that tree-hole communities from around the world appear to have several characteristics in common, most notably a limited number of species present in individual holes, and a stricking similarity in the taxa present. Further, the presence of plant-litter-shredding scirtid larvae seems essential as keystone decomposers in the system. Paradise and Dunson (1997) have demonstrated that, in tree holes in Pennsylvania, *Prionocyphon discoideus* and *Helodes pulchella* function as bottom-up facilitators of particle-feeding chironomid populations—specifically, under resource limitation, the latter grew larger in the presence of the scirtids. Kaufman *et al.* (1999) have shown that larvae of another very prominent tree-hole species, *Aedes triseriatus*, feed primarily on microorganisms actively filtered from the water column and grazed from plant-litter surfaces, and that larval presence can alter microbial community structure. Interestingly, Paradise (1999) has shown that the presence of scirtid larvae also influences the growth of this mosquito species, with the effects being either positive or negative, depending on resource level (low and high, respectively).

7.2.2 Pitcher plants

Pitcher plants occur in both the old and new worlds, where they are represented by species of the genera *Sarracenia* and *Nepenthes*, respectively. As aquatic environments, their habitable volume is relatively small and varies with species—ranging from 20 to 30 cm^3 in *Sarracenia* to several litres in the large *Nepenthes rajah* of northern Borneo. All are associated with a variety of arthropods, chiefly insects, but some vertebrates also make use of their fluid reservoirs. Many animal species have become resident in these chambers where they take advantage of trapped prey species, others, however, are herbivores and feed on pitcher plant tissues.

At least eight species of *Sarracenia* exist, primarily in the southeastern United States, but *S. purpurea* extends as far north as Labrador, and west to British Columbia. They occur in boggy locations characterized by saturated, low nutrient soils, low pH, and frequent fire, including floodplain terraces, sinkhole pond margins, *Sphagnum*-mat bogs, and springhead seepages. The pitchers are formed in tubular leaves that act as passive pitfall traps, but production of attractant chemicals is known in at least one species (Miles *et al.* 1975). Fresh pitchers are typically produced in the spring, but with a second batch in summer. Various downward-pointing hairs and waxy sections of the inner walls, together with low surface tension of the chamber fluid surface and the presence of the insect-paralyzing compound, coniine, serve to retain prey species (Folkerts 1999). Folkert (1999) produced a summary (Table 7.2) of the invertebrate

Table 7.2 Invertebrate species commonly associated with *Sarracenia* pitcher plants in the southeastern U.S.

Taxa	Sarracenia species	Relationship
Bombidae (several spp.)	Large-flowered spp.	Pollinators
Megachilidae (several spp.)	Small-flowered spp.	Pollinators
High diversity of arthropods (insects, spiders, millipedes, etc.)	Some specialization By different spp.	Prey
Culicidae: *Wyeomyia smithii*	*S. purpurea*	Prey consumers
Chironomidae: *Metriocnemus knabi*	*S. purpurea*	Prey consumers
Sarcophagidae: *Fletcherimyia fletcheri*	*S. purpurea*	Prey consumers
Sarcophagidae: *F. celerata*	Unknown host specificity	Prey consumers
Sarcophagidae: *F. jonesi*	Unknown host specificity	Prey consumers
Sarcophagidae: *F. rileyi*	Unknown host specificity	Prey consumers
Sarcophagidae: *Sarcophaga sarraceniae*	Unknown host specificity	Prey consumers
Sciaridae: *Bradysia macfarlanei*	Possibly all spp.	Prey consumers
Chloropidae: *Aphanotrigonum* sp.	*S. leucophylla* and tall forms	Prey consumers
Acari: *Sarraceniopus gibsoni*	*S. purpurea*	Prey consumers
Acari: *S. hughesi*	All except *S. purpurea*	Prey consumers
Sphecidae: *Isodontia mexicana*	Tall-form spp.	Capture interrupters
Araneae: Salticidae: *Phidippus* spp.		Capture interrupters
Araneae: Oxyopidae: *Peucetia viridans*		Capture interrupters
Other spiders and occasional Mantidae		Capture interrupters
Tortricidae: *Endothenia daeckeana*	*S. purpurea*	Herbivores on fruit
Tortricidae: *E. hebesana*	All spp. ?	Herbivores on fruit
Noctuidae: *Exyra semicrocea*	All spp.	Herbivores on pitcher
Noctuidae: *E. ridingsii*	*S. flava* only	Herbivores on pitcher
Noctuidae: *E. fax*	*S. purpurea* only	Herbivores on pitcher
Tortricidae: *Choristoneura parallela*	Tall-form spp.	Herbivores on pitcher
Aphidae: *Macrosiphum jeanae*	*S. purpurea*	Herbivores on pitcher
Diaspididae: ?*Aspidiotus* sp.	*S. psittacina/S. purpurea*	Herbivores on pitcher
Noctuidae: *Papaipema appassionata*	*S. purpurea*/possibly *S. flava*	Herbivores on rhizomes
Araneae: *Strotarchus piscatoria*		Predator of *Exyra* spp.
Acari: *Macroseius biscutatus*	All except *S. purpurea*	Predator of several spp.
Sarcophagidae: ?*Senotainia trilineata*		Parasite of *I. mexicana*
Chalicidae: ?*Eurytoma bicolor*		Parasite of *I. mexicana*
Tachinidae and Braconidae		Predators of several spp.

Source: From Folkerts 1999

species commonly associated with various species of *Sarracenia* which illustrates the many novel plant–animal interactions that have developed. The type of prey captured typically varies with species of *Sarracenia*, and is largely related to leaf morphology, for example: *S. leucophylla* (with its tall, nectar-producing pitchers) captures mostly dipterans and lepidopterans; *S. rubra* (smallest-leaved species) captures more small dipterans and wasps; *S. minor* captures predominantly ants; and *S. purpurea* traps more spiders than other species of *Sarracenia*.

Table 7.2 lists a number of invertebrate species that have become resident in *Sarracenia* by virtue of adaptation to the harsh conditions of the pitcher fluid (which include the presence of digestive enzymes, especially proteases). In addition, there are now records of megalopteran larvae (*Chauliodes* sp.), caddisfly larvae (*Oligostomis* sp., a phryganeid), and stonefly nymphs (*Leuctra duplicata* and *L. maria*) from *S. purpurea* (Hamilton *et al.* 1996; Turner *et al.* 1996 Hamilton *et al.* 1998). How these immature stages colonize the pitcher chambers is unknown, but migration during bog flooding events has been suggested. Most of these pitcher inhabitants profit from the food resource provided by the accumulation of the bodies of prey.

As was the case for tree holes, dipteran larvae are the most frequent inhabitants of these fluid-filled leaf chambers. Fish and Hall (1978) recorded three species from *Sarracenia purpurea*, the common pitcher plant of glacial peat bogs in North America (Figure 7.3). These species are the flesh fly *Fletcherimyia fletcheri*, the mosquito *Wyeomyia smithii*, and the chironomid *Metriocnemus knabi*. Fish and Hall found that only newly opened pitchers attract and capture insect prey, and that the latter slowly decompose as the pitcher ages, accompanied by a lowering of the pH of the fluid. The relative abundance of the resident insects in the pitcher is also related to pitcher age, as each of the three species consumes insect remains that are at different stages of decomposition. Specifically, the larvae of *Fletcherimyia* feed upon freshly caught prey floating on the fluid surface, those of *Wyeomia* filter feed on the partially decomposed material and microorganisms (bacteria and protoctists) in

Figure 7.3 Photographs of the pitcher plant *S. purpurea* showing the entire plant, the pitchers, and a longitudinal section of one pitcher with insect remains in its base.

the water column, and *Metriocnemus* larvae feed on the remains that collect on the bottom of the pitcher chamber. A seasonal succession in the pitcher plant fauna is thus produced, and interspecific competition for food is minimized. The feeding activities of these three species appear to speed up the release of nutrients (typically nitrogen and carbon dioxide) to the plant, whereas plant tissue processes remove potentially toxic metabolic wastes (ammonia and carbon dioxide), and release oxygen into the chamber fluid (Giberson and Hardwick 1999). In a field experiment, Kneitel and Miller (2002) manipulated both the abundances of resources (in the form of drowned ants) and predators (larvae of *W. smithii*), and found that the abundance of other pitcher inhabitants (mites, rotifers, protozoans, and bacteria) responded positively to resource addition. However, predator increase decreased rotifer abundance, yet increased bacteria abundance and species richness. These authors concluded that the latter indicated that a trophic cascade had occurred via a *Wyeomia*-rotifer-bacteria pathway.

Substantial life history work has been done on the two species of nematocerans in *S. purpurea*. For example, Kingsolver (1979) studied populations of *W. smithii* in northern Michigan and found significant differences in larval development rate, the number of generations per year, and larval mortality, all due primarily to microclimatic effects. He hypothesized that such mixed life history 'strategies' would be favoured in the uncertain microaquatic environment of the pitcher chamber. Lounibos *et al.* (1982) showed an interesting relationship between pitcher conditions and reproductive 'strategies' in *W. smithii* over a wide geographic range. In southern US populations, where density-dependent constraints on larvae were deemed to be perpetually severe, the adult female mosquitoes required a blood meal in order for their eggs to mature. However, in northern populations, which exhibit periods of larval growth that are density independent (the mean number of larvae/pitcher decreases with increasing latitude), haematophagy is lost—as reserves accumulated in the larval stages are sufficient for egg maturation in the adult.

Paterson and Cameron (1982) have shown a similarly complex life cycle for *Metriocnemus knabi*. Their data, based on a study in New Brunswick, Canada, suggested that there were two cohorts in the population. Each cohort had three generations every 2 years, emerging in May and August of one year and July of the next year. One cohort thus produced two generations in even-numbered years, and the other cohort produced two generations in odd-numbered years. Some intermixing of cohorts was possible during the midpoint of the July–August adult emergence period. This complex life cycle may represent an evolutionary adaptation in this species, or simply a pattern repeatedly reinforced by the seasonal characteristics of the host plant. Giberson and Hardwick (1999) summarized the phenology of all three common dipteran species in *S. purpurea* in relation to the latter's own life cycle (Figure 7.4).

Nepenthes is a genus of pitcher plant (Figure 7.5) that evolved in Indo-malaysia and which supports communities of organisms that live either in the pitcher fluid or on the upper walls of the chamber. Almost all of the 70 or so species derive nutrients from insect prey that has fallen into their pitchers—indeed, Moran and Moran (1998) have shown that specimens of *N. rafflesiana* deprived of such prey exhibit reduced chlorophyll, are smaller, and have fewer pitchers. The fauna resembles that seen in *Sarracenia* (even including a species of *Metriocnemus*), but includes some additional taxa. In total, the fauna comprises bacteria, protozoans, rotifers, nematodes, oligochaetes, crustaceans, arachnids, and insects. In addition, frog tadpoles may occur, and also some species of algae (primarily diatoms and desmids), although primary production is not a significant element in the food web (Beaver 1983). Beaver upheld Thienemann's (1932) division of the fauna into three subgroups, based on their frequency and abundance: (1) *Nepenthebiont*—species characteristic of the pitchers, and whose aquatic stages are not known from other habitats; (2) *Nepenthephil*—species that may occur at high densities in pitchers, but which are known to occur in other habitats; and (3) *Nepenthexene*—species that are only occasionally found in pitchers, and in low numbers, typically in

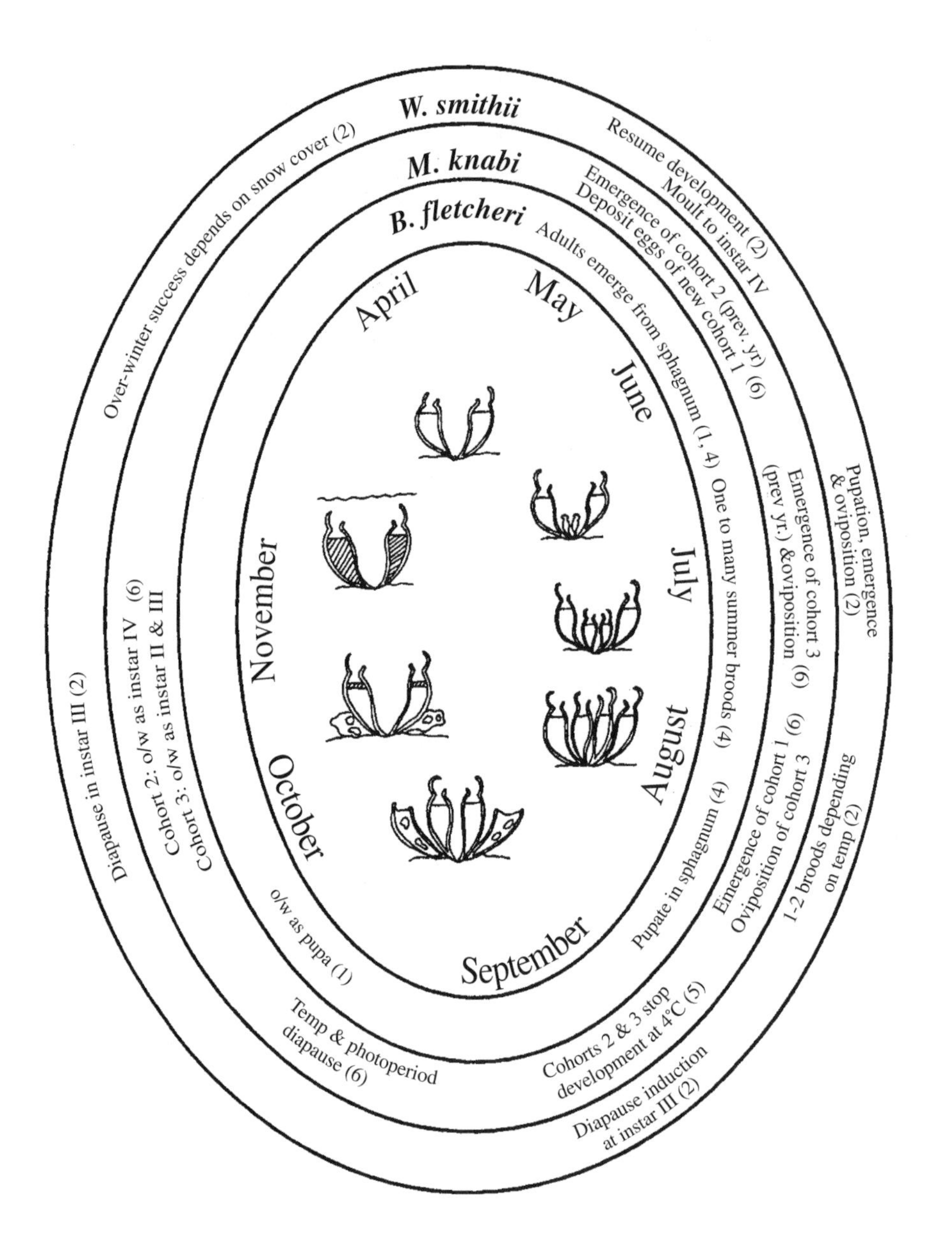

Figure 7.4 Summary of the phenology of the three main dipteran species that live in *S. purpurea*, set in the context of the latter's own life cycle (redrawn from Giberson and Hardwick 1999, and based on: [1] Farkas and Brust 1986b; [2] Farkas and Brust 1986a; [3] Evans and Brust 1972; [4] Forsyth and Robertson 1975; [5] Paterson 1971; [6] Paterson and Cameron 1982).

old pitchers where the other two subgroups are no longer present.

When pitchers first open, their fluid is sterile, but this becomes immediately colonized by bacteria. This is followed by colonization by animals, resulting in a rapid increase in both species and individuals during the first few weeks—which correspond with the period of maximum prey

Figure 7.5 Photographs of *N. rafflesiana* (left) and *N. ampullaria* (not to scale).

input. There then follows a slow decline until after 6 or 7 months, only a few individuals representing a few species persist. Local community composition is thought to depend on both colonization opportunity and biological interactions, with larger patches of plants likely providing greater stability and continuity. Insects, especially culicids, and arachnids (spiders and mites) comprise the bulk of the fauna. Mosquitoes are represented by at least eight genera and 94 species, but another 13 dipteran families are known from *Nepenthes* (including Chaoboridae, Ceratopogonidae, Chironomidae, Mycetophilidae, Phoridae, Syrphidae, and Sarcophagidae), yielding a total of at least 130 species of dipteran. Compilation of records of mosquitoes occurring in *Nepenthes* pitchers in different countries indicates that the different subgenera of *Tripteroides* appear to have radiated into different areas, which Beaver (1983) interpreted as indicating that speciation has followed a few initial colonization episodes of the pitcher plant habitat (Table 7.3). Other insects include odonates (Libellulidae), lepidopterans (Noctuidae and Psychidae), and ants. Crustaceans include harpacticoid copepods and small crabs (Beaver 1983).

Section 6.5.1 summarized the findings of Beaver's 1985 study on the structure of the food webs within species of *Nepenthes* from different localities. Beaver concluded that significant differences were evident that appeared to be related to the size of the country of origin (including its degree of spatial and temporal isolation), the size of the local pool of potential colonizers, and the number of *Nepenthes* species present.

7.2.3 Bromeliads

Approximately 2,000 species of bromeliad are known from tropical and warm temperate regions of the Americas, and they range from altitudes of over 4,000 m down to sea level. Tank bromeliads characteristically collect water in their leaf axils, a result of enlarged leaf bases that overlap tightly, and absorb nutrients from this water (Figure 7.6). Some bromeliads provide a relatively long-term, potentially more stable habitat for aquatic organisms because the plants are long-lived; some may live for more than 20 years and achieve a volumetric capacity of 1.3 litres (Benzing 1980).

Table 7.3 Numbers of mosquito species (*nepenthebiont* and *nepenthephil*) recorded from *Nepenthes* pitchers in different countries

Taxa	Country								
	Madag	**Seych**	**Sri Lan**	**Malaya**	**Sumat**	**Borneo**	**Java**	**Philip**	**New G**
Aedes						1	1		2
Armigeres				1					
Culex				8	1	6	1		
Toxorhynchites				4	2	1		1	
Tripteroides									
(*Polylepidomyia*)									2
(*Rachionotomyia*)			1	3	1	3	1		
(*Rachisoura*)									13
(*Tricholeptomyia*)								9	
(*Tripteroides*)				3	1	1		5	2
Uranotaenia	4	1	1	3	1	1	1		
No. of *Nepenthes* spp.	2	1	1	11	19	28	2	10	11

Note: Country abbreviations are: Madagascar, Seychelles, Sri Lanka, Sumatra, Philippines, New Guinea.
Source: From Steenis and Balgooy (1966) and Beaver (1983)

Due to their epiphytic, three-dimensional distribution, tank bromeliads represent perhaps the highest densities of phytotelmata found anywhere. For example, in a cloud forest in Colombia, Sugden and Robins (1979) recorded a mean density of 17.5 plants per m^2 of ground area. Assuming the volume of water retained by each plant to be approximately 250 cm^3, this suggests that the amount of water available for colonization by aquatic animals in such a location would be more than 50,000 litres per hectare.

Bromeliads may be subdivided into two categories based on their method of nutrition: *Dendrophilous bromeliads* generally grow in forested areas and collect nutrients leached by rainfall from tree canopies; their tanks thus contain a nutrient-rich soup. *Anemophilous bromeliads* tend to grow on cacti in deserts, or at the top of forest canopies, and here they obtain nutrients that are windborne.

Tank bromeliads exhibiting dendrophilous nutrition tend to have detritus-based food chains, whereas those with anemophilous nutrition support alga-based food chains (Frank 1983). Food resources for animals in bromeliads are partitioned as the water is not held in one single cavity but in several smaller ones among individual leaf axils; migration between axils is known to occur in the larvae of some mosquitoes (Frank and Curtis 1981). Barrera and Medialdea (1996) found that mosquito species originally from phytotelmata (such as *Ae. aegypti*) showed a greater resistence to starvation than species (such as *Culex quinquefasciatus*) originally from standing-water habitats.

Habitat selection is made by adult female insects, and is dependent on a number of habitat features important to particular species. Female mosquitoes, for example, select specific locations on temperature and humidity gradients that occur from ground level to the top of the forest canopy. Shape of the plant, reservoir volume, condition of the leaves, and presence of a flower spike also may be important factors.

Besides mosquitoes, other groups of Diptera commonly found in bromeliads include chironomids, ceratopogonids, psychodids, tipulids, syrphids, muscids, stratiomyids, phorids, tabanids, borborids, and anisopids. In addition, beetles (especially Scirtidae), hemipterans (Veliidae), many species of odonate, and some frogs may be found. Stonefly nymphs and caddisfly larvae are represented by a limited number of species (Fish 1983). Other invertebrates also occur (see below, and also Kitching 2000).

Corbet (1983) reported as many as 47 species (belonging to 17 genera) of odonate from phytotelmata, in general, with many of these living in

Figure 7.6 Photos of: (a) a terricolous bromeliad; (b) an epiphytic bromeliad.

ground-dwelling or epiphytic bromeliads. Phytotelmata seem to be more important for the Zygoptera than for the Anisoptera. The former are found in plants in the neotropical, Oriental and Pacific regions, and (though more rarely) in afrotropical and Australasian regions. The latter are found, regularly, only in plants in the Oriental region, and are less diverse. Odonate nymphs, because they can walk and climb, have wider options open to them than many of the other inhabitants of phytotelmata; *Roppaneura beckeri*, for example, can move between neighbouring plants of *Eryngium*.

It has been suggested that the Odonata may have evolved the habit of living in bromeliads through proximity of the latter to running water habitats (Calvert 1911). The ancestral habitat of the Bromeliaceae is the riparian forests along the Amazon River. This river floods seasonally, with water level fluctuating, vertically, as much as 10 m (see Chapter 8). At peak flood, water levels are close to the level of the lowest epiphytes. As bromeliads retain water for long periods of time, shallow-water, riverine odonates may have formed associations with these plants—which may have provided secure habitats in places or during times when rainfall was too patchy to maintain forest floor pools. Once established, such associations might well have allowed these species to extend their ranges into the forests, beyond the influence of river floods (Corbet 1983).

In a survey of the waters retained by *Vriesea inflata*, a bromeliad widely distributed in the Atlantic Forest of Brazil, Mestre *et al.* (2001) found a rich fauna dominated by beetles (especially scirtids), dipterans (10 families, and especially chironomids), and ants. Also present were turbellarians, nematodes, snails, oligochaetes, leeches, pseudoscorpions, harvestmen (Opiliones), spiders (18 families), centipedes, millipedes, amphipods, and isopods. There was a higher abundance of invertebrates in low-growing plants during the spring, but in the autumn they were more abundant in plants growing higher up—this was thought to be related to the degree of penetration of the forest canopy by rain (high in spring, low in autumn).

Armbruster *et al.* (2002) have pointed to the attractiveness of the physically discrete habitat presented by bromeliads, and its role in community studies. Their detailed analysis of 209 tank bromeliads in lowland Ecuador, yielded 11,219 individuals representing 354 morphospecies. The abundance distribution of these taxa was approximated by a log-series distribution, with 57% of the morphospecies being represented by a single specimen. The majority (62%) of the variation in taxon richness among plants was explained by plant volume, number of leaves, detritus content, and water volume. The authors concluded that both abiotic and biotic factors are important in influencing the structure of these communities. Janetzky and Vareschi (1993) have provided information on the physicochemical nature of the water retained by *Aechmea paniculigera* in Jamaica. Their study revealed the following ranges: conductivity 24–100 μS; phosphorus 78–200 $\mu g\,l^{-1}$; and oxygen concentrations that varied with time of day and water depth, according to the amount of detritus present and the degree of insolation. Bacterial densities were highly variable ($0.25–60 \times 10^3\ cm^{-3}$), and not correlated with detritus. The communities present comprised six species of testate protozoan, 19 species of rotifer, cyclopoid copepods (*Ectocyclops* and *Tropocyclops*), ostracods (*Metacypris* spp. and *Candonopsis*), unidentified cladocerans, and several insect species (two chironomids, the beetle *Cyphon* sp., two damselflies, and several mosquitoes). An endemic crab, *Metopaulius depressus*, uses these habitats as breeding sites. A later study, added the harpacticoids *Phyllognathopus viguieri*, *Epactophanes richardi*, and *Attheyella (Canthosella) mervini* (Janetzky *et al.* 1996). Colonization experiments showed that *Tropocyclops jamaicensis*, along with ostracods, and chironomid and mosquito larvae, invaded some plants only 2–3 weeks after they became available; after 13 weeks, 8 out of 10 plants contained populations of the copepod (Reid and Janetzky 1996).

Predator–prey associations have been observed within the minute volumes ($\sim 0.5\ cm^3$) of water found in taro axils. Mogi and Sembel (1996) found predatory larvae of the mosquito *Topmyia tipuliformis* alongside up to nine other taxa. These larvae did not appear to influence prey community organization through selective removal of particular species, although prey densities were lower in axils containing the predator. The study showed that inter-axil distribution of *To. tipuliformis* was aggregated in the first instar, but uniform in third and fourth instars where cannibalism became evident—a mechanism ensuring the survival of at least one individual under short food supply.

In a study of the ostracod genus *Elpidium* in Jamaican bromeliads, Little and Hebert (1996) found that most of the nine species exhibited very restricted distributions, together with low heterozygosity and marked gene-pool fragmentation. This suggested that population bottlenecks had occurred frequently, and that bromeliad habitats represent a source of high biodiversity and possibly ideal sites for extensive allopatric speciation.

7.2.4 Inflorescences

Not only do leaf axils retain sufficient water to support aquatic communities, some large flower bracts do so as well. The genus *Heliconia* (Zingiberales: Heliconiaceae), for example, has some 150 species native to the Neotropical region, but is also found in Indonesia, New Guinea, and Samoa. The floral parts consist of a series of alternately arranged, large bracts each of which surrounds 10–14 pairs of small flowers (Figure 7.7). The bracts of many species hold water which is derived from both rain and plant-transportation (in *H. caribaea* the mean volume per bract is 8.4 cm^3). Initially, the young flower buds are immersed in this water but as they mature they emerge and are held above the water (Machado-Allison *et al.* 1983).

In *H. caribea* in the northern, lowland tropical rainforests of Venezuela, new bracts open approximately every 6–7 days, and flowering occurs throughout the year although it peaks in the rainy season (June and July). The volume of the water reservoir increases with bract age as does the amount of organic carbon in the bract water. pH of the water is generally around 7.2. Machado-Allison *et al.* (1983) found the larvae of 15 species of insect in *H. caribea*. Of the eight most abundant species, three were mosquitoes (*Wyeomia ulocoma, Culex pleuristriatus*, and *Toxorhynchites haemorrhoidalis*), one a syrphid (*Quichuana* sp.), one a richardid (*Beebeomyia* sp.), and one a stratiomyid (*Merosargus* sp.). Also present were the chrysomelid beetles *Cephaloleia* and *Xenarescus*, the larvae of which feed by scraping the interior of the bract chamber. Densities of larvae of the three most abundant species, *Wyeomia, Culex*, and

Figure 7.7 Photo of the inflorescence of *H. caribaea* showing bracts and emerging flowers.

Quichuana, correlated positively with the amount of rainfall. Predators in these bracts seem rare, but intraspecific competition may occur. Adult females show distinct preferences for bracts of different ages when ovipositing. Bracts over 10 weeks old may be less stable habitats than younger ones because they tend to be damaged by vertebrates. Food input to the community seems to be chiefly from decomposition of the flowers and the inner-wall of the bract.

Thompson (1997) cites the interesting, facultatively aquatic, occurrence of nymphs of the Costa Rican spittlebugs *Mahanarva insignita* and *M. costaricensis* on several species of *Heliconia*. Spittlebug nymphs (Homoptera: Cercopidae) possess a ventral breathing tube on the abdomen which has evolved as an adaptation to living in semi-liquid spittle masses. Despite this 'pre-adaptation' to an aquatic existence, there were no previously substantiated examples of aquatic spittlebugs. Thompson found the two species of *Mahanarva* living along a gradient of inundation—some nymphs were submerged in the water-filled bracts, whereas others lived in drier conditions on the leaves, stems, or inflorescences.

7.2.5 Bamboo

Bamboo stumps are becoming recognized as common phytotelmata, with a wide distribution in the tropics and subtropics. In a comparison of the communities living in stump waters at low (200 m a.s.l.) and high (1,030–1,050 m) altitudes in North Sulawesi, Sota and Mogi (1996) found several similarities, but also differences. The lowland stumps yielded 38 taxa, including two predators, compared with 35 taxa, including two predators, at the higher site. The dominant detritivores were tipulids, scirtids, chironomids, culicids, and ceratopogonids. The predators were larvae of *Toxorhynchites*, which occurred in 67% of the lowland stumps, but only 28% of the highland stumps. Mean biomass per stump did not differ between the two sites, but the mean number of species per stump was significantly higher at the highland site, where variation in species composition among stumps was less. Where *Toxorhynchites* was present, the density and biomass of other culicids was significantly reduced. Other invertebrates present were nematodes, tubificid and naidid oligochaetes, leeches, harpacticoids, and libellulid odonates.

In a comparison of the faunas of bamboo stumps and tree holes, in Japan, Sota (1996) found a total of 21 taxa: 17 saprophagous dipterans, a scirtid, a nematode, a naidid, and *Toxorhynchites towadensis*, the single predator. Tree holes contained 20 taxa, and stumps 17, with 16 common to both. Taxon count and biomass were positively correlated with habitat volume. However, increased capacity in tree holes promoted more scirtids and *Culicoides*, whereas, in bamboo, chironomids and *Dasyhelea* were favoured. Another Japanese study examined interspecific competition between larvae of *Aedes albopictus* and *Tripteroides bambusa*, two common stump inhabitants (Sunahara and Mogi 1997). It showed that the outcome varied according to whether competition occurred over the short- or the long-term (single versus four cohorts, respectively), and that the presence of dead bamboo leaves in the stump water was a contributing factor.

Recently, Lozovei (1999) has reported another aquatic habitat provided by bamboo, the water present in the internode spaces of living plants. Access to bamboo stems in Parana, Brazil is through lateral holes bored by the noctuids *Eucalyptra fumida* and *E. barbara*, and the water they contain is derived from the plant itself. Seventeen species of dendricolous mosquito have been observed using such waterbodies.

7.2.6 Teasel axils

Even the very small volumes of water retained within the cup-like leaf bases of teasels (Dipsacaceae) support aquatic communities. Maguire (1959) showed that *Dipsacus sylvestris* in the vicinity of Ithaca, New York, contain a number of taxa, including tardigrades, oligochaetes, and nematodes (Table 7.4). Illoricate rotifers were by far the most ubiquitous inhabitants (found in 97% of axils), as were a number of ciliate groups. Most taxa, however, were more irregular in

Table 7.4 The frequencies of occurrence of aquatic taxa found in 186 teasel axils (113 plants) in New York State

Taxa	% occurrence
Rotifera (illoricate)	97
Ciliata	
(hypotrichs)	73
(holotrichs)	22
Colpoda	67
Cyclidium	49
Breslaua	18
Spathidium	15
Nematoda	43
Turbellaria (Rhabdocoela)	28
Tardigrada	8
Oligochaeta (*Aeolosoma*)	5
Algae (filamentous)	9
Unident cholorphytes	Common

Source: From Maguire (1959)

occurrence—for example, in one area tardigrades were present in 26.4% of plants, whereas in another (comprising 79 plants) none was found. Such inconsistencies are probably related to the 'chance' factor in passively dispersing organisms, combined with the extremely small size and isolation of the colonization 'target', and the limited time for which these waters persist. Kitching (2000) reports that the leaf bases of *Anglica sylvestris* (Apiaceae), a common weed of roadsides and woodland edges in Europe (and now feral in North America), also support aquatic communities.

7.3 Gastrotelmata

Although these habitats have not been well studied, at least four species of rotifer, representing the genera *Cephalodella*, *Lecane*, and *Macrotachela*, have been recorded from water-filled snail shells on Jamaica (Koste *et al.* 1995). Further, some insects with short life cycles, particularly mosquitoes, are known to breed in these habitats (Service 2000). For example, 67% of water-filled snail shells in Tanzania were found to contain larvae of *Ae. aegypti*, making them breeding habitats of significance in the spread of dengue fever (Trpis 1972).

7.4 Anthrotelmata

A wide variety of small, temporary water habitats has been artificially created as a result of human activities (see Chapter 1). However, whereas their existence is known, details of their inhabitants are less so. An exception, because they seem to be increasingly linked to the spread of disease-carrying dipterans, is rain-filled tyres, either abandoned in the landscape, or stockpiled for eventual reprocessing. Some of the latter are very large. For example, in a rural tyre dump (~300,000 tyres) in southeastern Illinois, Lampman *et al.* (1997) observed the following mosquito occurrences: (1) a high percentage of tyres containing *Aedes triseriatus* in an open-field situation (this species is more typically found in woodland tree holes); (2) a greater abundance of *Culex pipiens* than *Cx. restuans* in late-season collections; (3) a seasonal change in the distribution of *Ae. atropalpus* larvae in tyres from open field to edge of woods areas; and (4) the presence of *Ae. albopictus* as a major, late-season species. The rapid spread of *Ae. albopictus* (the Asian tiger mosquito), a species known to transmit a variety of pathogens that infect both humans and livestock, throughout the continental United States is cause for alarm (Moore 1999). Since its first detection in the east, in 1985, this species has spread and become established in 911 counties in 25 states, as far west as California and New Mexico. Its success is due to an ability to colonize diverse ecological settings (O'Meara *et al.* 1997), and tyre habitats are believed to have been a significant factor in this. Schreiber *et al.* (1996) reported on the effectiveness of the cyclopoid copepod *Mesocyclops longisetus* to control mosquito larvae in tyres. Adult copepods produced a greater than 90% reduction in the numbers of first and second instar larvae after 4 weeks, and a 90% reduction in third and fourth instars after 7 weeks. Copepod numbers in these tyres were significantly enhanced by the presence of leaf litter.

In a related habitat type, ground pools formed by the treads of bulldozers, in Indonesia, Mogi *et al.* (1999) recorded not only mosquitoes, but other insects and crustaceans. These unusual pools were created in the context of land that was in transition between deforestation and the development of rice field. Most of these pools were less than 1 m^2, and in full sunlight. Fourteen species of mosquito were present, overall, and their occurrence was related to water clarity: *Anopheles vagus* and *Culex vishnui* dominated clear pools, whereas *Cx. tritaeniorhynchus* and *Cx. gelidus* were dominant in turbid pools. Other insects commonly present were mayflies, dragonflies, chironomids, dytiscid beetles, notonectids, and veliids. Rarer were damselflies, corixids, hydrophilids, scirtids, ceratopogonids, and caddisflies.

Many mosquito species are inveterate colonizers of small 'container' habitat, as exemplified by *Ae. albopictus* in tyre water. In Florida, this species is also well known from flower-holding vases in cemeteries (Walker *et al*. 1996). In the urban area of Dakar, Senegal, anopheline and culicine mosquitoes abound in the more than 5,000 market-garden wells available as habitats (Robert *et al*. 1998). In Barcelona, there are approximately 1,300 public fountains and associated small reservoirs. Rieradevall and Cambra (1994) found that fountains without reservoirs, where there were hygropetric surfaces and, sometimes, accumulations of filamentous algae, had fewer invertebrates and typically those with short life cycles and semi-terrestrial adaptations (nematodes, ceratopogonids, chironomids [*Limnophyes*], and psychodids). Fountains with reservoirs supported grazing and detritivorous taxa, including chironomids (*Cricotopus* gr. *sylvestris*, and *Chironomus* spp.), ostracods, nematodes, oligochaetes, microturbellarians (*Prorhynchus*), and gastropods (*Physa, Lymnaea*, and hydrobiids). Medium to large reservoirs (up to 2 m deep) contained additional taxa (*Cloeon* and *Caenis* mayflies, a greater diversity of chironomids, and, where flow was sufficient, larvae of the caddisfly *Tinodes*); larger predators became more important also (e.g. *Lestes* and *Pyrrhosoma* odonates, notonectids, and mesoveliids).

7.5 Snowfields

A significant number of organisms is associated with mountain snowfields. The biota can be subdivided into two basic groups: (1) those species which actually live on the surface of the snow during some stage in their life cycle, either finding their food or a mate there; and (2) those species which are accidentally carried to high altitudes by air currents (perhaps during a dispersal phase in their life cycles), and which fall or settle onto the snow and subsequently die (Ashmole *et al*. 1983). The first group contains some forms that actively seek out snow as a moist environment, or become established after passive transport.

Two prominent components of the biota are insects and algae, but enchytraeid worms (which are somewhat amphibious in nature) and mites are often present also. Several insect species periodically migrate to snowfields to feed on the 'animal fallout' discussed earlier, but these are adult, non-aquatic forms, such as the muscid fly *Spilogona triangulifera* and the carabid beetle *Nebria gyllenhali* (Kaisila 1952). By far the most numerous hexapods on snow are the Collembola (springtails) which appear to migrate up through the snow to the surface during thaw and back down to the soil as the temperature drops. Leinaas (1981a) found that the extent of migration was species dependent. For example, in his study, *Hypogastrura socialis* moved up and down in the snow profile according to temperature and was the most abundant on the surface at air temperatures above 0°C. Species of *Isotoma*, in contrast, showed little vertical movement during winter and were found mostly in the coarse-grained bottom layer of snow—although a few individuals appeared on the snow surface during mild weather. A rich mix of species (e.g. *Entomobrya marginata* and *Lepidocyrtus lignorum*) tended to remain within the snow pack, particularly in areas of coarse-grained snow. As the guts of many species of Collembola have been found to contain only small amounts of complex plant residues, which are more abundant in the underlying soil, Leinaas suggested that these animals do not move into this nival habitat to seek food. They may be moving up into the snow in order to avoid compacting ice on the soil surface during mid winter, and also water-logging of the soil as the snow melts in late winter. Both Leinaas and Zettel (1984) have suggested that snow-surface activity may be important in terms of the dispersal of populations of springtails. Many species are hygrophilic and so dispersal, under the dry conditions of summer, may be impossible. By moving across snow surfaces, in winter, the springtails can cover considerable distances under optimal humidity.

As a physical adaptation to jumping on snow, Leinaas (1981b) has shown that a cyclomorphosis occurs in *Hypogastrura socialis*. In summer, the teeth on the furcula (abdominal jumping organ) are small, but in winter the teeth are larger and the surrounding cuticle becomes thickened which may improve the insect's grip on the snow surface. Some winter-active insects are protected from freezing when walking on snow by the presence of antifreeze agents in their bodyfluid, which cause a thermal hysteresis between the freezing-melting curves thus preventing the growth of ice crystals within the temperature range at which the insects are active. In the snow scorpionfly (*Boreus westwoodi*; Mecoptera), for example, there is a difference of 5–6°C between the melting point and the temperature of ice growth in the haemolymph, indicating the presence of antifreeze compounds. These compounds are likely to be proteins or glycoproteins, and in *B. westwoodi* they prevent the growth (not the formation) of ice crystals down to −7 °C, which covers the minimum temperature at which they are active (−2 °C) with a considerable margin of safety (Husby and Zachariassen 1980). There do not appear to be comparative values for the Collembola.

Craneflies (Tipulidae) also occur in snow, and adults of three species of *Chionea* have been found in the free air space under snow in Finland. The larvae are thought to live in tunnels inside damp grass hummocks where they are likely to be protected from severe frosts (Itamies and Lindgren 1985).

Considerable numbers of algae occur in snow where they frequently give a reddish tinge to the surface layers. Most common are species belonging to the Chlorophyceae (greens), Cyanophyceae (blue-greens), and Xanthophyceae (yellow-greens), but diatoms (Bacillarophyceae) may be present also. Gram-negative bacteria and microfungi may form associations with some types of snow-dwelling alga (Lichti-Federovich 1980).

Chlamydomonas nivalis is a unicellular, motile, green alga found commonly in snow fields in many parts of the world (e.g. US, Greenland, and Australia). The most common form found in snow is a resting stage which typically exhibits a red colour, due to the presence of carotenoids in cytoplasmic vacuoles (Marchant 1982); this colour may extend to depths of up to 10 cm into the snow. These resting cells are roughly spherical, 10–50 μm in diameter, and have a thick cell wall with a smooth outer surface. On the outside of the wall is a loose network of fibres that intermesh with encapsulated bacteria and surface debris. The bacteria, too, have a thick capsule wall and it is thought that this and the thick algal cell wall are important structures in protecting the cells from dehydration and freeze-thaw cycles as the snow melts (Weiss 1983). Hoham *et al.* (1983) have shown that a succession of species within the algal genus *Chloromonas* occurs, and that this may be due to different tolerance levels of individual species to environmental factors such as light intensity, carbon dioxide concentration, and temperature, factors that vary according to the age of a snowfield.

7.6 Marine littoral pools

Another type of temporary freshwater habitat is represented by brackishwater pools found in marine littoral areas. Because of their intermittent position between land and sea, the salt content, and therefore their freshwater content, often fluctuates widely. Ganning (1971) measured salinity changes in Swedish rockpools and correlated these with environmental changes. Figure 7.8 shows some of the factors that contributed to these changes. Ganning produced a classification scheme of rockpools based largely on salinity levels, but warned that these pools are not static and stable biotopes, but dynamic and changing, and consequently defy absolute categorization. His classification scheme spans permanent saltwater pools to permanent freshwater pools, however only two of his pool types are particularly relevant to this chapter. These are the *ephemeral saltwater pool* and the *brackishwater pool*. Ephemeral saltwater

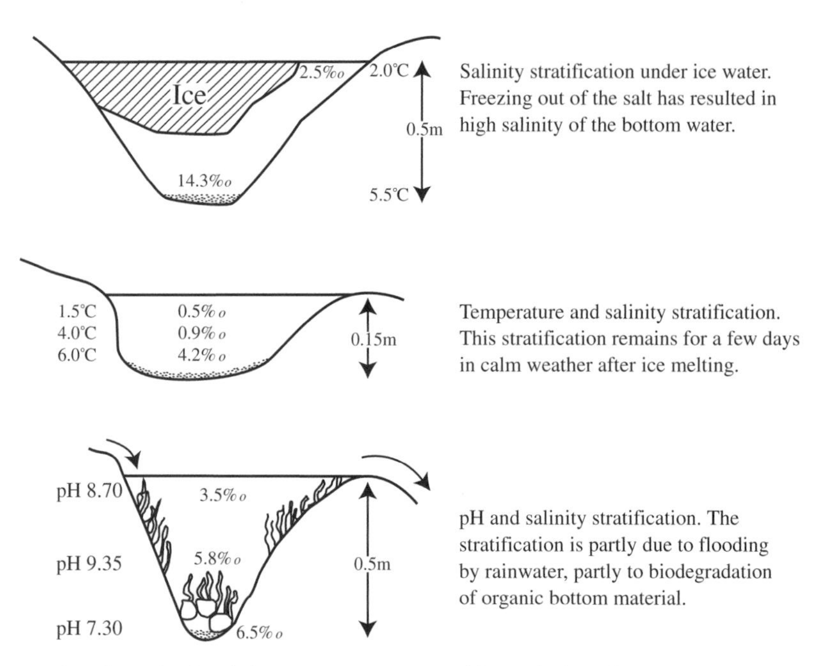

Figure 7.8 Changes in the salinity of rockpools due to various environmental factors (redrawn from Ganning 1971).

pools are generally of small volume and are situated high up on the shore. They receive saltwater input from splashing during strong winds at high tide. Physical and chemical parameters of the habitat vary greatly and the biota is limited. Figure 7.9(a) shows some of the characteristics of the community living in these pools in the Baltic. Brackishwater pools are also supplied by both rain and saltwater spray, but are generally lower on the shore. The distinction between these pools and the ephemeral saltwater pools is based chiefly on size, amount of bottom sediment, degree of salinity, and, though to a lesser extent, the presence of the filamentous alga *Enteromorpha intestinalis* (which is more abundant in brackishwater pools). The biota is somewhat richer in these latter pools (Figure 7.9(b)).

In a study of the fluctuations in salinity in salt marsh pools in Northumberland, England, Sutcliffe (1961) found a gradation in salinity. Some pools were flooded only by seawater at very high-spring tides, and their salinity was generally less than 15 ppt. Other pools were flooded more frequently and their salinity was generally greater than 15 ppt. A few species of euryhaline insect bred regularly in these pools: a caddisfly, a beetle, two species of midge, a mosquito, and a shore fly. In addition, during the summer, there were irregular influxes of adults of various species of hemipteran and beetle. However, these only colonized pools with salinities less than 10 ppt.

In a detailed study of the faunas of 14 rockpools along the north shore of the St Lawrence River, Canada, Williams and Williams (1976) found certain assemblages of animals to be characteristic of low and high salinities (Figure 7.10). Most of the insects were confined to lower salinities, for example, the dragonfly *Aeshna interrupta*, the caddisflies *Limnephilus tarsalis* and *Oecetis* sp., and the beetles. A few insect species, such as *Ephydra subopaca* and the marine chironomids *Halocladius* and *Cricotopus sylvestris*, were most abundant at salinities approaching full-strength seawater. In a survey of emerging adult insects from three coastal salt marshes on Prince Edward Island, Canada, 85% were found to be dipterans (Giberson *et al.*

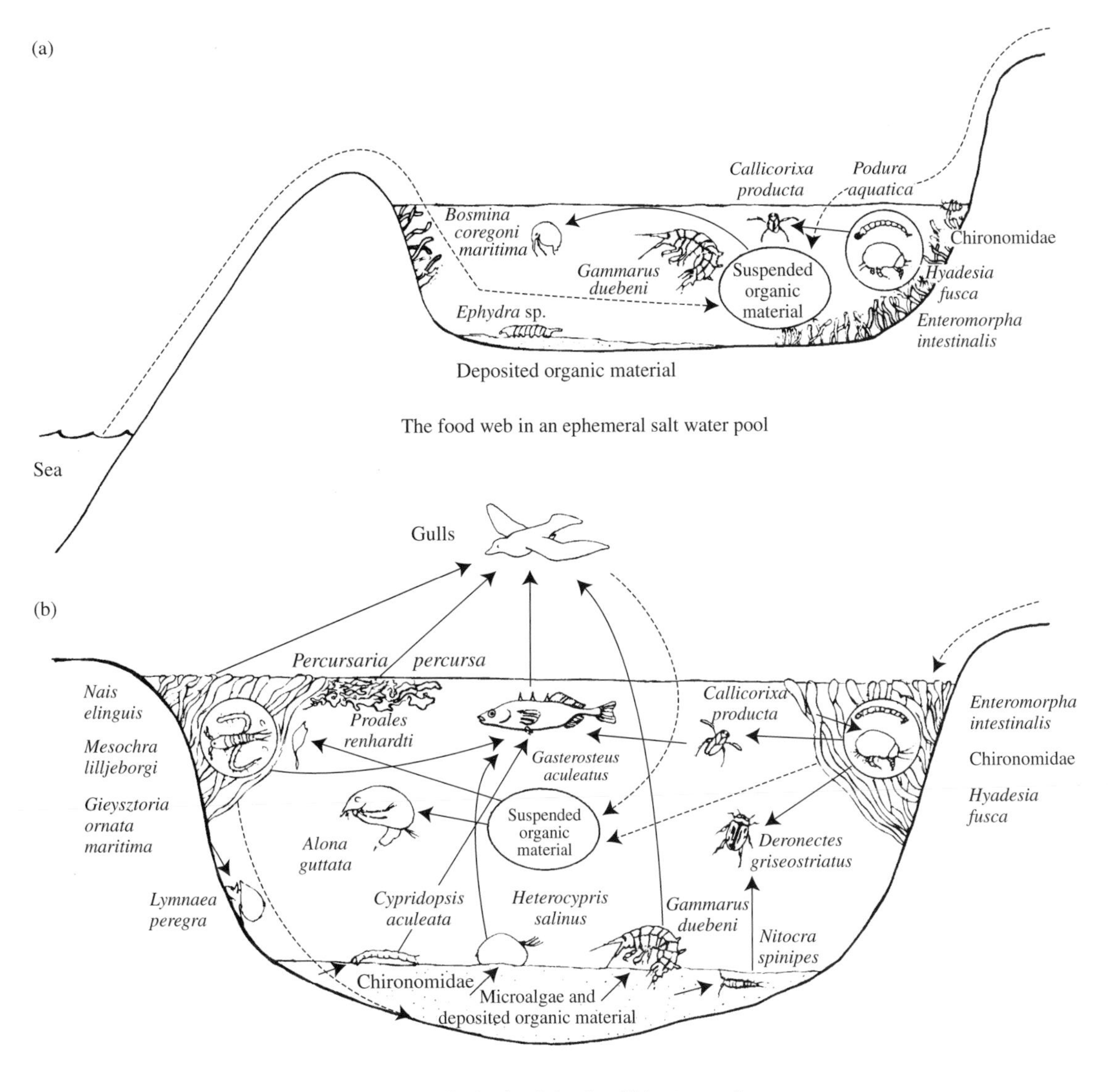

Figure 7.9 Examples of food webs in the two main types of rockpool in the Baltic region: (a) ephemeral salt water pool; (b) brackish water pool. (redrawn from Ganning 1971).

2001). Two-thirds of these were primarily chironomids, ceratopogonids, and culicids, with rare occurrences forming the bulk of the 43 species total. Salinity in the pools ranged from 11 to 27 ppt, and no consistent overall relationships were found between pool size and insect diversity or abundance. Yozzo and Diaz's study (1999) of tidal freshwater wetlands, largely along the Atlantic coast of the United States, characterized them as having high vascular plant diversity, but species-poor

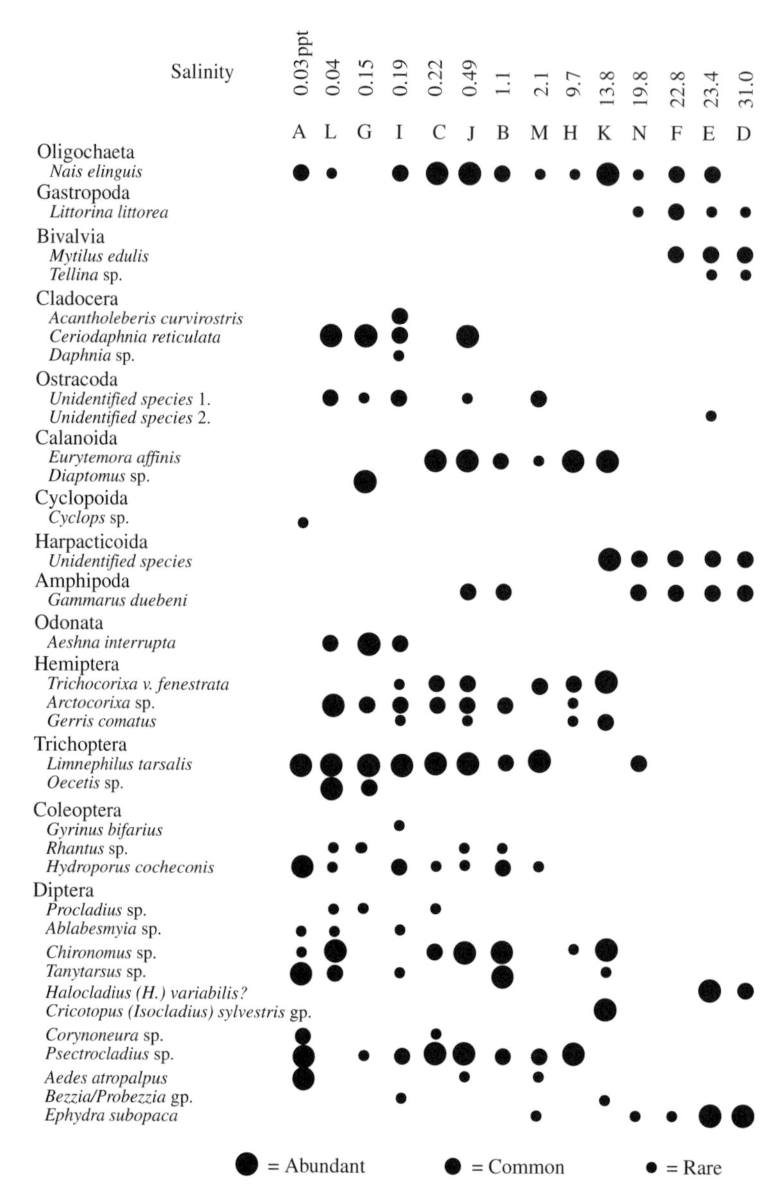

Figure 7.10 Relationship between salinity and the occurrence of invertebrates in 14 rockpools on the north shore of the St Lawrence River, Canada (redrawn from Williams and Williams 1976).

invertebrate communities. The macrofauna was reported to be dominated by tubificid, naidid, and enchytraeid oligochaetes, and chironomids, with a meiofauna comprising nematodes, ostracods, copepods, small naidids, and tardigrades.

Survival at high salinities requires a significant capacity for osmotic independence that may be achievable only by having, among other things, a body that tends to be impermeable to water and salts (Sutcliffe 1960). It has been suggested that osmotic regulation and the problems of respiring in saltwater (the ideal respiratory membrane is thin, of large surface area, and semipermeable) may require the evolution of such different physiological adaptations that few insects have been successful in doing so. This may have prevented widespread invasion of marine environments by insects. Marine insects are, for the most part, still restricted to 'bridging' habitats between land and sea (Cheng 1976). Marine littoral pools may therefore be the crucibles in which the evolution of improved physiological adaptations is taking place. This may also apply to inland saline lakes, ponds, and estuaries.

7.7 Reservoirs

In permanent, natural lakes, the littoral zone typically supports the most diverse and productive communities, comprising microorganisms, algae, macrophytes, invertebrates, and vertebrates. In regulated reservoirs, however, these communities are highly disturbed due to the vagaries of flow control. In some ways, the shores of reservoirs are analogous to the marine intertidal zone, but without predictable cycles of water level the former have a depauperate biota. For example, Prat *et al.*'s (1992) study of the communities of Spanish reservoirs showed them to be reduced to only highly resistant species. Filter-feeding species were found lacking in the Solina Reservoir, in Poland, leading Prus *et al.* (2002) to conclude that its community was not fully developed—the most common trophic groups present were plant-detritus feeders (especially oligochaetes, chironomids, mayflies, and caddisflies), and predators (chiefly leeches and mites). Hynes' study (1961) of Llyn Tegid in North Wales showed that its conversion to a water storage body in 1955 did not reduce littoral population densities *per se*, but dramatically altered community composition. The change from the lake's natural 2 m water level fluctuation to an imposed range of 5 m, together with related changes in substrate type, resulted in loss of sponges, flatworms, naidid oligochaetes, leeches, snails, amphipods, mites, and most aquatic insect groups. Densities were maintained by large increases in tubificid and enchytraeid worms, and chironomid larvae (Chironominae).

In the tropics, the increased rate of evaporation from reservoirs, may be compounded by heavy sediment loads resulting from riparian land degradation. In Lake Kariba, the water level can vary as much as 10 m, preventing the development of a recognizable littoral community, especially macrophytes, the latter severely affecting the presence of aquatic bird species, and also populations of large herbivores, such as buffalo, elephant, hippopotamus, and impala (Magadza 1995).

Similar trophically linked changes were observed by Zalewski *et al.* (1990a,b) during a long-term study of the Sulejów Reservoir, Poland. There, significant changes in water level prevented the development of typical littoral vegetation, which removed essential refuges for cladocerans that fed on phytoplankton (Moss 1990). Perch fry (*Perca fluviatilis*) were then able to feed freely on the cladocerans (reducing their biomass from $\sim$16 mg l^{-1} to $\sim$1–2 mg l^{-1}), which led to up to a 10-fold increase in phytoplankton, and a 2-fold increase in suspended materials. In addition, there was reduced growth and winter survival of pike-perch (*Stizostedion lucioperca*), as the fry of this species rely on large zooplankters as their first food.

Some changes similar to those seen in reservoirs are evident in both the littoral zone of large regulated rivers, and in natural river floodplains. In the former, there is commonly loss of stoneflies and mayflies, but enhancement of amphipods and simuliids. Some of these changes, however, have been related to factors other than water level change, for example, oxygen depletion, hypolimnetic water release from dams, and temperature modification (Ward and Stanford 1979; Blinn and Cole 1991). Aspects of natural river floodplains will be discussed in Chapter 8.

7.8 Deserts

Two groups of alga occur in deserts, Edaphic (or soil) algae and Lithophytic (or rock) algae. Friedmann *et al.* (1967) subdivided edaphic algae into three categories: (1) *Endedaphic algae* live in soil and occur in many deserts throughout the world. Frequently, they occur to depths of greater than 50 cm where they may be washed by rain and where they subsist at reduced metabolism for long periods (Schwabe 1963). Blue-green algae cultured in the laboratory from the Atacama Desert of northern Chile are slow to germinate and grow but are very resistant to desiccation and high levels of irradiation. These algae grow to form small, convex, cushion-like pellets of sand grains cemented together by the gelatinous sheaths of their filaments. It is thought that these structures shade the small area of soil beneath them thus preventing evaporation, and creating a minute water reservoir (Forest and Weston 1966). (2) *Epedaphic algae* live on the soil surface and are also widely distributed in deserts and semi-deserts throughout the world. In deserts of the southwestern United States, Cameron and Blank (1966) found three forms: (i) *Raincrusts*—which arise in shallow depressions in the soil where rainwater collects. These typically consist of a thin layer of algae that warps, curves upwards and finally breaks into fragments when dry. (ii) *Algal soil crusts*—which are soil-like in colour and seem to be more stable in structure, perhaps due to the presence of filamentous blue-green algae. (iii) *Lichen soil crusts*—which are the largest and thickest (up to 2–5 cm) of the formations. Here, fungal hyphae bind the algae into a lichen association and the resulting mass may become folded and stick up from the soil surface. Such crusts are known to reduce erosion via water and wind, conserve soil moisture, and also provide nitrogen for uptake by larger, rooted plants (Yair 1990; Evans and Johansen 1999). Despite such demonstrated and inferred importance, knowledge of the species composition of these microbiotic crusts is still very poor, although species richness is believed to be very rich (Flechtner *et al.* 1998). (3) *Hypolithic algae* live on the under-surfaces of stones and are found in both hot and cold deserts. The stones are typically translucent to some extent, thus light reaches the cells at intensities of between 30.0 and 0.06% of surface illumination (Vogel 1955). Because rainwater penetrates the soil most easily along the stone-soil boundary, small reservoirs of water persist under these rocks even after it has evaporated from the surrounding soil. The hypolithic community typically consists of filamentous and coccoid blue-green algae, and green algae (Friedmann *et al.* 1967). Hypolithic algae have become of considerable interest to those studying the possibility of extra-terrestrial life forms. For example, NASA researchers have a continuing programme that seeks to understand the biology of hypolithic species in various deserts around the world, including Death Valley National Park, the Gobi, Negev, and Atacama (Chile) deserts, and Antarctica (Landheim *et al.* 2004).

Lithophytic algae live in, rather than on, rocks. They are of two types: (1) *Chasmolithic algae* live in the fissures of rocks; and (2) *Endolithic algae* live within the rock matrix itself, colonizing spaces between particles. Again, the rocks tend to be somewhat translucent and porous. The latter feature is important in trapping and retaining moisture by capillary force, so as to provide the necessary water balance for the algae (Friedman 1964).

Provision of water to all these algal associations may be through rain, although in many cases it is derived from condensation of dew at night. Despite this, the algae are sometimes subjected to total loss of environmental water, in which case they are able to survive through physiological tolerance. Parker *et al.* (1969), for example, have shown some soil algae capable of surviving drying for over 60 years; Cameron *et al.* (1970) determined that some filamentous and coccoid blue-green algae could survive continuous high vacuum for 5 years; and Trainor (1962) has shown that the green alga, *Chlorella*, can survive desiccation at 130°C for 1 h.

In an analysis of the cryptoendolithic communities of the McMurdo Dry Valleys of Antarctica, Torres *et al.* (2003) identified more than 1,100 individual clones from two community types, one dominated by cyanobacteria, the other by lichens. No phylotypes were shared between them, and clone libraries showed a predominance of a relatively small number of phylotypes that, due to their relative abundance, likely represented the main primary producers in these communities. The lichen-dominated community contained three rRNA sequences, representing a fungus, a green alga, and a chloroplast that accounted for more than 70% of the clones. The cyanobacterium-dominated community proved to contain, in addition to cyanobacteria, a minimum of two other organisms thought to be important in community function. The latter comprised a member of the alpha-subdivision of the *Proteobacteria*, that may be capable of aerobic anoxygenic photosynthesis, and a distant relative of *Deinococcus*, a genus of highly stress-resistant bacteria. Again, survival of these organisms under winter conditions of −60°C, and summer conditions of −35°C to +3°C (with 15°C oscillations occurring within minutes, according to cloud cover, throughout the day), has prompted study of relationship with their environment as a potential analogue for conditions on Mars (McKay 1993).

7.9 Starfishes

One of the most unusual temporary marine habitats for an insect is seen in the caddisfly *Philanisus plebeius*. In New Zealand, eggs of this species develop in the coelomic cavity of the starfish *Patiriella regularis* (Winterbourn and Anderson 1980). Larvae migrate from the host soon after hatching and then are freeliving in the intertidal zone where they live among attached seaweeds and feed primarily on red algae (Rhodophyceae). It is not known exactly how the eggs are deposited in the starfishes or how the larvae leave, but it is hypothesized that oviposition occurs through the papular pores (abundant on the aboral surface of the host), and that newly hatched larvae may leave by this same route or through the wall of the stomach.

7.10 Dung

Stretching the broad definition of temporary aquatic habitats to near breaking point, there are aspects of the communities found in animal dung that would seem pertinent. Cattle dung produced in summer, when the animals are grazing on fresh grass, is quite liquid and has a high water content (73–89%; Hammer 1941). The fauna consists mostly of dipteran larvae, from a variety of families, and also beetles. Laurence (1954) studied cow pats in Hertfordshire, England and reported a diverse and abundant fauna (Table 7.5). Support for regarding this habitat type as a semiaquatic one is evident from the similarity between some of the taxa present and those comprising the summer terrestrial fauna of temporary ponds (Chapter 4; Figure 4.20). Taxa common to both habitats are larvae of the dipteran families Sphaeroceridae, Tipulidae, Sepsidae, Ceratopogonidae, and Chironomidae, as well as beetles belonging to the Hydrophilidae (scavenging waterbeetles) and Staphylinidae (rove beetles). Table 7.5 also provides a comparison with taxa found in cattle dung in the Sierra Nevada foothills of California (Merritt and Anderson 1977). Notable similarities include the presence of many of the same dipteran and beetle families, and identical genera (the nematocerans *Smittia* and *Psychoda; Scatophaga* [Scatophagidae]; and *Copromyza* [Sphaeroceridae]). Parasitic hymenopterans were not reported in the Hertfordshire study.

The dung habitat itself is a changing one, depending on intrinsic properties such as aging of the material itself, and extrinsic factors such as weather. After deposition, the pat begins to cool down to air temperature, and in summer, a crust forms on the surface, usually within 24 h. This crust separates from the more fluid dung underneath, and an air space is formed between the two. The crust appears to seal off the odour of the dung and the pat becomes less attractive to many adult flies. Much of the egg-laying must therefore occur soon after pat deposition, but not at night (Merritt and Anderson 1977). Winter-produced dung is coarser in texture, due to a change in cattle fodder, and tends not to crust over.

The fly larvae that inhabit cattle dung show many respiratory features in common with aquatic and semiaquatic species, and can be broadly categorized into: *amphipneustic* (spiracles on the prothorax and terminal segments)—many species; *apneustic* (no spiracles)—chironomids and ceratopogonids; and *metapneustic* (spiracles only on the terminal segment)—tipulids (Keilin 1944). An interesting seasonal change noted by Laurence (1954) was that larvae of species living in the less-aqueous, winter dung do not show the modification of the posterior end (i.e. spiracles borne on the end of a long, telescopic siphon) seen in some summer species.

In the Hertfordshire dung, a clear seasonal succession of species was evident (Figure 7.11) and

Table 7.5 Comparison of the insect groups found in cattle dung in Hertfordshire, England and the Sierra Nevada foothills of California

Taxa	England	California
Coleoptera		
Hydrophilidae	*Unident*	*Cercyon* (2[a]), *Sphaeridium* (3)
Scarabaeidae		*Aphodius* (6)
Staphylinidae	*Unident*	*Aleochara* (>4), *Apoloderus, Hypnonygrus, Leptacinus* (2), *Lithocaris, Oxytelus, Philonthus* (4), *Platystethus, Quedius*
Histeridae		*Peranus, Saprinus*
Rhizophagidae		*Monotoma*
Diptera—Nematocera		
Anisopodidae	*Anisopus*	
Cecidomyiidae		*Colpodia, Xylopriona*
Ceratopogonidae	*Culicoides*	*Forcipomyia* (2)
Chironomidae	*Smittia*	*Smittia*
Psychodidae	*Psychoda*	*Psychoda* (3)
Scatopsidae	*Scatopse*	*Regloclemina*
Sciaridae		*Lycoriella, Sciara*
Tipulidae		*Tipula*
Trichoceridae	*Trichocera*	
Diptera—Calypterate		
Anthomyiidae		*Calythea, Hylemya* (2),
Muscidae	*Dasyphora, Polietes* unident	*Pseudophaonia, Haematobia, Hydrotaea* (2), *Morellia, Musca, Myospila, Orthellia*
Calliphoridae	*Mesembrina*	
Sarcophagidae		*Ravinia* (3)
Scatophagidae	*Scatophaga*	*Scatophaga* (2)
Diptera—other families		
Sepsidae	*Unident*	*Saltella, Sepsis* (7)
Sphaeroceridae	*Copromyza, Limosina*	*Copromyza* (2), *Leptocera* (5), *Ischiolepta, Olinea*
Stratiomyiidae	*Unident*	*Microchrysa, Sargus* (2)
Empididae	*Unident*	*Drapetis*
Drosophilidae		*Drosophila*
Hymenoptera—parasitic		
Bethylidae		Unident
Braconidae		*Aphaereta, Aosbara, Idiasta, Pentapleura, Phaenocarpa*
Cynipidae		*Kleidotoma, Eucolia*
Diapriidae		*Phaenopria*
Figitidae		*Figites, Xyalophora*
Ichneumonidae		*Phygadeuon*

Note: [a]Numbers in brackets indicate the number of species found within the genus; the Hertfordshire inventory was based on a much smaller sample size than that used in the California study.
Source: From Laurence (1954) and Merritt and Anderson (1977)

few genera were present throughout the year. The genus *Limosina* (dung flies) is exceptional in that several species succeed one another throughout the year. Larvae of the Muscidae and Sepsidae (scavenger flies) were absent in winter, whereas larvae of *Trichocera* (winter gnats) and also *Smittia* (chironomids; not shown) occurred mainly in the winter and entered a diapause during the warm summer months. Larvae of *Copromyza* (dung flies) occurred primarily in winter and spring, whereas larvae of other dung flies, (*Scatophaga*) had two peaks, one in April–May, the other in September–December. Larvae of *Psychoda phalaenoides* had a very rapid rate of development and were very abundant in the late autumn, with another, lesser, peak in late spring. In California, the pasture ecosystem and the season were of primary importance in determining the diversity and numbers of insects that colonized pats. Differences in these measures among pastures were due mainly to a combination of interacting local and microclimatological factors (Merritt and Anderson 1977).

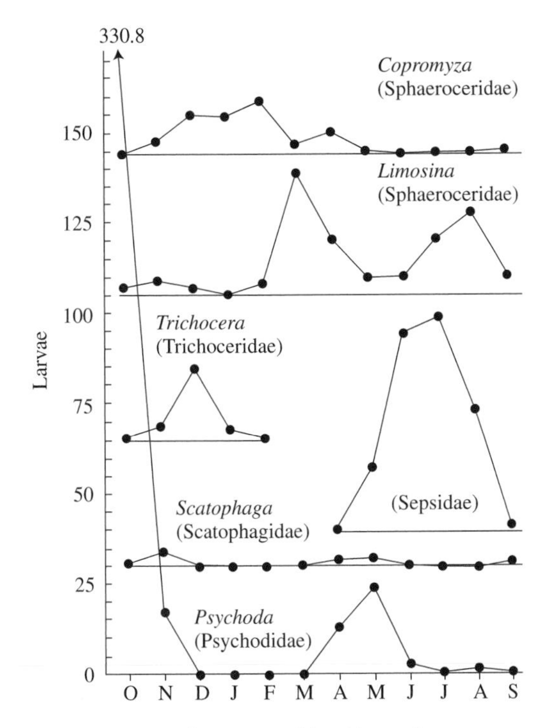

Figure 7.11 Seasonal succession of larval insects in cow pats (numbers of larvae are recorded as geometric means per pat sample of four 2.5-cm cores for each month; redrawn from Laurence 1954).

The rate of growth in some dung-inhabiting species is very high, some coprophagous muscid larvae, for example, mature in just 5 days in summer, and most summer-dipteran species have a development period of less than 40 days. In some of the stratiomyids (soldier flies) the period of development is long (60–100 days), as is the case for the chironomid *Camptocladius*. For species such as these, the pat probably disintegrates before development is complete. Periods of development may vary between species of the same genus as, for example, is seen in the scavenger flies *Sepsis* (Figure 7.12).

How numerous are these habitats and how significant are their faunas? Pat size varies between 13 and 45 cm in diameter, and a beast may deposit at least six per day. A single bullock may produce 26 kg of dung each day per 500 kg of body weight, or roughly 19 times its own weight in one year

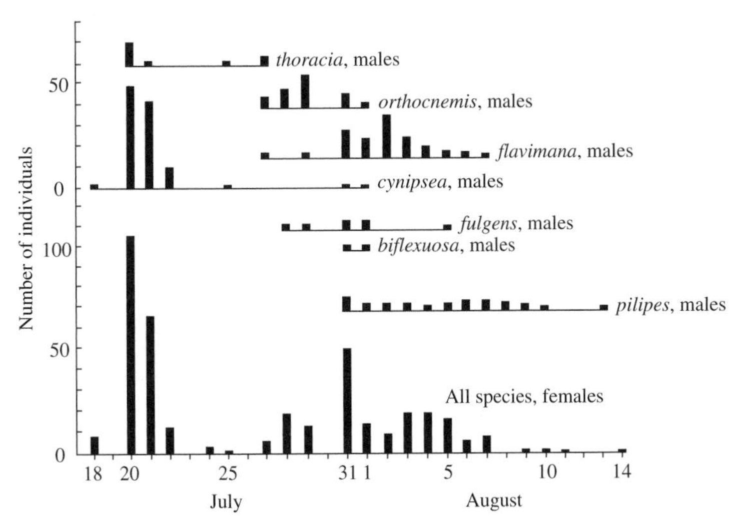

Figure 7.12 Emergence succession of species of *Sepsis* (Sepsidae) from dung deposited on the 5 July 1950 (redrawn from Laurence 1954).

(Henry and Morrison 1923). Approximately 1/80th of the weight of the dung is attributable to the animals living in it and so, in one year, a cow will leave in its dung enough material to support an insect population equal to 1/5th of its own weight (Laurence 1954).

7.11 Commonality among populations and communities

Attempting to seek trends in patterns of organization in the communities of organisms living in the 'less conventional' temporary waters covered in this chapter is difficult, and perhaps unwise. Nevertheless, at first sight, some of the smallest habitats, for example gastrotelmata and bamboo, would seem to support much reduced communities (if the latter term is even applicable), and very small populations compared with the shore of reservoirs and pools of the marine littoral zone. However, the former habitats are poorly known, and further study, particularly of the microbiota and meiofauna, may indicate otherwise. Moreover, under certain conditions some of these small temporary waters are important contributors to local biodiversity. For example, although the number of organisms and species living in individual *Nepenthes* pitcher plants is limited, the cumulative totals occurring in thousands of such pitchers within a relatively small forest are significant.

Despite the wide variety of habitat types exemplified above, it is possible to identify a central 'core' of taxa that is present in most. Apart from microorganisms (bacteria, algae, and protozoans), nematodes, and mites, the most prominent group in this core comprises nematoceran flies: specifically, chironomids, ceratopogonids, psychodids, and tipulids; hydrophilid beetles may also be present. The core becomes supplemented by other taxa, both in habitats that are more aquatic and those that are more terrestrial (Figure 7.13). In the more aquatic habitats, such as tree holes, pitcher plants, and bromeliad tanks, mosquitoes and scirtid beetles become major components of the fauna.

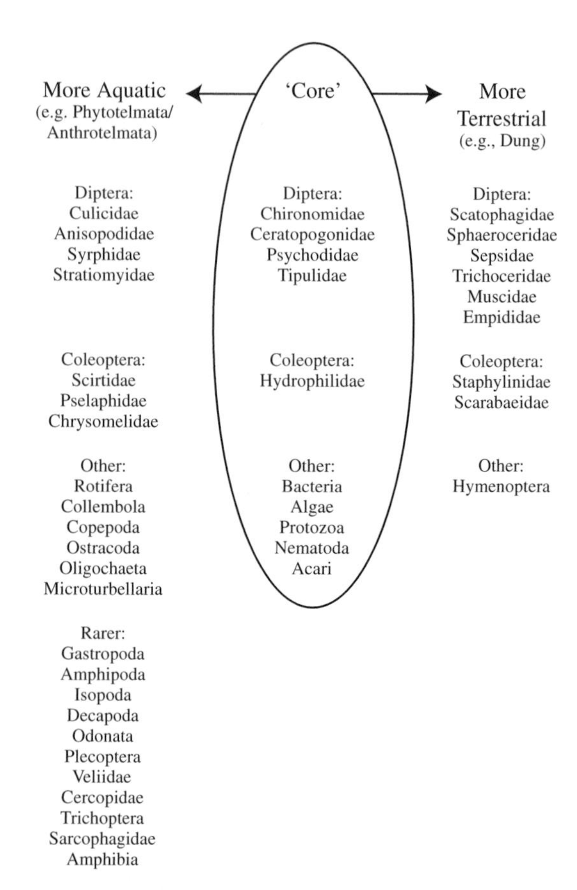

Figure 7.13 Comparison of trends occurring in the biotas of the less traditional temporary waters covered in this chapter. A central 'core' is identifiable that is present, but supplemented, both in habitats that are more aquatic and those that are more terrestrial.

Also present may be syrphid, anisopodid, and stratiomyid flies, pselaphid and chrysomelid beetles, springtails, microcrustaceans (copepods and/or ostracods), and oligochaetes. Rarer, are snails, odonates, caddisflies, stoneflies, hemipterans, amphipods, isopods, crabs, and amphibian larvae. In the more terrestrial habitats, such as cattle dung, the core nematoceran flies, while present, are overshadowed by brachyceran families, such as the Scatophagidae, Sphaeroceridae, Sepsidae, Muscidae, and Empididae, along with staphylinid beetles.

CHAPTER 8

Applied aspects of temporary waters

8.1 Introduction

The purpose of this chapter is to briefly examine those aspects of temporary waters that are largely associated with human activities. Many of these associations are very ancient. For example, use of temporary floodplain pools to trap fishes migrating from main river channels likely dates back to early, hunter-gatherer times; the rice paddy-culture method for fishes dates back almost two millenia; and rotation of grain and carp harvests in Europe dates back to the Middle Ages. Only a subset of temporary water types has been selected, and it is not intended that coverage be exhaustive, but rather that it emphasizes the application of temporary water properties and communities dealt with in other chapters of this book.

8.2 Aquaculture/agriculture rotation—an ancient art

A classic example of large-scale, applied use of temporary waters (see Mozley 1944) is to be seen in the Dombes ponds of France. The Dombes is a platform to the northeast of Lyon comprising long, fan-shaped morainic mounds created by Quaternary glaciers. The moraine itself is approximately 280 m high and was formed during the Riss glaciation. During the late Würm glaciation, substantial amounts of loess (wind-borne, calcareous silt: 0.002–0.05 mm in grain diameter) were deposited in the depressions between the morainic mounds (Figure 8.1). Post-glacial rains leached much of the loess creating decalcified clayey soils which provided a waterproof lining to these basins.

Up until the thirteenth century, only a few natural ponds (termed 'lescheres') existed but these were stocked with fishes. During the Middle Ages, because of a food demand for fish (for religious reasons), and the fact that the soil in the region was poor yet retained water, large numbers of artificial ponds were created. Making a pond simply entailed damming the ends of the depression between morainic mounds. This caused social problems, however, as the land so flooded frequently belonged to several landowners. To alleviate this, the *Rules of Villars* ('La Coutume de Villars') were proposed and have governed the use of the ponds from the fourteenth century until today (Truchelut 1982).

The rules proposed a rotation of use of the ponds: 2 years full, 1 year dry (Figure 8.2). Water for filling a pond ('evolage') generally comes either from a pond situated at higher elevation, or from a system of ditches which lead into the pond, and which collect rainwater. Water level in the pond is controlled by a sluice-gate ('thou'). When filled, each pond is then stocked with fishes, usually carp. After 2 years of evolage, the fishes are harvested and marketed, the profits going to the collective owners of the evolage. As soon as the pond bed is dry enough to be workable, it is ploughed (except for the central channels and the fishing area, which are never drained) and planted with cereals ('assec'). The grain is harvested at the end of the summer and the profits distributed among the owners of the assec—who may be different from the owners of the evolage. Stubble and roots left behind form a source of food for aquatic organisms when the pond is reflooded at the start of a new evolage.

Because of the assec/evolage time ratio, and the fact that the ponds are often arranged in a series, approximately one-third of the Dombes ponds are dry in 1 one year and two-thirds are flooded.

This rotational system has changed little since the fourteenth century apart from a reduction in the total surface area of the ponds (19,000 ha in 1862 to 11,000 ha at present), and occasional change in the relative duration of the evolage/assec stages (e.g. to 3:2 or 4:2).

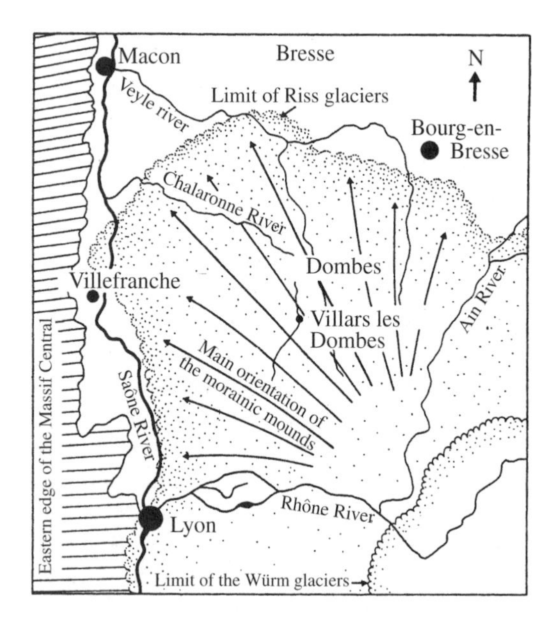

Figure 8.1 The origins of the Dombes ponds, France (based on information supplied by H. Tachet).

Typically, the Dombes region supports over 1,000 ponds each of surface area somewhat less than 10 hectares. Each pond is formed behind a dyke, made from compacted soil, and a channel runs through its middle and widens on the upstream side—this is the fishing area ('pêcherie') (Figure 8.3). The pêcherie and channel may be as much as 2 m deep but the bulk of the pond is between 0.5 and 0.8 m. Flow of water through the dyke is controlled by a conical plug connected via a threaded rod to the 'thou', the sluice-gate mechanism. A grating prevents fishes from escaping through an overflow in times of high water. To assist drainage during the assec, furrows are ploughed at right-angles to the channel with a unique pattern of 10–12 shallow furrows to one deep one (the 'billon' structure).

Fishes are usually harvested between September and April, and apart from the large ponds, there are smaller ones which contain the breeding stock. Mirror carp (*Cyprinus carpio*) is the predominant species cultivated (65% of the stock), followed by roach (*Rutilus rutilus*; 20%), tench (*Tinca tinca*; 10%), and pike (*Esox lucius*; 5%).

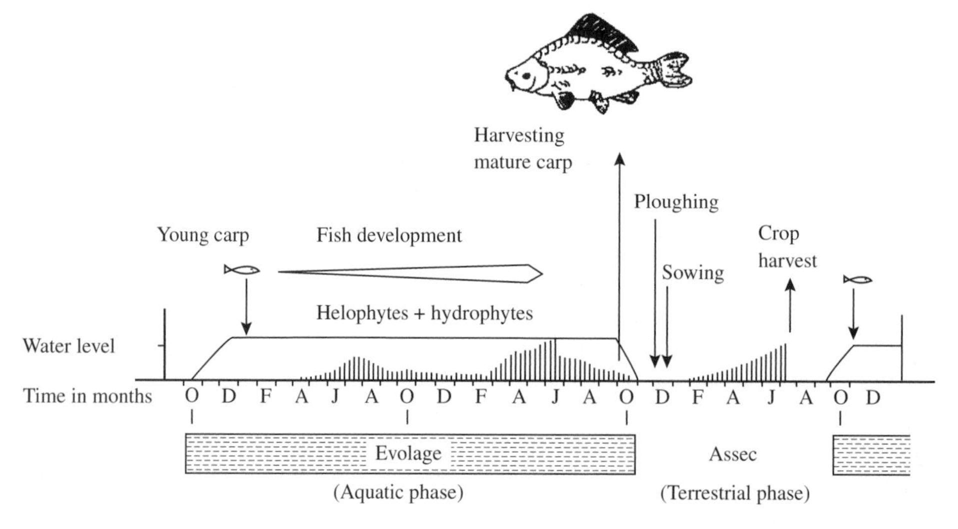

Figure 8.2 The traditional *evolage-assec* rotation in the Dombes ponds, France (based on information supplied by H. Tachet).

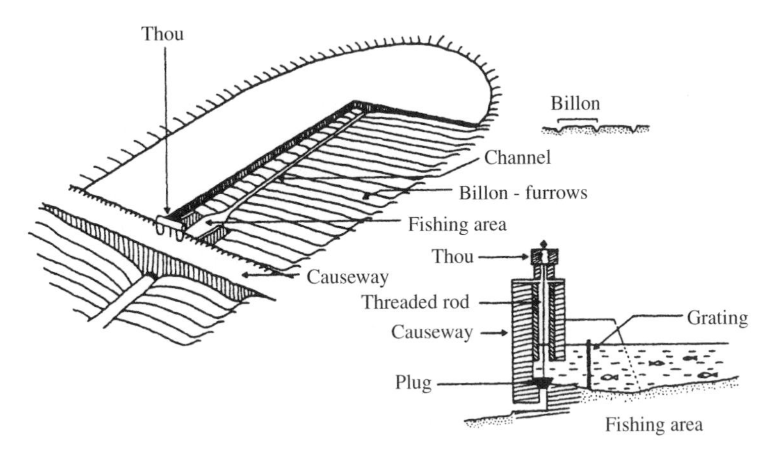

Figure 8.3 Simplified structure of a Dombes pond, together with a section through a 'thou' (based on information supplied by H. Tachet).

The Dombes system has been sustained biologically for centuries, and indeed improves soil of basically poor quality. The soil is allowed to lie fallow, under water, while the fishes are being raised. Aquatic plant debris from the evolage is mineralized when exposed to the air as the ponds are drained—if left submerged (in a non-degradable, acidic environment) much of this debris would accumulate on the pond bottom and its nutrients would not be released. Grant and Seegers (1985) have shown that in non-fertilized paddy fields, lowland rice uses ammonium-nitrogen present at the time of flooding and also nitrogen mineralized from soil organic matter. Experiments using a species of *Limnodrilus* taken from Philippine rice soil demonstrate that these worms readily increase the ammonium-nitrogen production, probably by promoting the breakdown of soil organic matter through ingestion of soil and excretion of compounds in various stages of mineralization. These nutrients then become available to the terrestrial crops in the assec and, in turn, their remains provide nutrients for the base of the aquatic food web (algae, macrophytes, invertebrates) upon which the fishes depend.

8.3 Floodplains and fisheries

Chapter 4 of the present volume provided a brief introduction to floodplain habitats, and emphasized that they tend to be more common in tropical and subtropical regions. In the latter, many cultures have come to rely heavily on the fishes that colonize these waters as a source of food. In the 1980s the 'flood pulse' concept was developed to underline the key role of the cyclical change between high and low water levels in structuring the ecology of large rivers and their floodplains (Junk *et al.* 1989). More recently, a comprehensive and interdisciplinary review of one of the world's largest floodplain systems, the central Amazon, has been published (Junk 1997). The aim of that volume was to consolidate existing knowledge, and to develop a theoretical framework for future ecological studies and for the development of methods for sustainable, multipurpose management. A particular focus was how the flood pulse affects the fish fauna. Coverage in this chapter will begin with some trophic considerations and follow with examples of the biology of fishes and fisheries practices in different floodplain systems.

8.3.1 Trophic considerations

The trophic dynamics of floodplain communities that fuel fish production have not been studied in great detail in the tropics, however, some generalities from studies in temperate regions likely apply. Merritt and Lawson (1992) emphasized that floodplains should be regarded as ecotones between aquatic and terrestrial habitats, and cited the conceptual model of riparian zones proposed by Gregory *et al.* (1991) which incorporated temporal and spatial patterns of hydrological and geomorphological processes. Merritt and Lawson's study focused on the role that floodplain macroinvertebrates play in the breakdown of leaf litter. Comparing floodplain and river communities in the Augusta Creek watershed, Michigan, they found that although the decomposer groups that operated in both systems were taxonomically different, their functional roles as leaf-litter detritivores were very similar. Major leaf-processing groups in the river included amphipods, stoneflies, mayflies, caddisflies, and dipterans, whereas on the floodplain this function was performed by semi-aquatic and terrestrial species of oligochaete (especially enchytraeids and lumbricids), snail, diplopod, isopod, dipteran (mainly tipulids), and beetle (Figure 8.4). Maximum decomposition is thought to occur in areas that are generally aerobic, yet have sufficient moisture, whereas it is less at drier sites, and least in permanently anaerobic sites (Brinson *et al.* 1981). Some riparian species have been shown to be prodigious leaf processors—for example, during 4 weeks in April–May, earthworms (*Lumbricus terrestris*) consumed leaves equivalent to 93.8% of the total annual leaf fall (mean consumption rate was 11.0 mg dry weight of leaves g^{-1} live weight of earthworms per day; Knollenberg *et al.* 1985). In Augusta Creek, itself, the bulk of the processing occurred in the autumn and winter, whereas on the floodplain it took place during spring. Nutrients thus liberated on wetted floodplain floors become available for aquatic species when pools form. Further, when connected to the main channel flow, a portion will enrich the nutrients within the river (Figure 8.4). As river water may also bring nutrients to riparian pools, there is thus a two-way exchange of nutrients between a river and its floodplain habitats (Junk *et al.* 1989). Moreover, it is widely held that floodplain forests are net importers of inorganic forms of nutrients and net exporters of organic forms—this transformation role having important implications for secondary production in downstream ecosystems (Sharitz and Mitsch 1993). At the same time, flooding provides colonization and dispersal opportunities for various plant and animals species. It is worth pointing out, also, that floodplain processes are little understood, indeed unknown, in many regions of the world. For example, in the riparian forests of Namibia species of fungus, millipede, and isopod—many of which are new to science—are thought to play an important role in both litter decomposition and nutrient cycling, ultimately releasing essential nutrients to riparian plants (Jacobson *et al.* 1995).

As noted in Chapter 4 (Section 4.4.6), flooding also influences the composition of riparian communities. For example, *Eucalyptus camaldulensis* (Red Gum) forests along the Murray River, Australia, and cypress (*Taxodium distichum*) wetlands of the southeastern United States are highly dependent on annual flooding (Mitsch *et al.* 1979; Bren 1988). Huffman and Forsythe (1981) recorded over 100 species of woody plants on floodplains of the central and eastern United States, all of which are capable of existing, to varying degrees, in both inundated and non-inundated conditions, but none actually requiring flooded, anaerobic soils for growth and reproduction. Nevertheless, hydroperiod-induced anaerobic soil conditions appear to be a powerful selection agent in determining floodplain plant community composition (see Sharitz and Mitsch 1993 for an example of zonal patterns in floodplain forest tree species). Such floodplain soils often have high clay and organic matter contents, leading to high fertility (especially of phosphorus, which binds to clay particles) that enhances growth of those plant species able to cope with poor soil aeration (Patrick 1981). Anaerobic soil conditions are created rapidly (within just a few days) when floodplain forests are inundated, but aerobic conditions return quickly following flood drawdown (Sharitz and Mitsch 1993).

In a study of macrophyte seed banks in a series of floodplain habitats, Abernethy and Willby (1999) found variations in germination success under different degrees of hydrological and management-related disturbance. For example, at

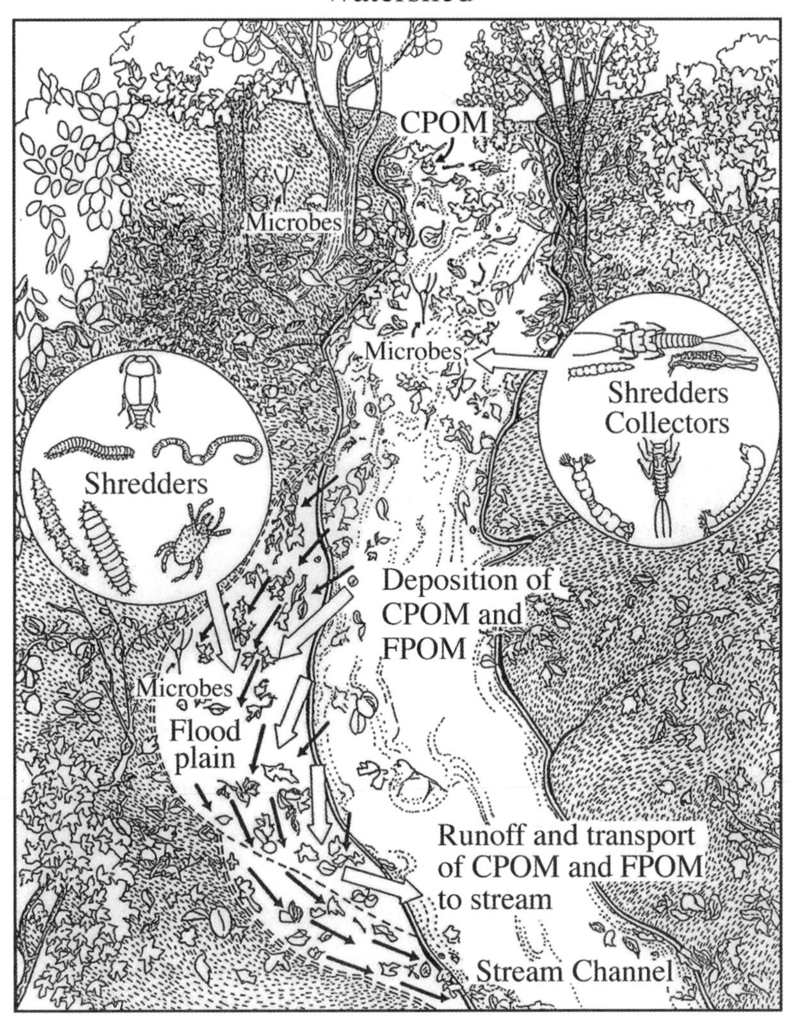

Figure 8.4 A conceptual model of the ecological processes and interactions occurring between a river and its floodplain (black arrows denote deposition of CPOM and FPOM onto the floodplain from the river; white arrows indicate movement of these materials from the floodplain into the river; reproduced with permission from Merritt and Lawson 1992).

permanently inundated sites, germination and species richness were low, with floodplain perennials and helophytes being most affected. Here, floodplain annual populations were maintained primarily through continued germination of a limited number of flood-tolerant species. However, in temporary backwaters, the most intensely disturbed sites, there were significant, species-rich seed banks representing largely floodplain annuals. Sites on the floodplain with intermediate inundation had a seed bank dominated by facultative amphibious, ruderal hydrophytes.

Phytoplankton community structure in floodplain lakes also tends to be hydrologically driven, often in association with macrophyte dynamics. For example, in Lake Camaleão in central Amazonia, diversity and abundance were strongly influenced by flooding from the large, white-water

River Solimões. The community comprised 262 taxa, especially euglenoids (81%), and was most rich at the end of the dry season which was characterized by high nutrient concentrations, much of which were derived from the decomposition of macrophytes (Ibañez 1998).

Macrophyte decomposition dynamics and phytoplankton, together with flood pulses undoubtedly influence riparian aquatic invertebrates, but quantitative studies linking these groups and processes are uncommon—although Harper *et al.* (1997) have suggested a dependence of invertebrate diversity in riparian pools on the quality of floodplain forests. In Redondo Lake, permanently connected to the Rio Negro, Brazil, Baptista (1999) has shown that invertebrate occurrence is linked to the high/low water cycle. During flooding, leaf litter (typically new and whole leaves) input is at its peak, but litter fragmentation increases markedly towards low water. During flood, shredder abundance equalled that of collectors, but the latter became dominant in the dry season. Scrapers occurred only when water levels were either rising or falling.

Smock (1994) observed and quantified the movements of invertebrates between stream channels and hardwood forest floodplains in the southeastern United States. There was a net output of individual animals from the floodplains to the streams, but a net input of biomass to the floodplains. Most of those invertebrates moving onto the floodplain did so via the discharge-related drift, however, most of the biomass accrued via individuals (typically Ephemeroptera, Trichoptera, and Megaloptera) crawling across the channel-floodplain boundary and was associated with habitat shifts resulting from life history events. In a later study, Smock (1999) re-emphasized the fact that such floodplains represent ecotones along a spatial and temporal continuum from continually flooded to terrestrial habitats. However, Brinson (1990) has cautioned that, in forested floodplains (or 'riverine forests'), flooding frequency alone is unlikely to explain the plant species mixture at a given site, but rather flooding in combination with the influence of soil texture and drainage on moisture regimes. Further, different stressors are likely to impose limits at the wet and dry ends of the moisture continuum. For example, at the wet end, forest development is limited by deep water, strong water movement, and anaerobic soils. At the dry end, a lack of 'supplementary' water from the river channel marks a transition to a true upland community. Such changes in forest composition affect the qualitative and quantitative nature of leaf-litter available for riparian and main-channel consumers, although as we have already noted, relatively few studies have addressed this issue (but see Bird and Kaushik 1981 for a comparison of leaf species processing in permanent streams). Forested wetlands occupy some 250 million ha, globally, and thus represent significant temporary water habitats that not only play a role in global biogeochemical cycles, but also support numerous freshwater fisheries (Lugo 1990).

Riparian sedge meadows are a characteristic feature of the landscape of Maine, United States, and during spring snow-melt, are inundated by water from the many small rivers that flow through them. Despite being short-lived, these wet meadows support distinct invertebrate communities whose species are derived from two sources, the floodplain and the rivers (Huryn and Gibbs 1999). Predominant among the floodplain taxa are *Aedes* mosquitoes, caddisflies (*Anabolia* and *Limnephilus*), and clams (*Pisidium*), which complete their entire life cycles on the floodplain. In contrast, the river-floodplain taxa have life cycles exhibiting both floodplain and river phases, and comprise primarily the mayfly genera *Leptophlebia*, *Siphlonisca*, and *Siphlonurus*. These mayflies undergo a rapid, but crucial, period of growth and development on the floodplain, and make up some 75% of the invertebrate biomass in that habitat. Alongside dytiscid beetles, some predatory caddisflies, dipterans and mites, and leeches, several vertebrate predators also migrate onto the floodplains, notably the common shiner (*Notropis cornutus*), the three-spine stickleback (*Gasterosteus aculeatus*), the chain pickerel (*Esox niger*), and the common white sucker (*Catostomus commersoni*). Common snipe (*Capella gallinago*) and black duck (*Anas rubripes*) also feed on the mayflies. Huryn and Gibbs suggest that given the rarity of unaltered, large river-floodplain habitats throughout the world, the pristine wetlands of Maine could serve as small-scale analogues for study of non-forested systems.

As the above examples show, invertebrates are quick to colonize floodplain waterbodies in order to make use of the food opportunities and environmental benefits (e.g. increased water temperature) available there. In turn, these invertebrates represent an important food source for vertebrate predators on floodplains. Aquatic macroinvertebrates provide food for adult fishes, as do terrestrial invertebrates falling from the forest canopy, and smaller taxa (e.g. microcrustaceans and chironomid larvae) are likely important to larval fishes. Observations suggest that many amphibians, and riparian species of bird, reptile and mammal also benefit from these invertebrate populations (Drobney and Fredrickson 1979)—although quantification of the energy that flows to these higher consumers is largely lacking (Smock 1999).

8.3.2 Fishes and fisheries

Floodplain fishes have best been studied in the tropics. Welcomme (1979) divided the fish species of tropical floodplains into two groups based on their behavioural responses to the fluctuating environment:

(1) those fishes that avoid the fluctuations by migrating to and from the main river channel, for example, species of Cyprinidae (minnows), Characoidei (characoids), and some species of Siluridae (catfishes) and Mormyridae; and
(2) those fishes that are able to survive the fluctuations, particularly that of dissolved oxygen, for example, species of Ophiocephalidae (snakeheads), Anabantidae (climbing perches), Osteoglossidae, Polypteridae (bichirs), Protopteridae (lungfishes), and most Siluridae.

Although there are many species in this latter category, very few of them are specifically adapted to survive complete loss of water from their habitat. If this occurs, then there is typically a huge fish kill. Only the lungfishes, some cyprinodonts, and some murrels (Ophicephalidae) appear to be able to survive—the cyprinodonts as 'annual' species (surviving as eggs), and the other two groups as cocooned adults (see Chapter 5).

Fish production and biomass tend to vary according to the flood characteristics of specific areas. Very high floods cause the main channel water to spread out over the floodplain and this enables fishes to forage farther afield. Feeding rate is greatest at times of inundation and the fishes store what they assimilate as fat, which lasts them through the lean dry period. Breeding, in most species, also coincides with the rise in water level, and significantly, a smooth and gradual rise, with a high, sustained peak produces higher numbers of young-of-the-year than lesser floods. Thus in good years, young fishes emerge at a time when microorganisms, plankton, benthos, and aquatic vegetation are abundant (Awachie 1981). Timing of migration onto floodplains is thus important for several reasons, and in some locations species have been observed to travel in an orderly sequence. For example, in the Mekong River, southeast Asia, fishes enter the monsoon-generated waters on the floodplain in a series of waves, with siluroids migrating first, and whitefishes later, and in the Niger River, West Africa, species of *Clarias*, *Distichodus*, *Citharinus*, and *Labeo* enter first whereas *Alestes*, *Tilapia*, mormyrids, *Schilbe*, and *Synodontis* migrate later (Soa-Leang and Dom Saveun 1955; Welcomme 1979). At low water on the floodplains of the Zaire River system, many juvenile fishes move into areas of floating meadows or marginal macrophytes to take advantage of the abundant populations of entomostracans and insects found there. This migration has attracted an intensive basket fishery comprising chiefly species of mormyrid, characoids (*Hydrocynus*, *Distichodus*, and *Citharinus*), siluroids (*Clarias*, *Synodontis*, and *Schilbe*), and tilapias (Lowe-McConnell 1991).

In the Rupununi savannas, located between the Amazon and Essequibo drainages (latitude 2–4°N), there is one well-defined rainy season each year. From May to August, these rains flood the savanna to a depth of 1–2 m and generate extensive aquatic plant growth. Lowe-McConnell (1991) found the fish fauna to comprise around 150 large species (61 characoids, 35 siluroids, 19 cichlids, 9 gymnotoids, and lesser numbers of 9 other families), together with numerous smaller (<5 cm) species. In the dry season, ponds within the savanna support 60–70 of the larger species, many of them different from those found in the river channels and pools. Considerable inter-year variation in population sizes has been observed, thought to reflect spawning success in the preceding year(s) which, in turn, is largely dependent on the way flooding occurs. For example, too little rain prevents fishes trapped in some of the ponds from being released, and may strand spawn. Refuge pond size appears to be a major factor in determining fish distribution. Many species are diurnally active within these very clear-water ponds, but hide in tree litter at night (e.g. cichlids and most of the characoids). Others, such as catfishes and gymnotoids, adopt a night-active behaviour. For some species, rock crevices and tree litter form an important, non-benthic refuge above oxygen-depleted bottom water, although many small fishes are capable of hiding in the leaf-litter that carpets these ponds (Lowe-McConnell 1991).

Many of the smaller, tropical floodplain fishes tend to have short life cycles, such that they may mature in one year and spawn at the next flood (Lowe-McConnell 1977). Larger species, such as those belonging to *Tilapia* and *Sarotherodon*, typically will not spawn until they are three or 4-years-old—however, in exceptionally dry years individuals may become stunted and mature in their first year (Dudley 1976). Many species have developed the capacity for rapid embryo development after their eggs are laid, and fry can emerge in as little as 16–24 h, as is the case for some nest-building and open-substrate spawners (e.g. *Labeo victorianus*); the larvae of mouthbrooding species may take up to 4–5 days (Welcomme 1979). Interestingly, and paralleling many invertebrate species in temporary waters, the eggs of some floodplain species, for example, *Arapaima gigas*, do not all mature at the same time (Lowe-McConnell 1991), presumably a bet-hedging strategy.

Floodplain fish assemblages can be extremely rich and diverse, for example, Petry *et al.* (2003a) recorded 139 species from a floodplain lake system in the central Amazon Basin. This assemblage proved not merely to be a random association of species, but rather one in which species were strongly correlated with various physical variables, macrophyte coverage, and habitat complexity. Interestingly, some species favoured shallow water areas low in dissolved oxygen, thought to provide refuge from piscivory—which also may influence assemblage structure (Rodríguez and Lewis 1997). In addition, in the upper Paraná River floodplain, Petry *et al.* (2003b) found that variation in assemblage composition was related to the degree of hydrological connectivity—with species richness, density, and biomass being significantly lower in connected lagoons than in disconnected ones. The coexistence of such high numbers of species on floodplains is believed to be possible, in part, because of marked morphological differences in species' trophic morphology (Hyslop 1986; Fugi *et al.* 2001). However, Junk *et al.* (1997) concluded that the fish fauna of the central Amazon floodplains exhibits largely *r*-strategy traits (rapid growth, early maturity, and high reproductive rates), as exemplified by the high numbers of characoid species present. Such traits maximize survival under a flood-pulse regime (Figure 8.5). In contrast, under such conditions, these authors argued that there should be fewer species present that exhibit the more specialized traits (e.g. complex behaviours linked to gathering specific foods, maintaining territories, parental care, etc.) characteristic of K-selected species (such as cichlids, callichthyids, and loricariids).

Many of these fish species are subject to heavy fishing by native peoples. However, this tends to be seasonal as, for example, during high water the fishes are spread out over a wide area of floodplain. Therefore, fishing effort tends to be concentrated when the flood recedes and the water starts to drain back into the main channel; migrating species, in particular, are caught in this way. Heavy fishing continues, during the dry season, in the basins of the floodplain, and also occurs at the beginning of the next flood cycle so as to again trap migrating species, this time on their way out of the main channel (Welcomme 1979). Often, the movement and habitation patterns of indigenous

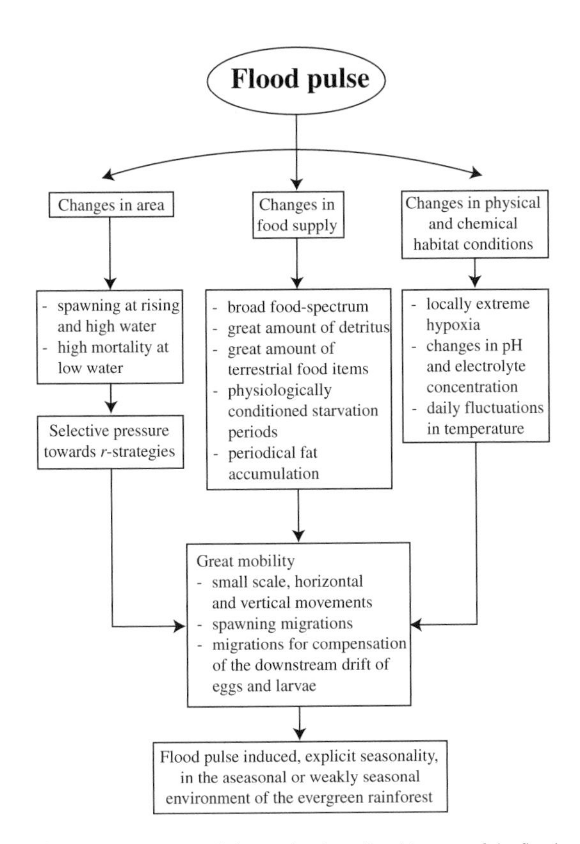

Figure 8.5 Summary of observed and predicted impacts of the flood pulse on fishes occurring on floodplains of the central Amazon (redrawn from Junk *et al.* 1997).

peoples are dictated by the seasonal availability of the fish resources.

The methods used to harvest fishes from floodplain waterbodies are highly varied according to local conditions, cultural traditions, and species' behaviour. Methods include an ingenious array of nets, traps, and other techniques, operated both from the shore and from boats (see Von Brandt 1984 for details). Frequently, there is modification of the floodplain to enhance fish production and capture. For example, installation of barrages and guide fences (constructed of brushwood, reeds, or earth) to corral exiting migrants, damming of pond outflows to retain fishes for harvesting later in the dry season, and planting of vegetation to attract spawning adults. More invasive modifications include regularizing the shape of lagoons, and excavating 'drain-in' fish ponds (typically 3–4 m wide channels, up to several kilometres long, and around 1.5 m deep). The latter may be extensive—for example, they cover 3% of the 1,000 km^2 surface area of the Ouémé River floodplain, West Africa. Similar drain-ins have been constructed on the Mekong floodplain, but these range between 20 and 100 m in length, and average 2–3 m deep. Excavation of floodplain soils to construct artificial islands for human occupancy has resulted in around 30,000 ponds in Bangladesh (Welcomme 1979). Floodplain ponds along the River Anambra, Nigeria, are extensively managed, particularly for species of *Clarias* (Ezenwaji 1998). In Zimbabwe, and other developing tropical countries, governments and various non-governmental organizations have, for some time, been actively encouraging the creation of 'backyard' fish ponds as a means of enhancing natural subsistence fish production that has failed due to over exploitation (Mandima 1995).

Table 8.1 lists floodplain river fish species, from several regions, that had high commercial value in the 1970s. Since that time, however, floodplain fisheries in many regions have been subject to overfishing, resulting in declining stocks—for example, species of the várzea, in the Amazon (Ruffino 1996). In other floodplains, productivity has been affected by changes to the environment—for example, low fish yield on the Mekong River, at Phu Thanh, Vietnam, has resulted from severe acid-sulphate soils disturbed following modification of the local hydrological regime (de Graaf and Chinh 2004). Further, in Cambodia, the Mekong floods back through the Tonle Sap into Grand Lac, a vast (11,000 km^2), shallow, seasonally fluctuating lake. Formerly, in the wet season (June–October), this lake typically extended into forest to support one of the world's largest freshwater fisheries. Now, however, and associated with forest felling, the lake is shallower and the fishery is in recession (Dussart 1974; Lowe-McConnell 1991). In the Sokoto-Rima floodplain, Nigeria, construction of the Goronyo dam significantly reduced the diversity of the juvenile fish community over at least a 100 km stretch downstream. Timing of the flood on the Sokoto-Rima plain is known to be important in determining its duration—if flooding

Table 8.1 Floodplain river fish species of significant commercial value[a]

Amazon	Niger	Indus	Mekong
Arapaima gigas	*Alestes dentex*	*Hilsa ilisha*	*Pseudosciaena soldado*
Myletus bidens	*Brachysynodontis batensoda*	*Notopterus chitala*	*Cirrhinus jullieni*
Prochilodus insignis	*Hydrocynus forskhalii*	*Catla catla*	*C. auratus*
Plecostomus spp.	*Sarotherodon niloticus*	*Cirrhinus mrigala*	*Ophicephalus micropeltes*
Brycon nattereri	*Labeo senegalensis*	*Labeo calbasu*	*Thynnichthys thynnoides*
Colossoma sp.	*Lates niloticus*	*L. rohita*	*Kryptopterus apogon*
Rhinosardinia sp.	*Bagrus bayad*	*Wallago attu*	*Macrones nemurus*
Prochilodus corimbata	*Mormyrus rume*	*Rita rita*	*Cyclocheilichthys enoplus*
Leporinus spp.	*Citharinus latus*	*Mystus* spp.	*Labeo chrysophekadion*
Plagioscion surinamensis		*Sarotherodon mossambicus*	*Ambassis wolffii*

[a] Only the top 10 or so species are given.

Source: After Fily and d'Aubenton (1966), Husain (1973), Meschkat (1975), Raimondo (1975).

takes place during the rains, the flood will extend over 3–4 months, however, if flooding does not begin until the dry season, increased evaporation and low humidity will result in a much shorter flood and rapid isolation of floodplain pools (Hyslop 1988).

In many parts of the world, floodplains are under cultivation by man during the dry phase and so such systems may parallel that of the Dombes ponds described earlier in this chapter. A summary of the main activities that occur on both natural and cultivated floodplains is given in Figure 8.6, and a scheme of nutrient and energy cycling in the system is given in Figure 8.7. Provided a natural community of plants and animals remains, these floodplain systems appear to be able to function under both regimes. However, man's increasing appetite for land for development is compromising these systems, particularly if it includes flood control, because cyclical flooding is the underlying factor upon which these complex ecosystems have evolved.

Although agricultural activities on floodplains, at traditional levels, appear to have relatively minor effects on water quality and fisheries, increases in human populations are accompanied by changes that have major effects. One of the most severe is water retention behind dams, as part of sophisticated irrigation systems for increasing crop production. Dams, by definition, change the patterns and timing of water flow, with downstream wetlands being compromised and losing their natural species. Loss of the natural flooding cycle fails to release nutrients into floodplain soils, necessitating the application of inorganic fertilizers. Such sudden availability of nutrients typically results in algal blooms which may both increase the numbers of algal-feeding fish species (e.g. *Tilapia* and *Labeo*) and those that can tolerate low oxygen levels (e.g. *Clarias*), and remove others—as has been the case in West Africa. In addition, continued application of inorganic fertilizers to floodplain soils, where flooding and the water table are low, leads to the accumulation of salts which render the soils unsuitable for agriculture (Ita 1993). In some countries, for example, India, dam construction has become so extensive as to significantly affect riverine fisheries, and shift attention to those species that thrive in impounded waters, such as carps (*Catla catla*, *Labeo rohita*, *L. fimbriatus*, and *Cirrhina mrigala*) (Hora and Pillay 1962; Ackermann *et al.* 1973; Lowe-McConnell 1991).

In Australia, the fish fauna of the largest floodplain river, the Murray-Darling system, has been severely compromised over the last 100 years from a combination of general habitat degradation, disease, and competition from introduced species. However, in its original state, the Murray represented one of the most variable rivers on earth,

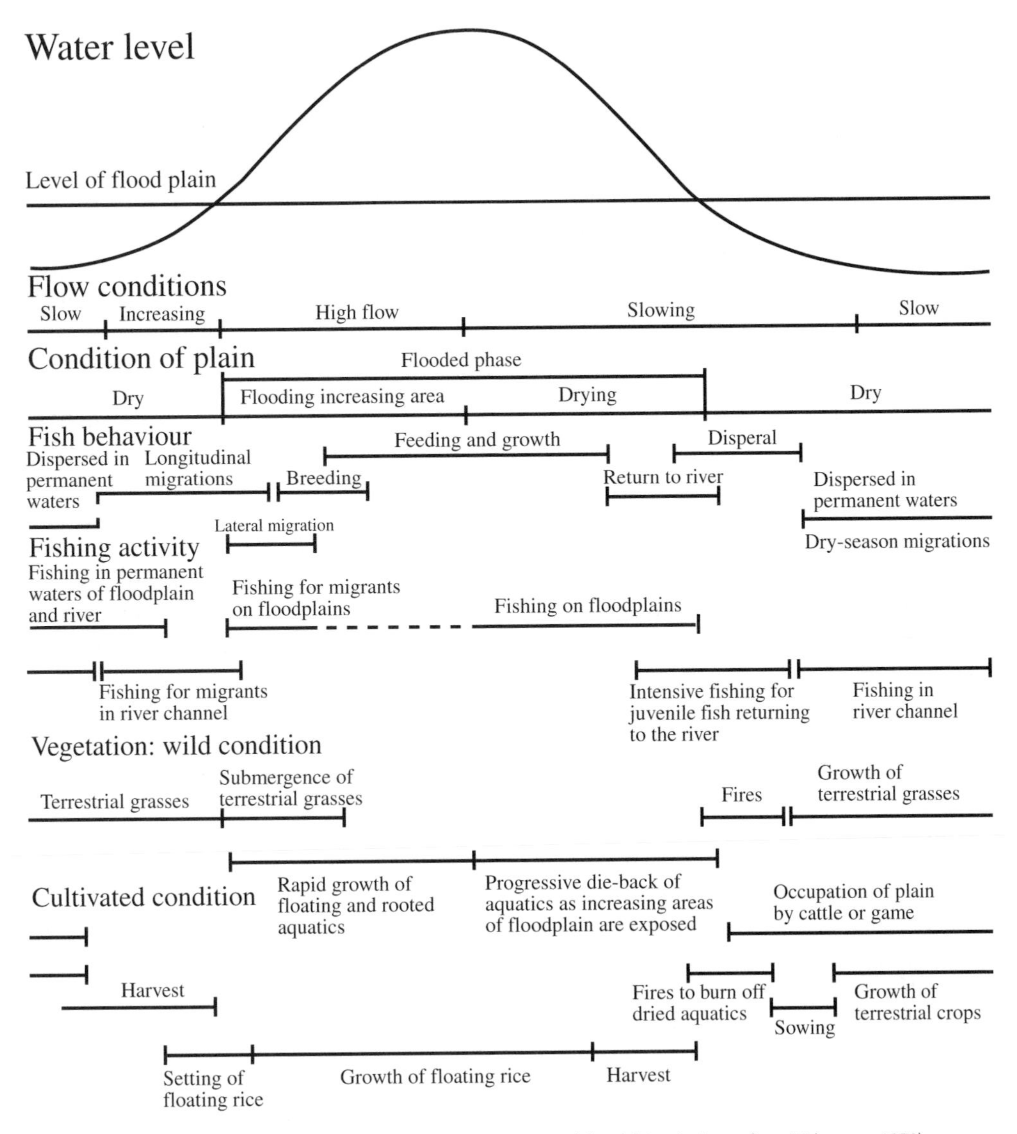

Figure 8.6 The main annual activities seen on natural and cultivated tropical floodplains (redrawn from Welcomme 1979).

and only some 30 species had been able to establish themselves in its floodplain-driven waters. Until recently, fisheries targetted Murray cod (*Maccullochella peelii peelii*), callop (*Macquaria ambigua*), and bony bream (*Nematalosa erebi*), but in 2003 the emphasis was shifted to non-native species (South Australia Fisheries Research Advisory Board 2004). All three of these native species exhibit remarkable tolerance of high water temperatures, high turbidity, and sluggish flow. For example, the callop ('Murray perch') is frequently associated with floodplain backwaters and billabongs, that may reach as high as 37°C; the bony bream prefers slow-flowing or still waters particularly in dry eucalypt-scrub country, where it breeds in the wet season; and the Murray cod inhabits silty, alluvial lowland channels (South Australian Research and Development Institute 2001; Native Fish Australia

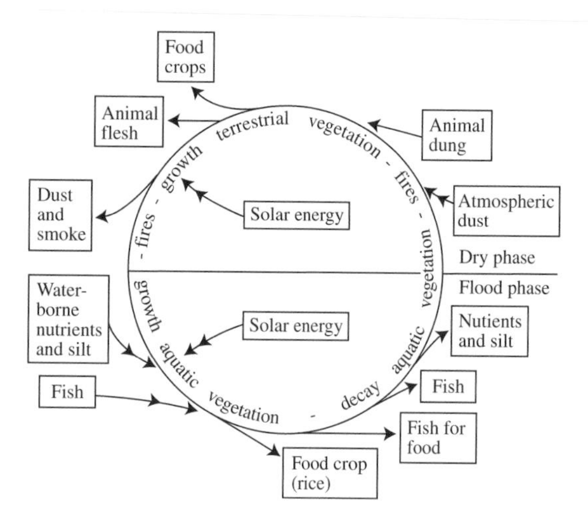

Figure 8.7 Energy and nutrient cycling in African floodplain systems (redrawn from Welcomme 1979).

2004). In Northern Australia, where the climate is monsoonal, 90% of the annual rainfall occurs between November and March and creates extensive floodplains (the Kakadu National Park) associated with the South, West and East Alligator rivers, the Wildman Rivers system, and Megela Creek. Beginning in May, these begin to dry out with the water contracting into lagoons and billabongs. These river systems support distinctive fish communities, including species found on the lower floodplain zone which supports major refuge habitats. Species commonly found in the billabongs and creeks include barramundi (*Lates calcarifer*), freshwater long-toms (*Strongylura* spp.), salmon-tailed catfishes (*Neosilurus* spp.), checkered rainbowfishes (*Melanotaenia splendida inornata*), and the Saratoga (*Scleropages* spp.). Backflow billabongs and floodplain billabongs are particularly important as nursery areas for the juveniles of many of the larger fish species, most of which breed at the beginning of the wet season when habitat area is significantly increased and there is a plentiful supply of planktonic and benthic foods (Tappin 2004).

8.4 Rice paddy fields

Included in the floodplain cultivation activities of humans (summarized in Figures 8.6 and 8.7 above), is the growing of rice, which is well suited to the hydrological regime of the floodplain environment. Rice fields are, in reality, temporary man-made aquatic habitats in which the hydrological regime is tightly controlled (Welcomme 1979). Typically grown in subtropical and tropical regions, rice is planted and harvested once or twice (occasionally three-times) a year in a range of irrigated, rain-fed lowland, flood-prone areas. Half the present world population is estimated to subsist wholly or partially on rice, with 90% of the world crop being grown and consumed in Asia (90–181 kg per person yr^{-1}); in terms of quantity produced, rice ranks with wheat and corn. Rice has been grown in China for an estimated 4,000 years, however, plant remains discovered in Spirit Cave on the Thailand-Myanmar border, suggest that its cultivation may be as old as 10,000 years BC. Since these times, the method of rice cultivation has diversified according to locality. However, in most Asian countries the traditional methods of hand-cultivation and harvesting have persisted. Basically, this entails preparing paddy fields with a simple plough, adding readily available fertilizer (typically dung or sewage), and smoothing the soil. Thirty to 50 day-old rice seedlings are then planted out to coincide with the flooding of the paddy fields either by natural (river flooding), or contrived (irrigation) means. Water is then retained and managed by a system of canals and dykes, but is typically drained before crop harvesting. During the dry phase, some farmers plant another crop, such as maize or soybean, to profit from the 'off season' (Food and Fertilizer Technology Centre 1994).

During the aquatic phase, many organisms other than rice colonize the paddy fields, and fishes from source rivers, in particular, are sometimes seen as being detrimental to the rice harvest. On the Niger River floodplains, for example, Matthes (1977) observed damage from three categories of fish behaviour: (1) fishes that feed extensively on rice plants—including *Alestes dentex*, *A. baremoze*, *Distichodus brevipinnis*, and *Tilapia zillii*; (2) fishes that feed occasionally on rice plants or their epiphytes—such as *Alestes nurse* and *Sarotherodon niloticus;* and (3) species that cause other damage, such as using rice stems for nest construction (e.g. species of *Heterotis* and *Gymnarchus*), or that uproot seedlings while searching for benthic food (e.g. species of *Clarias*, *Heterobranchus*, and *Protopterus*). Despite such problems, the practice of culturing fishes in paddy fields can be traced back to the Sichuan and Guizhou Provinces in China almost 2,000 years ago. However, due to a shift in government policy and also an increase in pesticide and chemical fertilizer use from the 1950s to late 1970s, rice-fish culture in China ceased. In the early 1980s, the practice was re-examined and underwent a rapid expansion in the 1990s, extending over an area of 1,528,027 hectares by 2001, and producing some 849,055 tons of fish per annum (Xiuzhen 2003). Elsewhere in the world, for example, Latin America, Africa, and other parts of Asia, fish culture in rice fields is practiced at a number of different levels. These range from the historical, non-managed level—where wild fishes entering during flooding are captured when the rice is harvested, to the intensively managed level—where selected species are stocked, fed, protected from water quality decline, and otherwise generally enhanced through paddy-field engineering.

Some of the advantages of the rice-fish culture system are: additional food and income in the form of fish flesh; control of certain mollusc and insect species that are injurious to rice plants; reduced risk of crop failure resulting from the integration of rice and fishes; continued flooding of the paddy fields together with rooting activity of the fishes helps to control weeds; and stirring up of soil nutrients by fishes makes them more available to the rice plants, leading to increased rice production. Potential disadvantages include the restricted use of pesticides, increase in the labour force compared with rice alone, and production of smaller fishes at harvest compared with traditional fish pond culture (International Centre for Aquaculture and Aquatic Environments 2004). Standard rice paddies also require some modification if they are to be used for effective fish culture. Foremost, is deepening of a section of the paddy so as to create a fish shelter and harvest area. However, no more than 10% of the paddy area should be modified so as not to affect rice production—although new forms of floating rice that have evolved on floodplain wetlands in southeast Asia (Abbasi 1997) may permit greater overall water depth.

Management practices differ according to region, fish species, and the number of fishes stocked (International Centre for Aquaculture and Aquatic Environments 2004). For example, Huat and Tan (1980) provide the following management sequence for carp (*Cyprinus carpio*) in Indonesia:

• Paddies have peripheral trenches 30–50 cm deep, dikes 25 cm high, with bamboo pipes and screens at water inlets and outlets. Five days after planting out rice seedlings, 1 cm long fry are stocked into the paddy at a rate of 60,000 per hectare. Roughly two weeks later, the paddy is drained for weeding, and the fishes take refuge in the trenches. The paddy is refilled and, about 2 weeks later, the rice is weeded again. At this time, fishes 3–5 cm are harvested and sold to fish pond owners for culture to market size, whereas larger fishes (8–10 cm) are restocked into the paddy at a rate of 1,000–2,000 ha^{-1}. Fishes and rice are grown for an additional 1.5–2 months. At harvest, the carp weigh between 50 and 75 g, and are 14–16 cm. Total fish yield from such a system is between 75 and 100 kg ha^{-1}.

Cruz and de la Cruz (1991) provide details of a modified management protocol that enhanced carp production during 75-day experimental trials in Indonesia; these trials took place during the long fallow period between two rice cropping seasons:

• Paddy dikes were raised to allow a minimum water depth of 40 cm. Standing rice stubble remained in the paddy, and loose rice straw was stacked in small mounds. Fingerlings, weighing between 53 and 111 g were stocked at densities of 1,500, 3,000, and 4,500 per ha^{-1} in 9 paddies. Each density was replicated three-times. Supplemental feeding at 5% of the fish biomass per day (feed composition: 72% rice bran, 20% copra meal, 8% soybean meal) was adjusted weekly. After 35 days, the paddies were drained and the fishes sampled to determine growth. They were then refilled and restocked with the same fishes. The hydroperiod continued for an additional 40 days before the final

Table 8.2 Results of carp-culture trials in rice paddies in Indonesia

Parameter	Stocking density[a]		
	1,500	3,000	4,500
Mean fish weight (g)	355.1	303.8	257.2
Gross yield (kg ha^{-1})	508.0	913.3	1,044.4
Survival (%)	95.0	100.0	89.7
Profit ($)	356.0	519.5	574.2
Rate of return (%)	65.5	47.4	45.1

[a] Fishes per hectare.

Source: After Cruz and de la Cruz (1991).

harvest. The results of these trials are shown in Table 8.2. Clearly, on a per hectare basis, alternating a fish crop with a rice crop generated significantly greater fish yields (5–10 times) than when the two species were grown simultaneously. Why should this be so? There is the obvious difference that the alternating protocol included supplemented food (most of which was a waste product of the rice harvest), but the stubble, roots, and straw persisting from the rice phase would also have provided nutrients for aquatic microorganisms and invertebrates that, in turn, would enhance the food base for the carp. Draw-down intervals between the fish and rice phases would provide an alternation of aerobic and anaerobic conditions in the paddy soil likely to increase the availability of phosphates and potassium, both essential to plant growth. In reality, the alternating protocol produced markedly greater fish yields, likely by simulating some of the conditions found in natural intermittent ponds.

Kirk and Kronzucker (2000) have reviewed the processes by which rice roots growing in flooded soils acquire nitrogen, and cautioned that a balance should be struck between the rapid uptake of fertilizer-based nitrogen that is broadcast on crop fields, and the more gradual uptake of nitrogen in the soil. Rice plants are capable of absorbing very high rates of fertilizer nitrogen (up to 10 kg ha^{-1} d^{-1}) if its application coincides with high plant demand. However, root system and canopy development typically peak before this occurs resulting in loss of nitrogen via ammonia volatilization and immobilization by microorganisms. Kronzucker *et al.* (2000) have suggested that lowland rice growing in flooded soil may absorb substantial amounts of nitrate formed by nitrification in the rhizosphere, which otherwise will be lost via denitrification in the soil bulk. Plant growth and yield are typically improved when plants are able to obtain their nitrogen through a mixture of nitrate and ammonium (Gill and Reisenauer 1993). When rice culture is alternated with that of fish, considerable nitrogen is available as excreted ammonia, which can then be converted to ammonium by attaching to hydrogen ions found in the soil. Nitrate also becomes available in the paddy soil through the biological and chemical transformation of ammonium—however, the rate at which this happens is influenced by the amount of oxygen available through organic matter management and irrigation regime.

Over time, rice management practices have changed, often as a result of switching rice varieties, and the paddy field habitat has changed, too. For example, in the Philippines, the introduction of the IR8 strain (a semi-dwarf, high-yield variety) in 1962, necessitated intensive use of inorganic fertilizers and pesticides. But while this doubled annual rice production, it did so at a considerable cost to the environment, decimating fish and amphibian populations during the 1970s. Other practices are more environmentally friendly, such as encouraging the fast-growing, floating macrophyte, *Azolla*, to grow among rice plants, such that when it dies, its breakdown products supply nitrogen to the rice.

Interestingly, some species that have evolved in natural ponds are now being used as biological control agents in rice culture. For example, in Japan, the problem of annual weeds infesting paddy fields has been countered by introduction of several species of tadpole shrimp (*Triops*). This practice began in the 1920s, and it has been shown that at a density of 25–80 animals m^{-2}, the typical 500 man-hours required to weed 1 ha could be reduced to only 20 h (Matsunaka 1976; Yonekura 1979). Weed control is brought about via shrimp activity which agitates the soil surface, uproots weed seedlings, and creates turbid water (which compromises photosynthesis), and by consumption of developing weed buds. In Californian rice fields, tadpole shrimp have the potential to harm developing, directly seeded rice plants, however, this can be avoided if seedlings are transplanted into the fields at a larger size (Grigarick *et al.* 1961).

One final point to consider is the effect of converting significant areas of the world's natural wetlands to managed rice fields. Elphick (2000) examined the functional equivalency between rice fields and semi-natural wetlands in California in terms of habitats for aquatic birds. Parameters compared included food abundance, perceived predation threat, foraging performance, and allocation of time to different behaviours. The study concluded that flooded rice fields provide equivalent foraging habitat, and, because of reduced predation threat, may actually represent a safer habitat for birds.

CHAPTER 9

Habitats for vectors of disease

9.1 Introduction

This chapter considers the many disease-carrying organisms, chiefly invertebrates, that are supported by temporary water habitats. Blood flukes (*Schistosoma* spp.), for example, cause schistosomiasis in humans, and are spread by snails that live in temporary ponds, marshes, and drains. The nematode *Dracunculus medinensis* (Guinea worm) is a parasite of the human skin, and has as its intermediate host a copepod that lives in small ponds and step wells. Of course, many of the best known and most serious diseases, such as malaria, yellow fever, and elephantiasis, are spread by temporary water-dwelling mosquitoes. The most notable disease vector and pest species are those belonging to the genera *Anopheles, Culex, Aedes, Psorophora, Haemagogus,* and *Sabethes* (Service 2000). For example, populations of *Anopheles neivai* in Colombia, *An. bellator* and *An. homunculus* in Brazil and Trinidad, and *An. cruzii* in Brazil are proven carriers of human malarias. *Anopheles bellator* is also known to carry the nematode *Wuchereria bancrofti*; all of these mosquitoes breed in phytotelmata (Frank 1983). The biology of such species will be presented alongside a discussion of the eradication/control methods that are aimed at these organisms and their habitats.

9.2 Why temporary waters?

Why is it that so many of the world's most human-centric diseases are associated with temporary waterbodies? At least part of the answer must lie in the need for close association of humans with both natural and culturally created water supplies. As a consequence, human disease organisms are best served by also associating themselves with these waters, and by harnessing the vector properties of many of the invertebrates that live in them. As some vector species have spread beyond the boundaries of natural temporary waters (e.g. rockpools, and phytotelmata) to colonize those in even greater proximity to human hosts (e.g. rain-filled cooking vessels and livestock troughs), disease transmission efficiencies and infection rates have risen. For example, *Anopheles dirus* is the main vector of malaria in parts of southeastern Asia, where it typically lives in forests, feeding on primates. However, when humans move into these forested areas they fuel transmission conditions by increasing the blood meals available, and by creating many more small container habitats for larval development (Pates and Curtis 2005).

9.3 Temporary water-facilitated diseases—who are the vectors?

Disease vectors may be either biological or mechanical. A *biological vector* is a vector for which there is evidence of a long-standing ecological relationship with the disease organism. Typically, the vector species is persistently infected, and is often a required host for an essential part of the pathogen's life cycle. In contrast, a *mechanical vector* transmits the pathogen, but the latter remains unaltered while on or in the vector. The vast majority of temporary water-living vector species act as biological vectors.

The main human diseases spread by vector species that live in temporary waters are given in Table 9.1. The most serious of these is malaria. It is caused through infection by species belonging to

Table 9.1 Prominent human diseases spread by vector species that inhabit temporary waters (after various sources, but especially WHO)

Disease	Organism	Biogeography	Vector	Habitat
Malaria	Protozoa: *Plasmodium* spp.	Global, but 80–90% in Africa	Anopheline mosquitoes	Various temporary waters
Schistosomiasis (Bilharzia)	Trematoda: *Schistosoma* spp.	Global, but mostly in Africa and S. China	Snails: *Bulinus, Biomphalaria, Oncomelania*	Ditches, swamps, etc.
Japanese encephalitis	Arbovirus	South, southeast and east Asia,	Culicine mosquitoes	Flooded rice fields
Lymphatic filariasis (includes elephantiasis)	Nematoda: *Wucheraria bancrofti Brugia malayi, B. timori*	*bancroftian*: Mexico, Central and South America, Africa, India, S-E Asia, S-W Arabia	Culicine and Anopheline mosquitoes	Various temporary waters
		brugian: India, S-E Asia	Culicine mosquitoes	Various temporary waters
Yellow fever	Arbovirus	Mexico, Central and South America, sub-Saharan Africa, S-W Arabia	Culicine mosquitoes	Phytotelmata, container habitats
Dengue	Arbovirus	Mexico, Central and South America, Africa, India, S-E Asia, S-W Arabia, South Pacific	Culicine mosquitoes	Especially container habitats
Dracunculiasis (Guinea worm)	Nematoda: *Dracunculus* sp.	India, S-E Asia, sub-Saharan Africa, S-W Arabia	Copepods	Puddles, stagnant pools
Leishmaniasis	Protozoa: *Leishmania* spp.	Mexico, Central and South America, Africa, S-E Asia, S-W Arabia	Psychodidae	Various temporary waters
Onchocerciasis	Nematoda: *Onchocerca volvulus*	West and Central Africa, some parts of South America, Arabian peninsula	Simuliidae	Typically permanent rivers, but sometimes steep canals and dam spillways

the protozoan genus *Plasmodium*, of which *P. vivax* is perhaps the most widespread. *P. vivax* is a parasite of human liver and red blood cells, and is spread through the bite of many species of anopheline mosquito. It occurs in temperate as well as tropical climates, and is relatively benign (McKenzie *et al.* 2002). Other forms of malaria, in humans, are caused by *P. falciparum* (the most deadly, and found in the subtopics and tropics), *P. malariae* (broadly tropical, but with a spotty geographical distribution), and *P. ovale* (primarily tropical West Africa), whereas other species in the genus may be specific to non-human primates and birds (Olsen 1974). The symptoms of malaria are well-known, expressed as chills, recurrent fever, weakness and emaciation. It affects hundreds of millions of people, directly, through debilitation but many more, indirectly, through losses in labour and efficiency. The disease has been known for at least 2,500 years, and Hippocrates is credited with having made the connection between the occurrence of malaria and the proximity to swamps and stagnant waterbodies.

Mosquitoes that carry malarial parasites include the genera *Culex, Aedes, Culiseta, Psorophora, Armigeres, Mansonia*, and *Anopheles*, but only the latter is responsible for spreading human malaria. However, there are about 40 species of *Anopheles* that can transmit *Plasmodium vivax* alone, and these include species found in Europe and North America. Survival of the mosquitoes when their larval habitats dry up is often dependent on gravid females. Females of *Anopheles gambiae*, the most important of the African malaria vectors, frequently aestivate in a dormant state in crevices (e.g. in walls, in dry wells, and in rodent burrows) during the long dry season. Omer and Cloudsley-Thompson (1968) showed that, in the Sudan, 77.1% of female *An. gambiae* in this condition were engorged with blood (90.6% of it human), and that their ovaries developed very slowly during the drought so that when the rains came the females were ready to oviposit.

Schistosoma is a genus of blood fluke (Trematoda) that causes the condition known as schistosomiasis, or bilharzia, in humans. Various species are common throughout many parts of Africa, South America, and southern China where their vector hosts are freshwater snails. Snail species suitable for the transmission of schistosomiasis (generally of the family Planorbidae: *Biomphalaria, Bulinus*) are common and pick up the ciliated larvae from a variety of shallow waterbodies such as storage ponds, marshes, streams, and banana drains (Sturrock 1974). These miracidia secrete histolytic enzymes from special glands and these enable them to bore into the snails' tissue. Once inside the snail, each miracidium absorbs food from the host, and after 2 weeks produces daughter sporocysts. After a further 2–4 weeks the cysts produce cercariae which then leave the snail and swim about in the water until they come into contact with the final host, often man. The usual method of entry into the body is by the animal forcing its way through the skin. The most important species are *Schistosoma mansoni, S. haematobium* (both occurring in the Middle East and much of southern and equatorial Africa; *S. mansoni* also occurs in parts of the West Indies and South America), and *S. japonicum* (common in the Far East). In *S. mansoni*, perhaps the most studied species, adults live in those portal system veins that drain the colon. The eggs are laid in small veins and develop into miracidia, some of which find their way into the intestine and thence into water when the host defecates near a pond or marsh. Symptoms of the disease include rashes, fever, inflamation, pain, and damage to internal organs such as intestine, liver, spleen, lungs, bladder, spinal cord, and brain. In tropical Africa, *S. haematobium* is chiefly carried by *Bulinus* snails and is commonly endemic in areas of standing waters which disappear in the dry season. It is clear that not only can the snails survive drought but also that the schistosomes within them survive and are infective when the ponds refill after rain (Beadle 1981). *S. mansoni* is known to survive in aestivating *Biomphalaria* snails in Africa and Brazil (Barbosa and Barbosa 1958; Jordan and Webbe 1969). It seems likely that the larvae stop their development when the snails go into aestivation, and, in this state, they can survive for up to 7 months in snails not in water (Richards 1967). *S. japonica* is known to be transmitted by *Oncomelania* snails (Bulimidae) found in eastern Asia and nearby Pacific islands.

Another serious condition caused by trematodes is fascioliasis. This is primarily a disease of the liver of cattle, sheep and other grazing animals, but it may also affect humans. Again, the intermediate hosts are snails, typically those living on the wet mud at the edges of small ponds, seeps, and marshes (e.g. *Stagnicola bulimoides* and *Fosaria modicella*). They particularly thrive in temporary pools where they are capable of surviving the dry period by aestivation (Olsen 1974). Eggs are deposited in the water in the faeces of grazing animals, and only hatch after contact with the water. Again, it is the miracidium stage in the life cycle that seeks out a snail host. Development in the snail culminates in cercariae which emerge and encyst to form metacercariae. They are grazed off herbage by the final host. In *Fasciola hepatica*, the young fluke emerges from its cyst in the duodenum below the opening of the bile duct, where the bile triggers emergence. The metacercariae initially feed on the mucosal lining and eventually migrate to the liver. The flukes mature and begin to lay eggs anytime from 9 to 15 weeks later, depending on the host species. Symptoms of the disease include loss of weight and condition, swelling, breathing difficulties, toxemia, and often death.

Another form of fascioliasis is caused by *Fasciolopsis buski*, a fluke common in eastern Asia and the southwest Pacific where it may affect as many as 10 million people, causing intestinal and toxic symptoms. The life cycle is similar to that of *F. hepatica* but *F. buski* lives in the duodenum. After a sporocyst generation and two generations of rediae, cercariae are produced which encyst on the surfaces of aquatic vegetation such as water caltrop and water chestnut. When the seed pods or bulbs of these plants are harvested and eaten, they are invariably first peeled using the teeth. It is during this process that the metacercarial cysts are ingested (Jones 1967).

Paragonimus westermani, the oriental lung fluke, infects humans and a number of domestic and wild animals. It has two aquatic intermediate hosts. The snail (e.g. *Pomatiopsis lapidaria*) is penetrated by miracidia which swim about in marshy seepages and floodplain pools after having been deposited in faeces. Sporocysts develop in the snail and produce rediae which migrate into the lymph system of the liver. The rediae contain mature cercariae which emerge in the late afternoon and early evening to coincide with the peak activity time of their second intermediate host, either freshwater crayfishes or crabs. In these secondary hosts, the cercariae encyst around the heart where metacercariae are produced but are not released until the crustacean is eaten by the definitive host which is, in many cases, a human.

Japanese encephalitis is caused by an arbovirus (a virus in which the life cycle includes transmission by arthropods) spread by culicine mosquitoes, especially *Culex tritaeniorrhynchus* and *Cx. gelidus*. Prevalent in southern and eastern Asia, it is closely linked to irrigated rice production areas which provide vector larval habitats. The disease manifests itself in periodic outbreaks, often of epidemic proportion, and in which children suffer high mortality. The virus is amplified in pigs, however, rapid build-up of vector population densities, resulting from favourable local climate conditions, causes a spill-over into the human population for access to blood meals. Migratory birds are believed to facilitate spread of the virus over large distances (FAO 2004).

Lymphatic filariasis results from infestations of nematode worms, especially *Wuchereria bancrofti* and *Brugia malayi* (Table 9.1). The former causes either mumu or elephantiasis, depending on the severity of the infection, and also a lung condition known as tropical pulmonary eosinophila (TPE). Mumu is a painful swelling of the lymphatic ducts, whereas elephantiasis (which develops in about 5% of infected people) is a widespread blockage of lymphatic channels from any distal portion of the body, causing extreme swelling of the legs and genitals (WHO 1993). Bancroftian filariasis (*W. bancrofti*) is found in most tropical regions (Africa, India, southeast Asia, South America, the Caribbean, and Pacific Islands), and is transmitted to humans by any of several species of mosquito, but particularly those of the genera *Culex*, *Aedes*, *Anopheles*, and *Mansonia* that breed in temporary waters (Jones 1967). Microfilariae are ingested by female mosquitoes, as they take a blood-meal from an infected person, and penetrate the stomach lining to lodge in the insect's thoracic muscles. Here they transform into robust, infective larvae which then migrate (some 8–14 days later) to the mosquito's proboscis in readiness for transfer back to a human host. In certain regions of the tropics, virtually all the inhabitants of a particular village or community may have larvae in their blood (Jones 1967; Dreyer *et al.* 2000). *Brugia malayi* and *B. timori* are found in southeast Asia, and are responsible for brugian filiariasis, spread by species of *Mansonia* mosquitoes (Pradeep Kumar *et al.* 1992). Lymphatic filariasis is seldom fatal.

A number of other nematode parasites are associated with temporary waterbodies. For example, *Dracunculus medinensis* is a parasite of human subcutaneous tissues, and occurs in many parts of the tropics, particularly in Africa and Asia where it is known as the Guinea worm. It is typically a parasite of rural populations in arid regions, where household water is obtained from step wells and small ponds (Olsen 1974). The larval nematode lives in copepods such as those of the *Mesocyclops leuckarti*-group and *Thermocyclops hyalinus*, together with other species of *Mesocyclops*,

Eucyclops, and *Macrocyclops*. If a copepod becomes infected it tends to sink and, in wells, this stage seems to coincide with the period of drought—this may increase the probability of infection by concentrating the infective larvae in a relatively small volume of water (Croll 1966). Infection of people occurs when these copepods, containing fully developed third-stage larvae, are swallowed in drinking water. Over the course of 10–14 months, the immature worms migrate through the intestinal wall, grow (up to 1 m in the case of females), and mate. Males subsequently die, but females make their way into subcutaneous areas, particularly in the legs. The presence of these females produces inflamation and painful blisters, which the host tends to bathe. This process bursts the blisters and releases vast numbers of young nematode larvae into the water where they are consumed by copepods, to begin the cycle again.

Although more typically associated with fast, permanent waters, the simuliid vectors of *Onchocerca volvulus* are occasionally found in steep, intermittently flowing canals, ditches, and dam spillways. This parasitic nematode is the cause of onchocerciasis, or river blindness, and is widespread in west and central Africa, parts of South America, and the Arabian peninsula (Table 9.1). Onchocerciasis is a non-fatal disease that affects an estimated 18–40 million people worldwide. Approximately 2 million people have been blinded by the disease, and about 85 million people, in 35 countries, live in regions where this disease is endemic (Carter Centre 1999). The worm larvae are taken into the blackfly host during a blood-feeding episode, and there they mature to their infective stage within 6–10 days. A subsequent blood meal transfers them back into a human host where they mature and mate. Female worms produce thousands of microfilarial larvae each day, by vivipary. Larvae take up to one year to mature, and may be found encapsulated with females in fibrous-coated nodules (onchocercomas) throughout the body. Blindness results from the microfilariae using ocular tissues as migration routes, and through invasion of the conjunctiva and cornea (Rowe and Durand 1998).

Yellow fever results from infection by an arbovirus which, in its most common, urban form, is transmitted by *Aedes aegypti*. Jungle (sylvatic) yellow fever is transmitted by additional culicine species, for example, those belonging to the genera *Sabethes* and *Haemagogus* in the Americas, and several species of *Aedes* in Africa. The disease, which is acute and often fatal, is not known from Asia, and its global prevalence (potentially from 40°N to 40°S) has been successfully checked during the last 50 years by effective vaccines—although sporadic epidemics still occur in South America and Africa (WHO 2004a). In the past, yellow fever was particularly common in port cities and was spread by trading sailing ships, sick seamen, and its mosquito vector which often bred in drinking-water barrels on board (Jones 1967). The disease's primary vector, *Ae. aegypti*, is a peridomestic species found in abundance in towns and cities where it flies only a few hundred metres from its larval habitats. The latter are typically phytotelmata (especially tree holes), but have come to include a wide range of man-made container habitats, such as rain vats, and water-filled cans, jars, urns, and discarded tyres. The eggs of *Ae. aegypti* can resist desiccation for up to a year, and are typically laid on the damp margins of waterbodies as they are drying. They hatch when they are flooded by water that is deoxygenated (Womack 1993).

Nowadays, in tropical Africa, jungle yellow fever is maintained and circulated chiefly among animal populations but is sporadically passed on to humans. In addition to *Ae. aegypti*, it can be transmitted by 5 other species, and a further 11 possible vector species. Of these 17 species, 12 belong to the genera *Aedes* or *Eretmapodites* and all, except for those of the *Aedes dentatus* group, either normally or occasionally have larvae capable of being reared in common types of phytotelmata, particularly tree holes and leaf axils. Availability of water-filled leaf axils has been shown to be influenced by agriculture, as crops such as banana, plantain, pineapple, and cocoyam form particularly good water reservoirs and, for example, are the principle habitats of *Aedes simpsoni* (Pajot 1983).

Ae. aegypti is responsible for transmitting yet another arbovirus, dengue, or 'break-bone fever'. The pattern of infection is very similar to that of yellow fever, with old trading ships acting as breeding places for the mosquito. As a new port was reached, the virus got ashore either in the blood of an infected sailor, or in the salivary glands of an infected mosquito. It then became established through local susceptible hosts and local species of *Aedes*, and an epidemic would typically follow. Reservoirs of both these diseases exist in populations of monkey in many tropical forests. In recent years, dengue has become a major international public health concern, currently endemic in more than 100 countries in Africa, the Americas, the eastern Mediterranean, southeast Asia, and the western Pacific. Whereas dengue has a low mortality rate, a more severe complication, dengue haemorrhagic fever (DHF), appeared in the Philippines and Thailand in the 1950s and has caused high mortality among children throughout southeast Asia, where it has become a leading cause of hospitalization and death. In 1981, it reached the Caribbean region, with subsequent outbreaks in, for example, Cuba, Mexico, and Puerto Rico (WHO 2004a).

Figure 9.1 (a) An example of a used-tyre dump in Ohio, prime breeding habitat for *Ae. albopictus* (Asian tiger mosquito), which colonized the state in 1996; (b) close-up of habitat (reproduced with permission of R. Large, Ohio State Office, US Environmental Protection Agency); (c) a female *Ae. albopictus* emerging from its pupal skin at the water surface (copyright 2005 Dexter Sear iovision.com; with permission).

Fairly recently, a secondary dengue vector species in Asia, *Aedes albopictus*, has become established in the United States, several Caribbean and Latin American countries, parts of Europe and Africa, and this has been deemed a direct consequence of the international trade in used tyres—an adopted larval habitat (Figure 9.1; WHO 2004b).

Related to dengue are the various forms of viral encephalitis which break out periodically in the United States, Canada, Australia, Japan, and other countries. These affect the human central nervous system and are potentially very dangerous. Many birds carry enough of the virus in their blood to act as reservoirs for the disease, and transmission to humans is effected by quite a number of mosquito species, chiefly those belonging to the genera *Culex, Anopheles, Aedes*, and *Psorophora* (Jones 1967).

Other important human diseases spread by insects associated with temporary waters or moist habitats during some stage in their life cycles include leishmaniasis (Kala-azar of most tropical and subtropical regions) transmitted by phlebotomine sandflies (Psychodidae); other forms of filariasis (in African rainforests) transmitted by biting midges (Ceratopogonidae); Loa loa (in Africa) another form of lymphatic filariasis, but transmitted by tabanid flies (Tabanidae); pinkeye, conjunctivitis, and trachoma (in the New World), transmitted by eye flies (Chloropidae); and trypanosomiasis (African sleeping sickness), transmitted by tsetse flies (Muscidae) (Jones 1967; Lainson and Shaw 1971; Bannister *et al.* 1996; Ottesen 2000).

9.4 Economic and humanitarian costs

Table 9.2 summarizes the humanitarian and economic statistics (where known) for the major diseases spread by temporary water-dwelling vectors. Most of the figures are estimates subject to errors introduced by: (1) the impossibility of obtaining accurate counts of mortality and, especially, nonlethal infections in remote regions; and (2) the fundamental difficulty in estimating monetary losses due to reduced efficiency in the workplace. Despite these shortcoming, however, the indicated orders of magnitude of both the financial tallies and human suffering are formidable and of global concern.

Ironically, as tropical, third-world nations are solving, and being helped to solve, issues of basic supply of drinking and irrigation waters, they are also often increasing the extent and diversity of habitats available for breeding disease vector species. For example, dengue is a mosquito-borne infection very much on the rise in recent years. Together with its complication, dengue haemorrhagic fever, it now affects 50 million people each year, and 2.5 billion people (2/5ths of the world's population) are now considered to be at risk. In 2001, more than 600,000 cases were reported from the Americas alone, representing more than twice the number of infections recorded from that region in 1995. The disease is also spreading to new regions, and explosive outbreaks are occurring, for example, 390,000 cases (including 670 of DHF) in Brazil in 2001. This unprecedented spread is linked to the expanding distributions of both the virus (actually comprising four closely related forms) and its mosquito vectors (WHO 2004b). In particular, *Ae. aegypti*, the most important vector, is flourishing in its preferred urban environment where it breeds in household water-storage tanks and inadequately drained street-scapes. Further, it is unfortunate that whenever, in warm climates, water-related disasters strike (such as severe river flooding, hurricanes, or the south Asian tsunami of 2004), added to the problems of loss of life and property is the inevitability of disease. Many diseases, such as malaria, yellow fever, dengue, and arboviruses (Nasci and Moore 1998) follow the retreating waters as large numbers of temporary vector-breeding habitats are created on floodplains and coastal lands.

Of course, many of these disease problems are not just contemporary, some, such as malaria, have been associated with mankind for a very long time. Virulent forms of malaria are thought to have evolved at least 10,000 years ago (around the dawn of agriculture)—as evidenced from the analysis of human genes that confer a reduced risk from malarial infection—although less virulent forms were likely in existence well before that time (Pennisi 2001). Throughout this association, such diseases have had a major influence on human history and endeavours. For example, in parts of

Table 9.2 Humanitarian and economic statistics for global diseases spread by vector species that inhabit temporary waters

Disease	Affected No.	Annual data and/or forcast	Comments
Malaria	2.4 billion (>40% of world population); 1.4 million deaths p.a.	16% growth p.a. 0.8 million children under 5 yrs old die each year	Ranked 3rd globally infectious disease; expenditure for treatment, prevention, research: ~$84 million p.a.; economic costs vast and difficult to estimate
Schistosomiasis	>200 million infected; seldom fatal, but it is chronically debilitating	88 million children infected p.a.	80% of transmission is in sub-Saharan Africa
Japanese encephalitis	30–50,000 cases p.a.; 13,000 deaths in 2002	range is expanding (e.g. to N. Australia)	Leading cause of viral encephalitis in Asia; suspected econ. loss $10s of millions p.a.
Lymphatic filariasis	1.1 billion at risk; 120 million infected	Most infections are acquired in childhood	Endemic in 80 countries; economic loss $1 billion p.a. in India alone
Yellow fever	200,000 cases p.a.; ~30,000 deaths p.a.; >0.5 billion at risk	1998–2003 highest rates since 1948	Although restricted to Africa and South America, suitable environments and vector species are present in other countries
Dengue/ Haemorrhagic fever	20 million cases p.a.; 24,000 deaths p.a.; 2.5 billion at risk	Dengue and DHF have steadily increased in incidence/distribution since 1960s	Occur in >100 countries; risk is higher in urban, peri-urban and rural areas of tropics and subtropics; economic cost in 1998 ~$418 million in S-E Asia alone
Dracunculiasis	0.5–1 million p.a.	Appears on decline as 75,223 cases in 2000 vs 3.6 million in 1986	Endemic in sub-Saharan Africa, especially Sudan, Nigeria, Ghana; greatest decreases in India
Leishmaniasis	2.4 million DALYs[a] 59,000 deaths p.a.	Recent drug resistance reported, requiring use of more toxic chemicals	Endemic in 88 countries on 4 continents; visceral form can be fatal
Onchocerciasis	50% of men over 40 years old blinded in some W. African areas	90% of the cases in Africa; also in Latin America and Yemen	In the 1970s, economic losses estimated at $30 million p.a.; *ivermectin* is now an effective drug for killing microfiliariae

[a] DALYs represent disability-adjusted life years, a non-monetary economic measure of impact lost to disease.
Source: After various sources, but especially FAO (2004), UNESCO (2004), WHO (1999, 2004a,b,c).

Asia, malaria has affected cultural and economic issues for more than 4,000 years, sometimes contributing to the success and wellbeing of nation states (Kidson and Indaratna 1998; UNESCO 2004). In 1910, resumption of work on the Panama Canal was made possible only after malaria (and yellow fever) was controlled in the area—in 1906, out of 26,000 project workers, more than 21,000 had to be hospitalized for malaria (Marshall 1913). At the beginning of the twenty-first century, more people are dying from malaria than 40 years ago, and the disease is seen very much as a 're-emerging' threat (Guerin *et al.* 2002), despite the attempts of long-term control programmes (Anon 2005).

Some of the temporary water-facilitated diseases that have become more prevalent in recent years—or perhaps we are now more knowledgeable of their specific affects, include a variety of arboviruses. Many of these infect human hosts outside the traditionally disease-prone tropics and subtropics, striking at populations in major urban areas with high standards of living. By way of examples, Table 9.3 lists some of the contemporary diseases transmitted to people in North America, via mosquitoes that breed in temporary woodland habitats. In particular, the West Nile virus (WNV), first detected in New York City in 1999 but first isolated in Uganda in 1937, is spread by a number of species, foremost among which, in terms of the

Table 9.3 Examples of diseases transmitted to humans, in North America, via mosquitoes that breed in temporary woodland habitats[a]

Habitat	Mosquito species	Disease	Vector status[b]
Temporary woodland ponds	*Aedes vexans*	West Nile virus; Eastern Equine Encephalomyelitis; La Crosse virus	Bridge vector Epidemic/bridge vector Potential vector
	Aedes (Och.) stimulans	Jamestown Canyon virus	Primary vector
	Aedes (Och.) communis	Jamestown Canyon virus	Primary vector
	Aedes (Och.) canadensis	West Nile virus; La Crosse virus	?Bridge vector; potential vector
Tree holes	*Aedes (Och.) triseriatus*	West Nile virus; La Crosse virus	Bridge vector; primary vector
Artificial containers, including tree holes, pools with leaf-litter	*Cx. restuans*	West Nile virus	Primary vector
Wetland margins/ floodplain pools	*Cx. salinarius*	West Nile virus; Eastern Equine Encephalomyelitis; St. Louis Encephalitis	Bridge/primary vector Potential vector Potential vector
	Cx. tarsalis	Western Equine Encephalomyelitis; St. Louis Encephalitis; West Nile virus	Primary vector Primary vector Primary vector

[a] Based on information derived from various sources, including: Wood *et al.*, 1979; Walter Reed Biosystematics Unit, 1997; Virginia Department of Health, 2003.

[b] Primary (or enzootic) vectors typically feed on non-human hosts, for example, in the case of WNV, on birds; Bridge vectors feed on numerous animal species, including humans, and serve as a bridge for the virus to enter other species.

greatest number of populations infected, are *Culex pipiens*, *Cx. restuans*, and *Cx. tarsalis* (Turell *et al.* 2001). The first two species breed mostly in urban and rural habitats, such as storm sewer catch basins, waste lagoons, and other eutrophic or organically polluted waters, whereas *Cx. tarsalis* breeds in containers plus a variety of natural habitats. However, there are a number of other mosquitoes (*Aedes vexans*, *Ae. (Ochlerotatus) canadensis*, *Ae. (Och.) triseriatus*, and *Cx. salinarius*) that carry the disease but which live in cleaner temporary waters, including woodland ponds, or associated habitats, such as tree holes, riparian pools, and wetland margins. Further, these, together with other woodland species (e.g. *Ae. (Och.) communis* and *Ae. (Och.) stimulans*) are known to spread other diseases to humans, including both Eastern and Western Equine Encephalomyelitis, and the La Crosse, Jamestown Canyon, and St Louis Encephalitis viruses. The presence of suitable vector species, existence of a pool of infectious individuals, and the availability of suitable local aquatic habitats for the vectors are all factors in the equation of the spread of such diseases. At present, however, there appears to be a relatively low risk of disease from natural, woodland pool populations, compared with mosquitoes living in other habitats, such as containers and permanent/semi-permanent open-water habitats. Nevertheless, the increasing trend of global warming could well escalate the role of woodland populations—perhaps through creation of more, and warmer, intermittent ponds. Intensified scientific study of vector populations in such temporary waters is required before potential infection rates can be estimated and informed control measures can be applied.

9.5 Eradication vs control: vector vs habitat

The history of mankind's attempts to avoid being infected with diseases spread by temporary water vectors, and especially by biting insects, is a long one. There are examples of significant success (e.g. the near elimination of malaria in North America), of continuous struggle (e.g. the World Health

Organization/World Health Assembly's 1955 attempt to eradicate malaria worldwide), and of failure (e.g. the current escalation of dengue and DHF). Control of many of these diseases typically involves identification of the specific vector, and reduction of either its numbers or its habitat. For vectors such as freshwater snails and mosquitoes, this may involve application of molluscicides and insecticides, respectively, drainage of wet areas of land, introduction of predators (e.g. dytiscid beetles or fishes), or controlled raising or lowering of water levels at crucial times in the vector's life cycle. Of course, interruption of the parasite–human transmission cycle, and immunization against parasites are also part of the control arsenal. Rarely, nowadays, is one control mechanism employed alone, as integrated programmes have proved to be more successful, and also somewhat less harmful to natural environments.

Early examples of using a single control mechanism (e.g. vector habitat removal) include deforestation around towns in southern Brazil, which proved to be effective control for *Anopheles* because it removed the epiphytic bromeliads that were providing habitats for the larvae. Although flight ranges of *Anopheles bellator* and *An. cruzii* are known to be in excess of 2 km, very few adults crossed the 1 km deforested zone. Similarly, in Trinidad, the incidence of malaria decreased after the bromeliads of shade trees around cacao plantations were destroyed by herbicides (Smith 1953). However, some of the other insects that live in bromeliads are known to be important pollinators of both crops and natural vegetation so that widespread destruction of phytotelmata is counterproductive (Frank 1983).

During the first 20–30 years of the twentieth century, malaria was rampant in the United States, with close to 6 million cases being reported annually. In 1947, the National Malaria Eradication Programme, a cooperative effort between the US Federal Public Health Service and health agencies in 13 southeastern states began spraying DDT inside rural buildings in counties known to have had malaria in previous years. More than 4.5 million premises had been sprayed by 1949, and the country was effectively declared to be free of malaria as a significant health problem (CDC 2004a). The occasional outbreaks of malaria that now occur in the United States are typically attributed to infected people returning from countries in which malaria is still common.

The goal of the WHO/World Health Assembly's 1955 malaria eradication programme (which did not include Africa due to reasons of high transmission rates and lack of infrastructure) was to cut back mosquito populations to levels where transmission of the *Plasmodium* parasite to humans was interrupted—rather than to try to eradicate all vectors. The programme provides an example of an early attempt at integrated control as it comprised house spraying with residual insecticides (especially DDT), treatment of people with antimalarial drugs, and accurate surveillance. The programme was successful in countries with temperate climates and/or seasonal transmission patterns, but other countries made little progress (e.g. Nicaragua, Haiti, Indonesia, and Afghanistan), or had initial success followed by a return of the disease once control efforts faltered (e.g. India and Sri Lanka). Ultimately, eradication was given up in favour of control (CDC 2004b).

Even within integrated control programmes, some single methods tend to predominate due to reasons of cost, feasibility, and local conditions. Destruction of vector populations with chemicals is one such technique, and has led to vector species, particularly mosquitoes, becoming quite quickly resistant to these poisons. Indeed, it is now acknowledged that mosquitoes have developed resistance to all of the major groups of chemicals, including DDT, at a time when there is a shortage of new insecticides available. As a consequence, WHO has adopted a global strategy of using insecticides on a selective basis, for example, primarily for indoor spraying, impregnating bednetting, and larviciding (WHO/SEARO 1997). To make matters worse, many of the parasites also have developed resistance to chemicals designed to prevent them from surviving within the human host. For example, in the 1960s strains of *P. falciparum* in South America and southeast Asia became resistant to chloroquine and spread to most of the world, except Africa. In 1980, chloroquine-resistant *P. falciparum* appeared in coastal regions of Kenya and Tanzania, and had spread to most of the continent by the end of that decade (Collins and Paskewitz 1995).

Realization that integrated control is the optimum way forward, together with newly emerging technologies, has produced a number of new approaches to disease control. Fundamental to these is a detailed understanding of the biology of vector species, including their physiology, behaviour, ecology, and genetics. Greater knowledge is also required of the other organisms with which they interact in both their natural habitats (e.g. woodland pools and phytotelmata) and adopted ones (e.g. cisterns and ricefields).

Knowledge of vector behaviour can be central to developing strategies for disease control. For example, Pates and Curtis (2005) have pointed out that the effectiveness of indoor residual chemical spraying against *Anopheles* mosquitoes depends on whether the adult mosquitoes rest indoors (i.e. exhibit endophilic behaviour)—a trait that not only varies among species, but also among populations. Some individuals have an immediate avoidance response to irritant insecticides (e.g. DDT or pyrethroids), and the possibility exists that the continuous presence of these chemicals in houses has promoted the evolution of such avoidance (exophilic) behaviour, or, even worse, a 'bite and run' behaviour—as has been shown in populations of *Anopheles gambiae* in the Tanga Region of Tanzania (Gerold 1977).

Knowledge of oviposition behaviour is crucial to mosquito larval control as it helps in identifying breeding sites, which can then be treated with larvicides. Mosquito females are known to be selective in where they lay their eggs, and may avoid waters containing predators, such as notonectids, fishes, and tadpoles (Ritchie and Laidlaw-Bell 1994). Further, container-habitat laying species, such as *Ae. aegypti*, show a dispersal pattern driven by the search for suitable oviposition sites. For example, in Puerto Rico a release-and-recapture study has shown that *Ae. aegypti* females were recaptured significantly more in houses in which breeding containers were placed compared with houses with no suitable containers (Edman *et al.* 1998). The increased dispersion of adults resulting from females seeking out specific waters for their larvae is thought to be an important factor in the spread of diseases, such as dengue (Reiter *et al.* 1995; Pates and Curtis 2005). In Bangladesh, female *An. dirus*, a malaria vector, lay their eggs at the waterline of small temporary pools. Here, the embryos develop and remain viable for up to 2 weeks. Synchronized hatching follows heavy rainfall, and waves of biting adults can emerge in as little as 5–6 days. During very dry conditions, larvae exhibit the unusual behaviour of abandoning drying pools and crawling as much as 0.5 m in search of adjacent pools (Rosenberg 1982).

For development of effective control protocols in general, there is no substitute for sound knowledge of the basic biology of vector species. For example, while the application of a thin film of oil to the surface of temporary waters effectively kills the air-breathing larvae and pupae of most mosquito species, the larvae and pupae of the genera *Mansonia* and *Coquillettidia* (both vectors of filariasis and arboviruses) have respiratory tubes that allow them to pierce aquatic vegetation and so extract oxygen without surfacing (Service 2000).

9.5.1 New technologies for surveying vector populations and habitats

Habitat characterization and modelling have been used to estimate vector population parameters, such as abundance and distribution. For example, using satellite imagery Rejmankova *et al.* (1998) surveyed the larval habitats of *Anopheles vestitipennis* and *An. punctimacula* in Belize and identified eight types based on hydrology and dominant life forms. A subsequent discriminant function for *An. vestitipennis* correctly predicted the presence of larvae in 65% of sites, and correctly predicted the absence of larvae in 88% of sites. A similar analysis for *An. punctimacula* correctly predicted 81% of the sites for the presence of larvae, and 45% for the absence of larvae. Dale *et al.* (1998) have pointed to the potential benefits from combining remote sensing analyses with Geographic Information Systems (GIS) to minimize disease risk. They cite examples from subtropical Queensland, Australia where the salt marsh mosquito *Aedes vigilax*, and the freshwater species *Culex annulirostris* act as vectors of human arboviruses. Risk of contracting the diseases is modelled based on knowledge of the breeding habitats of the vectors in a localized area. This is subsequently related to computer-assisted analysis of remotely sensed data in order to map the potential temporary water breeding sites of *Cx. annulirostris*. This can then act as a guide for vector control at critical times, such as following heavy summer rainfall, or when there is a disease outbreak. In addition, mapping techniques such as colour infrared aerial photography can be used to identify areas of salt marsh where the eggs and larvae of *Ae. vigilax* are likely to be found, and to produce detailed water distribution patterns under mangrove forest canopy, again to identify larval breeding habitats.

9.5.2 Habitat control

Removal or modification of vector habitats, where practicable, is an obvious first line of defence against diseases associated with temporary waters. Often such habitats are natural ones, such as marshes or phytotelmata. However, there are many examples of man-made structures which create artificial temporary waters that not only support but often enhance populations of vector species. For example, particularly in tropical regions, low flow regimes associated with the downstream areas of dams are particularly prone to creating vector habitats in the form of pools in the river bed. Often, these become transmission foci for malaria vectors (*Anopheles* spp.) and for schistosomiasis (via species of *Biomphalaria* and *Bulinus*). Where reduced flow in coastal areas results in salt intrusion into estuaries, brackish-water vector populations are enhanced (e.g. *Anopheles sundaicus*, *An. melas*, *An. merus*, and *An. albimanus*) (FAO 2004). Other consequences of dam construction include the ideal, fastwater habitat for blackflies (onchocerciasis vectors) created by spillways, and the encouragement of aquatic macrophytes which support snails. Some of these problems can be reduced by modification of the engineering structures and/or operational procedures that created them. For example, blackflies can be removed by constructing two spillways that can be used alternately, thus not allowing the larvae to complete their development into adulthood (Pike 1987). Also, physical removal of macrophytes has been shown to reduce snail populations. Alternating wet and dry phases in paddy fields, together with synchronized cropping of rice harvests, can interfere with vector species life cycles, and creation of self-draining standing waters in Zimbabwe has significantly reduced schistosomiasis transmission (Chimbale *et al*. 1993). Further, modification of timing of seasonal rice-culturing procedures and water management, have been effective in controlling *Culex tritaeniorhynchus*, the principal vector of the Japanese encephalitis virus in Japan (Takagi *et al*. 1995). Replacement of traditional, open irrigation canals and ditches with more modern water-delivery methods, such as sprinklers, drip-, or subsurface-irrigation is also effective in reducing the amount of standing water available as vector habitat (USAID 1975; Worthington 1983).

Russell (1999) has pointed out the vector-enhancing problems inherent in constructing wetlands for the expressed purpose of 'polishing' urban drainage and storm water by reducing contaminants before they are discharged into rivers. He cites these engineered waters, in Australia, as major contributors to mosquito population increases and thence the transmission of arboviruses and malaria. In particular, Russell states that the peri-urban siting of such wetlands and their design (typically shallow with dense vegetation) greatly favour mosquito production and contact with humans. However, with some re-engineering, for example, deeper basins with cleaner, steeper margins, more open water, vegetation control, and water management (e.g. aeration and sprinkler systems, together with planned flooding and draining regimes), mosquito populations can be curbed. Russell admits that such measures may go against the preferred objectives and operation of these constructed wetlands, but argues that mosquito management needs to be an integral part of modern wetland design and maintenance.

Even less ambitious engineering projects, such as shallow wells can contribute to vector increase. For example, 37 out of 79 wells (47%) in three geographically diverse regions of Greece were found to support populations of sandflies (Psychodidae). Two species of *Phlebotomus* (*P. tobbi* and *P. neglectus*) are known to transmit visceral leishmaniasis (Chaniotis and Tselentis 1996). Similarly, in Senegal, wells dug by market-gardeners in the Dakar area are known to support large populations of *Anopheles arabiensis*, a member of the *An. gambiae* complex, and vector of malaria (Awono-Ambene and Robert 1999). In southern Tanzania, *An. arabiensis* is more predominant in waters close to cattle (Charlwood and Edoh 1996). In Sri Lanka, the ancient, small-tank-based irrigation network for rice production is still widely practiced, and a survey of these village-associated waters yielded 12 species of anopheline mosquito. The most abundant were the malaria vectors *Anopheles varuna* and *An. culicifacies*; the latter being observed to switch breeding habitats from streambed pools to tank bed and drainage area pools during the pre-monsoon period (Amerasinghe *et al.* 1997).

Recently, significant spread of mosquito vectors has been discovered associated with the practice of shipping used tyres around the world to re-treading plants. In 1979, the first recorded occurrence of *Ae. albopictus* outside Asia and Australasia was made close to a rubber factory in Albania (Adhami and Reiter 1998). Many such transfers have included human parasites, such as the Eastern Equine Encephalitis (EEE) virus isolated from *Ae. albopictus* in Florida in 1991, the Potosi and Cache Valley viruses isolated from *Ae. albopictus*, and the La Crosse virus isolated from *Ae. triseriatus* in Illinois in 1994/1995, all collected at used tyre sites (Mitchell *et al.* 1998). In a survey of arboviruses associated with mosquitoes in nine Florida counties in 1993, 21 (48.8%) of the 43 virus strains detected were isolated from mosquitoes collected at waste tyre sites (Mitchell *et al.* 1996). These authors warned of the obvious high potential of such sites to act as vector production foci, and nuclei for the spread of diseases.

For those vector species, chiefly mosquitoes, that breed in urban, container-type habitats, relatively simple precautions can be taken, such as covering the water surface to prevent access by egg-laying females, placing screening over cesspits and wells, creating soakaways to remove waste water from the ground surface, and removing potential rainwater-collecting garbage. In larger habitats, such as irrigation projects, Amerasinghe *et al.* (1995) have emphasized the need to determine significant associations between the abundance of individual vector species and specific physicochemical parameters of the water in which they live. Such information (e.g. a positive correlation between *An. culicifacies* and phosphate, dissolved oxygen, and temperature; but a negative relationship between *An. nigerrimus* and the last two parameters) may then be useful in predicting which vector populations are likely to develop as, for example, forest land in Sri Lanka is cleared for rice production.

9.5.3 Vector control by biological means

A variety of natural and introduced predator species have been tested as potential vector control agents.

Crustaceans—primarily copepods and notostracans

Notostracans commonly co-habit temporary lentic waters with mosquito larvae. Fry *et al.* (1996) evaluated the potential of *Triops longicaudatus* to control *Culex tarsalis* in shallow ponds in California, and found that a high density of these shrimp had a significant negative impact on larval populations. Shrimp populations persisted in 94% of these ponds, but unfortunately also had a negative affect on non-target chironomid larvae. The best form of field inoculation was deemed to be as drought-resistant eggs.

Predation of newly hatched mosquito larvae by cyclopoid copepods was first documented by Hurlburt (1938). Since that time, a number of experimental trials have been made. For example, in the 1980s application of *Mesocyclops aspericornis* to crab holes in Rongaroa (French Polynesia) proved to be effective against *Aedes aegypti* and *Ae. polynesiensis* (producing a 76% reduction in adults of the latter), but not against species of *Culex* (Riviere *et al.* 1987). A related species, *Mesocyclops longisetus*, has proved similarly effective against *Ae. aegypti* breeding in cisterns, water drums, and other domestic containers in Honduras (Marten *et al.* 1994). Marten (1990) conducted field trials in which *Macrocyclops albidus* was released into water-filled tyres at a dump in Louisiana. High populations of *Aedes albopictus* present at the start of the trials were virtually eliminated within two months. Around 1990, the New Orleans Mosquito Control Board began exploring the use of cyclopoids for mosquito control on a large scale, creating a culture facility capable of producing 1 million adult *Mesocyclops longisetus* and *Macrocyclops albidus* per month (Marten *et al.* 1994). An interesting community-based approach was tested using *M. longisetus* in northeastern Mexico. Here, *Ae. aegypti* breeds prolifically in metal drums, discarded tyres, and cemetery vases. The programme involved inoculating these habitats with female *M. longisetus* and having the local community trained to rescue and reinoculate the copepods before the drums, in particular, were cleaned and refilled. Community participation proved to be good, with all peridomestic drums still retaining copepods after 4 months. The average reduction of larvae was 37.5% for the drums, 67.5% for the flower vases, and 40.9% for the tyres; however, the latter two habitats proved more difficult to manage as they desiccated more readily (Gorrochoteguiescalante *et al.* 1998). Blaustein and Margalit (1994) have pointed to the potential of the common temporary pond copepod *Acanthocyclops viridis* for control of several mosquito species, but particularly *Ae. aegypti.*

Other mosquitoes

Several mosquito genera contain species that prey on other mosquitoes, for example, *Toxorhynchites*, *Psorophora*, and *Ochlerotatus*, and thus the potential for their use in vector control exists. However, experimental studies have yielded different degrees of success. For example, in a study of the facultative predator *Anopheles barberi* on its treehole-dwelling prey, *Aedes triseriatus*, Nannini and Juliano (1998) found only a limited potential for reducing the latter's populations—and then only during a short period in mid-summer. In contrast, in a field experiment in Louisiana, Focks *et al.* (1982) showed that stocking water-filled car tyres, plastic buckets, and discarded paint cans with one or two first-instar larvae of *Tx. rutilus* resulted in an overall control of populations of *Ae. aegypti* and *Cx. quinquefasciatus* of around 74%. Some problems have been noted with the use of *Toxorhynchites* as vector-control agents, in that at high prey densities predator searching efficiency decreases through an interference effect (Hubbard *et al.* 1988). However, Collins and Blackwell (2000) have reviewed the evidence for using *Toxorhynchites* in biological control, and conclude that there has been some success, but at a low level. As a single *Toxorhynchites* larva may eat over 5,000 first instar prey larvae before pupating, the impact of these predators is potentially large. Future improvements are seen to depend on better knowledge of the general biology of these predators, especially of their oviposition site selection and cues, which may then be better matched to those of target vector species. Brown *et al.* (1996) explored the possibility of integrating *Toxorhynchites* with another predator, *Mesocyclops aspericornis*. Introduction of the latter alongside naturally occurring *Tx. speciosus* resulted in a compatible predator pair capable of reducing *Aedes notoscriptus* and *Cx. quinquefasciatus* populations in tyre habitats—specifically, only 51% of tyres containing both predators supported larvae (at a median density of 4 larva l^{-1}), compared with 97% of tyres (with a median density of 43 larvae l^{-1}) where predators were absent.

Other insects

Many of the insects that co-occur with mosquitoes in temporary waters, such as dytiscid beetles, odonates, and notonectids, prey on their larvae and pupae. However, they are largely ineffective in reducing prey populations because the latter have rapid, well synchronized development that is very often completed before the predator populations become fully established (Collins and Washino 1985). Interestingly, *Culiseta longiareolata* is known to strongly avoid ovipositing in habitats containing the predator *Notonecta maculata*, and is repelled, by chemical residues, for up to 8 days after the predator has been removed (Blaustein *et al.* 2004).

Fishes

Whereas there has been some success with using fishes as vector predators, typically they cannot be used in waterbodies with short hydroperiods (~1 month or less). However, fishes have proved particularly useful in controlling mosquito larvae (especially *Ae. aegypti*) in large cisterns and domestic drinking-water containers. Small species or small individuals of larger species are best, and include the mosquito-fish *Gambusia affinis*, *Tilapia nilotica*, *Clarias fuscus*, and species of *Macropodus* (Neng *et al.* 1987). In the United States, fathead minnows (*Pimphales promelas*) and *G. affinis* are recommended species for stocking drainage ditches, intermittent ponds, and ornamental pools in order to reduce the spread of West Nile virus; individuals are capable of consuming around 300 mosquito larvae per day. Although survival of *G. affinis* was initially restricted to the southern states, cold-tolerant strains are now available that allow populations to successfully overwinter in regions with air temperatures as low as −30°C (Keeton Industries 2005).

In Quangzhou County, China, addition of common carp (*Cyprinus carpio*) and grass carp (*Ctenopharyngoden idella*) to rice fields (at stocking densities of 6,000–9,000, and 150–1,500 per hectare, respectively) has been shown to significantly reduce the populations of *Anopheles sinensis*, the main vector of local malaria, and *Cx. tritaeniorhynchus*, the vector of Japanese encephalitis (Neng *et al.* 1995). However, Blaustein and Karban (1990) have demonstrated that the larvae of *Cx. tarsalis* developed faster and had a higher survival in Californian rice-field enclosures that contained *G. affinis*, compared with fishless enclosures. It was thought that the mosquito-fish reduced cladoceran populations that were in competition with the larvae for food, and mosquito-eating notonectid populations were also lower in the *Gambusia* enclosures (Blaustein 1992).

Amphibians

Many temporary waters support populations of frogs, toads, and salamanders, and many of these have omnivorous larval stages whose diets include mosquito larvae. While direct predation is likely the primary means of prey removal, other mechanisms may operate, including competition. For example, Mokany and Shine (2002) have shown that in ponds in southeastern Australia, two tadpole-mosquito associations are common. In the *Crinia signifera* (common eastern froglet)-*Aedes australis* system, direct physical interactions suppressed the mosquito populations, but this effect disappeared when densities were lowered. In contrast, in the *Limnodynastes peronii* (striped marsh frog)-*Culex quinquefasciatus* system, the tadpoles suppressed the mosquitoes even when the two species were separated by a physical partition, suggesting that chemical or microbiological cues may be at work. In terms of biological control, larvae of *Hyla septentrionalis*, the giant Cuban tree frog, have been used to reduce mosquito larval populations living in water containers in the Bahamas (Spielman and Sullivan 1974).

In addition to predator introduction, a variety of other biological techniques have been explored, for example: (1) infection with natural vector parasites (especially microsporidia and bacteria; see below); (2) displacement with other mosquito species (e.g. the displacement of *Ae. aegypti* by *Ae. albopictus* in the southern United States, although both are highly successful disease vectors; see Rai 1991); (3) production of genetically modified mosquito strains (e.g. competitive transgenic strains capable of driving disease-refractory genes into wild vector populations; see review by Rai 1999; and Alphey *et al.* 2002); (4) use of growth-regulating hormones (hormonomimetics; see Staal 1975); and (5) use of plant extracts (that may act as either toxicants, growth and reproduction inhibitors, repellents, or oviposition deterrents; see review by Sukumar *et al.* 1991). In Hawaii, the preference of certain mosquito species for taking blood meals from cats, has led to the suggestion that increasing urban cat populations might slow down the transmission rate of dengue—a 'zooprophylactic' solution.

Natural infections of mosquito species with microsporidia (Protozoa) have control possibilities. Microsporidia are intracellular parasites in which the infective stage, the spore, injects its contents into host cells. Here, multiplication takes place eventually resulting in the production of more spores. Spores are protected by a thick coating during their subsequent infective stage, and may (e.g. in species of *Amblyospora*) or may not (e.g. in species of *Edhazardia* and *Culicospora*) require transfer to an intermediate host (typically a copepod) in order to complete their development (Vossbrinck *et al.* 2004). Andreadis (1999) recorded infection rates by *Amblyospora stimuli* in adult female *Aedes stimulans* of between 1.0% and 9.6%, with an annual rate of transovarian transmission to larval populations ranging from 1.3% to 5.9%. Meiospore infections in F-1 generation larvae were significantly correlated with infections in parental-generation females. This suggests that the infection rate of larvae could be increased to possible control levels if methods could be developed to facilitate the transmission of the parasite to a larger portion of the female population—likely via release of infected copepods. Comiskey *et al.* (1999) have noted the potential for another parasitic protozoan, the gregarine *Ascogregarina taiwanensis*, to reduce adult size and egg production in *Ae. albopictus* breeding in tyre dumps.

Among the fungi, although several species (e.g. from the genera *Tolypocladium*, *Coelomomyces*, *Culicinomyces*, and *Leptolegnia*) are known to infect mosquitoes, only the oomycete *Lagenidium giganteum* appears to have had its potential to control mosquitoes studied in any detail. This species is commonly found, in a vegetative state, on submerged plant materials and dead insect carcasses, and has a motile, highly infective spore stage that appears to have high specificity for mosquito larvae. It will infect and kill a wide variety of freshwater-breeding mosquito species in waters that are between 16 and 32°C. There is an oospore phase in the life cycle that can survive drying and can be cultured in bulk. These oospores survive naturally for many years in soil, but require soaking for several weeks before they reactivate. When used as an operational mosquito control agent (e.g. in California and Florida), it has to be activated before being sprayed onto water surfaces. Studies show a good rate (up to 50%) of mosquito mortality after spraying, and the fungus has the added benefit of becoming permanently established in vector habitats after just one application (Kerwin *et al.* 1986, 1994). Scholte *et al.* (2004) have reviewed the topic of entomopathogenic fungi and conclude that while there is considerable potential for vector control, the advent of Bti (see below) tended to curtail exploration of other biological control agents.

Bacteria are now an accepted form of biological control agent for vectors. The best known is *Bacillus thuringiensis*, a rod-shaped, aerobic bacterium which, in the 1920s, was discovered to have insecticidal properties. Currently, there are 34 subspecies known, and about 40,000 strains. *Bacillus thuringiensis* subspecies *Kurstaki* is a strain that has been used to control spruce budworm in Canada since 1987 (Canada 1996). *Bacillus thuringiensis* subspecies *israelensis* (Bti) is a strain first isolated from a stagnant pond in the Nahal Besor Desert river basin in the Negev Desert of Israel in 1976 (Goldberg and Margalit 1977). It has since been shown to have significant larvicidal properties for controlling biting insects, and is especially toxic to many species of *Culex*, *Anopheles*, and *Aedes* mosquito (Chui *et al.* 1995). A related species, *Bacillus sphaericus* (Bs), shows toxicity towards mainly species of *Culex* and *Anopheles*. Formulated insecticides based on Bti and Bs have in fact been available since 1981 and 1987, respectively, and have been used in vector control programmes in many areas. For example, in Cameroon and the Ivory Coast in 1992 a large-scale trial using Bs to reduce the number of *Culex* mosquitoes, via control of their larvae, proved successful provided that the majority of breeding sites were treated three times each year. In southern India, Bs forms part of an ongoing, integrated control programme (alongside spreading polystyrene beads over the water surface in cess pits, and chemotherapy of sufferers) to reduce *Culex* mosquitoes and reduce filariasis in rural areas. In West Africa, onchocerciasis has been combatted since 1985 by an integrated control programme against *Simulium damnosum*, wherein use of Bti, carbamates, and pyrethroids is alternated (Neilsen-LeRoux and Silva-Filha 1996). Although Bti and Bs, and their genetically engineered descendents (e.g. Soltes-Rak *et al.* 1995) are still under evaluation, their role in control programmes is assured, particularly as the development of resistance by mosquitoes appears to be very much slower than to chemical insecticides, such as organophosphates (Becker and Ludwig 1993). However, while a much lauded property of *Bacillus* species control is high specificity towards target vector species, typically mosquitoes and blackflies (e.g. Neilsen-LeRoux and Silva-Filha 1996), studies of the effects on other inhabitants of temporary waters are rare. An exception is that of Pont *et al.* (1999) who found that application of Bti in temporary marshland reduced the density of chironomid larvae by 38%, and affected adult emergence. Moreover, Bti has been shown to be ineffective against some types of mosquito, for example, the abovirus-spreading *Coquillettidia perturbans* in Minnesota (Sjogren *et al.* 1986). Blaustein and Margalit (1991) have suggested that the reason why Bti may be less effective in some habitats might be due to the presence of bacteria-feeding crustaceans. Specifically, they showed that mortality of *Ae. aegypti* larvae exposed to Bti decreased when the fairy shrimp *Branchipus schaefferi* and the ostracod *Cypridopsis vidua* were

Table 9.4 An example of a recommended integrative control programme for controlling mosquito populations in the State of New Jersey[a]

Major component	Sub-component	Details
Surveillance	Larval surveillance	Extensive sampling of a wide range of aquatic habitats for the presence of pest species during developmental stages
	Adult surveillance	Measures the size of adult mosquito populations using standardized methods (e.g. 1 min landing rates; light traps)
	Virus surveillance	Assesses the size of vector populations by testing specimens for presence of virus, on a weekly basis; findings are disseminated to all control agencies in the state
Source reduction	Sanitation	Routine de-snagging of waterways to restore flow, catch basin cleaning, removal of tyres and containers
	Water management: Freshwater wetlands	Using Best Management Practices and surveillance data, agencies conduct various water-management activities (e.g. improvement of stormwater facilities)
	Salt marshes	Employs knowledge of tidal marsh ecology and improvement of drainage channels (e.g. removal of ditch plugs; creation of tidal runoff channels)
Chemical control	Larviciding	Largely ground-based application of methoprene, temephos, Bti, Bs, and petroleum oils to kill larvae in large concentrations
	Adulticiding	Largely pyrethroids and malathion, used when biting populations reach critical levels; applied through ultra-low-volume sprays via well-calibrated dispensers
	Airspray programme	Largely for controlling saltmarsh mosquitoes in their larval stages, using mostly 'biorational' pesticides (e.g. growth regulatory hormones, bacteria, viruses, fungi)
Biological control		Primarily via introduction of fishes, especially *Gambusia affinis*, but also fathead minnows, sunfishes, killifishes
Education	Continuing education	Directed towards operational workers to maintain control skills
	Public education	Directed towards the general public to teach mosquito biology and encourage the practice of prevention techniques; schools
Cooperation	With Government	Control agencies regularly work with municipalities to eliminate breeding sites, and to develop management plans
	With Private Enterprise	Likewise with industries, especially to remove container habitats, waste sites, and poor drainage
Progress		Ongoing evaluation to meet the goals of the American Mosquito Control Association, especially the Pesticide Environmental Stewardship Programme

[a] Adapted from the New Jersey Mosquito Control Association 1998.

present. In contrast, Perich *et al.* (1990) advocate the integration of Bti-use, alongside introduction of the predatory planarian *Dugesia dorotocephala*, against *Ae. tritaeniorhynchus*.

9.6 Conclusions

It is clear from the preceding sections of this chapter that when it comes to combatting the many diseases associated with temporary waters, *elimination* is rarely possible (without severe damage to the environment), and the operative word is *control*. The evidence shows that, typically, no single method can be totally effective as, for example, many species of mosquito become resistant to chemicals quite quickly, and so an integrative programme of control must be adopted—where a combination or succession of techniques are intelligently applied. By way of illustration, the recommended integrative control programme for controlling mosquito populations in the State of New Jersey, is summarized in Table 9.4. New Jersey comprises some 22,560 km^2, and has a population of around 8.5 million people, making it the most densely populated state in the United States. It supports a variety of natural and man-made environments that support over 60 species of mosquito, and has practiced responsible mosquito management since 1914; control is now mandated by state law. The latter assigns the control of vector and pest species to mosquito control commissions in each county, and these function as autonomous units of local government. Tax levies provide the operating funds on a county by county basis. The county commissions have the powers of local boards of health in dealing with mosquito matters, and have the right of entry onto private and public properties, and may issue abatement notices whenever needed (New Jersey Mosquito Control Association 1998). The necessary components of responsible control are deemed to include: surveillance of populations of both the vectors and diseases; vector source reduction through Best Management Practices (BMP); use of both chemical and biological control agents; development of education programmes aimed at field operatives and the public; cooperation with various levels of government and industry; and ongoing programme evaluation to meet the goals of national agencies, particularly with regard to environmental stewardship.

Clearly, such an idealized programme may simply be untenable in many other countries, because of lack of funding and infrastructure, and also because in the latter the diseases may affect a far greater proportion of the population, be more debilitating, result in far more fatalities, and be historically well established. Nevertheless, such programmes provide useful models towards which control efforts can be directed.

CHAPTER 10

Importance and stewardship of temporary waters

10.1 Introduction

It is becoming clear that many temporary waters are repositories for species that do not occur elsewhere, that reach their maximum abundance in these habitats, or that enrich, genetically, metapopulations encompassing both permanent and temporary habitats. This realization comes at a time when temporary waters are being destroyed or altered at a rapid rate. Using data gathered for the better-documented wetland habitats, its is clear that despite the existence of several international agreements—such as the Ramsar Convention, which lists more than 1,000 sites of international importance that cover nearly 800,000 km^2—many of these types of environment have either been lost or are under threat (Turner *et al.* 2003a). Such threats come from a variety of sources, ranging from regional resource exploitation and harvesting, to urban and agricultural encroachment, to the effects of global climate change. In northwestern Europe, for example, intensification of agriculture together with industrial development have been responsible for a 60% reduction of wetland area (European Environment Agency 1999). In the United States, where the extent of alteration of freshwater habitats through, particularly, channelization is very great, loss of many invertebrate species has reached such a critical level that, for molluscs at least, major steps have been taken towards their conservation. These include: identification of endangered species and initiation of recovery programmes; development of techniques for relocating species or creating new habitat; and declaring certain sanctuary areas, as has been done, for example, in Tennessee (Stansbery and Stein 1971; Clarke 1981).

This chapter will examine, via case histories, some of these problems and describe attempts to ameliorate them through protective legislation. Arguments will be made for the suitability of temporary waters as ideal testing sites for contemporary hypotheses in *ecology*. The chapter will conclude by evaluating some existing management practices. While the latter are aimed, again, largely at wetlands, parallels can be drawn, and similar principles applied to many other temporary waters.

10.2 Role in the natural environment

Many temporary ponds and pools represent stages in the hydroseral succession of wetlands to more terrestrial habitats. This results from an excess production and accumulation of organic (largely plant) matter faster than it can be degraded. At a certain point in this transition, previously permanent ponds contain water for only part of the year (i.e. they become intermittent) and eventually only on an irregular basis (i.e. they become episodic). The process is entirely natural and as the water regime changes so do both the aquatic/emergent vegetation and the invertebrate fauna (see Wrubleski 1987). Unfortunately, in many rural areas, the increased pressure of agriculture often leads to large-scale land drainage in an attempt to bring so-called 'marginal' wetlands into cultivation. To do this, below-ground tile systems are installed with the expressed purpose of lowering the local groundwater table (GWT). Such practices destroy temporary water habitats and their associated biotas quickly and permanently, and should be discouraged. In Sweden, it is estimated that 30–60% of present-day arable land in the main agricultural regions has been enabled as a result of subsurface pipe-drainage of wetlands (Löfroth 1991).

As well as their role in hydroseral succession, temporary waters/wetland ecosystems have been shown to provide a number of valuable 'goods' and 'services' to human society—helping to further dispel the idea that they are areas of wasteland and/or sources of disease. Table 10.1 shows the breadth of these benefits, which range, in the jargon of public policy managers, from those with direct *economic* value (e.g. water abstraction and flood retardation) to those of *non-use* values. The latter includes the hotly debated component, *existence* value, which claims that some level of satisfaction is to be derived from knowing that certain features of the natural environment continue to exist (Turner *et al.* 2003a).

Table 10.1 Categories of wetland benefit to human society[a]

Services	Goods
Flood control	Water abstraction
Prevention of saline intrusion	Forest resources
Storm protection/windbreak	Agricultural resources
Sediment removal	Wildlife resources
Toxicant removal	Forage resources
Nutrient removal	Fisheries
Groundwater recharge	Mineral resources
Erosion control	Water transport
Wildlife habitat	Tourism/recreation
Fish habitat	Aquaculture
Toxicant export	Research sites
Shoreline stabilization	Education sites
Microclimate stabilization	Fertilizer production
Macroclimate stabilization	Energy production
Biological diversity provision	
Wilderness value provision	
Aesthetic value provision	
Cultural value provision	
Historical value provision	
Existence value provision	

[a] Largely from Turner *et al.* (2003a).

10.2.1 Biodiversity, rare species, and habitat loss

The cyclical nature of many temporary waters creates habitats that are quite distinct from those found in permanent waters. The former support biotas containing many elements that either are not found in other habitat types, or have their highest populations in temporary waters. In the United Kingdom, for example, the fairy shrimp, *Chirocephalus diaphanus*, and the tadpole shrimp, *Triops cancriformis*, are restricted to temporary lentic waters because of their physiological requirement of a dry-phase in their life cycles. Both species are geographically, and often temporally, rare in Britain. *C. diaphanus* is known from only one site in Wales and about 12 in England—for example, from ponds in the New Forest, the southwest, Cambridgeshire, and Sussex. The most northern record is from near York, in 1862, but most sightings lie south of a line from the Severn Estuary to the Wash (Bratton and Fryer 1990). *Triops cancriformis* has been recorded from only about 10 localities over the past 200 years and is currently known only from a single pool in the New Forest (Bratton 1990). Their populations are able to persist by means of a rapid life cycle and eggs that are both drought-resistant and viable over many years, and that hatch within hours of a pond basin filling. Both species are now protected by the Wildlife and Countryside Act 1981. In an ordination analysis of the invertebrate communities of 39 permanent and temporary ponds in Oxfordshire, Collinson *et al.* (1995) found that the caddisfly *Limnephilus auricula* and the corixid bug *Callicorixa praeusta* emerged as indicator species of the temporary ponds. Although *C. praeusta* is not known exclusively from temporary waters (Savage 1989), it prefers open water and may benefit from the fishless environment of temporary ponds. Collinson *et al.* (1995) also found that although temporary and permanent ponds displayed similar species rarity indices, four of the five highest rarity index scores were from temporary or 'semi-permanent' sites. Notable and Red Data Book species include the snail *Lymnaea glabra*, the damselfly *Lestes dryas*, and the waterbeetles *Graptodytes flavipes*,

Agabus uliginosus, *Haliplus furcatus*, *Dryops similaris*, *Helophorus strigifrons*, *H. nanus*, and *H. longitarsus*.

In terms of contributing to the United Kingdom's overall invertebrate biodiversity, temporary waters are of considerable importance. This importance is not just in terms of presence/absence statistics, but also may be manifest through maximization of the gene pool of species that occur in both temporary and permanent waters. For example, species that have populations in both habitat types might be expected, because of increased fitness demands, to have greater genetic diversity than species inhabiting only permanent ponds—indeed, the highest degree of genetic differentiation for some crustacean species (e.g. *Daphnia magna*, in Rhode Island, US) occurs in small temporary ponds (Korpelainen 1986), suggesting that such populations are vital to the gene pool of certain species. Such increased diversity may be crucial to the survival of species faced with possible future changes to global environments, as may result from climate warming (Hogg and Williams 1996). The consequences of decreased genetic diversity include the extinction of locally adapted populations with possible loss of alleles from a species' gene pool—this, in turn, may further reduce the species' ability to track future environmental change. For example, genetic variability in the anostracan *Branchinecta sandiegonensis* (endemic to temporary pools along the coast of San Diego County, US) is very low—thought to be a consequence of low gene flow and founder effects resulting from habitat fragmentation and a lack of potential vectors for dispersal of their cysts (Davies *et al.* 1997). Such species are especially at risk from human activities—in the case of *B. sandiegonensis*, San Diego County has already lost 90–95% of its historic vernal pool habitat (Goettle 2000).

In the global warming scenario, it is forecast that the impact will be most severe in northern latitudes (Hengeveld 1990). As genetic diversity is often lowest at the edges of a species' range, especially in northern latitudes (Sweeney *et al.* 1992), such populations will be most likely to experience the greatest environmental change in the next few decades, yet may be among the least able to adapt. These potential consequences led Hogg *et al.* (1995) to urge aquatic biologists and conservationists to consider the evolutionary as well as the ecological consequences of habitat alteration. Once genetic diversity has been lost for a given species it may not easily be regained, even if pristine environmental conditions are restored.

Using insects as an example, it is clear that, on a global scale, species diversity in all habitats is dropping (between 100,000 and 500,000 species are predicted to be lost over the next 300 years; Mawdsley and Stork 1995) and the world fauna is becoming more homogeneous. Factors contributing to this include increased abundance and spread of a small number of native species that thrive in habitats disturbed by human activity; the attraction of invading exotic species to native ecosystems; human-induced species extinctions and population declines; and the overarching and poorly understood synergistic effects of global change (Samways 1996). Fossil evidence from the Quaternary indicates that, in the past, it has been possible for insects to track climatic changes, gradually altering their geographical ranges and thus preventing species extinctions (Coope 1995). However, the rate of human-induced climate change is far faster (e.g. mean global air temperature is predicted to increase by 1.5–4.5°C over the next 20 years; IPCC 1990) than even insects appear to be able to cope with. Presumably, the effects on the other main component of the temporary water biota, the crustaceans, with their typically lower dispersal abilities are likely to be far more severe. Not only will crustacean extinctions occur, but also the population characteristics of surviving species are likely to be altered. Virtually no studies have been done to assess such changes in freshwater crustaceans—although a large-scale field manipulation to mimic global warming showed an increase in growth, smaller size at maturity, and precocial breeding in a population of the amphipod *Hyalella azteca* (Hogg and Williams 1996). Assessing such differences/evolutionary change among and within invertebrate populations typically has involved use of morphological and ecological characters. However, it has recently been shown (Müller *et al.* 2000) for *Gammarus fossarum*, in Europe, that morphological traits were 10-times less effective as genetic characters (enzyme loci) in revealing population variance. These findings suggest that molecular techniques should be applied when the status of threatened temporary water species is being evaluated, as observed morphological stasis may belie the level of genetic differentiation, which perhaps may lead to an alternative conservation strategem.

Quite apart from the direct effects of global climate change on species populations, there will be many serious effects as a consquence of how mankind responds to the problem of reduced water retention on the planet surface as a result of elevated temperatures. Again, using temporary water crustaceans as examples, Table 10.2 summarizes some of these impacts and indicates whether they are likely to have a positive or negative affect. Most of the effects are seen, intuitively, as being negative although manipulation experiments are required for confirmation. Only one impact, that of an increase in irrigation networks needed for agriculture under a warmer climate, is likely to be beneficial (see Caspers and Heckman 1981).

By way of underlining our lack of knowledge of precisely how temporary water biotas are likely to respond to global changes, Table 10.3 compares what is known for insect and crustacean populations. Only a very few studies (cited in the table legend) address, or come close to addressing, the likely responses by crustaceans, and all point to an escalation of the listed variables as a result of human-induced changes to temporary waters—which largely parallels the insect responses. The severity and speed of response are, however, largely unknown and require considerable research and managerial input.

Table 10.2 Example of environmental impacts likely to be associated with mankind's attempts to conserve water resources at risk fom global climate change

Impact	*Likely to affect temporary water crustaceans*	
	Positively	**Negatively**
Changed magnitude and seasonality of run-off regimes	?	√
Accompanying altered nutrient loading, for example, release from shoreline soils due to increased water level change	?	√
Accompanying limited habitat availability at low flow		√
Reduction in phytoplankton diversity, for example, increased phosphorus loading + higher temperatures promote competitively superior cyanobacteria over diatoms		√
Loss of wetlands due to water extraction and tillage		√
Disconnection of GWT from shallow basins	?	√
Increase in amount of emergent vegetation due to lowered water table	?	?
Increasing distance among habitats due to landscape fragmentation		√
Reduced availability of waterfowl as dispersal agents		√
Increasing salinity due to evaporation and reduced run-off	?	√
Shifts in riparian vegetation (qualitative and quantitative), for example, increased temperature favours emergence of weed species from wetland seed banks; also fire-resistant forms predominate because of increasingly dry conditions	?	?
Creation of dams for increased water retention		√
Increased irrigation networks for agriculture	√	?
Increased proximity of agricultural chemicals		√

Source: Based on Covich *et al.* (1997) and Meyer *et al.* (1999).

Table 10.3 Comparison of the responses of insects and crustaceans to human-induced changes to ecosystems (insect responses in part based on Samways 1996)

Variable	Insects		Crustaceans	
	Natural	**Human-induced**	**Natural**	**Human-induced**
Population surges	Common	Common	Common	Increasing[a]
Range increase	Rare	Common	Rare	Increasing[b]
Population fragmentation (via landscape fragmentation, habitat loss, etc.)	Rare	Very common	Rare	Increasing[c]
Population crashes	Fairly common	Common	Fairly common	Increasing[d]
Species extinctions	Occasional	Increasingly more common and widespread	Occasional	Strong potential[e] + actual?
Loss/reduction of keystone species	Rare	Increasing	Rare	Strong potential[f]
Shifts in life history traits	Occasional	Increasing	Occasional	Increasing[g]
Competition/predation from invading species	Occasional	Increasing	Occasional	Increasing[h]
Loss of genetic variability	Rare	Increasing	Rare	Strong potential[i]

Examples, but not necessarily all from temporary waters.
Source: [a]Patalas (1975); [b]Maude (1988); Mills *et al.* (1993); [c]Goettle (2000); [d]Hobbs and Hall (1974); [e]Collinson *et al.* (1995); [f]Neckles *et al.* (1990); [g]Abdullahi (1990), Gallaway and Hummon (1991), Hogg and Williams (1996); [h]Berrill (1978), Garvey and Stein (1993), Lehman and Caceres (1993); [i]Korpelainen (1986).

10.2.2 Temporary waters as research sites

The ubiquitous nature and small size of many temporary waterbodies make them ideal subjects for pure and applied research studies. Perhaps the greatest ecological utility that these waters represent is their potential to contribute to our understanding of community ecology and ecosystem function. In a very thoughtful assessment of the practices and problems associated with past and present study of biological communities, Putman (1994) concluded that there are two basic approaches. First, in what has been termed the *reductionist* approach, individual, single-species populations and their controlling factors are studied in isolation before being compared with similar data for other populations within the same system. Ultimately, relationships among these focal populations are sought in an attempt to re-assemble a picture of the community. This approach generally suffers from consideration of only a few predominant (although not necessarily 'key') species, and only the perceived major interactions. In addition, it has been argued strongly that separate cause–effect relationships are unlikely to ever be reassembled into a functioning whole (Peters 1991). Second, is the *holistic* approach in which an attempt is made to study a community in its entirety. Unfortunately, the sheer magnitude of this undertaking necessitates an imposed simplification of the system which often allows only qualitative or semiquantitative measurement of patterns and processes. In spite of such limitations, many apparently widely applicable underlying principles of community organization have been proposed and tested. However, with very few exceptions (e.g. Carpenter and Kitchell 1984), a high degree of quantification in community dynamics is generally lacking from these top-down approaches. May (1999) has echoed the problems associated with 'linear' approaches to studying population and community dynamics, pointing out that non-linearities inherent in such dynamical systems render it essentially impossible to take complex systems apart in order to study them piece by piece in a controlled and comparative way.

Temporary water systems would seem to have properties that lend themselves to the advancement of population and community ecology, and ecosystem function. First, many of their communities are less complex than those typically found in permanent waterbodies, yet they appear to exhibit most of the structural (e.g. a wide range of biota comprising many major taxa) and functional (e.g. producers, various levels of consumers, and decomposers) properties seen in other communities. Second, the natural variation in community structure, and presumably in function, seen along the continuum from episodic to intermittent waterbodies would seem to be an ideal testing ground for questions of community trophic efficiency, predator–prey dyamics, and competition. Trophic dynamics continue to be a central theory of ecology (Fretwell 1987; Cohen *et al.* 1990; Pimm *et al.* 1991), however, they are particularly difficult to study because of the spatial and temporal complexity of most natural food webs (e.g. Tavares-Cromar and Williams 1996; May 1999). The empirical simplicity of food webs in episodic waterbodies, in particular, lends them to such study.

Given the high degree of specificity of many temporary water-dwelling species to these habitat types, together with the oftentimes isolated nature of individual waterbodies (e.g. container habitats), it would seem that some temporary waters provide ideal habitat-island models for the experimental analysis of *metapopulation dynamics theory*. Within a metapopulation (spatially isolated sub-populations, or patches, where individuals interact more strongly within patches than between patches; Levins 1970), examination of the characteristics (e.g. genetic relatedness, dispersal ability) of populations from a series of adjacent and distant waterbodies should provide an ideal platform for testing this theory.

Further, many species that are restricted to temporary waters have buoyant populations yet live alongside species that occur in a wider range of freshwater habitat types and are more widely distributed. Clearly such species have very different traits and requirements, not least of which are relative migrational/colonizational abilities. How these might relate to their interactions within a single habitat, might form the basis of some interesting questions on their respective roles in community structure and function, particularly if these roles were to be compared, in the case of cosmopolitan species, with those in other habitat types. The answers to such questions are likely to be of interest to ecologists in general.

Evolutionary biologists may also find temporary waters intriguing research venues. For example, in exploring the classic phenomenon of why guppies (*Poecilia reticulata*) have not speciated on Trinidad, despite great potential to do so, one of the main reasons is seen as their ability to inhabit temporary waters (Endler 1995). It is argued that small pools, ditches, and wetlands do not persist long enough for reproductive isolation to occur (Magurran 1999). Yet, there remains the hypothesis, put forward by W.D. Williams (1988; see Chapter 1), that ponds which dry out periodically collectively represent a very ancient habitat type, and possibly an alternative site for the origin of some forms of life.

10.3 Management and conservation

The purpose of this section is to examine some of the biological consequences, primarily for invertebrate animals, of environmental management practices that involve temporary water habitats. The primary examples chosen are where drainage channels have been created and/or modified for agricultural and urban purposes, and the construction of urban wetlands. Two case-study datasets will be considered in detail, one from Ontario, Canada, the other from southeastern Australia. Based on these and other examples, management recommendations are offered that may allow preservation of species diversity alongside habitat alteration. The focus is on invertebrates because of their fundamental and pervasive importance in all global ecosystems, particularly through their roles in food webs, cycling of nutrients, and maintenance of soil structure and fertility (Wells *et al.*, 1983).

10.3.1 Concerns—replacing natural streams with drainage ditches, and 'cleaning' ponds

The economics of modern agriculture demand maximum use of land area, which frequently requires improved drainage. Typically, this involves laying below-ground tiles which drain into larger underground drains or surface-excavated ditches. By definition, to be successful this practice must result in a substantially altered local drainage pattern. Previously subsurface water may thus be brought to the surface if newly created ditches are dug below the level of the local groundwater table, or if they are designed as sumps for tiles. Conversely, in some instances, tiles and drains that cut across small, natural streams often drain them completely so that their discharge is removed to the subsurface. Such practices have dramatic consequences not only for the aquatic biota living in wetland areas but also for those larger animals, especially birds, whose food supply stems from aquatic communities. If it is unrealistic to hope that all wetlands can be protected by legislation, an alternative strategy is to manage drainage protocols, using sound ecological principals. The following outlines some of the biological changes that take place when natural streams are replaced by drainage ditches, and examines the link between faunal diversity and ditch hydrology.

Moser Creek was a small intermittent stream, in Waterloo County, Ontario (Figure 10.1), which flowed for about 7 months of the year (October to April). It was about 400 m long and ran through a pasture field where it had cut a uniform channel some 50 cm deep and 80 cm wide in the clay-loam soil. Nine small springs supplied most of its flow but lateral seepage (interflow) from the adjacent land was important also. At its lower end, it emptied into a slow-flowing, meandering silty stream fed mostly from agricultural drainage, which is a tributary of the Nith River. As part of a local drainage improvement scheme, Moser Creek was replaced by an open channel about 3 m wide and 1.5 m deep. Using a dragline bucket, the excavation crossed Moser Creek in two places, allowing the creek to drain into it. The ditch also received water from a newly laid field drainage tile at its uppermost end. At its lower end, the ditch was joined to the Nith tributary. Excavated materials were dumped on Moser Creek and bulldozed flat thus effectively obliterating the old stream bed. As the bed elevation in the new channel was some 1 m deeper than in the old one, flow continued throughout the summer, although only at a rate of 2–3 $cm\,s^{-1}$.

The qualitative changes that took place in the fauna after the natural stream had been replaced by the ditch are shown in Table 10.4. Some 60 taxa were found in Moser Creek, representing many of the major invertebrate groups found in running waters (Williams and Feltmate 1992). One year after the ditch had been completed, 40 taxa were present. Of the 86, total, taxa collected from these two habitats, only 13 (15.1%) were common to both. Thus the majority of the original fauna was lost, in particular the flatworms, larger crustaceans, mayflies, caddisflies and many of the cooler-water chironomids. However, some new taxa colonized the ditch, notably beetles and warmer-water associated chironomids. In terms of the trophic structure of the two surface communities, all major feeding groups (*sensu* Merritt and Cummins 1988) were present in both (Figure 10.2), however, the numbers of predators, scrapers and shredders were considerably reduced in the ditch community.

Clearly, what happened in the above case was the replacement of a cool, intermittent running water habitat by a warmer, slower-flowing but permanent one. The original community was dominated by detritus-feeding species but a considerable diversity of predators was present also. Most of the original taxa would have been considered *r*-selected, that is they exhibited features characteristic of species found in unstable habitats (Pianka 1970; Williams 1987). The ditch community was also dominated by detritivores and included taxa that might be considered more K-selected (e.g. the elmid beetle *Dubiraphia* and the chironomid

Figure 10.1 Moser Creek, an intermittent stream in Ontario, Canada, seen in: (a) winter; (b) spring spate; (c) early summer low flow conditions; and (d) summer dry phase.

Eukiefferiella), although a number of opportunistic forms (e.g. the hydrophilid and dytiscid beetles) persisted. How ecologically efficient or stable the newly composed trophic structure of the ditch community became is unknown.

Development of the fauna in this ditch, over the first year of its existence was decribed by Williams and Hynes (1977). They found that the colonizing species came from four sources: drift from the upstream remains of Moser Creek; upstream movement from the Nith Tributary; oviposition by aerial adults from nearby habitats; and from the newly exposed subsurface tiles which drained into the ditch. At first, colonization and extinction rates

Table 10.4 Qualitative change in the fauna after a small, intermittent stream in Ontario had been replaced by an agricultural drainage ditch.

Taxonomic group	Moser Creek	Ditch
Tricladida	*Fonticola velata*	—
Nematoda	*	*
Oligochaeta	Enchytraeidae	—
	?*Sparganophilus* sp.	—
	Tubifex tubifex	*Tubifex tubifex*
Gastropoda	*Physa gyrina*	*Physa gyrina*
	Lymnaea humilis	—
	Helisoma sp.	—
	Gyraulus parvus	—
Ostracoda	*Candona stagnalis*	*Candona stagnalis*
Copepoda	*Cyclops vernalis*	*Cyclops vernalis*
	Attheyella nordenskioldii	—
Amphipoda	*Crangonyx minor*	*Crangonyx minor*
	Crangonyx setodactylus	—
	Hyalella azteca	—
Decapoda	*Fallicambarus fodiens*	—
Acari	*Hydryphantes* sp.	—
	Hydrachna sp.	—
	Oribatida	*
Collembola	*Isotomurus* sp.	*Isotomurus* sp.
	—	*Sminthurides* sp.
Insecta:		
Ephemeroptera	*Paraleptophlebia ontario*	—
	Leptophlebia sp.	—
	Siphlonurus marshalli	—
Hemiptera	*Gerris remigis*	—
	—	*Gerris buenoi*
	Sigara sp.	*Sigara* sp.
	Mesovelia sp.	—
Trichoptera	*Ironoquia punctatissima*	—
	Limnephilus sp.	—
Coleoptera		
Haliplidae	*Peltodytes* sp.	—
Hydrophildae	*Helophorus orientalis*	*Helophorus orientalis*
	Anacaena limbata	—
	Tropisternus sp.	—
	Hydrobius sp.	—
	—	*Enochrus* sp.
		Laccobius sp.
Dytiscidae	*Hydroporus wickhami*	—
	—	*Hydroporus* sp.
	Rhantus sp.	—
	—	*Agabus* sp.
Heteroceridae	—	*Heterocerus* sp.
Elmidae	—	*Dubiraphia* sp.
Scirtidae	—	*Cyphon* ?*variabilis*
Georyssidae	—	*

Table 10.4 (*Continued*)

Taxonomic group	Moser Creek	Ditch
Diptera:		
Chironomidae		
Tanypodinae	*Natarsia* sp.	—
	Psectrotanypus sp.	—
	Pentaneura sp.	—
	—	? *Trissopelopia* sp.
Diamesinae	*Diamesa* sp.	—
Orthocladiinae	*Diplocladius* n. sp.	—
	Trissocladius sp.	—
	Orthocladius n. sp.	—
	Acricotopus sp.	—
	Pseudosmittia sp.	—
	Paraphaenocladius sp.	—
	—	*Cricotopus* sp.
	—	*Eukiefferiella* sp.
Chironominae		
Tanytarsini	*Micropsectra* sp.	*Micropsectra* sp.
	—	*Rheotanytarsus* sp.
Chironomini	*Chironomus* sp.	*Chironomus* sp.
	—	*Cryptochironomus* sp.
	—	*Dicrotendipes* sp.
	—	*Stictochironomus* sp.
Diptera:		
Tipulidae	*Tipula* ?*ignoblis*	—
	Tipula cunctans	—
	Limnophila sp.	—
	Hexatoma sp.	—
	—	*Prionocera* sp.
Culicidae	*Aedes vexans*	—
Ceratopogonidae	*Bezzia/Probezzia* grp.	*Bezzia/Probezzia* grp.
	—	*Stilobezzia* sp.
	—	*Dasyhelea* sp.
Simuliidae	*Simulium* sp.	—
	—	*Simulium vittatum*
Tabanidae	—	*Chrysops* sp.
Ephydridae	*Ephydra subopaca*	*Ephydra subopaca*
	Hydrellia sp.	—
	—	*
Psychodidae	—	*Psychoda alternata*
	—	*Pericoma* sp.
Sphaeroceridae	*Leptocera* spp.	—
Sepsidae	*	—
Muscidea	—	*Limnophora aequifrons*
Pisces	*Culaea inconstans*	—
	Semotilus atromaculatus	—
Total taxa	60	40

'—' indicates absence of a particular taxon, '*' indicates that the taxonomic group was represented but the species was not identified.

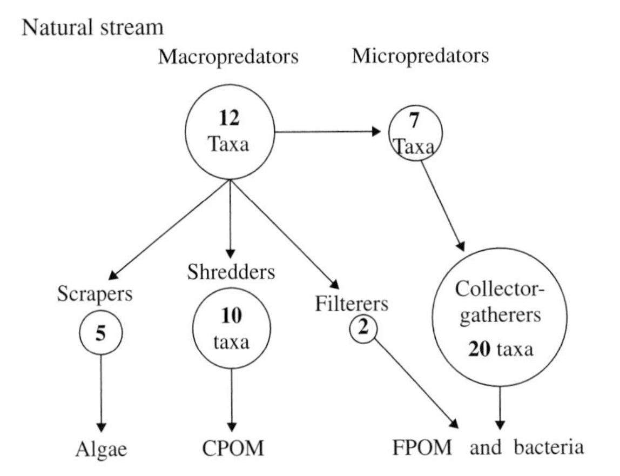

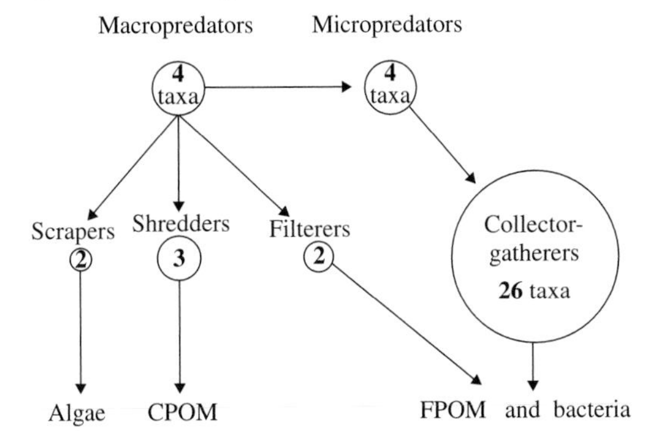

Figure 10.2 Comparison of the trophic structures of: (a) Moser Creek, an intermittent stream in Ontario, Canada; and (b) the agricultural drainage ditch that replaced it. No quantities have been assigned to the various links, and the size of the circles simply reflects the number of taxa present in a particular feeding group.

were high, but after 109 days these began to drop and converge, suggesting that a more stable community was being achieved. However, after 373 days of monitoring (in May), the colonization rate picked up again, likely due to an influx of adults (especially beetles) that typically disperse to warm, shallow bodies of water in this region in the spring (Fernando and Galbraith 1973).

Replacement of natural aquatic systems by drainage schemes has, on a much larger scale, been responsible for unprecedented loss of invertebrate diversity. For example, as a direct result of impoundment and channelization of water courses in the southeastern United States, 40–50% of the formerly rich (>1,000 species) freshwater molluscan fauna has become either extinct or endangered (Standsbery 1971). Populations of some of the North American species of freshwater crayfish have also declined due to changes in drainage pattern (Hobbs and Hall 1974), although a few (e.g. *Procambarus clarkii* and *Fallicambarus fodiens*) find drainage ditches to be acceptable habitats (Penn 1943; LaCaze 1970; Maude and Williams 1983).

In the United Kingdom, extinctions of the damselflies *Lestes dryas* and *Coenagrion armatum* are thought to have been due to the lowering of water tables as a consequence of 'land improvement' (Moore 1976, 1980). Non-aquatic invertebrates may be affected also. For example, draining of the English fens in 1847/8 resulted in the loss of *Lycaena d. dispar* the Large Copper butterfly, and the distribution of the Mole cricket (*Gryllotalpa gryllotalpa*) has been dramatically reduced by drainage of wetlands and meadows (Duffey 1968; Wells *et al.* 1983).

There is evidence of a relationship between diversity of species and ditch hydrology. For example, the faunal communities of three agricultural drainage ditches (within a 2 km radius) in southern Ontario were shown in Table 3.1 (Chapter 3). Although broadly comparable in their physical and chemical characteristics, these ditches differed in the time for which they held water. These differences were due to a combination of local topography and farming practice. The table shows a clear trend of decrease in faunal diversity with increase in the length of the dry period. Taxa that were lost included oligochaetes, some of the mites, caddisflies, elmid beetles, and several dipterans. Whereas some cool-adapted chironomids were lost but warm-adapted forms gained between the 11 month- and 8 month-flowing ditches, respectively, a 6 month dry period was sufficient to remove all midges other than *Pseudosmittia*, a genus known to be semi-terrestrial. Other taxa occurring in the drier ditches included several species of *Helophorus* (beetles), *Fonticola velata* (flatworm), *Crangonyx minor* (amphipod), and *Fallicambarus fodiens* (crayfish). All of these are known from temporary waters and have either behavioural, physiological, or phenological adaptations to deal with loss of water from their habitat (Williams and Hynes 1976). Although 31 taxa was the highest richness recorded in any one ditch, together the three ditches supported 52 different taxa. Trophically, these three communities were, as in the case of the ditch that replaced Moser Creek, dominated by detritivores (Figure 10.2). Further, compared with local, natural intermittent streams, the ditch communities were low in predators.

From the conservation perspective of promoting maximum faunal richness and diversity, creation and maintenance of ditches with a range of water-holding capacities would seem to represent a sound management strategy. How practical, from an engineering perspective, this would be remains to be seen as many factors, such as soil type, slope, surrounding land use, etc., would have to be considered. Nevertheless, in many regions hydrologically different ditches do exist in close proximity—although such diversity is likely more an artifact of local conditions rather than a result of environmental planning.

In the event that an undertaking is made to incorporate ecological considerations into the design of engineered ditches, once the latter are physically in place a sound management programme is necessary for sustained success. In some cases, conflict may arise between conservation and land use. For example, in the Elbe River Valley between Hamburg and Niedersachsen is an extensive (~170 km^2) fruit-growing area. Throughout this region is a network of ditches that were originally dug to drain the land, and these support a very diverse flora and fauna (Caspers and Heckman 1981). In the 1960s and early 1970s, many of these ditches were replaced by underground tiles and pumping stations in order to lower the groundwater table (GWT) further to make the soil suitable for the cultivation of small fruit trees. Other ditches have been allowed to fill as a result of normal eutrophication processes or by dumping. Present fruit-growing technology acknowledges that ditches are in fact desirable because they improve the microclimate, especially in terms of preventing frost damage to spring blossoms. Extensive application of pesticides, however, has resulted in the elimination of large numbers of aquatic arthropods species (particularly mites, mayflies, dragonflies, caddisflies, and beetles), although several others have undergone population explosions (e.g. dipterans). As in the Canadian examples, loss of species from these ditches was most noticable among predators. The Elbe ditches contain species that are rapidly disappearing from the German landscape (Korneck *et al.* 1977). Because these habitats are artificial, their persistence, together with that of their biota, depends on continued management (Caspers and Heckman 1981). Loss of species, due to inappropriate farming practices, from the relatively rich communities of lowland ditches in the Netherlands have also been recorded (Higler and Repko 1981). These latter ditches, totalling approximately 300,000 km, represent an aquatic biotope of considerable importance in northwestern Europe. Macroinvertebrate richness has been shown to be closely linked to the spatial-physical structure of each ditch, particularly that associated with aquatic vegetation, and to the way in which the latter is managed (Higler 1984; Higler and Verdonschot 1989).

Part of the reason for the paucity of conservation and management of drainage system habitats in countries such as Canada and Australia (see below) may be that such engineered structures are historically recent, compared with the situation in Europe, coupled with much greater land areas which have made conservation seem less urgent. In Canada, as in Europe, ditches are particularly common alongside major and minor roads in both urban and rural settings. These are designed with the primary purpose of removing water and snow from the road surface and, typically, empty into nearby streams and rivers. Any features of local topography that lessen the amount of water that can be transported are mechanically removed, both initially and on a routine basis (as part of operational maintenance), if necessary. Habitat stability is therefore often not achieved and a climax community cannot be reached. Further, the proximity to vehicles results in many pollutants entering both the ditches and receiving rivers. Common among these are road de-icing salts, gasoline and oil, and various heavy metals associated with particles of tyre rubber, for example, cadmium and zinc. In rural Australia, roadside ditches are uncommon owing to the generally arid climate and flat topography (annual run-off represents ony 13% of the total annual precipitation; W.D. Williams 1981). However, in urban areas, run-off from paved surfaces (parking lots, roads, etc.) results in the entry of vehicle-associated compounds (e.g. oil, gasoline, and lead) into urban waterways and eventually into natural systems (Australian Capital Territory Electricity and Water, unpublished data).

Attainment of natural climax communities is also prevented by management practices in other types of temporary waters. For example, thousands of ponds, both permanent and temporary, in the United Kingdom are subject to 'pond-management programmes'. Often annual in occurrence, these events involve groups of professionals or volunteers dredging, removing silt, and general 'cleaning-up' pond basins in comparative ignorance of the ecological consequences (Biggs *et al.* 1994). Whereas the removal of human-related debris is laudable, the removal of silt, for example, disrupts natural hydroseral succession. The devolution of temporary ponds is a normal consequence as this succession proceeds. To keep a balance, managers may wish to create new ponds at a younger hydroseral stage—especially where natural replenishment is prevented due to adverse land practices. This would ensure a supply of ponds representative of all stages in the succession, together with the survival of those taxa and communities that specialize in each stage or habitat type. This includes inhabitants of episodic ponds in their last throwes of existence, which may support semi-terrestrial forms like sepsid, sphaerocerid and some ceratopogonid (e.g. *Dasyhelea*, *Culicoides*, *Stilobezzia*) and chironomid (e.g. *Georthocladius*, *Gymnometriocnemus*, *Pseudosmittia*, *Lapposmittia*, *Limnophyes*, *Paraphaenocladius*, *Smittia*) dipterans, together with scirtid (helodid), hydraenid, and heterocerid beetles. To dredge 'terminal' ponds back to new pond status would deny habitat to such taxa.

10.3.2 Concerns—biological consequences of urban wetland construction

In urban areas, too, the transformation of naturally pervious rural land to predominantly impervious land forms, such as roofs and roads, often overwhelms the capacity of natural watercourses. Typically, the urbanization process results in:

- altered hydrologic regimes
- deterioration of water quality
- increased sedimentation rates, particularly during development phases (Cordery 1976).

To accommodate these changes, engineered structures (e.g. concrete channels, urban wetlands, reservoirs, and balancing ponds) may be constructed to improve flow rates, control discharge and sedimentation rates, and improve water quality. The conservation potential of these anthropogenic systems is often limited because of poor water quality, however, their impact on the aquatic habitats they may be designed to protect should be considered. For example, most urban development in Australia occurs in previously cultivated areas which are characterized by geologically 'old', deeply weathered soils rich in clay-sized particles (Olive and Walker 1982). Because such material is easily transported following disturbance of surface cover (Norris 1991), provision is often made for sediment retention structures (e.g. settling ponds and artificial lakes). The following outlines the effects of such a development on the Murrumbidgee River, Australian Capital Territory, and discusses the consequences of this project for benthic invertebrate populations.

The Murrumbidgee River is part of Australia's largest river system, running some 1,600 km through the south-eastern part of the country before entering the Murray River. Virtually the entire 60 km length contained within the A.C.T. has been gazetted as nature reserve (Anon 1988). Within this stretch the Murrumbidgee consists chiefly of pools and areas of shifting sand interspersed with riffles. Riparian land use is primarily agricultural (livestock), although more recently the growth of Canberra has led to increased urbanization and associated water-control structures within the Murrumbidgee catchment (Hogg and Norris 1991). Tuggeranong Creek is a small urban stream that flows through southern Canberra before discharging into the Murrumbidgee. In an effort to minimize the effects of urban development on the Murrumbidgee, several settling ponds and an artificial lake were constructed along the length of Tuggeranong Creek to improve water quality and intercept sediment before it entered the Murrumbidgee. The settling ponds and 70 ha 'Lake Tuggeranong' were created using a series of small dams along the length of the creek. The lake basin was further graded by earth movers, and together with the downstream portion of the dam spillway, was denuded of surface vegetation cover. Despite having a mean discharge an order of magnitude smaller than the Murrumbidgee River, the land-servicing activities (e.g. clearing for development) in the Tuggeranong catchment led to an often hundred-fold increase in suspended solids in the Murrumbidgee downstream of its confluence with Tuggeranong Creek. Most of this material, especially particles <125 μm in diameter, settled immediately downstream of Tuggeranong Creek. Other chemical parameters were similar upstream and downstream of Tuggeranong Creek with the exception of nitrate and nitrite levels. Such increases are typical of cleared catchments because of topsoil erosion and the loss of associated nutrients, especially nitrates (Martin and Pierce 1980).

Benthic samples collected during the construction period were lower in both species (54 vs 23; Table 10.5) and animal numbers downstream of the confluence—with the exception of the oligochaetes *Nais communis/variabilis* and the chironomid *Cryptochironomus* sp., almost all species had lower densities downstream. No new species colonized the reach downstream of the confluence. Immature stages of predaceous taxa (especially odonates, tanypodine chironomids, and ceratopogonids) were almost entirely absent in samples taken downstream of Tuggeranong Creek, although the tanypod *Procladius* sp. did show some recovery 35 km downstream. Sensitivity of predaceous species to anthropogenic disturbance has been well documented in other studies on Australian freshwaters (e.g. Watson *et al.* 1982). Decreased densities of some species, such as the caenid mayfly *Tasmanocoenis tillyardi*, were somewhat more surprising—as it is a common inhabitant of the depositing zones of rivers throughout Australia. Its partial recovery 35 km downstream may have reflected the severity of conditions immediately downstream of Tuggeranong Creek. Across Australia, alteration of stream discharge is known to have threatened at least five species of Odonata (Key 1978).

The effectiveness of the sediment-retention structures in Tuggeranong Creek remains largely unknown—recovery of *T. tillyardi* and some other invertebrates had not been realized 6 years after construction. The absence or low densities of such species downstream of Tuggeranong Creek, combined with the limited dispersal capabilities of the adults of some species, could have long-term consequences for the survival of species within the affected area. Habitat loss occurring as a result of the genesis of Lake Tuggeranong, the sediment retention structures, etc. downstream of Tuggeranong Creek may have exacerbated existing limits to gene flow among such populations.

In many parts of the developing world where unique communities of wetland invertebrates occur, threats from adjacent conurbations or water resource development schemes are very real (Wells *et al.* 1983) and of great concern, particularly as little or no protective legislation is in place. In addition, such schemes also threaten human populations as a result of disease-vector enhancement (Russell 1999; and Chapter 9).

10.3.3 Forestry and other plant issues

Chapter 3 examined some of the ways in which the surrounding trees may influence woodland temporary ponds and streams, primarily through effects of shading and leaf litter input. In addition, Chapter 4 (Section 4.2.4) gave examples of how forestry practices affect wetland function, including both increasing the incidence of low flow events (resulting from pre-planting drainage), and stabilizing flow (depending on the rate of vegetation re-growth). A major and rapid upset occurs, however, when timber is harvested, as the resulting loss of litter input combined with increased insolation changes the food base for the aquatic communities, that is, from detritus-based to algal-based. Such recurring events, even though their

Table 10.5 Summary of species collected at Control, Recovery, and Impact sites along the Murrumbidgee River, Australia (1987–88) (extent of broken line indicates site occurrence; based on data supplied by I.D. Hogg)

Taxonomic group	Species	Control	Recovery	Impact
Oligochaeta				
Naididae	*Nais communis/variabilis*	......	......	......
Tubificidae	*Branchiura sowerbyi*	......	......	......
	Limnodrilus hoffmeisteri	......	......	......
Gastropoda	Unident sp.	......	......	......
Bivalva				
Corbiculidae	*Corbiculina australis*	......	......	......
Ostracoda	*Candonocypris* sp.	......	......	......
Copepoda	*Eucyclops speratus*	......	......	......
	Paracyclops fimbriatus	......	......	......
	Macrocyclops albidus	......		
Acari				
Hygrobatidae	*Australiobates violaceus*	......	......	
Hydrodromidae	*Hydrodroma* sp.	......	......	
Oribatida	*Hydrozetes* sp.	......		
Ephemeroptera				
Baetidae	*Baetis* sp.	......		
Caenidae	*Tasmanocoenis tillyardi*	......	......	
Leptophlebiidae	*Atalophlebia* sp.	......		
	Nousia sp.	......		
Oniscigastridae	*Tasmanophlebia* sp.	......		
Odonata				
Coenagrionidae	*Ischnura heterostica*	......		
Lestidae	*Austrolestes cingulatus*	......		
Gomphidae	*Austrogomphus ochraceus*	......		
	Hemigomphus sp.	......		
Corduliidae	*Cordulephya pygmaea*	......		
Plecoptera				
Gripopterygidae	*Dinotoperla bassae*	......		
Hemiptera				
Corixidae	Unident sp.	......	......	......
Notonectidae	Unident sp.	......	......	......
Trichoptera				
Ecnomidae	*Ecnomus* sp.	......	......	
	E. pansus	......	......	
	E. continentalis	......	......	
Leptoceridae	*Oecetis* sp.	......		
	Triaenodes sp.	......		
Hydroptilidae	*Hellyethira simplex*	......	......	
	H. ?malleoforma	......	......	
Coleoptera				
Dytiscidae	Unident sp.	......	......	......
Helminthidae	Unident sp.	......	......	......
Hydrophilidae	Unident sp.	......	......	......

Table 10.5 (*Continued*)

Taxonomic group	Species	Control	Recovery	Impact
Diptera				
Chironomidae				
Orthocladiinae	*Cricotopus* sp. 1	......	......	......
	Cricotopus sp. 2	......	......	......
	Unident sp.	......		
	Unident sp.	......		
Chironominae				
Chironomini	*Cryptochironomus* sp.	......	......	......
	Chironomus sp.	......	......	......
	Dicrotendipes sp.	......	......	......
	Polypedilum oresitotrophus	......	......	......
	Unident sp.	......		
Tanytarsini	*Cladotanytarsus* sp.	......	......	......
	Rheotanytarsus sp.	......		
	Riethia sp.	......	......	......
	Tanytarsus sp.	......		
Tanypodinae	*Procladius* sp.	......	......	......
	P. paludicola	......		
	Coelopynia pruinosa	......		
	Paramerina sp.	......		
Ceratopogonidae	Unident sp.	......		
Empididae	Unident sp.	......	......	......
Total taxa		54	29	23

frequency may be measured in decades, cause significant shifts in aquatic community structure and function.

A number of studies have examined the integrity of plant communities in artificially created or restored wetlands compared with natural wetlands. Galatowitsch and van der Valk (1996), for example, report that thousands of wetland restorations have taken place in the mid-continental US. In this region, wetlands tend to re-vegetate naturally once appropriate hydrological regimes have been restored. Comparing 10 restored wetlands with 10 adjacent natural wetlands in northern Iowa, they tested the 'efficient-community' hypothesis, which states that: all plant species that can become established and survive under the environmental conditions prevailing at a site will eventually be found growing there and/or will be found in its seed bank. After 3 years of post-restoration study, the mean species count for the restored wetlands was 27 (58.7%), compared with 46 for the natural wetlands. Although the richness of emergent species was similar in the two wetland types, there were many differences, including, in the restored condition: the under- (e.g. sedge meadow) or over-representation (e.g. submerged aquatics) of certain guilds of species; elevational differences in species distributions and abundances; and poorer seedbanks (both in terms of species and seed numbers). Many of these differences were seen as resulting from differential dispersal abilities, causing concern that natural restoration processes might be undesirably protracted.

Similar findings have been made for manufactured wetlands, especially where assumptions are made that the latter will eventually come to physically resemble natural wetlands and thus will serve as functional replacements. In a comparison of created versus natural wetlands in Pennsylvania,

Campbell *et al.* (2002) found that vegetation species richness and total cover were both greater in natural wetlands, in particular created sites supported a greater proportion of upland species. Furthermore, in created wetlands ranging in age from 2 to 18 years, there were significant differences in soils (less organic matter), bulk densities (greater), soil matrix chroma (higher), and rock fragment content (higher). Even the older sites could not be said to be trending towards the reference wetland condition. If such differences exist within the wetland plant base, then the establishment of other components (e.g. invertebrates, amphibians, reptiles) that constitute natural wetlands may also be compromised.

While manufactured wetlands are often to replace natural ones previously destroyed, there is an increasing trend to construct artificial wetlands largely for the purpose of treating wastewater. For example, construction of urban wetlands in Queensland has been proposed as a means of treating municipal wastewater. Nine pilot wetlands were studied for their performance efficiency and nutrient bioaccumulation abilities (Greenway and Woolley 1999). Located in different geographical locations (tropical, subtropical, and arid), each wetland had a different configuration and developed a variety of macrophyte types and species, through planting or self colonization. In general, many wastewater parameters showed decreases in these systems, including Biochemical Oxygen Demand (17–89% reduction), suspended solids (14–77%), total nitrogen (18–86%), and reactive phosphorus (~13%). Many of the submerged and floating macrophytes (e.g. *Ceratophyllum* and duckweed) proved to be high nutrient bioaccumulators, compared with emergent species. However, emergent species had a much greater biomass and were therefore able to store more nutrients per unit area of wetland. The faunas associated with these constructed wetlands were not reported.

10.4 Triumphs, failures, and 'works in progress'

In parts of Europe and North America, the degradation of wetlands has been identified as a matter of concern, and both government and private organizations have initiated restoration programmes (e.g. the North American Wetlands Conservation Act of 1989, and its subsequent amendments; and the European Community Framework Directive for Water, 2000). These measures take account of the close relationship between aquatic ecosystems and the hydrology of the watershed, in the context of local land use, and range from the re-introduction of meanders to the planting of appropriate riparian trees and restocking of fauna. However, because of existing and oftentimes extensive environmental deterioration, for example, groundwater contamination and flow alteration, recovery may be slow (Higler and Verdonschot 1993). Interestingly, the initial stimulus, especially in North America, for such policy has stemmed largely from the plight of migratory birds and reduced waterfowl habitat, rather than a desire to protect wetlands *per se*. Historical records indicate that in the United States, in 1776, the contiguous 48 states contained around 89 million hectares of wetlands. By the mid-1970s, this had been reduced to just under half that area (43 million ha). Since 1980, however, it is claimed that there has been no net loss of wetlands (Council on Environmental Quality 1995). This has been due largely to a much greater appreciation of US wetlands as highly sensitive habitats that support a wide diversity of plant and animal life, together with passage of a Clean Water Act (1972), which included a section that regulated wetland use. Unfortunately, the act has caused great controversy as much of the nation's remaining wetlands is on private land and no compensation is available for development prohibition. Further, activities that may potentially impact wetland areas require that the landowner obtain a permit before work may begin.

In Australia, the lower Cooper Creek and Coongie Lakes region of Queensland (~2 million ha) represents one of the world's largest freshwater Ramsar sites. It includes a discharge pattern that is among the most variable on the planet, typifying the 'boom-and-bust' cycle typical of dryland rivers. Such a flow regime has promoted naturally high taxonomic diversity within the system, yet the scarcity of water in the region has attracted potential development. In 1995, for example, a consortium proposed to extract (42,000 ML p.a.) and store (15,000 ML) water for cotton farming in the area. The Queensland Minister for Natural Resources was subsequently persuaded, by wetland ecologists, to reject the proposal on the grounds of major environmental damage (Boulton 2000).

Such ministerial decisions and protective legislation are crucial to conservation, however, they are not always durable or enforceable, and sometimes succumb to private-interest groups. For example, *Lepidurus packardi*, the vernal pool tadpole shrimp, is known only from California's Central Valley and from a few locations in the San Francisco Bay area. Because loss of vernal pool habitat in California's Central Valley is estimated at between 65% and 90% of its former extent—due to urbanization of flat lands (slope $\leq$ 3–4%) close to metropolitan areas—in 1994 this species was designated as endangered and was protected by federal law (Goettle 2000). In 1996, agricultural and business groups asked for delisting of two of the four species of endangered vernal pool shrimp, and proceeded to sue the US Fish and Wildlife Service over the matter, claiming that protection of the shrimp was being used as a surrogate for land-use control (Evans 2000). Another example of lack of public sympathy for the fate of notostracans is the fact that in California rice fields, *Triops* are looked on as pest species as they are reported to uproot and eat young rice plants (Fry and Mulla 1992).

In some cases, judgement based on habitat size rather than importance has failed to protect species. For example, small, isolated wetlands, such as are found in the southeastern Coastal Plain of the United States, lack the legal protection given to riparian and lacustrine wetlands. Indeed, the US Army Corps of Engineers permits wetlands less than 0.12 ha to be filled on an *ad hoc* basis, and requires only a minimal review of circumstances if they do not exceed 1.2 ha (Federal Register 1996). As a result of uncontrolled agricultural conversion and regional development, and in apparent contrast to the view of the Federal Government (Council on Environmental Quality 1995), loss of these habitats and their unique faunas is escalating (Kirkman *et al.* 1999).

Abbasi (1997) provides some discouraging statistics on regional wetlands: 80,000 ha of bog in Ireland have been drained since 1946; 35% of wetlands in Tunisia have been drained or built upon in the last century; 40% of coastal wetlands in Brittany have disappeared since the mid-1970s. In addition, one-third of the Banados del Este in Uruguay has been drained for agriculture—despite it being designated a Ramsar site and a UNESCO Biosphere Reserve; another Ramsar and World Heritage site, Lac Ichkeul, in Tunisia has been compromised through dam building on its input rivers; and gold mining in the Paraguay River basin has resulted in mercury pollution in the Pantanal.

Hails (1997) showcases many wetland areas in Asia that are under threat from human activities, including inadequate management. For example, the Mai Po marshes represent a mosaic of some 2,000 ha of wetlands in the northwestern corner of Hong Kong. Their diversity has resulted from almost 100 years of land-use changes which have managed to maintain a 'wet' nature to the landscape, and provide habitat for a remarkable diversity of wildlife. However, the existence of the area has been significantly compromised by organic pollution in nearby Deep Bay, and by in-filling to support local development, consequently there is uncertainty over the area's long-term future (Young 1997). Tasek Bera, Malaysia's first Ramsar site, is one of only two major freshwater lake systems on the Malay Peninsular. The lake (6,150 ha) links to a much larger (61,383 ha) catchment which serves as a natural hydrological buffer in flood control, flow regulation, water purification, and water supply. The system is, however, under threat from a number of sources, including increased pesticide and fertilizer run-off, inputs of domestic pollution, clearing of vegetation for navigation, and collection of species for the aquarium trade. These problems have been exacerbated by ineffective cross-sectoral coordination among management agencies. In 1995, however, Ramsar Convention edicts came into play, with a primary initiative being the development of an integrated management plan (D'Cruz 1997).

The Sundarbans, the largest mangrove system in the world (over 1 million ha on the India/Bangladesh border), and upon which depend more than 300,000 people, may be regarded as a wetland that has attained sustainability through implementation of principles of 'wise use' which developed over a 125-year period. Management involves a 20-year exploitation cycle of the tree resources, which are divided into 20 compartments, each of which is harvested once in the cycle. Harvesting of leaves for thatching, fishing, and collections of other resources (honey and shells) take place on shorter cycles and are also closely regulated to ensure that their exploitation is sustainable (Hussain 1997).

Occasionally, serendipitous events may benefit or even restore wetlands. For example, at Lac Fetzara, in Algeria, changes made by local hydraulic engineers to avert flooding of the nearby steelworks at Annaba, resulted in the inadvertent restoration of the lake and wetland complex. Improved management of the outlet sluices, now supplies sufficient water for irrigation, an extended grazing season, flood protection, reinstated wildfowl habitat, and hunting opportunities (Stevenson *et al.* 1989).

The Kävlinge River Project in Sweden can be considered an example of a 'work in progress'. In 1990, it was initiated to create wetlands and riparian buffer zones in southeastern Scania. The river runs through a heavily cultivated region where natural wetlands had been eroded over several hundred years to a current cover of less than 1%. The main objective of the project was to reduce the nutrient load to the coast via the creation of 300 ha of ponds and a further 210 ha of buffer zones, to be done alongside changes in agricultural practices and upgraded waste-water treatment. Interestingly, the project was largely a local and regional initiative (a 'bottom-up' agreement) to implement a national policy for wetland creation. Key elements that have aided the project's progress were seen as:

- the existence of a clear overarching policy objective to which all involved were committed
- the devotion and persistence of key personnel (e.g. officials and politicians) in participating municipalities
- commitment among municipalities to finance the implementation of the policy
- well defined and interactive channels of communication between the authorities and the landowners
- the presence of an independent mediating agent between the parties involved (Söderqvist and Lindahl 2003).

The UK's Norfolk and Suffolk Broads serve as an example of extensive (~300 km^2), open-access, and multipurpose wetlands that have been managed to varying degrees for centuries. Early management comprised alterations to optimize the land for harvesting reeds and sedges, and for cattle grazing, and these traditional uses have resulted in a distinctive landscape. In recent times, however, much greater and varied uses have developed, in particular those associated with water abstraction, navigation, and flood control. In response to these oftentimes conflicting demands, together with environment preservation issues, a series of management schemes culminated in the establishment of the Broads Authority in 1989. The BA is charged with coordinating the management of this network of rivers and shallow lakes surrounded by fens and drained marshes, which represent the only UK habitat for rare animals and many of the 250 species of plants that it supports. The Authority has the following core mandates:

- to conserve the wildlife of the Broads
- to conserve the cultural heritage of the Broads, as well as promoting an understanding and enjoyment of these lands
- to protect the interests of navigation.

Oftentimes, these objectives involve complex political, economic, and environmental trade-offs, made more difficult recently by European Union Directives (e.g. the Birds and Habitat Directive)—which feature a somewhat static approach to nature protection. It is clear, however, that the Broads represent a semi-natural/human system that continues to evolve, requiring continuous adjustments to support both scientific and managerial goals. To this end, the BA is required to prepare a 'Broads Plan' every 5 years, into which all stakeholders (government, local people, scientists, and special interest groups) will be allowed input (Turner *et al.* 2003b).

In the neotropics, despite the daunting challenge of dealing with a complex of geographical, ecological, and socio-political issues, compounded by incomplete knowledge of wetland dynamics, a pilot project on Esteros del Ibera in Argentina is proving useful for the management of Latin American wetlands (Loiselle *et al.* 2004). By means of remote sensing information, *in situ* hydrological and microclimate sensors, hydrological and ecological modelling approaches based on (limited) historical data, energy-based models, and ecological-economic indicators of wetland resource use, local and regional decision-makers are being provided with tools to help formulate sustainable wetland management plans.

Gains are also being made on much smaller scales. For example, 'Best Development Practices' have been drawn up to conserve vernal pool-breeding amphibians in residential and commercial developments in several northeastern US states. These practices encourage the public to become aware of these habitats, educate communities as to their intrinsic worth, and recommend practical measures to achieve conservation (Calhoun and Klemens 2002). 'Pond Action', a non-profit organization founded in 1988 has surveyed many small ponds in the United Kingdom (up to 40% of which were thought to be temporary), and concluded that pond management practices in that country were frequently based on a range of myths about pond properties. Among these misconceptions were:

- drying out is disastrous for ponds
- ponds should be at least 2 m deep
- all classical pond zones (from shallow margins to deep open water) should be present and maintained
- the bigger the pond the better
- ponds should not be shaded by trees
- ponds should be periodically dredged to prevent them becoming choked with vegetation
- steps should be taken to minimize water level fluctuations
- ponds should be protected from livestock
- ponds are entirely self-contained ecosystems with little interaction with their surroundings (Biggs *et al.* 1994).

A highly encouraging recent outcome with regard to temporary pools has been the call for the global conservation of these habitats, and its adoption at the 8th Ramsar Conference, in 2002 (Grillas *et al.* 2004).

10.5 Conclusions and recommendations

In terms of the potential of, for example, man-made drainage channels and manufactured wetlands to be habitats for the persistence and conservation of temporary freshwater species, the conclusion would seem to be 'high' in theory, but 'moderate to low' in practice. Although, as seen in the southern Ontario study (Section 10.3.1), replacing natural wetlands with a range of ditch types can maintain a reasonable level of species diversity, man-made habitats support somewhat different communities. These communities are not only *structurally* but also *functionally* different, particularly, in the case of invertebrates, with respect to predatory species, the latter shift seeming to be universally characteristic of man-made freshwaters. In the case of the Murrumbidgee River Protection Scheme, relatively little initial consideration was given to the conservation of the invertebrate component of the old and new water courses, although to have the protection, from urbanization, of a major river as an engineering goal is, indeed, laudable. However, ecological input must be sought early on in such undertakings so that potential biological problems can be identified and solutions found. Even the high diversity measured in the Ontario ditches was achieved serendipitously, through creation of a range of habitat types due to local differences in hydrology, rather than to any constructural plan. Clearly, there is a need for much greater discussion among ecologists, planners, and engineers.

Adoption of ecological guidelines for the management of man-made waterbodies can make economic as well as conservation sense. For example, for centuries in the Dombes region of southern France, large, constructed clay 'ponds' have been used for alternately reaping harvests of fishes and grain (Section 8.2). These shallow waters were created by damming the ends of depressions between moraines and water levels are controlled by a series of source ponds, ditches, and sluice gates. Some 11,000 ha of ponds currently exist in this region. Many tropical floodplains are, similarly, under crop cultivation during the dry phase (Williams 1987).

The following premises should be accepted when making management recommendations for the safeguarding of temporary waters:

(1) wetland features, such as temporary ponds, pools, streams, bogs, and floodplains are not 'wasted' areas of land, they are natural components of many regional and global environments and represent distinct habitats for many species—some that are not found elsewhere, others that are most numerous there;
(2) there is a strong likelihood that temporary waters contribute to maximizing the gene pool of species that occur in both temporary and permanent waters, and this increased diversity may be crucial to the survival of species facing future changes to global environments—aquatic biologists and conservationists must therefore consider the evolutionary as well as the ecological consequences of habitat alteration;
(3) evidence shows that wetland drainage results in significant loss of both aquatic and wetland-associated terrestrial species, sometimes on an enormous scale.

For temporary ponds, *per se*, Biggs *et al.* (1994) have identified three factors that are strongly associated with ponds that support diverse plant and animal assemblages, and also a high proportion of rare species:

- good water quality
- variety in physical shape and structure—areas with a mosaic of pools, both permanent and temporary, each supporting a range of mesohabitats (emergent vegetation, riparian plant debris, submerged plants) are particularly desirable
- proximity or high connectivity to other wetlands, so that metapopulations may develop.

These properties are likely applicable to many other types of temporary water habitat.

Pragmatic and specific recommendations to achieve such goals include:

(1) when dealing with pristine natural systems, wherever possible using minimal-intervention management protocols;
(2) where intervention is necessary, involving ecologists in all engineering projects that involve alteration of habitats, starting at the planning stage;
(3) if the rate and extent of intervention cannot be reduced by adoption of alternative, ecologically sound technologies, then (e.g. in the case of drainage channels) creating an overall diversity of channels within a given area, particularly with respect to hydrological regime;
(4) stabilizing littoral zones;
(5) maintaining natural inundation patterns, which may necessitate management of regional groundwater dynamics;
(6) mimicking the natural frequency, duration, and seasonal patterns of drying in water basins and channels;
(7) increasing the diversity of microhabitats within individual drainage channels through:
(a) providing a variety of natural substrate types on the bed; and (b) for lotic bodies, creating a variety of current regimes (including laminar and turbulent), for example, via modification of the channel course and installation of in-stream devices;
(8) restoring natural riparian vegetation, including both annuals, and perennial shrubs and mature trees;
(9) controlling excess growth of aquatic macrophytes, at least initially, until natural shading takes over;
(10) establishing sanctuary habitats within the general area affected by the disruption. These will both maintain adequate gene flow among populations and allow the persistence of rare and endangered species;

(11) determining the ecological requirements of rare and endangered species (including dispersal abilities) and, where necessary, enacting legislation to protect them;
(12) preventing entry of agricultural chemicals, especially fertilizers and pesticides, and those pollutants reaching roadside ditches from passing vehicles or snow removal practices;
(13) rather than abandoning sites after intervention, continuing to monitor the ecological processes that develop within newly modified or created water bodies, such that the colonization success of the biota, establishment of energy budgets and food webs, and the resilience potential of the communities can be assessed;
(14) raising public awareness of the importance of temporary water habitats, through formal and informal education programmes, including schools and the media.

A final point worth making is that temporary water biologists, should not only be advocating the need for further basic research into these habitats, but also be prepared to provide answers to some of the human-health problems associated with their inhabitants. The role of many temporary waters as disease-vector breeding sites is well known, and the details were summarized in Chapter 9. While many of these diseases are most prevalent in tropical and subtropical areas, the spread of forms such as West Nile virus and Equine Encephalomyelitis are giving cause for great concern by health-control agencies in more temperate countries. Thus, even where there is sympathy for the conservation of temporary water habitats, fear for public health has resulted in such current practices as the wholesale drainage of standing waters, and mosquito larviciding, which are crude control measures in the extreme. There is an urgent need to replace these with more environmentally friendly alternatives, based on sound ecological knowledge of natural population control mechanisms. If such alternatives cannot be provided, then there is the very real risk that any recent, hard-earned, protective legislation for wetlands may be quickly overturned to protect the public health.

References

Chapter 1

Bayly, I.A.E. (1967). The general biological classification of aquatic environments with special reference to those in Australia. In *Australian Inland Waters and their Faunas: Eleven Studies* (ed. A.H. Weatherly), pp. 78–104. Australian National University Press, Canberra.

Blaustein, L. and S.S. Schwartz. (2001). Why study ecology in temporary pools? *Israel Journal of Zoology* **47**: 303–12.

Boulton, A.J. and M.A. Brock. (1999). *Australian Freshwater Ecology, Processes and Management*. Gleneagles Publishing, Adelaide, Australia.

Bradley, C. and A.G. Brown. (1997). Modelling of hydrological processes in a floodplain wetland. In *Groundwater/Surface Water Ecotones: Biological and Hydrological Interactions and Management Options* (eds J. Gibert, J. Mathieu, and F. Fournier), pp. 102–10. Cambridge University Press, Cambridge.

Cole, G.A. (1979). *Textbook on Limnology*. C.V. Mosby Company, San Louis.

Colless, D.H. (1957). Notes on the culicine mosquitoes of Singapore. III. Larval breedong places. *Annals of Tropical Medicine and Parasitology* **51**: 102–16.

Comín, F.A. and W.D. Williams. (1994). Parched continents: our common future? In *Limnology Now: A Paradigm of Planetary Problems* (ed. R. Margalef), pp. 473–527. Elsevier, Amsterdam.

Dahl, T.E. and C.E. Johnson. (1991). *Wetlands Status and Trends in the Conterminous United States, Mid–1970s to Mid–1980s*. US Department of the Interior Fish and Wildlife Service, Washington DC.

Decksbach, N.K. von (1929). Zur Klassifikation der Gewässer von astatischen Typus. *Archiv für Hydrobiologie* **20**: 349–406.

Duigan, C.A. and A.T. Jones. (1997). Pond conservation symposium: introduction. *Aquatic Conservation: Marine and Freshwater Ecosystems* **7**: 87–9.

Elgmork, K. (1980). Evolutionary aspects of diapause in freshwater copepods. In *Evolution and Ecology of Zooplankton Communities* (ed. W.C. Kerfoot), pp. 411–17. University of New England Press, Hanover.

Fernando, C.H. (1980). Some important implications for tropical limnology. In *Proceedings of the 1st Workshop on Promotion of Limnology in Developing Countries* (eds S. Mori and I. Ikusuma), pp. 103–7. International Society for Limnology, Kyoto.

Fernando, C.H. and J. Holcík. (1989). Origin, composition and yield of fish in reservoirs. *Archive für Hydrobiologie* **33**: 637–41.

Finlayson, C.M. and A.G. Van der Valk. (eds). (1995). *Classification and Inventory of the World's Wetlands*. Kluwer Academic Publishers, Dordrecht.

Giudicelli, J. and M. Bournard. (1997). Invertebrate biodiversity in land–inland water ecotonal habitats. In *Biodiversity in Land–Inland Water Ecotones* (eds J.-B. Lachavanne and R. Juge), *Man and the Biosphere Series* **18**: 143–60. Parthenon Publishing Group, New York.

Gordon, N.D., T.A. McMahon, and B.L. Finlayson. (1993). *Stream Hydrology: An Introduction for Ecologists*. John Wiley & Sons, Chichester, UK.

Hammer, U.T. (1986). *Saline Lake Ecosystems of the World*. W. Junk Publishers, Dordrecht.

Hebert, P.D.N. (1999). Variable environments and evolutionary diversification in inland waters. *Advances in Molecular Ecology* **306**: 267–90.

Heddinghaus, T.R. and P. Sabol. (1991). A review of the Palmer Drought Severity Index and where do we go from here? In *Proceedings of the Seventh Conference on Applied Climatology*, American Meteorological Society, Boston, pp. 24–26.

Hershey, A.E., L. Shannon, G.J. Niemi, A.R. Lima, and R.R. Regal. (1999). Prairie wetlands of south–central Minnesota: effects of drought on invertebrate communities. In *Invertebrates of Freshwater Wetlands of North America: Ecology and Management* (eds D.P. Batzer, R.B. Rader, and S.A. Wissinger), pp. 515–41. John Wiley & Sons, New York.

Hinton, H.E. (1968). Reversible suspension of metabolism and the origin of life. *Proceedings of the Royal Society B* **171**: 43–56.

International Society for Salt Lake Research. (2001). http://www.isslr.org/biblio/biblio.htm.

Johansson, S.R. (1999). Death and doctors: medicine and elite mortality in Britain from 1500 to 1800. In *Cambridge Group for the History of Population and Social Structure*. Working Paper No. **7**: 1–37.

Junk, W.J. (1993). Wetlands of tropical South America. In *Wetlands of the World: Inventory, Ecology and Management*, Vol. 1 (eds D. Whigham, D. Dykyjová, and S. Hejny), pp. 679–739. Kluwer Academic Publishers, Dordrecht.

Keeley, J.E. and P.H. Zedler. (1998). Characterization and global distribution of vernal pools. In *Ecology, Conservation, and Management of Vernal Pool Ecosystems* (eds C.W. Witham, E.T. Bauder, D. Belk, W.R. Ferren Jr., and R. Ornduff), pp. 1–14. *Proceedings of the 1996 Conference of the California Native Plant Society*. Sacramento, CA.

Klimowicz, H. (1959). Tentative classification of small water bodies on the basis of the differentiation of the molluscan fauna. *Polskie Archwm Hydrobiologie* **6**: 85–104.

Laird, M. (1988). *The Natural History of Larval Mosquito Habitats*. Academic Press, London.

Lake, P.S., I.A.E. Bayly, and D.W. Morton. (1988). The phenology of a temporary pond in western Victoria, Australia, with special reference to invertebrate succession. *Archiv für Hydrobiologie* **115**: 171–202.

Mawson, D. (1950). Occurrence of water in Lake Eyre, South Australia. *Nature* **166**: 667–8.

Mitsch, W.J. and J.G. Gosselink. (2000). *Wetlands*, 3rd edn. John Wiley & Sons, New York.

Mozley, A. (1944). Temporary ponds, neglected natural resource. *Nature* **154**: 490.

Pearsall, W.H. (1921). The development of vegetation in the English Lakes, considered in relation to the general evolution of glacial lakes and rock basins. *Proceedings of the Royal Society B* **92**: 259–84.

Pichler, W. (1939). Unsere derzeitige Kenntnis van der Thermik Kleiner Gewässer Thermische Kleingewassertypen. *Internationale Revue der gesamten Hydrobiologie* **38**: 231–42.

Poff, N.L. and J.V. Ward. (1990). Physical habitat template of lotic systems: recovery in the context of historical pattern of spatiotemporal heterogeneity. *Environmental Management* **14**: 629–45.

Puckridge, J.T., F. Sheldon, K.F. Walker, and A.J. Boulton. (1998). Flow variability and the ecology of large rivers. *Marine and Freshwater Research* **49**: 55–72.

Rackham, O. (1986). *The History of the Countryside*. J.M. Dent and Sons Ltd., London.

Schmitt, W.L. (1971). *Crustaceans*. University of Michigan Press, Ann Arbor, MI.

Schram, F.R. (1986). *Crustacea*. Oxford University Press, Oxford.

Spray, S.L. and K.L. McGlothlin. (eds). (2004). *Wetlands*. Rowman and Littlefield Publishers, Lanham, MD.

Stemberger, R.S. (1995). Pleistocene refuge areas and postglacial dispersal of copepods of the northeastern United States. *Canadian Journal of Fisheries and Aquatic Sciences* **52**: 2197–210.

Styczynska-Jurewicz, E. (1966). Astatic-water bodies as characteristic habitat of some parasites of man and animals. *Verhandlungen des Internationalen Vereinigung für theoretische und angewandte Limnologie* **16**: 604–11.

Tasch, P. (1969). Branchiopoda. In *Treatise on Invertebrate Paleontology* (ed. R.C. Moore), *Part R. Arthropoda 4*, Vol. 1, R128–R191. Geological Society of America and the University of Kansas, Lawrence, Kansas.

Vincent, W.F. and C. Howard-Williams. (1986). Antarctic stream ecosystems: physiological ecology of a blue-green algal epilithon. *Freshwater Biology* **16**: 209–33.

Warner, B.G. and C.D.A. Rubec (eds). (1997). *The Canadian Wetland Classification System*. National Wetland Working Group, Wetland Research Centre, University of Waterloo, Ontario, Canada.

Way, C.M., D.J. Hornbach, and A.J. Burky. (1980). Comparative life history tactics of the sphaeriid clam, *Musculium partumeium* (Say), from a permanent and temporary pond. *American Midland Naturalist* **104**: 319–27.

Whigham, D., D. Dykyjová, and S. Hejny. (eds). (1993). *Wetlands of the World: Inventory, Ecology and Management*, Vol. 1. Kluwer Academic Publishers, Dordrecht.

Wiggins, G.B., R.J. Mackay, and I.M. Smith. (1980). Evolutionary and ecological strategies of animals in annual temporary pools. *Archiv für Hydrobiologie* **58** (Suppl.): 97–206.

Williams, D.D. (1987). *The Ecology of Temporary Waters*. Croom Helm, London.

Williams, M. (1993). *Wetlands: A Threatened Landscape*. Institute of British Geographers Special Publication. Blackwell Publishing, Oxford.

Williams, P., J. Biggs, G. Fox, P. Nicolet, and M. Whitfield. (2001). History, origins and importance of temporary ponds. *Freshwater Forum* **17**: 7–15.

Williams, W.D. (1964). A contribution to lake typology in Victoria, Australia. *Verhandlungen des Internationalen Vereinigung für theoretische und angewandte Limnologie* **15**: 158–68.

Williams, W.D. (1981). Inland salt lakes: an introduction. *Hydrobiologia* **81/82**: 1–14.

Williams, W.D. (1988). Limnological imbalances: an antipodean viewpoint. *Freshwater Biology* **20**: 407–20.

Yavercovski, N., P. Grillas, G. Paradis, and A. Thiéry. (2004). Biodiversity and conservation issues. In *Mediterranean Temporary Pools*. Vol. 1: *Issues Relating to Conservation, Functioning and Management* (eds P. Grillas, P. Gauthier, N. Yavercovski, and C. Perennou), pp. 11–33. Station Biologique de la Tour du Valat, La Sambuc, Arles, France.

Chapter 2

Amatali, M.A. (1993). Climate and surface water hydrology. In *Freshwater Ecosystems of Suriname* (ed. P.E. Ouboter), pp. 29–51. Kluwer Academic Publishers, Dordrecht.

Beaver, R.A. (1983). The communities living in *Nepenthes* pitcher plants: fauna and food webs. In *Phytotelmata: Terrestrial Plants as Hosts for Aquatic Insect Communities* (eds J.H. Frank and L.P. Lounibos), pp. 129–59. Plexus Publishing Inc., Medford, NJ.

Biggs, J., P.J. Williams, A. Corfield, M.A. Whitfield, C.J. Barr, and C.P. Cummins. (1996). *Pond Survey 1996. Stage I Scoping Study*. Pond Action and Institute for Terrestrial Ecology, Oxford.

Bowen, R. (1982). *Surface Water*. Wiley-Interscience, New York.

Bradley, R.S., H.F. Diaz, J.K. Eischeid, P.D. Jones, P.M. Kelly, and C.M. Goodess. (1987). Precipitation fluctuations over northern hemisphere land areas since the mid-19th century. *Science* **237**: 171–5.

Brooks, R.T. and M. Hayashi. (2002). Depth–area–volume and hydroperiod relationships of ephemeral (vernal) forect pools in southern New England. *Wetlands* **22**: 247–55.

Bull, L.J. and M.J. Kirkby. (2002). Dryland river characteristics and concept. In *Dryland Rivers: Hydrology and Geomorphology of the Semi-Arid Channels* (eds L.J. Bull and M.J. Kirkby), pp. 3–15. John Wiley & Sons, Chichester, UK.

Carson, M.A. and E.A. Sutton. (1971). The hydrologic response of the Eaton River basin, Quebec. *Canadian Journal of Earth Sciences* **8**: 102–15.

Davis, S.N. and R.J.M. DeWeist. (1966). *Hydrogeology*. John Wiley & Sons, New York.

Everard, M., R. Blackham, K. Rouen, W. Watson, A. Angell, and A. Hill. (1999). How do we raise the profile of ponds. *Freshwater Forum* **12**: 32–43.

Gooch, J.L. and D.S. Glazier. (1991). Temporal and spatial patterns in mid-Appalachian springs. *Memoirs of the Entomological Society of Canada* **155**: 29–49.

Horton, R.E. (1933). The role of infiltration in the hydrogeologic cycle. *American Geophysical Union Transactions* **14**: 446–60.

Jacobson, P.J., K.M. Jacobson, and M.K. Seely. (1995). *Ephemeral Rivers and their Catchments: Sustaining People and Development in Western Namibia*. Desert Research Foundation of Namibia, Windhoek.

Johnson, R. (1998). The forest cycle and low river flows: a review of U.K. and international studies. *Forest Ecology and Management* **109**: 1–7.

Larson, D.L. (1995). Effects of climate on numbers of Northern Prairie wetlands. *Climatic Change* **30**: 169–80.

Manabe, S. and R.T. Wetherald. (1986). Reduction in summer soil wetness induced by an increase in atmospheric carbon dioxide. *Science* **232**: 626–8.

Matthews, J.V. (1979). Tertiary and Quaternary environments: historical background for an analysis of the Canadian insect fauna. *Memoirs of the Entomological Society of Canada* **108**: 31–86.

Nanson, G.C., S. Tooth, and A.D. Knighton. (2002). A global perspective on dryland rivers: perceptions, misconceptions and distinctions. In *Dryland Rivers: Hydrology and Geomorphology of the Semi-Arid Channels* (eds L.J. Bull and M.J. Kirkby), pp. 17–54. John Wiley & Sons, Chichester, UK.

Odum, H.T. (1957). Trophic structure and productivity of Silver Springs, Florida. *Ecological Monographs* **27**: 55–112.

Poiani, K.A. and W.C. Johnson. (1991). Global warming and prairie wetlands. *BioScience* **41**: 611–18.

Reid, G. K. (1961). *Ecology of Inland Waters and Estuaries*. Van Nostrand Reinholt, New York.

Resh, V.H. (1982). Age structure alteration in caddisfly populations after habitat loss and recovery. *Oikos* **38**: 280–4.

Rex, R.W. (1961). Hydrodynamical analysis of circulation and orientation of lakes in northern Alaska. In *Geology of the Arctic* (ed. G.O. Raasch), pp. 1021–43. University of Toronto Press, Toronto.

Schindler, D.W., P. Jefferson Curtis, B.R. Parker, and M.P. Stainton. (1996). Consequences of climatic warming and lake acidification for Uvb penetration in North American boreal lakes. *Nature* **379**: 705–8.

Schneider, S.H. (1989). The greenhouse effect: science and policy. *Science* **243**: 771–81.

Shannon, J., R. Richardson, and J. Thornes. (2002). Modelling event-based fluxes in ephemeral streams.

In *Dryland Rivers: Hydrology and Geomorphology of the Semi-Arid Channels*, (eds L.J. Bull and M.J. Kirkby), pp. 127–72. John Wiley & Sons, Chichester, UK.

Taylor, C.C. (1972). Medieval moats in Cambridgeshire. In *Archeology and the Landscape* (ed. P.J. Fowler), pp. 237–49. J. Barker, London.

Tilly, L.J. (1968). The structure and dynamics of Cone Spring. *Ecological Monographs* **38**: 169–97.

Webb, D.W., M.J. Wetzel, P.C. Reed, L.R. Phillippe, and M.A. Harris. (1995). Aquatic biodiversity in Illinois springs. *Journal of the Kansas Entomological Society* **68**: 93–107.

Weitkamp, W.A., R.C. Graham, M.A. Anderson, and C. Amrhein. (1996). Pedogenesis of a vernal pool entisol–alfisol–vertisol catena in southern California. *Journal of the Soil Science Society of America* **60**: 316–23.

Williams, D.D. and I.D. Hogg. (1988). Ecology and production of invertebrates in a Canadian coldwater spring-springbrook system. *Holarctic Ecology* **11**: 41–54.

Williams, D.D. and H.B.N. Hynes. (1976a). The ecology of temporary streams I. The faunas of two Canadian streams. *Internationale Revue der gesamten Hydrobiologie* **61**: 761–87.

Williams, P., J. Biggs, G. Fox, P. Nicolet, and M. Whitfield. (2001). History, origins and importance of temporary ponds. *Freshwater Forum* **17**: 7–15.

Yan, N.D., W. Keller, N.M. Scully, D.R.S. Lean, and P.J. Dillon. (1996). Increased UV-B penetration in a lake owing to drought-induced acidification. *Nature* **381**: 141–3.

Chapter 3

Allan, S.A. and D.L. Kline. (1998). Larval rearing water and pre-existing eggs influence oviposition by *Aedes aegypti* and *A. albopictus* (Diptera: Culicidae). *Journal of Medical Entomology* **35**: 943–7.

Baird, D.J., L.R. Linton, and R.W. Davies. (1987). Life-history flexibility as a strategy for surviving in a variable environment. *Functional Ecology* **1**: 45–8.

Bärlocher, F., R.J. Mackay, and G.B. Wiggins. (1978). Detritus processing in a temporary vernal pool in southern Ontario. *Archive für Hydrobiologie* **81**: 269–95.

Batzer, D.P., R.B. Rader, and S.A. Wissinger. (eds). (1999). *Invertebrates in Freshwater Wetlands of North America*. John Wiley & Sons, New York.

Bazzanti, M., M. Seminara, and S. Baldoni. (1997). Chironomids (Diptera: Chironomidae) from three temporary ponds of different wet phase duration in central Italy. *Journal of Freshwater Ecology* **12**: 89–99.

Boix, D., J. Sala, and R. Moreno-Amich. (2001). The faunal composition of Espolla Pond (N.E. Iberian Peninsula): the neglected biodiversity of temporary waters. *Wetlands* **21**: 577–92.

Boulton, A.J. and P.J. Suter. (1986). Ecology of temporary streams—an Australian pespective. In *Limnology in Australia* (eds P. DeDeckker and W.D. Williams), pp. 131–27. CSIRO/Dr. W. Junk, Melbourne/Dordrecht.

Boulton, A.J., C.G. Peterson, N.B. Grimm, and S.G. Fisher. (1992). Stability of an aquatic macroinvertebrate community in a multiyear hydrologic disturbance regime. *Ecology* **73**: 2192–207.

Brönmark, C. and L.-A. Hanson. (1998). *The Biology of Ponds and Lakes*. Oxford University Press, Oxford.

Busnardo, M.J., R.M. Gersberg, R. Langis, T.L. Sinicrope, and J.B. Zedler. (1992). Nitrogen and phosphorus removal by wetland mesocosm subjected to different hydroperiods. *Ecological Engineering* **1**: 287–307.

Butler, J.L. (1963). Temperature relations in shallow ponds. *Proceedings of the Oklahoma Academy of Science* **43**: 90–5.

Cameron, G.N. and T.W. LaPoint. (1978). Effects of tannins on the decomposition of Chinese tallow leaves by terrestrial and aquatic invertebrates. *Oecologia* **32**: 349–66.

Cole, G.A. (1968). Desert limnology. In *Desert Biology* (ed. G.W. Brown), pp. 423–86. Academic Press, New York.

Corti, D., S.L. Kohler, and R.E. Sparks. (1997). Effects of hydroperiod and predation on a Mississippi River floodplain invertebrate community. *Oecologia* **109**: 154–65.

Daborn, G.R. and H.F. Clifford. (1974). Physical and chemical features of an aestival pond in western Canada. *Hydrobiologia* **44**: 43–59.

Davidsson, T.E. and L. Leonardson. (1998). Seasonal dynamics of denitrification activity in two water meadows. *Hydrobiologia* **364**: 189–98.

Dent, C.L., J.D. Schade, N.B. Grimm, and S.G. Fisher. (2000). Subsurface influences on surface biology. In *Streams and Ground Waters* (eds J.B. Jones and P.J. Mulholland), pp. 381–402. Academic Press, Boston, MA.

Dick, J.T.A., S.E. Faloon, and R.W. Elwood. (1998). Active brood care in an amphipod: influences of embryonic development, temperature and oxygen. *Animal Behaviour* **56**: 663–72.

Dieterich, M. and N.H. Anderson. (1998). Dynamics of abiotic parameters, solute removal and sediment retention in summer dry headwater streams of western Oregon. *Hydrobiologia* **379**: 1–15.

Duff, J.H. and F.J. Triska. (2000). Nitrogen biogeochemistry and surface–subsurface exchange in streams. In *Streams and Ground Waters* (eds J.B. Jones and P.J. Mulholland), pp. 197–220. Academic Press, Boston, MA.

Eisenberg, J.N., W.K. Reisen, and R.C. Spear. (1995). Dynamic model comparing the bionomics of two isolated *Culex tarsalis* (Diptera: Culicidae) populations: sensitivity analysis. *Journal of Medical Entomology* **32**: 98–106.

Eriksen, C.H. (1966). Diurnal limnology of two highly turbid puddles. *Verhandlungen des Internationalen Vereinigung für theoretische und angewandte Limnologie* **16**: 507–14.

Esteban, G., A. Baltanás, and B. Finlay. (2002). Saline ponds and secretive ciliates. *Planet Earth*. UK Natural Environment Research Council, Autumn: 10–11.

Feminella, J.W. (1996). Comparison of benthic macroinvertebrate assemblages in small streams along a gradient of flow permanence. *Journal of the North American Benthological Society* **15**: 651–69.

Fernando, C.H. (1958). The colonisation of small freshwater habitats by aquatic insects. I. General discussion, methods and colonisation in the aquatic Coleoptera. *Ceylon Journal of Science (Biological Sciences)* **1**: 117–54.

George, M.G. (1961). Diurnal variations in two shallow ponds in Delhi, India. *Hydrobiologia* **18**: 265–73.

Grimm, N.B., S.G. Fisher, and W.L. Minckley. (1981). Nitrogen and phosphorus dynamics in hot desert streams of the southwestern USA. *Hydrobiologia* **83**: 303–12.

Guiral, D., R. Arfi, M. Bouvy, M. Pagano, and L. Saint-Jean. (1994). Ecological organization and succession during natural recolonization of a tropical pond. *Hydrobiologia* **294**: 229–42.

Harrison, R.G. (1980). Dispersal polymorphisms in insects. *Annual Review of Ecology and Systematics* **11**: 95–118.

Herbst, D.B. (1988). Comparative population ecology of *Ephydra hians* Say (Diptera: Ephydridae) at Mono Lake (California) and Albert Lake (Oregon). *Hydrobiologia* **158**: 145–66.

Hershey, A.E., L. Shannon, G.J. Niemi, A.R. Lima, and R.R. Regal. (1999). Prairie wetlands of south-central Minnesota: effects of drought on invertebrate communities. In *Invertebrates in Freshwater Wetlands of North America* (eds D.P. Batzer, R.B. Rader, and S.A. Wissinger), pp. 515–41. John Wiley & Sons, New York.

Hynes, H.B.N. (1975). The stream and its valley. *Internationale Vereinigung für theoretische und angewandte Limnologie* **19**: 1–15.

Jeffries, M.J. (1994). Invertebrate communities and turnover in wetland ponds affected by drought. *Freshwater Biology* **32**: 603–12.

Jeffries, M.J. (2001). Modeling the incidence of temporary pond microcrustacea: the importance of dry phase and linkage between ponds. *Israel Journal of Zoology* **47**: 445–58.

Jensen, T., D.A. Carlson, and D.R. Barnard. (1999). Factor from swamp water induces hatching in eggs of *Anopheles diluvialis* (Diptera: Culicidae) mosquitoes. *Environmental Entomology* **28**: 545–50.

Johansson, F. (1993). Intraguild predation and cannibalism in odonate larvae: effects of foraging behaviour and zooplankton availability. *Oikos* **66**: 80–7.

Juliano, S.A. and T.L. Stoffregren. (1994). Effects of habitat drying on size at, and time to, metamorphosis in the tree hole mosquito *Aedes triseriatus*. *Oecologia* **97**: 369–76.

Kaushik, N.K. and H.B.N. Hynes. (1968). Experimental study on the role of autumn-shed leaves in aquatic environments. *Journal of Ecology* **56**: 229–43.

Lake, P.S. (1995). Of flood and droughts: river and stream ecosystems of Australia. In *Ecosystems of the World 22: River and Stream Ecosystems* (eds C.E. Cushing, K.W. Cummins, and G.W. Minshall), pp. 659–94. Elsevier, Amsterdam.

Landin, J. (1980). Habitats, life histories, migration, and dispersal by flight of two water beetles, *Helophorus brevipalpis* and *H. strigifrons* (Hydrophilidae). *Holarctic Ecology* **3**: 190–201.

Laurila, A. and J. Kujasalo. (1999). Habitat duration, predator risk and phenotypic plasticity in common frog (*Rana temporaria*) tadpoles. *Journal of Animal Ecology* **68**: 1123–32.

Magnusson, A.K. and D.D. Williams. (2006). The relative role of natural temporal and spatial variation versus biotic influences in shaping the physicochemical environment of intermittent ponds—a case study *Archiv für Hydrobiologie* (in press).

Maher, M. and S.M. Carpenter. (1984). Benthic studies of waterfowl breeding habitats in south-western New South Wales. II. Chironomid populations. *Australian Journal of Marine and Freshwater Research* **35**: 97–110.

McLachlan, A.J., P.R. Morgan, C. Howard-Williams, S.M. McLachlan, and D. Bourn. (1972). Aspects of the recovery of a saline African lake following a dry period. *Archive für Hydrobiologie* **70**: 325–40.

Merritt, R.W. and D.L. Lawson. (1981). Adult emergence patterns and species distribution and abundance of Tipulidae in three woodland floodplains. *Environmental Entomology* **10**: 915–21.

Merritt, R.W. and D.L. Lawson. (1992). The role of leaf litter macroinvertebrates in stream–floodplain dynamics. *Hydrobiologia* **248**: 65–77.

Morin, P.J., S.P. Lawler, and E.A. Johnson. (1988). Competition between aquatic insects and vertebrates: interaction strength and higher order interactions. *Ecology* **69**: 1401–9.

Nilsson, A.N. and B.W. Svensson. (1994). Dytiscid predators and culicid prey in two boreal snowmelt pools differing in temperature and duration. *Annales Zoologiae Fennici* **31**: 365–76.

Paradise, C.J. and W.A. Dunson. (1998). Relationship of atmospheric deposition to the water chemistry and biota of treehole habitats. *Environmental Toxicology and Chemistry* **17**: 362–8.

Podrabsky, J.E., T. Hrbek, and S.C. Hand. (1998). Physical and chemical characteristics of ephemeral pond habitats in the Maracaibo basin and Llanos region of Venezuela. *Hydrobiologia* **362**: 67–78.

Poff, N.L. and J.V. Ward. (1989). Implications of streamflow variability and predictability for lotic community structure: a regional analysis of streamflow patterns. *Canadian Journal of Fisheries and Aquatic Sciences* **46**: 1805–18.

Schneider, D.W. and T.M. Frost. (1996). Habitat duration and community structure in temporary ponds. *Journal of the North American Benthological Society* **15**: 64–86.

Schneller, M.V. (1955). Oxygen depletion in Salt Creek, Indiana. *Investigations on Indiana Lakes and Streams* **4**: 163–75.

Sposito, G. (1989). *The Chemistry of Soils*. Oxford University Press, Oxford.

Stearns, S.C. (1976). Life-history tactics: review of the ideas. *Quarterly Review of Biology* **51**: 3–47.

Suberkropp, K. and M.J. Klug. (1976). Fungi and bacteria associated with leaves during processing in a forest stream. *Ecology* **57**: 707–19.

Suter, P.J., P.M. Goonan, J.A. Beer, and T.B. Thompson. (1994). The response of chironomid populations to flooding and drying in floodplain wetlands of the Lower River Murray in South Australia. In *Chironomids: From Genes to Ecosystems* (ed. P.S. Cranston), pp. 185–97. C.S.I.R.O. Publications, Melbourne.

Tabacchi, E., H. Décamps, and A. Thomas. (1993). Substrate interstices as a habitat for larval *Thraulus bellus* (Ephemeroptera) in a temporary floodplain pond. *Freshwater Biology* **29**: 429–39.

Therriault, T.W. and J. Kolasa. (2001). Desiccation frequency reduces species diversity and predictability of community structure in coastal rock pools, Israel. *Journal of Zoology* **47**: 477–89.

Tibby, J., K. Sabbe, and N. Roberts. (2000). Effects of depth, salinity, and substrate on the invertebrate community of a fluctuating tropical lake. *Ecology* **81**: 164–82.

Tugel, A.J. and A.M. Lewandowski. (eds). (1999). *Soil Biology*. Natural Resources Conservation Service, Soil Quality Institute, Ames, Iowa.

Vaas, K.F. and M. Sachlan. (1955). Limnological studies on diurnal fluctuations in shallow ponds in Indonesia. *Verhandlungen des Internationalen Vereinigung für theoretische und angewandte Limnologie* **12**: 309–19.

Valett, H.M., S.G. Fisher, N.B. Grimm, and P. Camill. (1994). Vertical hydrologic exchange and ecological stability of a desert stream ecosystem. *Ecology* **75**: 548–60.

Webster, J.R., J.B. Wallace, and E.F. Benfield. (1995). Organic processes in streams of the eastern United States. In *Ecosystems of the World 22: River and Stream Ecosystems* (eds C.E. Cushing, K.W. Cummins, and G.W. Minshall), pp. 117–87. Elsevier, Amsterdam.

Wegis, M.C., W.E. Bradshaw, T.E. Davison, and C.M. Holzapfel. (1997). Rhythmic components of photoperiodic time measurement in the pitcher-plant mosquito, *Wyeomia smithii*. *Oecologia* **110**: 32–9.

Whitney, R.J. (1942). Diurnal fluctuations of oxygen and pH in two small ponds and a stream. *Journal of Experimental Biology* **19**: 92–9.

Williams, D.D. (1983). The natural history of a nearctic temporary pond in Ontario with remarks on continental variation in such habitats. *Internationale Revue der gesamten Hydrobiologie* **68**: 239–54.

Williams, D.D. (1996). Environmental constraints in temporary fresh waters and their consequences for the insect fauna. *Journal of the North American Benthological Society* **15**: 634–50.

Williams, D.D. (2002). Temporary water crustaceans: biodiversity and habitat loss. In *Modern Approaches to the Study of Crustacea* (eds E. Escobar-Briones and F. Alvarez), pp. 223–33. Kluwer Academic/Plenum Publishers, New York.

Williams, W.D., A.J. Boulton, and R.G. Taaffe. (1990). Salinity as a determinant of a salt lake fauna: a question of scale. *Hydrobiologia* **197**: 257–66.

Young, F.N. and J.R. Zimmerman. (1956). Variations in temperature in small aquatic situations. *Ecology* **37**: 609–11.

Zamoramunoz, C. and B.W. Svensson. (1996). Survival of caddis larvae in relation to their case material in a group of temporary and permanent pools. *Freshwater Biology* **36**: 23–31.

Chapter 4

Abbasi, S.A. (1997). *Wetlands of India: Ecology and Threats. Vol.1. The Ecology and the Exploitation of Typical South Indian Wetlands.* Discovery Publishing House, New Delhi.

Abbasi, S.A. and P.K. Mishra. (1997). *Wetlands of India: Ecology and Threats. Vol. 2. Asia's Largest Lake (Chilka) Ecology, Threats and Imperatives.* Discovery Publishing House, New Delhi.

Abbasi, S.A., N. Abbasi, and K.K.S. Bhatia. (1997). *Wetlands of India: Ecology and Threats. Vol. 3. The Wetlands of Kerala.* Discovery Publishing House, New Delhi.

Alcocera, J., A. Lugo, E. Escobar, and M. Sánchez. (1997). The macrobenthic fauna of a former perennial and now episodically filled Mexican saline lake. *International Journal of Salt Lake Research* **5**: 261–74.

Alcocera-Durand, J. and E.G. Escobar-Briones. (1992). The aquatic biota of the now extinct lacustrine complex of the Mexico Basin. *Freshwater Forum* **1**: 171–83.

Alkins-Koo, M. (1989/90). The aquatic fauna of two intermittent streams in the southwestern peninsula, Trinidad. *Living World Journal of the Trinidad and Tobago Field Naturalists' Club* **1989–1990**: 36–42.

Allan, D.G., M.T. Seaman, and B. Kaletja. (1995). The endorheic pans of South Africa. In *Wetlands of South Africa* (ed. G.I. Cowan), pp. 75–102. Department of Environmental Affairs and Tourism, Pretoria.

Andrushchyshyn, O., A.K. Magnusson, and D.D. Williams. (2003). Ciliate populations in temporary freshwater ponds: seasonal dynamics and influential factors. *Freshwater Biology* **48**: 548–64.

Aragno, M. and B. Ulehlova. (1997). Microbial diversity and functions in land–inland water ecotones. In *Biodiversity in Land–Inland Water Ecotones* (eds J.-B. Lachavanne and R. Juge), pp. 81–108. Parthenon Publishing Group, New York.

Arts, M.T., E.J. Maly, and M. Pasitschniak. (1981). The influence of *Acilius* (Dytiscidae) predation on *Daphnia* in a small pond. *Limnology and Oceanography* **26**: 1172–5.

Ayeni, J.S.O. (1977). Waterholes in Tsavo National Park, Kenya. *Journal of Applied Ecology* **14**: 369–78.

Bagnold, R.A. (1954). The physical aspects of dry deserts. In *Biology of Deserts* (ed. J.C. Cloudsley-Thompson), pp. 7–12. Institute of Ecology, London.

Ball, I.R., N. Gourbault, and R. Kenk. (1981). The planarians (Turbellaria) of temporary waters in eastern North America. *Life Sciences Contributions, Royal Ontario Museum* **127**: 1–27.

Bambach, R.K., C.R. Scotese, and A.M. Ziegler. (1980). Before Pangea: the geographies of the Paleozoic World. *American Scientist* **68**: 26–38.

Bärlocher, F. (1985). The role of fungi in the nutrition of stream invertebrates. *Botanical Journal of the Linnean Society* **91**: 83–94.

Barlocher, F., R.J. Mackay, and G.B. Wiggins. (1978). Detritus processing in a temporary vernal pool in southern Ontario. *Archiv für Hydrobiologie* **81**: 269–95.

Basta, J. (1997). Contribution to knowledge of the ground-beetle fauna (Coleoptera: Carabidae) in the environs of Brno. *Acta Museum Moraviae Scientiarum Biologica* **82**: 207–13.

Batzer, D.P., R.B. Rader, and S.A. Wissinger. (eds). (1999). *Invertebrates in Freshwater Wetlands of North America.* John Wiley & Sons, New York.

Bauder, E., D.A. Kraeger, and S.C. McMillan. (1998). *Vernal Pools of Southern California; Recovery Plan.* US Fish and Wildlife Service, Portland, OR.

Bayly, I.A.E. (1969). The body fluids of some centropagid copepods: total concentration and amounts of sodium and magnesium. *Comparative Biochemistry and Physiology* **28**: 1403–9.

Bayly, I.A.E. (1982). Invertebrate fauna and ecology of temporary pools on granite outcrops in southwestern Australia. *Australian Jounal of Marine and Freshwater Research* **33**: 599–606.

Bazzanti, M., S. Baldoni, and M. Seminara. (1996). Invertebrate macrofauna of a temporary pond in Central Italy; composition, community parameters and temporal succession. *Archiv für Hydrobiologie* **137**: 77–94.

Beaver, R.A. (1985). Geographical variation in food web structure in *Nepenthes* pitcher plants. *Ecological Entomology* **10**: 241–8.

Bechara, J.A. (1996). The relative importance of water quality, sediment composition and floating vegetation in explaining the macrobenthic community structure of floodplain lakes (Parana River, Argentina). *Hydrobiologia* **333**: 95–109.

Belk, D. (1981). Patterns in anostracan distribution. In *Vernal Pools and Intermittent Streams* (eds S. Jain and P. Moyle), pp. 168–72. Institute of Ecology, University of California, Davis, Publication Number 28.

Benenati, P., J.P. Shannon, and D.W. Blinn. (1998). Desiccation and recolonization of phytobenthos in a regulated desert river: Colorada River at Lee's Ferry, Arizona, USA. *Regulated Rivers—Research and Management* **14**: 519–32.

Benson, R.H. (1961). Ecology of ostracode assemblages. In *Treatise on Invertebrate Paleontology. Part Q: Arthropoda 3, Crustacea, Ostracoda.* pp. 56–63. Geological

Society of America. University of Kansas Press, Lawrence.

Bentley, P.J. (1966). Adaptations of Amphibia to arid environments. *Science* **152**: 619–23.

Bilton, D.T. (1988). A survey of aquatic Coleoptera in central Ireland and in the Burren. *Bulletin of the Irish Biogeographical Society* **11**: 77–94.

Blackstock, T.H., C.A. Duigan, D.P. Stevens, and M.J.M. Yeo. (1993). Vegetation zonation and invertebrate fauna in Pant-y-llyn, an unusual seasonal lake in South Wales, UK. *Aquatic Conservation: Marine and Freshwater Ecosystems* **3**: 253–68.

Blanc, M., J. Daget, and F. D'Aubenton. (1955). Recherches hydrobiologiques dans le bassin du Moyen–Niger. Bulletin d'Institute Francais Afrique Noire (A) **17**: 619–746.

Blaustein, L., J. Friedman, and T. Fahima. (1996). Larval *Salamandra* drive temporary pool community dynamics: evidence from an artificial pool experiment. *Oikos* **76**: 392–402.

Blom, C. (1999). Adaptations to flooding stress: from plant community to molecule. *Plant Biology* **1**: 261–73.

Boix, D., J. Sala, and R. Moreno-Amich. (2001). The faunal composition of Espolla Pond (N.E. Iberian Peninsula): the neglected biodiversity of temporary waters. *Wetlands* **21**: 577–92.

Bonecker, C.C. and F.A. Lansactoha. (1996). Community structure of rotifers in two environments of the upper River Parana floodplain: Brazil. *Hydrobiologia* **325**: 137–50.

Bonetto, A.A. (1975). Hydraulic regime of the Parana River and its influence on ecosystems. *Ecological Studies* **10**: 175–97.

Bonetto, A.A., C. Pignelberi, and E. Cordoviola. (1969). Limnological investigations on biotic communities in the Middle Arana River Valley. *Verhandlungen Internationale Vereinigung für Theoretische und Angewandte Limnologie* **17**: 1035–50.

Borowitzka, L.J. (1981). The microflora: adaptations to life in extremely saline lakes. In *Salt Lakes* (ed. W.D. Williams), pp. 33–46. Junk, The Hague.

Boulton, A.J. and P.S. Lake. (1992a). The macroinvertebrate assemblages in pools and riffles in two intermittent streams (Werribee and Lerderderg Rivers, southern central Victoria). *Occasional Papers of the Museum of Victoria* **5**: 55–71.

Boulton, A.J. and P.S. Lake. (1992b). Benthic organic matter and detritivorous macroinvertebrates in two intermittent streams in southeastern Australia. *Hydrobiologia* **241**: 107–18.

Boulton, A.J. and P.J. Suter. (1986). Ecology of temporary streams—an Australian perspective. In *Limnology in Australia* (eds P. De Deckker and W.D. Williams), pp. 313–27. CSIRO/Dr. W. Junk, Melbourne/Dordrecht.

Boulton, A.J., C.G. Peterson, N.B. Grimm, and S.G. Fisher. (1992). Stability of an aquatic macroinvertebrate community in a multiyear hydrological disturbance regime. *Ecology* **73**: 2192–207.

Bragg, A.N. (1944). Breeding habits, eggs and tadpoles of *Scaphiopus hurterii*. *Copeia* **1944**: 230–41.

Bratton, J.H. (1990). Seasonal pools. An overlooked invertebrate habitat. *British Wildlife* **2**: 22–9.

Bratton, J.H. and G. Fryer. (1990). The distribution and ecology of *Chirocephalus diaphanus* Prévost (Branchiopoda: Anostraca) in Britain. *Journal of Natural History* **24**: 955–64.

Bretschko, G. (1973). Benthos production of a high-mountain lake: nematoda. *Verhandlungen des Instituts für Vereinforschung in Limnologica* **18**: 1421–8.

Brinck, P. (1949). Studies on Swedish stoneflies. *Opuscula Entomologica (Supplementum)* **11**: 1–250.

Brinkhurst, R.O. and M.J. Austin. (1978). Assimilation by aquatic oligochaetes. *Internationale Revue der gesamten Hydrobiologie* **63**: 863–8.

Brinkhurst, R.O. and S.R. Gelder. (1991). Annelida: Oligochaeta and Branchiobdellida. In *Ecology and Classification of North American Freshwater Invertebrates* (eds J.H. Thorp and A.P. Covich), pp. 401–35. Academic Press, New York.

Brown, B.T., L.E. Stevens, and T.A. Yates. (1998). Influences of fluctuating river flows on bald eagle foraging behaviour. *Condor* **100**: 745–8.

Brown, K.M. (1991). Mollusca: Gastropoda. In *Ecology and Classification of North American Freshwater Invertebrates* (eds J.H. Thorp and A.P. Covich), pp. 285–314. Academic Press, New York.

Brtek, J. (1974). Zwei *Streptocephalus* Arten aus Afrika undeinige notizen zur gattung *Streptocephalus*. Annotationes Zoologicae Botanicae, Bratislava **96**: 1–9.

Bunn, S.E. and P.M. Davies. (1992). Community structure of the macroinvertebrate fauna and water quality of a saline river system in southwestern Australia. *Hydrobiologia* **248**: 143–60.

Burky, A.J. (1983). Physiological ecology of freshwater bivalves. In *The Mollusca* (ed. W.D. Russell-Hunter), pp. 281–327. Academic Press, New York.

Burky, A.J., D.J. Hornbach, and C.M. Way. (1985). Comparative bioenergetics of permanent and temporary pond populations of the freshwater clam, *Musculium partumeium* (Say). *Hydrobiologia* **126**: 35–48.

Bushdosh, M. (1981). Behavioural adaptations to spatially intermittent streams by the longfin dace, *Agosia chrysogaster* (Cyprinidae). In *Vernal Pools and Intermittent Streams* (eds S. Jain and P. Moyle), pp. 69–75. Institute of Ecology, University of California, Davis, Publication Number 28.

Byrne, R.A. (1981). *Ecological Comparisons of Three Water Bodies in the Burren District of County Clare*. M.Sc. Thesis, University of Dublin.

Camarasa Belmonte, A.M. and F. Segura Beltrán. (2001). Flood events in Mediterranean ephemeral streams (ramblas) in Valencia region, Spain. *Catena* **45**: 229–49.

Campbell, S., J. Gunn, and P. Hardwick. (1992). Pant-y-llyn—the first Welsh turlough? *Earth Science Conservation* **31**: 3–7.

Carey, T.G. (1967). Some observations on distribution and abundance of the invertebrate fauna. *Fisheries Research Bulletin, Zambia* **3**: 22–4.

Case, T.J. and R.K. Washino. (1979). Flatworm control of mosquito larvae in rice fields. *Science* **206**: 1412–14.

Caspers, H. and C.W. Heckman. (1981). Ecology of orchard drainage ditches along the freshwater section of the Elbe Estuary. Biotic succession and influences of changing agricultural methods. *Archiv für Hydrobiologie* **43**: (Suppl.): 347–486.

Castillo, M.M. (2000). Influence of hydrological seasonality on bacterioplankton in two neotropical floodplain lakes. *Hydrobiologia* **437**: 57–69.

Closs, G.P. and P.S. Lake. (1994). Spatial and temporal variation in the structure of an intermittent-stream food web. *Ecological Monographs* **64**: 1–21.

Cole, G.A. (1966). Contrasts among calanoid copepods from permanent and temporary ponds in Arizona. *American Midland Naturalist* **76**: 351–68.

Cole, G.A. (1968). Desert limnology. In *Desert Biology* (ed. G.W. Brown), pp. 423–86. Academic Press, New York.

Colless, D.H. (1957). Notes on the culicine mosquitoes of Singapore. III. Larval breeding places. *Annals of Tropical Medicine and Parasitology* **51**: 102–16.

Collins, F.H. and R.K. Washino. (1979). Factors affecting the density of Culex tarsalis and Anopheles freeborni in northern California rice fields. *Proceedings of the California Mosquito Control Association* **46**: 97–8.

Collins, T.W. (1967). Oxygen-uptake, shell morphology and desiccation of the fingernail clam, *Sphaerium occidentale* Prime. Ph.D. Thesis, University of Minnesota, Minneapolis, MN.

Collinson, N.H., J. Biggs, A. Corfield, M.J. Hodson, D. Walker, M. Whitfield, and P.J. Williams. (1995). Temporary and permanent ponds: an assessment of the effects of drying out on the conservation value of aquatic macroinvertebrate communities. *Biological Conservation* **74**: 125–33.

Conrad, R. (1993). Mechanisms controlling methane emission from wetland rice fields. In *Biogeochemistry of Global Change: Radiatively Active Trace Gases* (ed. R.S. Oremland), pp. 336–52. Chapman and Hall, New York.

Convey, P. and W. Block. (1996). Antarctic Diptera: ecology, physiology and distribution. *European Journal of Entomology* **93**: 1–13.

Corbet, P.S. (1983). Odonata in phytotelmata. In *Phytotelmata: Terrestrial Plants as Hosts for Aquatic Insect Communities* (eds J.H. Frank and L.P. Lounibos), pp. 29–54. Plexus Publications Incorporated, Medford, NJ.

Courtney, W.R. and J.D. Williams. (2005). Snakeheads (Pisces, Channidae): a biological synopsis and risk assessment. US Geological Survey. http://fisc.er.usgs.gov/Snakehead_circ_1251/html/abstract.html.

Covich, A.P. and J.H. Thorp. (1991). Crustacea: introduction and Peracarida. In *Ecology and Classification of North American Freshwater Invertebrates* (eds J.H. Thorp and A.P. Covich), pp. 665–89. Academic Press, New York.

Coxon, C.E. (1987). An examination of the characteristics of turloughs, using multivariate statistical techniques. *Irish Geography* **20**: 24–42.

Cronin, G., K.D. Wissing, and D.M. Lodge. (1998). Comparative feeding selectivity of herbivorous insects on water lilies: aquatic vs. semiterrestrial insects and submerged vs. floating leaves. *Freshwater Biology* **39**: 243–57.

Crowe, J. (1975). The physiology of cryptobiosis in tardigrades. *Memoire dell'Istituto Italiano di Idrobiologia* **32**: 37–59.

Curtis, B.A. (1991). Freshwater macroinvertebrates of Namibia. In *The Status and Conservation of Wetlands in Namibia* (eds R.E. Simmons, C.J. Brown, and M. Griffin). *Madoqua* **17**: 163–87.

Curtis, B.A., K.S. Roberts, M. Griffin, S. Bethune, C.J. Hay, and H. Kolberg. (1998). Species richness and conservation of Namibian freshwater macroinvertebrates, fish and amphibians. *Biodiversity and Conservation* **7**: 447–66.

Dana, G.L. (1981). *Artemia* in temporary alkaline ponds near Fallon, Nevada with a comparison of its life history strategies in temporary and permanent habitats. In *Vernal Pools and Intermittent Streams* (eds S. Jain and P. Moyle), pp. 115–25. Institute of Ecology, University of California, Davis, Publication Number 28.

Dance, K.W. and H.B.N. Hynes. (1979). A continuous study of the drift in adjacent intermittent and permanent streams. *Archiv für Hydrobiologie* **87**: 253–61.

Danks, H.V. (1990). Arctic insects: instructive diversity. In *Canada's Missing Dimension: Science and History in the Canadian Arctic Islands* (ed. C.R. Harrington), pp. 444–70. Canadian Museum of Nature, Ottawa.

Davis, G.M. (1982). Historical and ecological factors in the evolution, adaptive radiation, and biogeography of freshwater molluscs. *American Zoologist* **22**: 375–95.

Davies, R.W. (1991). Annelida: leeches, polychaetes, and acanthobdellids. In *Ecology and Classification of North American Freshwater Invertebrates* (eds J.H. Thorp and A.P. Covich), pp. 437–79. Academic Press, New York.

Davies, R.W. and T.B. Reynoldson. (1976). A comparison of the life-cycle of *Helobdella stagnalis* (Linn. 1758) (Hirudinoidea) in two different geographical areas in Canada. *Journal of Animal Ecology* **45**: 457–70.

Delorme, L.D. (1991). Ostracoda. In *Ecology and Classification of North American Freshwater Invertebrates* (eds J.H. Thorp and A.P. Covich), pp. 691–722. Academic Press, New York.

Delucchi, C.M. and B.L. Peckarsky. (1989). Life history patterns of insects in an intermittent and a permanent stream. *Journal of the North American Benthological Society* **8**: 308–21.

del Rosario, R. B. and V.H. Resh. (2000). Invertebrates in intermittent and perennial streams: is the hyporheic zone a refuge from drying? *Journal of the North American Benthological Society* **19**: 680–96.

Dieterich, M. and N.H. Anderson. (2000). The invertebrate fauna of summer-dry streams in western Oregon. *Archiv für Hydrobiologie* **147**: 273–95.

Driver, E.A. (1977). Chironomid communities in small prairie ponds: some characteristics and controls. *Freshwater Biology* **7**: 121–33.

Ducharme, A. (1975). Informe tecnico de biologia pesquera (Limnologia). *Pubblicazione Progress Dessarrollo Pesca Contribucion* INDERENA/FAO, Columbia **4**: 1–42.

Duigan, C. and D.G. Frey. (1987). *Eurycercus glacialis* in Ireland (Cladocera, Chydoridae). *Internationale Revue gesamten Hydrobiologie* **72**: 235–49.

Ebert, T.A. and M.L. Balko. (1987). Temporary pools as islands in space and time: the biota of vernal pools in San Diego, Southern California, USA. *Archiv für Hydrobiologie* **110**: 101–23.

Eckroth, M.C. and R.O. Brinkhurst. (1996). Tenagodrilus musculus, a new genus and species of Lumbiculidae (Clitellata) from a temporary pond in Alabama. *Hydrobiologia* **334**: 1–9.

Eder, E., W. Hödl, and R. Gottwald. (1997). Distribution and phenology of large branchiopods in Austria. *Hydrobiologia* **359**: 13–22.

Edwards, J.A. (1980). An experimental introduction of vascular plants to the maritime Antarctic. *British Antarctic Survey Bulletin* **49**: 73–80.

Ehrenfeld, D.W. (1970). *Biological Conservation.* Holt, Rinehart and Winston, Toronto.

Elam, D.R. (1998). Population genetics of vernal pool plants: theory, data and conservation implications. In *Ecology, Conservation, and Management of Vernal Pool Ecosystems* (eds C.W. Witham, E.T. Bauder, D. Belk, W.R. Ferren, and R. Ornduff), pp. 180–9. California Native Plant Society, Sacramento, CA.

Erman, N.A. and D.C. Erman. (1995). Spring permanence, Trichoptera species richness, and the role of drought. *Journal of the Kansas Entomological Society* **68**: 50–64.

Euliss, N.H., D.A. Wrubleski, and D.A. Mushet. (1999). Wetlands of the Prairie Pothole Region: invertebrate species composition, ecology, and management. In *Invertebrates in Freshwater Wetlands of North America* (eds D.P. Batzer, R.B. Rader, and S.A. Wissinger), pp. 471–514. John Wiley & Sons, New York.

Fahd, K., L. Serrano, and J. Toja. (2000). Crustacean and rotifer composition of temporary ponds in the Donana National Park (SW Spain) during floods. *Hydrobiologia* **436**: 41–9.

Felsenstein, J. (1976). The theoretical population genetics of variable selection and migration. *Annual Review of Genetics* **10**: 253–80.

Felton, M., J.J. Cooney, and W.G. Moore. (1967). A quantitative study of the bacteria of a temporary pond. *Journal of General Microbiology* **47**: 25–31.

Fenchel, T. (1975). The quantitative importance of the benthic microfauna of an Arctic tundra pond. *Hydrobiologia* **46**: 445–64.

Fittkau, E.J. (1975). Productivity biomass and population dynamics in Amazonian water bodies. *Ecological Studies* **11**: 289–311.

Foissner, W. (1998). An updated compilation of world soil ciliates (Protozoa: Ciliophora), with ecological notes, new records, and descriptions of new species. *European Journal of Protistology* **34**: 195–235.

Fox, I. and I. Garcia-Moll. (1962). *Echiniscus molluscorum,* new tardigrade from the faeces of the land snail, *Bulimulus exilis* (Gamlin) in Puerto Rico (Tardigrada: Scutechiniscidae). *Journal of Parasitology* **48**: 177–181.

Frith, H.J. (1959). Ecology of wild ducks in inland Australia. In *Biogeography and Ecology in Australia* (eds A. Keast, R.L. Crocker, and C.S. Christian), pp. 383–95. Monographiae Biologicae 3, Junk, The Hague.

Fritz, K.M. and W.K. Dodds. (2002). Macroinvertebrate assemblage structure across a tallgras prairie stream landscape. *Archiv für Hydrobiologie* **154**: 79–102.

Frost, T.M. (1978). The impact of the freshwater sponge *Spongilla lacustris* on a sphagnum bog pond. *Verhandlungen Internationale Vereinigung für Theoretische und Angewandte Limnologie* **20**: 2368–71.

Frost, T.M. (1991). Porifera. In *Ecology and Classification of North American Freshwater Invertebrates* (eds J.H. Thorp and A.P. Covich), pp. 95–124. Academic Press, New York.

Fryer, G. (1996). Diapause, a potent force in the evolution of freshwater crustaceans. *Hydrobiologia* **320**: 1–14.

Fukuda, T., O.R. Willis, and D.R. Barnard. (1997). Parasites of the Asian tiger mosquito and other container-inhabiting mosquitoes (Diptera: Culicidae) in northcentral Florida. *Journal of Medical Entomology* **34**: 226–33.

Fumanti, B., S. Alfinito, and P. Cavacini. (1995). Floristic studies on freshwater algae of Lake Gondwana, Northern Victoria Land (Antarctica). *Hydrobiologia* **316**: 81–90.

Gafny, S. and A. Gasith. (1999). Spatially and temporally sporadic appearance of macrophytes in the littoral zone of Lake Kinneret, Israel: taking advantage of a window of opportunity. *Aquatic Botany* **62**: 249–67.

Garcia-Gil, L.J., C.M. Borrego, X. Vila, L. Baneras, and J. Colomer. (1992). Banyoles: visit to the aquatic ecosystems of Lake Banyoles karstic area. *S.I.L. Congress, Barcelona, Mid-Congress Excursions* **14**: 1–13.

Gasith, A. and V.H. Resh. (1990). Streams in mediterranean climate regions: abiotic influences and biotic responses to predictable seasonal events. *Annual Review of Ecology and Systematics* **30**: 51–81.

Gathman, J.P., T.M. Burton, and B.J. Armitage. (1999). Coastal wetlands of the upper Great Lakes: distribution of invertebrate communities in response to environmental variation. In *Invertebrates in Freshwater Wetlands of North America* (eds D.P. Batzer, R.B. Rader, and S.A. Wissinger), pp. 949–94. John Wiley & Sons, New York.

Geddes, M.C. (1976). Seasonal fauna of some ephemeral saline waters in western Victoria with particular reference to *Parartemia zietziana* Sayce (Crustacea: Anostraca). *Australian Journal of Marine and Freshwater Research* **27**: 1–22.

Geddes, M.C. (1980). The brine shrimps *Artemia* and *Parartemia* in Australia. In *The Brine Shrimp Artemia*, Vol. 3 (eds G. Persoone, P. Sargeloos, D. Roels, and E. Jaspers), pp. 57–65. Universa Press, Wetteren, Belgium.

Gomez, R. (1995). *Función de los humedales en la dinámica de nutrientes (N y P) de una cuenca de características áridas: Experiencias en el sureste ibérico*. Ph.D. Thesis, Universitat de Murcia.

Gonzalez-Jimerez, E. (1977). The capybara. *World Animal Review* **21**: 24–30.

Good, J.A. and F.T. Butler. (2001). Turlough pastures as a habitat for Staphylinidae and Carabidae (Coleoptera) in south-east Galway and north Clare, Ireland. *Bulletin of the Irish Biogeographical Society* **25**: 74–94.

Gopal, B. (2003). Perspectives on wetland science, application and policy. *Hydrobiologia* **490**: 1–10.

Gutierrezyurrita, P.J., G. Sancho, M.A. Bravo, A. Baltanas, and C. Montes. (1998). Diet of the red swamp crayfish *Procambarus clarkii* in natural ecosystems of the Donana National Park temporary freshwater marsh (Spain). *Journal of Crustacean Biology* **18**: 120–7.

Hails, A.J. (ed.). (2003). *Wetlands, Biodiversity and the Ramsar Convention* Vol. 4. http:www.ramsar.org/lib_bio_5.htm.

Hall, D.L., R.W. Sites, E.B. Fish, T.R. Mollhagen, D.L. Moorhead, and M.R. Willig. (1999). Playas of the Southern High Plains: the macroinvertebrate fauna. In *Invertebrates in Freshwater Wetlands of North America* (eds D.P. Batzer, R.B. Rader, and S.A. Wissinger), pp. 635–66. John Wiley & Sons, New York.

Hall, D.W. and D.D. Fish. (1974). A baculovirus from the mosquito *Wyeomyia smithii*. *Journal of Invertebrate Pathology* **23**: 383–8.

Hamer, M.L. and L. Brendonck. (1997). Distribution, diversity and conservation of Anostraca (Crustacea: Branchiopoda) in southern Africa. *Hydrobiologia* **359**: 1–12.

Hansson, L.-A., H.J.G. Dartnall, J.C. Ellis-Evans, H. MacAlister, and L.J. Tranvik. (1996). Variation in physical, chemical and biological components in the subantarctic lakes of South Georgia. *Ecography* **19**: 393–403.

Harper, P.P. and H.B.N. Hynes. (1970). Diapause in the nymphs of Canadian winter stoneflies. *Ecology* **51**: 425–7.

Harrison, L. (1922). On the breeding habits of some Australian frogs. *Australian Journal of Zoology* **3**: 17–34.

He, J.B., G.M. Bogeman, H.G. van de Steeg, J. Rijinders, L. Voesenek, and C. Blom. (1999). Survival tactics of *Ranunculus* species in river floodplains. *Oecologia* **118**: 1–8.

Heckman, C.W. (1998). The seasonal succession of biotic communities in wetlands of the tropical wet-and-dry climatic zone: V. Aquatic invertebrate communities in the Pantanal of Mato Grosso, Brazil. *Internationale Revue gesamten Hydrobiologie* **83**: 31–63.

Heitkamp, U. (1982). Untersuchungen zur Biologie, Ökologie und Systematik limnischer Turbellarien periodischer und perennierender Kleingewässer Südniedersachsens. *Archiv für Hydrobiologie* **64** (Suppl.): 65–188.

Hering, D. and H. Plachter. (1997). Riparian ground beetles (Coleoptera: Carabidae) preying on aquatic invertebrates: a feeding strategy in alpine floodplains. *Oecologia* **111**: 261–70.

Hershey, A.E., L.J. Shannon, G.J. Lima, and R.R. Regal. (1999). Prairie wetlands of south-central Minnesota, effects of drought on invertebrate communities. In *Invertebrates in Freshwater Wetlands of North America* (eds D.P. Batzer, R.B. Rader, and S.A. Wissinger), pp. 515–40. John Wiley and Sons, New York.

Hinton, H.E. (1968). Reversible suspension of metabolism and the origin of life. *Proceedings of the Royal Society of London B*. **171**: 43–57.

Hobbs, H.H. (1981). The crayfishes of Georgia. *Smithsonian Contributions to Zoology* **318**: 1–549.

Hobbs, H.H. (1991). Decapoda. In *Ecology and Classification of North American Freshwater Invertebrates* (eds J.H. Thorp and A.P. Covich), pp. 823–58. Academic Press, New York.

Holland, R.F. and S.K. Jain. (1981). Spatial and temporal variation in plant species diversity of vernal pools. In *Vernal Pools and Intermittent Streams* (eds S. Jain and P. Moyle), pp. 198–209. Institute of Ecology, University of California, Davis Publication No. 28.

Howard-Williams, C. and G.M. Lenton. (1975). The role of the littoral zone in the functioning of a shallow tropical lake ecosystem. *Freshwater Biology* **5**: 445–59.

Husband, B.C. and S.C.H. Barrett. (1998). Spatial and temporal variation in population size of Eichhornia paniculata in ephemeral habitats: implications for metapopulation dynamics. *Journal of Ecology* **86**: 1021–31.

Hynes, H.B.N. (1982). New and poorly known Gripopterygidae (Plecoptera) from Australia, especially Tasmania. *Australian Journal of Zoology* **30**: 115–58.

Hyslop, E.J. (1987). The growth and feeding habits of *Clarias anguillaris* during their first season in the floodplain pools of the Sokoto-Rima river basin, Nigeria. *Journal of Fish Biology* **30**: 183–92.

Innes, D.J. (1991). Geographic patterns of genetic differentiation among sexual populations of *Daphnia pulex*. *Canadian Journal of Zoology* **69**: 995–1003.

Ismail, M. (1990). The Sundarbans of Bangladesh. In *Ecosystems of the World 15: Forested Wetlands* (eds A.E. Lugo, M. Brinson, and S. Brown), pp. 357–86. Elsevier, Amsterdam.

Jaas, J. and B. Klausmeier. (2000). Atlas and bibliography of the first state and county records for anostracans (Crustacea: Branchiopoda) of the contiguous United States. *Milwaukee Public Museum, Contributions in Biology and Geology* **94**: 1–158.

Janetzky, W., W. Koste, and E. Vareschi. (1995). Rotifers (Rotifera) of Jamaican inland waters. A synopsis. *Ecotropica* **1**: 31–40.

Jeffries, M.J. (2001). Modeling the incidence of temporary pond microcrustacea: The importance of dry phase and linkage between ponds. *Israel Journal of Zoology* **47**: 445–58.

Jhingran, V.G. (1991). Lake Chilka. In *Fish and Fisheries of India*. Hindustan Publishing Corporation, New Delhi.

Johnson, R. (1998). The forest cycle and low river flows: a review of the U.K. and international studies. *Forest Ecology and Management* **109**: 1–7.

Joly, P. and A. Morand. (1997). Amphibian diversity and land–water ecotones. In *Biodiversity in Land–Inland Water Ecotones* (eds J.-B. Lachavanne and R. Juge), pp. 161–82. *Man and Biosphere Series*, Vol. 18. The Parthenon Publishing Group, New York.

Joly, P. and O. Grolet. (1997). Colonization dynamics of new ponds, and the age structure of colonizing alpine newts, *Triturus alpestris*. *Acta Oecologia* **17**: 599–608.

Juliano, S.A. (1985). Habitat associations, resources, and predators of an assemblage of *Brachinus* (Coleoptera: Carabidae) from southeastern Arizona. *Canadian Journal of Zoology* **63**: 1683–91.

Junk, W.J. and B.A. Robertson. (1997). Aquatic invertebrates. In *The Central Amazon Floodplain* (ed. W.J. Junk), pp. 279–98. Springer, Berlin.

Kaplan, R.H. (1981). Temporal heterogeneity of habitats in relation to amphibian ecology. In *Vernal Pools and Intermittent Streams* (eds S. Jain and P. Moyle), pp. 143–54. Institute of Ecology, University of California, Davis, Publication Number 28.

Kaster, J.L. and J.H. Bushnell. (1981). Cyst formation by *Tubifex tubifex* (Tubificidae). *Transactions of the American Microscopical Society* **100**: 34–41.

Kathman, R.D. and S.F. Cross. (1991). Ecological distribution of moss-dwelling tardigrades on Vancouver Island, British Columbia, Canada. *Canadian Journal of Zoology* **69**: 122–9.

Kaufman, M.G., E.D. Walker, T.W. Smith, R.W. Merritt, and M.J. Klug. (1999). Effects of larval mosquitoes (*Aedes triseriatus*) and stemflow on microbial community dynamics in container habitats. *Applied and Environmental Microbiology* **65**: 2661–73.

Kaushik, N.K. and H.B.N. Hynes. (1971). The fate of dead leaves that fall into streams. *Archiv für Hydrobiologie* **68**: 465–515.

Kehl, C. (1997). Die Hornmilbenzönosen (Acari, Oribatida) unterschiedlich stark degradierter Moor-standorte in Berlin und Brandenburg. Ph.D. Thesis, Free University of Berlin.

Kenk, R. (1944). The freshwater triclads of Michigan. *Miscellaneous Publications of the Museum of Zoology, University of Michigan* **60**: 1–44.

Kisielewska, G. (1982). Gastrotricha of two complexes of peat bogs near Siedlce. *Fragmenta Faunistica* **27**: 39–57.

Kitchell, J.F., D.E. Schindler, B.R. Herwig, D.M. Post, M.H. Olson, and M. Oldham. (1999). Nutrient cycling at the landscape scale: the role of diel Foraging migrations by geese at the Bosque del Apache National Wildlife Refuge, New Mexico. *Limnology and Oceanography* **44**: 828–36.

Kohler, S.L., D. Corti, M.C. Slamecka, and D.W. Schneider. (1999). Prairie floodplain ponds—mechanisms affecting invertebrate community structure. In *Invertebrates in Freshwater Wetlands of North America* (eds D.P. Batzer, R.B. Rader, and S.A. Wissinger), pp. 711–30. John Wiley & Sons, New York.

Kolasa, J. (1991). Flatworms: Turbellaria and Nemertea. In *Ecology and Classification of North American Freshwater Invertebrates* (eds J.H. Thorp and A.P. Covich), pp. 145–71. Academic Press, New York.

Korinek, V. and P.D.N. Hebert. (1996). A new species complex of *Daphnia* (Crustacea, Cladocera) from the Pacific northwest of the United States. *Canadian Journal of Zoology* **74**: 379–93.

Kozloff, E.N. (1990). *Invertebrates*. Saunders Publishing, Philadelphia, PA.

Kramer, D.L., C.C. Lindsey, G.E.E. Moodie, and E.D. Stevens. (1978). The fishes and the aquatic environment of the central Amazon basin, with particular reference to respiratory patterns. *Canadian Journal of Zoology* **56**: 717–29.

Kushner, D.J. (1978). Life in high salt and solute concentrations: halophilic bacteria. In *Microbial Life in Extreme Environments* (ed. E.J. Kushner), pp. 317–68. Academic Press, London.

Lahr, J. (1998). An ecological assessment of the hazard of eight insecticides used in Desert Locust control, to invertebrates in temporary ponds in the Sahel. *Aquatic Ecology* **32**: 153–62.

Lahr, J., A.O. Diallo, K.B. Ndour, A. Badji, and P.S. Diouf. (1999). Phenology of invertebrates living in a Sahelian temporary pond. *Hydrobiologia* **405**: 189–205.

Laird, M. (1988). *The Natural History of Larval Mosquito Habitats*. Academic Press, London.

Lake, P.S. (1982). Ecology of the macroinvertebrates of Australian upland streams—a review of current knowledge. *Bulletin of the Australian Society of Limnology* **8**: 1–16.

Lake, P.S., I.A.E. Bayly, and D.W. Morton. (1989). The phenology of a temporary pond in western Victoria, Australia, with special reference to invertebrate succession. *Archiv für Hydrobiologie* **115**: 171–202.

Lang, J.W. (1989). Sex determination. In *Crocodiles and Alligators* (ed. C.A. Ross), pp. 120–1. Facts on File, New York.

Larson, D.J. (1985). Structure in temperate predaceous diving beetle communities (Coleoptera: Dytiscidae). *Holarctic Ecology* **8**: 18–32.

Laurila, A. and T. Aho. (1997). Do female common frogs choose their breeding habitat to avoid predation on tadpoles? *Oikos* **78**: 585–91.

Leitch, W.G. and R.M. Kaminski. (1985). Long-term wetland waterfowl trends in Saskatchewan grassland. *Journal of Wildlife Management* **49**: 212–22.

Leopold, L.B., M.B. Wolman, and J.P. Miller. (1964). *Fluvial Processes in Geomorphology*. W.H. Freeman, San Francisco.

Lindeman, D.H. and R.G. Clark. (1999). Amphipods, land use impacts, and lesser scaup (*Aythya affinis*) distribution in Saskatchewan wetlands. *Wetlands* **19**: 627–38.

Linder, F. (1959). Notostraca. In *Freshwater Biology*, (2nd edn) (ed. W.T. Edmondson), pp. 572–6. John Wiley & Sons, New York.

Linhart, Y.B. (1988). Intrapopulation differentiation in annual plants. III. The contrasting effects of intra- and interspecific competition. *Evolution* **42**: 1047–64.

Linhart, Y.B. and M.C. Grant. (1996). Evolutionary significance of local genetic differentiation in plants. *Annual Review of Ecology and Systematics* **27**: 237–77.

Linton, L.R. and R.W. Davies. (1987). An energetics model of an aquatic predator and its application to life-history optima. *Oecologia* **71**: 552–9.

Linton, L.R., R.W. Davies, and F.J. Wrona. (1983). The effect of water temperature, ionic content and total dissolved solids on *Nephelopsis obscura* and *Erpobdella punctata* (Hirudinoidea: Erpobdellidae). II. Reproduction. *Holarctic Ecology* **6**: 64–8.

Lippert, B.E. and D.L. Jamieson. (1964). Plant succession in temporary ponds at Willamette Valley, Oregon. *American Midland Naturalist* **71**: 181–97.

Lott, D. (2001). Ground beetles and rove beetles associated with temporary ponds in England. *Freshwater Forum* **17**: 40–53.

Lude, A., M. Reich, and H. Plachter. (1999). Life strategies of ants in unpredictable floodplain habitats of Alpine rivers (Hymenoptera: Formicidae). *Entomologia Generalis* **24**: 75–91.

Mackay, R.J. and J. Kalff. (1973). Ecology of two related species of caddisfly larvae in the organic substrates of a woodland stream. *Ecology* **54**: 499–511.

Madoni, P. (1996). The contribution of ciliated Protozoa to plankton and benthos biomass in a European rice-field. *Journal of Eukaryotic Microbiology* **43**: 193–8.

Mahoney, D.L, M.A. Mort, and B.E. Taylor. (1990). Species richness of calanoid copepods, cladocerans and other branchiopods in Carolina Bay temporary ponds (South Carolina, USA). *American Midland Naturalist* **123**: 244–58.

Main, A.R., M.J. Littlejohn, and A.K. Lee. (1959). Ecology of Australian frogs. In *Biogeography and Ecology in Australia* (eds A. Keast, R.L. Crocker, and C.S. Christian), pp. 398–411. *Monographiae Biologicae 3*, Junk, The Hague.

Main, B.Y., V.T. Davies, and M.S. Harvey. (1985). *Zoological Catalogue of Australia, Vol. 3: Arachnida: Mygalomorphae, Araneomorphae in Part; Pseudoscorpionida; Amblypygi and Palpigradi.* Bureau of Flora and Fauna, Canberra.

Maly, E.J., S. Schoenholtz, and M.T. Arts. (1980). The influence of flatworm predation on zooplankton inhabiting small ponds. *Hydrobiologia* **76**: 233–40.

Marchant, R. (1982a). Seasonal variation in the macroinvertebrate fauna of Billabongs along Magela Creek, Northern Territory. *Australian Journal of Marine and Freshwater Research* **33**: 329–42.

Marchant, R. (1982b). Life spans of two species of tropical mayfly nymph (Ephemeroptera) from Magela Creek, Northern Territory. *Australian Journal of Marine and Freshwater Research* **33**: 173–9.

Margulis, L. and K.V. Schwartz. (1988). *Five Kingdoms*, 2nd edn. W.H. Freeman and Co., New York.

Marmonier, P., A.-M. Bodergat, and S. Doledec. (1994). Theoretical habitat templets, species traits, and species richnesses: ostracods (Crustacea) in the Upper Rhone River and its floodplain. *Freshwater Biology* **31**: 341–55.

Martens, K. (1988). Seven new species and two new subspecies of *Sclerocypris* Sars 1924 from Africa, with new records of some other Megalocycprids (Crustacea: Ostracoda). *Hydrobiologia* **162**: 243–73.

Martens, K. (1996). On *Korannacythere* gen. nov. (Crustacea, Ostracoda), a new genus of temporary pool limnocytherids from southern Africa, with the description of three new species and a generic reassessment of the Limnocytherinae. *Bulletin de l'Institut Royal des Sciences Naturelles de Belgique* **66**: 51–72.

Matthews, J.V. (1979). Tertiary and Quaternary environments: historical background for an analysis of the Canadian insect fauna. In *Canada and its Insects* (ed. H.V. Danks), pp. 31–86. Memoirs of the Entomological Society of Canada, Ottawa.

Mattox, N.T. (1959). Conchostraca. In *Freshwater Biology*, 2nd edn (ed. W.T. Edmondson), pp. 577–86. John Wiley & Sons, New York.

Mayhew, W.H. (1968). Biology of desert amphibians and reptiles. In *Desert Biology* (ed. G.W. Brown), pp. 195–356. Academic Press, New York.

McKee, P.M. and G.L. Mackie. (1980). Desiccation resistance in *Sphaerium occidentale* and *Musculium securis* (Bivalvia: Sphaeriidae) from a temporary pond. *Canadian Journal of Zoology* **58**: 1693–6.

McLachlan, A.J. (1981). Food resources and foraging tactics in tropical rain pools. *Zoological Journal of the Linnean Society* **71**: 265–77.

McLachlan, A.J. and M.A. Cantrell. (1980). Survival strategies in tropical rainpools. *Oecologia* **47**: 344–51.

McLachlan, A.J. and R. Ladle. (2001). Life in the puddle: behaviour and life-cycle adaptations in the Diptera of tropical rain pools. *Biological Reviews* **76**: 377–88.

McMahon, R.F. (1983). Physiological ecology of freshwater pulmonates. In *The Mollusca* (ed. W.D. Russell-Hunter), pp. 359–430. Academic Press, New York.

McMahon, R.F. (1991). Mollusca: Bivalvia. In *Ecology and Classification of North American Freshwater Invertebrates* (eds J.H. Thorp and A.P. Covich), pp. 315–99. Academic Press, New York.

Meintjes, S. (1996). Seasonal changes in the invertebrate community of small shallow ephemeral pans at Bain's Vlei, South Africa. *Hydrobiologia* **317**: 51–64.

Mellor, M.W. (1979). A study of the salt lake snail *Coxiella* (Smith)1844, *sensu lato*. B.Sc. (Hons) Thesis, University of Adelaide, Australia.

Menge, B.A. and J.P. Sutherland. (1976). Species diversity gradients: synthesis of the roles of predation, competition, and temporal heterogeneity. *American Naturalist* **110**: 350–69.

Merritt, R.W. and K.W. Cummins (eds). (1984). *An Introduction to the Aquatic Insects of North America.* Kendal/Hunt Publishers, Dubuque, Iowa.

Minckley, W.L. and W.E. Barber. (1979). Some aspects of the biology of the longfin dace, a cyprinid fish characteristic of streams in the Sonoran desert. *Southwestern Naturalist* **15**: 459–64.

Moghraby, A.I. (1977). A study on diapause of zooplankton in a tropical river—The Blue Nile. *Freshwater Biology* **7**: 77–117.

Mol, J.H. (1996). Impact of predation on early stages of the armoured catfish *Hoplosternum thoracatum* (Siluriformes-Callichthyidae) and implications for the syntopic occurrence with other related catfishes in a neotropical multi-predator swamp. *Oecologia* **107**: 395–410.

Monakov, A.V. (1969). The zooplankton and the zoobenthos of the White Nile and adjoining waters in the Republic of Sudan. *Hydrobiologia* **33**: 161–85.

Moore, W.G. (1970). Limnological studies of temporary ponds in southeastern Louisiana. *Southwestern Naturalist* **15**: 83–110.

Moorhead, D.L., D.L. Hall, and M.R. Willig. (1998). Succession of macroinvertebrates in playas of the Southern High Plains, USA. *Journal of the North American Benthological Society* **17**: 430–42.

Moreno, J.L., A. Millan, M.L. Suarez, M.R. Vidal-Abarca, and J. Velasco. (1997). Aquatic Coleoptera and Heteroptera assemblages in waterbodies from ephemeral coastal streams ('ramblas') of south-eastern Spain. *Archiv Hydrobiologie* **141**: 93–107.

Moreno, J.L., M. Aboal, M.R. Vidal-Abarca, and M.L. Suárez. (2001). Macroalgae and submerged macrophytes from fresh and saline waterbodies of ephemeral streams ('ramblas') in semiarid southeastern Spain. *Marine and Freshwater Research* **52**: 891–905.

Morgan, C.I. (1977). Population dynamics of two species of Tardigrada, *Macrobiotus hufelandii* (Schultze) and *Echiniscus* (*Echiniscus*) *testudo* (Doyère), in roof moss from Swansea. *Journal of Animal Ecology* **46**: 263–79.

Morgan, W.T. and R.W. Merritt. (1992). Conspicuous by absence? Protozoa in the diet of Aedes triseriatus. *Vector Control Bulletin, North Central States* **1**: 80.

Morley, A.W., T.E. Brown, and D.V. Koontz. (1985). The limnology of a naturally acidic tropical water system in Australia I. General description and wet season characteristics. *Verhandlungen Internationale Vereinigung für Theoretische und Angewandte Limnologie* **22**: 2125–30.

Myers, M.J. and V.H. Resh. (1999). Spring-formed wetlands of the arid West. In *Invertebrates in Freshwater Wetlands of North America* (eds D.P. Batzer, R.B. Rader, and S.A. Wissinger), pp. 811–28. John Wiley & Sons, New York.

Nelson, D.R. (1991). Tardigrada. In *Ecology and Classification of North American Freshwater Invertebrates* (eds J.H. Thorp and A.P. Covich), pp. 501–21. Academic Press, New York.

Nicholas, W.L. (1984). *The Biology of Free-living Nematodes*. Clarendon Press, Oxford.

Nilsson, A.N. (1986). Community structure in the Dytiscidae (Coleoptera) of a northern Swedish seasonal pond. *Annals Zoologica Fennici* **23**: 39–47.

Olsson, T.I. (1981). Overwintering of benthic macroinvertebrates in ice and frozen sediment in a North Swedish river. *Holarctic Ecology* **4**: 161–6.

Ortega, M., M.L. Suárez, M.R. Vidal-Abarca, R. Gómez, and L. Ramírez-Díaz. (1991). Aspects of postflood recolonization of macroinvertebrates in a 'Rambla' of southeast Spain ('Rambla del Moro'): Segura River Basin. *Verhandlungen Internationale Vereinigung für Theoretische und Angewandte Limnologie* **24**: 1994–2001.

Otto, C. (1976). Habitat relationships in the larvae of three Trichoptera species. *Archiv für Hydrobiologie* **77**: 505–17.

Ouboter, P.E. (1993). *The Freshwater Ecosystems of Suriname*. Kluwer Academic Publishers, Dordrecht.

Outridge, P. (1987). Possible causes of high species diversity in tropical Australian freshwater macrobenthic communities. *Hydrobiologia* **150**: 95–107.

Panter, J. and A. May. (1997). Rapid changes in the vegetation of a shallow pond in Epping Forest, related to recent droughts. *Freshwater Biological Association Freshwater Forum* **8**: 55–64.

Peckham, V. (1971). Notes on the chironomid midge *Belgica antarctica* Jacobs at Anvers Island in the maritime Antarctic. *Pacific Insects Monographs* **25**: 145–66.

Petersen, B. (1951). The tardigrade fauna of Greenland. *Meddelelser om Groenland* **150**: 1–94.

Petrov, B. and D.M. Cvetkovic. (1997). Community structure of branchiopods (Anostraca, Notostraca and Conchostraca) in the Banat province of Yugoslavia. *Hydrobiologia* **359**: 23–8.

Pizarro, H., I. Izaguirre, and G. Tell. (1996). Epilithic algae from a freshwater stream at Hope Bay, Antarctica. *Antarctic Science* **8**: 161–7.

Poff, N.L. and J.V. Ward. (1989). Implications of streamflow variability and predictability for lotic community structure: a regional analysis of streamflow patterns. *Canadian Journal of Fisheries and Aquatic Sciences* **46**: 1805–18.

Poinar, G.O., Jr. (1991). Nematoda and nematomorpha. In *Ecology and Classification of North American Freshwater Invertebrates* (eds J.H. Thorp and A.P. Covich), pp. 249–83. Academic Press, New York.

Poiani, K.A. and W.C. Johnson. (1991). Global warming and prairie wetlands. *BioScience* **41**: 611–8.

Pollock, M.M., R.J. Naiman, and T.A. Hanley. (1998). Plant species richness in riparian wetlands—a test of biodiversity theory. *Ecology* **79**: 94–105.

Pontin, R.M. (1989). Opportunistic rotifers: colonizing species of young ponds in Surrey, England. *Hydrobiologia* **186/187**: 229–34.

Por, F.D. (1995). *The Pantanal of Mato Grosso (Brazil)*. Kluwer Academic Publishers, Dordrecht.

Praeger, R.L. (1932). The flora of the turloughs: a preliminary note. *Proceedings of the Royal Irish Academy(B)* **41**: 37–45.

Price, P.W. (1984). Alternative paradigms in community ecology. In *A New Ecology: Novel Approaches to Interactive*

Systems (eds P.W. Price, C.N. Slobodchikoff, and S. Gaud), pp. 353–83. John Wiley & Sons, New York.

Pugh, P.J.A. (1996). Edaphic oribatid mites (Cryptostigmata, Acarina) associated with an aquatic moss on sub-antarctic South Georgia. *Pedobiologia* **40**: 113–7.

Puri, S.K. (2003). Biodiversity Profile of India. http://www.wcmc.org.uk/igcmc/main.html.

Rai, H. and G. Hill. (1984). Primary production in the Amazonian aquatic ecosystems. In *The Amazon: Limnology and Landscape Ecology of a Mighty Tropical River and its Basin* (ed. H. Sioli), pp. 311–35. W. Junk Publishers, Dordrecht.

Ramazzotti, G. and W. Maucci. (1983). Il Phylum Tardigrada, III deizione riveduta e aggiornata. *Memoire dell'Istituto Italiano di Idrobiologia* **41**: 1–1012.

Rasmussen, J.B. and J.A. Downing. (1988). The spatial response of chironomid larvae to the predatory leech *Nephelopsis obscura*. *American Naturalist* **131**: 14–21.

Reiss, F. (1976). Charakterisierung zentralamazonischer Seen aufgrund ihrer Makrobenthosfauna. *Amazoniana* **6**: 123–34.

Reiss, F. (1977). Qualitative and quantitative investigations on the macrobenthic fauna of Central Amazon lakes. I. Lago Tupe, a blackwater lake on the lower Rio Negro. *Amazoniana* **6**: 203–35.

Reynolds, J.D. (1983). 'Algal paper' on Inishmore, Aran Islands, Co. Galway. *Irish Naturalists Journal* **21**: 50.

Reynolds, J.D. (1996). *The Conservation of Aquatic Systems*. Royal Irish Academy, Dublin.

Reynolds, J.D., C. Duigan, F. Marnell, and A. O'Connor. (1998). Extreme and ephemeral water bodies in Ireland. In *Studies in Irish Limnology* (ed. P.S. Giller), pp. 67–99. The Marine Institute, Brunswick Press, Dublin.

Rhodes, M. (1950). Viability of dried bacterial cultures. *Journal of General Microbiology* **4**: 450–6.

Richards, K.J., P. Convey, and W. Block. (1994). The terrestrial arthropod fauna of the Byers Peninsula, Livingston Island, South Shetland Islands. *Polar Biology* **14**: 371–9.

Robertson, B.A. and E.R. Hardy. (1984). Zooplankton of Amazonian lakes and rivers. In *The Amazon: Limnology and Landscape Ecology of a Mighty Tropical River and its Basin* (ed. H. Sioli), pp. 337–52. W. Junk Publishers, Dordrecht.

Roessler, E.W. (1995). Review of Colombian Conchostraca (Crustacea)—ecological aspects and life cycles—families Lynceidae, Limnadiidae, Leptestheriidae and Metalimnadiidae. *Hydrobiologia* **298**: 125–32.

Romani, A.M. and S. Sabater. (1997). Metabolic recovery of a stromatolitic biofilm after drought in a Mediterranean stream. *Archiv für Hydrobiologie* **140**: 261–71.

Rzoska, J. (1961). Observations on tropical rainpools and general remarks on temporary waters. *Hydrobiologia* **17**: 265–87.

Rzoska, J. (1974). The Upper Nile swamps, tropical wetland study. *Freshwater Biology* **4**: 1–30.

Rzoska, J. (1984). Temporary and other waters. In *Sahara Desert* (ed. J.L. Cloudsley-Thompson), Pergamon Press, Oxford.

Sanders, H.L. (1968). Marine benthic diversity: comparative study. *American Naturalist* **102**: 243–82.

Sands, A. (1981). Algae of vernal pools and intermittent streams. In *Vernal Pools and Intermittent Streams* (eds S. Jain and P. Moyle), pp. 66–8. Institute of Ecology, University of California, Davis Publication No. 28.

Sassaman, C., M.A. Simovich, and M. Fugate. (1997). Reproductive isolation and genetic differentiation in North American species of *Triops* (Crustacea: Branchiopoda: Notostraca). *Hydrobiologia* **359**: 125–47.

Schneider, D.W. and T.M. Frost. (1996). Habitat duration and community structure in temporary ponds. *Journal of the North American Benthological Society* **15**: 64–86.

Schubart, C.D., R. Diesel, and S.B. Hedges. (1998). Rapid evolution to terrestrial life in Jamaican crabs. *Nature* **393**: 363–5.

Scott, D.A. (1989). *A Directory of Asian Wetlands*. IUCN, Gland, Switzerland, and Cambridge, UK.

Segura, F.S. (1990). *Las ramblas valencianas*. Ph.D. Thesis, Universitat de Valencia.

Sheath, R.G. and J.A. Hellebust. (1978). Comparison of algae in the euplankton, tychplankton and periphyton of a tundra pond. *Canadian Journal of Botany* **56**: 1472–83.

Simpson, T.L. and P.E. Fell. (1974). Dormancy among the Porifera: gemmule formation and hatching in freshwater and marine sponges. *Transactions of the American Microscopical Society* **92**: 544–77.

Sioli, H. (1975). Amazon tributaries and drainage basins. *Ecological Studies* **10**: 199–213.

Sliva, L. and D.D. Williams. (2005). Exploration of riffle-scale interactions between abiotic variables and microbial assemblages in the hyporheic zone. *Canadian Journal of Fisheries and Aquatic Sciences* **62**: 276–90.

Slobodkin, L.B. and P.E. Bossert. (1991). The freshwater Cnidaria—or Coelenterates. In *Ecology and Classification of North American Freshwater Invertebrates* (eds J.H. Thorp and A.P. Covich), pp. 125–43. Academic Press, New York.

Smith, I.M. and D.R. Cook. (1991). Water mites. In *Ecology and Classification of North American Freshwater Invertebrates* (eds J.H. Thorp and A.P. Covich), pp. 523–92. Academic Press, San Diego.

Smock, L.A. (1999). Riverine floodplain forests of the southeastern United States: invertebrates in an aquatic-terrestrial ecotone. In *Invertebrates in Freshwater Wetlands of North America* (eds D.P. Batzer, R.B. Rader, and S.A. Wissinger), pp. 137–66. John Wiley & Sons, New York.

Steiner, T.M. and W.F. Loftus. (1997). *Reptiles and Amphibians of the Everglades National Park*. Florida National Parks and Monuments Association, Public Affairs Office, Homestead, FL.

Stocker, Z.S.J. and H.B.N. Hynes. (1976). Studies on the tributaries of Char Lake, Cornwallis Island, Canada. *Hydrobiologia* **49**: 97–102.

Stout, J.D. (1984). The protozoan fauna of a seasonally inundated soil under grassland. *Soil Biology and Biochemistry* **16**: 121–5.

Strayer, D.L. and W.D. Hummon. (1991). Gastrotricha. In *Ecology and Classification of North American Freshwater Invertebrates* (eds J.H. Thorp and A.P. Covich), pp. 173–85. Academic Press, New York.

Stromberg, J.C., R. Tiller, and B. Richter. (1996). Effects of groundwater decline on riparian vegetation of semi-arid regions: the San Pedro River, Arizona. *Ecological Applications* **6**: 113–31.

Suberkropp, K. and M.J. Klug. (1980). The maceration of deciduous leaf litter by aquatic hyphomycetes. *Canadian Journal of Botany* **58**: 1025–31.

Sublette, J.E. and M.S. Sublette. (1967). The limnology of playa lakes on the Llano Estacado, New Mexico. *Southwestern Naturalist* **12**: 369–406.

Swanson, G.A. (1988). Aquatic habitats of breeding waterfowl. In *Ecology and Management of Wetlands Vol. 1. Ecology of Wetlands* (ed. D.D. Hook), pp. 228–67. Timber Press, Oregon.

Szkutnik, A. (1986). Freshwater Gastrotricha of Poland. VI. Gastrotricha of small astatic water bodies with rush vegetation. *Fragmenta Faunistica* **30**: 251–66.

Tasch, P. (1969). Branchiopoda. In *Treatise on Invertebrate Paleontology* (ed. R.C. Moore), Part R (4), Vol. 1, pp. 128–91. Geological Society of America, Boulder, CO.

Taylor, B.E., D.A. Leeper, M.A. McClure, and A.E. DeBiase. (1999). Carolina Bays: ecology of aquatic invertebrates and perspectives on conservation. In *Invertebrates in Freshwater Wetlands of North America* (eds D.P. Batzer, R.B. Rader, and S.A. Wissinger), pp. 167–96. John Wiley & Sons, New York.

Thiéry, A. (1991). Multispecies coexistence of branchiopods (Anostraca, Notostraca and Spinicaudata) in temporary ponds of Chaouia Plain (western Morocco): sympatry or syntopy between usually allopatric species. *Hydrobiologia* **212**: 117–36.

Thiéry, A. and A. Cazaubon. (1992). Epizootic algae and Protozoa on freshwater branchiopods (Anostraca, Notostraca and Spinicaudata) in Moroccan temporary ponds. *Hydrobiologia* **239**: 85–91.

Thomas, D.L. and J.B. McClintock. (1996). Aspects of the population-dynamics and physiological ecology of the gastropod *Physella cubensis* (Pulmonata: Physidae) living in a warm-water stream and ephemeral pond habitat. *Malacologia* **37**: 333–48.

Timms, B.V. (1983). A study of benthic communities in some shallow saline lakes of western Victoria, Australia. *Hydrobiologia* **105**: 165–77.

Timms, B.V. (1997). A comparison between saline and freshwater wetlands on Bloodwood Station, the Paroo, Australia, with special reference to their use by waterbirds. *International Journal of Salt Lake Research* **5**: 287–313.

Timms, B.V. and A.J. Boulton. (2001). Typology of arid-zone floodplain wetlands of the Paroo River (inland Australia) and the influence of water regime, turbidity, and salinity on their aquatic invertebrate assemblages. *Arch für Hydrobiologie* **153**: 1–27.

Towns, D.R. (1985). Limnological characteristics of a South Australian intermittent stream, Brown Hill Creek. *Australian Journal of Marine and Freshwater Research* **36**: 821–37.

Townsend, C.R., M.R. Scarsbrook, and S. Doledec. (1997). The Intermediate Disturbance Hypothesis, refugia, and biodiversity in Streams. *Limnology and Oceanography* **42**: 938–49.

Tsai, S.C. and E.F. Legner. (1977). Exponential growth in culture of the planarian mosquito predator, *Dugesia dorotocephala*. *Mosquito News* **37**: 474–8.

Van Buskirk, J. (2002). A comparative test of the adaptive plasticity hypothesis: relationships between habitat and phenotype in anuran larvae. *American Naturalist* **160**: 87–102.

Van Helsdingen, P.J. (1996). The spider fauna of some Irish floodplains. *Irish Naturalists' Journal* **25**: 285–93.

Vekhoff, N.V. (1997). Large branchiopod Crustacea (Anostraca, Notostraca, Spinicaudata) of the Barents Region of Russia. *Hydrobiologia* **359**: 69–74.

Vidal-Abarca, M.R., M.L. Suárez, and L. Ramírez-Díaz. (1996). Ramblas/wadis. In *Management of Mediterranean Wetlands. III. Case Studies 2* (eds C. Morillo and J.L. Gonzalez), pp. 17–38. MEDWET, Union Europea, Ministero de Medio Ambiente.

Vincent, W.F. and C. Howard-Williams. (1986). Antarctic stream ecosystems: physiological ecology of a blue-green algal epilithon. *Freshwater Biology* **16**: 209–33.

Visscher, P.K., R.S. Vetter, and R. Orth. (1994). Benthic bees? Emergence phenology of *Calliopsis pugionis* (Hymenoptera: Andrenidae) at a seasonally flooded site. *Annals of the Entomological Society of America* **87**: 941–5.

Walker, E.D., D.L. Lawson, R.W. Merritt, W.T. Morgan, and M.J. Klug. (1991). Nutrient dynamics, bacterial populations, and mosquito productivity in tree hole ecosystems and microcosms. *Ecology* **72**: 1529–46.

Walker, T.D. and P.A. Tyler. (1979). *A Limnological Survey of the Magela Creek System, Alligator Rivers Region, Northern Territory*. First and Second Interim Report of the Supervising Scientist, Alligator Rivers Region, Sydney.

Wallwork, J.A. (1983). *The Distribution and Diversity of Soil Fauna*. Academic Press, London.

Warwick, R.M. and R. Price. (1979). Ecological and metabolic studies on free-living nematodes from an estuarine mudflat. *Estuary Coastal Marine Science* **9**: 257–71.

Watson, G.F., M. Davies, and M.J. Tyler. (1995). Observations on temporary waters in northwestern Australia. *Hydrobiologia* **299**: 53–73.

Wauthy, G.M., M. Noti, and M. Dufrene. (1989). Geographic ecology of soil oribatid mites in deciduous forests. *Pedobiologia* **33**: 399–416.

Way, C.M., D.J. Hornbach, and A.J. Burky. (1980). Comparative life history tactics of the sphaeriid clam, *Musculium partumeium* (Say), from a permanent and temporary pond. *American Midland Naturalist* **104**: 319–27.

Welcomme, R.L. (1979). *Fisheries Ecology of Floodplain Rivers*. Longman, London.

Welling, C.H., R.L. Pederson, and A.G. van der Valk. (1988). Temporal patterns in recruitment from the seedbank during drawdowns in a prairie wetland. *Journal of Applied Ecology* **25**: 999–1007.

Wetlands International (2003). *A Case Study of Loktak Lake*. http://www.ramsar.org/wurchbk4cs5.doc.

Whitley, L.S. (1982). Aquatic oligochaeta. In *Aquatic Insects and Oligochaetes of North and South Carolina* (eds A.R. Brigham, W.U. Brigham, and A. Gnilka), pp. 2.1–2.29. Midwest Aquatic Enterprises, Mahomet, ILL.

Wiggins, G.B. (1973). A contribution to the biology of caddisflies (Trichoptera) in temporary pools. *Life Sciences Contributions of the Royal Ontario Museum* **88**: 1–28.

Wiggins, G.B. and R.J. Mackay. (1978). Some relationships between systematics and trophic ecology in nearctic aquatic insects, with special reference to Trichoptera. *Ecology* **59**: 1211–20.

Wiggins, G.B., R.J. Mackay, and I.M. Smith. (1980). Evolutionary and ecological strategies of animals in annual temporary pools. *Archiv für Hydrobiologie* **58** (Suppl.): 97–206.

Wilbur, H.M. (1987). Regulation of structure in complex systems: experimental temporary pond communities. *Ecology* **68**: 1437–52.

Wilbur, H.M. and J.P. Collins. (1973). Ecological aspects of amphibian metamorphosis. *Science* **182**: 1305–14.

Williams, D.D. (1983). The natural history of a nearctic temporary pond in Ontario with remarks on continental variation in such habitats. *Internationale Revue gesamten Hydrobiologie* **68**: 239–54.

Williams, D.D. (1993). Changes in freshwater meiofauna communities along the groundwater-hyporheic water ecotone. *Transactions of the American Microscopical Society* **112**: 181–94.

Williams, D.D. (1999). Why are there so few insects in the sea? *Trends in Entomology* **2**: 63–70.

Williams, D.D. and B.W. Coad. (1979). The ecology of temporary streams III. Temporary stream fishes in southern Ontario, Canada. *Internationale Revue gesamten Hydrobiologie* **62**: 761–87.

Williams, D.D. and B.W. Feltmate. (1992). *Aquatic Insects*. CAB International, Wallingford, Oxford, UK.

Williams, D.D. and H.B.N. Hynes. (1976). The ecology of temporary streams I. The faunas of two Canadian streams. *Internationale Revue gesamten Hydrobiologie* **61**: 761–87.

Williams, D.D. and H.B.N. Hynes. (1977). The ecology of temporary streams. II. General remarks on temporary streams. *Internationale Revue gesamten Hydrobiologie* **62**: 53–61.

Williams, D.D. and N.E. Williams. (1998). Invertebrate communities from freshwater springs: what can they contribute to pure and applied ecology? In *Studies in Crenobiology* (ed. L. Botosaneanu), pp. 251–61. Backhuys, Leiden.

Williams, D.D., L.G. Ambrose, and L.N. Browning. (1995). Trophic dynamics of two sympatric species of riparian spider (Araneae: Tetragnathidae). *Canadian Journal of Zoology* **73**: 1545–53.

Williams, D.D., C. Nalewajko, and A.K. Magnusson. (2005). Temporal variation in algal communities in an intermittent pond. *Journal of Freshwater Ecology* **20**: 165–70.

Williams, W.D. (1968). The distribution of *Triops* and *Lepidurus* (Branchiopoda) in Australia. *Crustaceana* **14**: 119–26.

Williams, W.D. (1980). *Australian Freshwater Life*. Macmillan, London.

Williams, W.D. (1981). Running water ecology in Australia. In *Perspectives in Running Water Ecology* (eds M.A. Lock and D.D. Williams), pp. 367–92. Plenum, New York.

Williams, W.D. (1983). *Life in Inland Waters*. Blackwell Scientific Publishers, Oxford. 252pp.

Williams, W.D. (1984). Chemical and biological features of salt lakes on the Eyre Peninsula, South Australia, and an explanation of regional differences in the fauna of Australian salt lakes. *Verhandlungen Internationale Vereinigung für Theoretische und Angewandte Limnologie* **22**: 1208–15.

Williams, W.D. (1985). Biotic adaptations in temporary lentic waters with special reference to those in semi-arid regions. In *Perspectives in Southern Hemisphere Limnology* (eds B.R. Davies and R.D. Walmsley), pp. 85–110. Junk, The Hague.

Williams, W.D. and R.T. Buckney. (1976). Stability of ionic proportions in five salt lakes in Victoria, Australia. *Australian Journal of Marine and Freshwater Research* **27**: 367–77.

Williamson, C.E. (1991). Copepoda. In *Ecology and Classification of North American Freshwater Invertebrates* (eds J.H. Thorp and A.P. Covich), pp. 787–822. Academic Press, New York.

Winfield, T.P., T. Cass, and K.B. MacDonald. (1981). Small mammal utilization of vernal pools, San Diego County, California. In *Vernal Pools and Intermittent Streams* (eds S. Jain and P. Moyle), pp. 161–7. Institute of Ecology, University of California, Davis, Publication Number 28.

Wirth, W.W. and J.L. Gressitt. (1967). Diptera: Chironomidae (midges). *Antarctic Research Series* **10**: 197–203.

Wissinger, S.A., A.J. Bohonak, H.H. Whiteman, and W.S. Brown. (1999). Subalpine wetlands in Colorado. In *Invertebrates in Freshwater Wetlands of North America* (eds D.P. Batzer, R.B. Rader, and S.A. Wissinger), pp. 757–90. John Wiley & Sons, New York.

Wolverton, B. and R.C. McDonald. (1976). Don't waste waterweeds. *New Scientist* **71**: 318–20.

Wrubleski, D.A. (1987). Chironomidae (Diptera) of peatlands and marshes in Canada. In *Aquatic Insects of Peatlands and Marshes in Canada* (eds D.M. Rosenberg and H.V. Danks), pp. 141–61. Memoirs of the Entomological Society of Canada 140, Ottawa.

Young, R. (1976). *Tanymastix stagnalis* (Linn.) in County Galway, new to Britain and Ireland. *Proceedings of the Royal Irish Academy (B)* **76**: 369–78.

Yozzo, D.J. and R.J. Diaz. (1999). Tidal freshwater wetlands: invertebrate diversity, ecology, and functional significance. In *Invertebrates in Freshwater Wetlands of North America* (eds D.P. Batzer, R.B. Rader, and S.A. Wissinger), pp. 889–918. John Wiley & Sons, New York.

Yu, H.S., E.F. Legner, and R.D. Sjogren. (1974). Mass release effects of Chlorohydra viridissima (Coelenterata) on field populations of *Aedes nigromaculis* and *Culex tarsalis* in Kern County. *Entomophaga* **19**: 409–20.

Zedler, P.H. (1981). Microdistribution of vernal pool plants of Kearny Mesa, San Diego County. In *Vernal Pools and Intermittent Streams* (eds S. Jain and P. Moyle), pp. 185–97. Institute of Ecology, University of California, Davis Publication No. 28.

Zimmerman, J.C., L.E. DeWald, and P.G. Rowlands. (1999). Vegetation diversity in an interconnected ephemeral system of north-central Arizona, USA. *Biological Conservation* **90**: 217–28.

Chapter 5

Alderdice, D.F. (1972). Responses of marine poikilotherms to environmental factors acting in concert. In *Marine Ecology* (ed. O. Kinne), pp. 1659–722. Wiley-Interscience, London.

Alekseev, V.R. and Y.I. Starabogatov. (1996). Types of diapause in Crustacea: definitions, distribution, evolution. *Hydrobiologia* **320**: 15–26.

Alpert, P. and M.J. Oliver. (2002). Drying without dying. In *Desiccation and Survival in Plants: Drying Without Dying* (eds M. Black and H.W. Pritchard), pp. 3–43. CAB International, Wallingford, Oxford.

Anderson, M. and A.K. Bromley. (1987). Sensory systems. In *Aphids, the Biology, Natural Enemies and Control* (eds A.K. Minks and P. Harrewijn), pp. 153–62. Elsevier, Amsterdam.

Anderson, N.H. (1967). Life cycle of a terrestrial caddisfly, *Philocasca demita* (Trichoptera: Limnephilidae) in North America. *Annals of the Entomological Society of America* **60**: 320–3.

Armbruster, P., W.E. Bradshaw, A.L. Steiner, and C.M. Holzapfel. (1999). Evolutionary responses to environmental stress by the pitcher-plant mosquito, *Wyeomyia smithii*. *Heredity* **83**: 509–19.

Arts, M.T., E.J. Maly, and M. Pasitschniak. (1981). The influence of *Acilius* (Dytiscidae) predation on *Daphnia* in a small pond. *Limnology and Oceanography* **26**: 1172–5.

Ball, I.R., N. Gourbault, and R. Kenk. (1981). The planarians of temporary waters in eastern North America. *Contributions to the Life Sciences, Royal Ontario Museum* **127**: 1–27.

Ball, S.L. (2002). Population variation and ecological correlates of tychoparthenogenesis in the mayfly, *Stenonema femoratum*. *Biological Journal of the Linnean Society* **75**: 101–23.

Barigozzi, C. (1980). Genus *Artemia*: problems of systematics. In *The Brine Shrimp Artemia* (eds G. Persoone, P. Sorgeloos, O. Roels, and E. Jaspers), pp. 147–53. Vol. 3, Universa Press, Wetteren, Belgium.

Barlocher, F., R.J. Mackay, and G.B. Wiggins. (1978). Detritus processing in a temporary vernal pool in southern Ontario. *Archiv für Hydrobiologie* **81**: 269–95.

Batzer, D.P. and V.H. Resh. (1992). Wetland management strategies that enhance waterfowl habitats can also control mosquitoes. *Journal of the American Mosquito Control Association* **8**: 117–25.

Beadle, L.C. (1981). *The Inland Waters of Tropical Africa*. Longman, London.

Beament, J.W.L. (1961). The waterproofing mechanism of arthropods II. The permeability of the cuticle of some aquatic insects. *Journal of Experimental Biology* **38**: 277–90.

Belcher, J.H. (1970). The resistance to desiccation and heat of the asexual cysts of some freshwater Prasinophyceae. *British Phycology* **5**: 173–7.

Benjamin, S.N. and W.E. Bradshaw. (1994). Body size and flight activity effects on male reproductive success in the pitcherplant mosquito (Diptera: Culicidae). *Annals of the Entomological Society of America* **87**: 331–6.

Beukema, J.J. (2002). Survival rates, site fidelity and homing ability in territorial *Calopteryx haemorrhoidalis* (Vander Linden) (Zygoptera: Calopterygidae). *Odonatologica* **31**: 9–22.

Bewley, J.D. (1979). Physiological aspects of desiccation tolerance. *Annual Review of Plant Physiology* **30**: 195–238.

Bilton, D.T., J.R. Freeland, and B. Okamura. (2001). Dispersal in freshwater invertebrates. *Annual Review of Ecology and Systematics* **32**: 159–81.

Bishop, J.A. (1967). Some adaptations of *Limnadia stanleyana* King (Crustacea: Branchiopoda, Conchostraca) to a temporary freshwater environment. *Journal of Animal Ecology* **36**: 599–609.

Bishop, J.A. (1974). The fauna of temporary rain pools in eastern New South Wales. *Hydrobiologia* **44**: 319–23.

Bishop, J.E. and H.B.N. Hynes. (1969). Upstream movements of the benthic invertebrates in the Speed River, Ontario. *Journal of the Fisheries Research Board of Canada* **26**: 279–98.

Blaustein, L. and B.P. Kotler. (1997). Differential development rates of two colour morphs of *Bufo viridis* tadpoles in desert pools. *Israel Journal of Zoology* **43**: 205–7.

Bohle, H.W. (1969). Untersuchungen über die Embryonalentwicklung und die embryonale Diapause bei *Baetis vernus* Curtis und *Baetis rhodani* (Pictet) (Baetidae, Ephemeroptera). *Zoologische Jahresbericht des Abteilung Anatomie und Ontogenie der Tiere* **86**: 493–557.

Bohonak, A.J. (1999). Effects of insect-mediated dispersal on the genetic structure of postglacial water mite populations. *Heredity* **82**: 451–61.

Bohonak, A.J. and H.H. Whiteman. (1999). Dispersal of the fairy shrimp Branchinecta coloradensis (Anostraca): effects of hydroperiod and salamanders. *Limnology and Oceanography* **44**: 487–93.

Bohonak, A.J., B.P. Smith, and M. Thornton. (2004). Distributional, morphological and genetic consequences of dispersal for temporary pond water mites. *Freshwater Biology* **49**: 170–80.

Boulton, A.J. (1989). Over-summering refuges of aquatic macroinvertebrates in two intermittent streams in Victoria. *Transactions of the Royal Society of South Australia* **113**: 23–34.

Boulton, A.J., E.H. Stanley, S.G. Fisher, and P.S. Lake. (1992). Over-summering strategies of macroinvertebrates in intermittent streams in Australia and Arizona. In *Aquatic Ecosystems in Semi-arid Regions: Implications for Resource Management* (eds R.D. Robarts and M.L. Bothwell), pp. 227–37. National Hydrological Research Institute Symposium Series 7, Environment Canada, Saskatoon.

Bouvet, Y. (1977). Adaptations physiologiques et compartementales des *Stenophylax* (Limnephilidae) aux eaux temporaires. In *Proceedings of the Second International Symposium on Trichoptera* (ed. M.I. Crichton), pp. 117–19. Junk, The Hague.

Bowen, S.T., K.N. Hitchner, and G.L. Dana. (1981). *Artemia* speciation: ecological isolation. In *Vernal Pools and Intermittent Streams* (eds S. Jain and P. Moyle), pp. 102–14. Institute of Ecology, University of California, Davis, Publication No. 28.

Bradshaw, W.E. and C.M. Holzapfel. (1983). Life cycle strategies in *Wyeomyia smithii*: seasonal and geographic adaptations. In *Diapause and Life Cycle Strategies in Insects* (eds V.K. Brown and I. Hodek), pp. 167–85. W. Junk Publishers, The Hague.

Bradshaw, W.E., C.M. Holzapfel, C.A. Kleckner, and J.J. Hard. (1997). Heritability of developmental time and protandry in the pitcher-plant mosquito, *Wyeomyia smithii*. *Ecology* **78**: 969–76.

Brendonck, L. (1996). Diapause, quiescence, hatching requirements: what can we learn from large branchiopods (Crustacea: Branchiopoda: Anostraca, Notostraca, 'Conchostraca'). *Hydrobiologia* **320**: 85–97.

Brendonck, L. and G. Persoone. (1996). Biological, ecological characteristics of large freshwater branchiopods from endorheic regions and consequences for their use in cyst-based toxicity tests. In *Progress in Standardisation of Aquatic Toxicity Tests* (eds A.M.V.M. Soares and P. Calow), pp. 7–35. Lewis Publishers, USA.

Brendonck, L. and B.J. Riddoch. (1999). Wind-borne short-range egg dispersal in anostracans (Crustacea: Branchiopoda). *Biological Journal of the Linnean Society* **67**: 87–95.

Briers, R.A., H.M. Cariss, and J.H.R. Gee. (2002). Individual and population movements of adult stream insects. *Published Abstract at the 2002 Meetings of the North American Benthological Society*, Pittsburgh.

Briers, R.A., J.H.R. Gee, H.M. Cariss, and R. Geoghegan. (2004). Inter-population dispersal by adult stoneflies detected by stable isotope enrichment. *Freshwater Biology* **49**: 425–31.

Broch, E.S. (1965). Mechanism of adaptation of the fairy shrimp, *Chirocephalopsis bundyi* Forbes to the temporary pond. *Cornell University Agriculture Experimental Station Memoir* **392**: 1–48.

Brown, L.R. and L.H. Carpelan. (1971). Egg hatching and life history of a fairy shrimp *Branchinecta mackini* Dexter (Crustacea: Anostraca) in a Mohave Desert playa (Rabbit Dry Lake). *Ecology* **52**: 41–54.

Burks, B.D. (1953). The Mayflies, or Ephemeroptera, of Illinois. *Illinois Natural History Survey Bulletin* **26**: 1–216.

Butler, M.G. (1984). Life histories of aquatic insects. In *The Ecology of Aquatic Insects* (eds V.H. Resh and D.M. Rosenberg), pp. 24–55. Praeger Scientific, New York.

Cao, Y., D.D. Williams, and N.E. Williams. (1998). How important are rare species in aquatic community ecology and bioassessment? *Limnology and Oceanography* **43**: 1403–9.

Castle, W.A. (1928). An experimental and histological study of the lifecycle of *Planaria velata*. *Journal of Experimental Biology* **51**: 417–76.

Chirkova, Z.N. (1973). Observations on the survival of cladocerans of the genus *Ilyocryptus* (Macrothricidae) in moist ground. *Inf. Byull. Biol. vnutr. Vad.* (*Information Bulletin of the Inland Water Biological Institute*) **17**: 37–9.

Chodorowski, A. (1969). The desiccation of ephemeral pools and the rate of development of *Aedes communis* larvae. *Polskie Archwm Hydrobiologie* **16**: 79–91.

Christensen, B.M. (1978). *Dirofilaria immitis*: effect on the longevity of *Aedes trivittatus*. *Experimental Parasitology* **44**: 116–23.

Claussen, D.L., R.A. Hopper, and A.M. Sanker. (2000). The effects of temperature, body size, and hydration state on the terrestrial locomotion of the crayfish *Orconectes rusticus*. *Journal of Crustacean Biology* **20**: 218–23.

Clegg, J.S. (1981). Metabolic consequences of the extent and disposition of the aqueous intracellular environment. *Journal of Experimental Zoology* **215**: 303–13.

Clifford, H.F. (1966). The ecology of invertebrates in an intermittent stream. *Investigations of Indiana Lakes and Streams* **7**: 57–98.

Clinton, S.M., N.B. Grimm, and S.G. Fisher. (1996). Response of a hyporheic invertebrate assemblage to drying disturbance in a desert stream. *Journal of the North American Benthological Society* **15**: 700–12.

Cohen, D. (1966). Optimizing reproduction in a randomly varying environment. *Journal of Theoretical Biology* **12**: 119–29.

Colbourne, J.K., T.J. Crease, L.J. Weider, P.D.N. Hebert, F. Dufrene, and A. Hobaek. (1998). Phylogenetic evolution of a circumarctic species complex (Cladocera: *Daphnia pulex*). *Biological Journal of the Linnean Society* **65**: 347–65.

Collins, N.C. and G. Stirling. (1980). Relationships among total dissolved solids, conductivity, and osmosity for five *Artemia* habitats (Anostraca: Artemiidae). *Great Basin Naturalist* **40**: 131–8.

Compton, S.G. (2002). Sailing with the wind: dispersal by small flying insects. In *Dispersal Ecology* (eds J.M. Bullock, R.E. Kenward, and R.S. Hails), pp. 113–33. Blackwell Publishing, Oxford.

Corbet, P.S. (1963). *A Biology of Dragonflies*. Quadrangle Books, Chicago, IL.

Corbet, P.S. (1980). Biology of Odonata. *Annual Review of Entomology* **25**: 189–217.

Crocker, D. and D. Barr. (1968). *Handbook of the Crayfish of Ontario*. University of Toronto Press, Toronto.

Crowe, J.H. (1971). Anhydrobiosis: an unsolved problem. *American Naturalist* **105**: 563–73.

Crowe, J.H. and L.M. Crowe (1982). Induction of anhydrobiosis: membrane changes during drying. *Cryobiology* **19**: 317–28.

Daborn, G.R. (1971). Survival and mortality of coenagrionid nymphs (Odonata: Zygoptera) from the ice of an aestival pond. *Canadian Journal of Zoology* **49**: 569–71.

Danilevskii, A.S. (1965). *Photoperiodism and Seasonal Development of Insects*. Oliver and Boyd, Edinburgh.

Danks, H.V. (1971). Overwintering of some north temperate and Arctic Chironomidae. II. Chironomid biology. *Canadian Entomologist* **103**: 1875–910.

Danks, H.V. (1987). Insect dormancy: an ecological perspective. *Biological Survey of Canada Monograph Series* No. 1, Ottawa.

Davey, K.G. (1956). The physiology of dormancy in the sweetclover weevil. *Canadian Journal of Zoology* **34**: 86–98.

Davy-Bowker, J. (2002). A mark and recapture study of water beetles (Coleoptera: Dytiscidae) in a group of semi-permanent and temporary ponds. *Aquatic Ecology* **36**: 435–46.

Davidson, J. (1932). Resistance of the eggs of Collembola to drought conditions. *Nature* **29**: 867.

Davies, C.P., M.A. Simovich, and S.A. Hathaway. (1997). Population genetic structure of a Californian endemic branchiopod, *Branchinecta sandiegoensis*. *Hydrobiologia* **359**: 149–58.

Delucchi, C.M. (1989). Movement patterns of invertebrates in temporary and permanent streams. *Oecologia* **78**: 199–207.

Denlinger, D.L. (1978). The developmental response of flesh flies (Diptera: Sarcophagidae) to tropical seasons. Variation in generation time and diapause in East Africa. *Oecologia* **35**: 105–7.

Denver, R.J., N. Mirhadi, and M. Phillips. (1998). Adaptive plasticity in amphibian metamorphosis: response of *Scaphiopus hammondii* tadpoles to habitat desiccation. *Ecology* **79**: 1859–72.

Dieterich, M. and N.H. Anderson. (1995). Life cycles and food habits of mayflies and stoneflies from temporary streams in western Oregon. *Freshwater Biology* **34**: 47–60.

Dwyer, G. and R.S. Hails. (2002). Manipulating your host: host–pathogen population dynamics, host dispersal and genetically modified baculoviruses. In *Dispersal Ecology* (eds J.M. Bullock, R.E. Kenward, and R.S. Hails), pp. 173–93. Blackwell Publishing, Oxford.

Eckblad, J.W. (1973). Population studies of three aquatic gastropods in an intermittent backwater. *Hydrobiologia* **41**: 199–219.

Elgmork, K. (1980). Evolutionary aspects of diapause in freshwater copepods. *Special Symposium Volume, American Society of Limnology and Oceanography* **3**: 411–7.

Erman, N.A. (1981). Terrestrial feeding migration and life history of the stream-dwelling caddisfly *Desmona bethula* (Trichoptera: Limnephilidae). *Canadian Journal of Zoology* **59**: 1658–65.

Evans, K.W. and R.A. Brust. (1972). Induction and termination of diapause in *Wyeomyia smithii* (Diptera: Culicidae), and larval survival studies at low and subzero temperatures. *Canadian Entomologist* **104**: 1937–50.

Evans, J.H. (1958). The survival of freshwater algae during dry periods. Part I. Investigation of algae of five small ponds. *Journal of Ecology* **46**: 148–67.

Everitt, D.A. (1981). An ecological study of an Antarctic freshwater pool with particular reference to Tardigrada and Rotifera. *Hydrobiologia* **83**: 225–37.

Farner, D.S. (1961). Comparative physiology: photoperiodicity. *Annual Review of Physiology* **23**: 71–96.

Fernando, C.H. (1958). The colonization of small freshwater habitats by aquatic insects. I. General discussion, methods and colonization in the aquatic Coleoptera. *Ceylon Journal of Science (Biological Sciences)* **1**: 117–54.

Fernando, C.H. and D.F. Galbraith. (1973). Seasonality and dynamics of aquatic insects colonizing small habitats. *Verhandlungen Internationale Vereinigung für Theoretische und Angewandte Limnologie* **18**: 1564–75.

Fischer, Z. (1967). Food composition and food preference in larvae of *Lestes sponsa* (L.) in astatic water environments. *Polskie Archwm Hydrobiologie* **14**: 59–71.

Fox, H.M. (1949). On *Apus*: its rediscovery in Britain, nomenclature and habits. *Proceedings of the Zoological Society London* **119**: 693–702.

Freeland, J.R., L.R. Noble, and B. Okamura. (2000). Genetic consequences of the metapopulation biology of a facultative sexual freshwater invertebrate. *Journal of Evolutionary Biology* **13**: 383–95.

Fryer, G. (1974). Attachment of bivalve molluscs to corixid bugs. *Naturalist* **28**: 18.

Fryer, G. (1996). Diapause, a potent force in the evolution of freshwater crustaceans. *Hydrobiologia* **320**: 1–14.

Gaff, D.F. (1977). Desiccation tolerant plants of southern Africa. *Oecologia* **31**: 95–109.

Geddes, M.C. (1976). Seasonal fauna of some ephemeral saline waters in western Victoria with particular reference to *Parartemia zietziana* Sayce (Crustacea: Anostraca). *Australian Journal of Marine and Freshwater Research* **27**: 1–22.

Gekko, K. and S.N. Timasheff. (1981). Thermodynamic and kinetic examination of protein stabilization by glycerol. *Biochemistry* **20**: 4677–86.

Gislason, G.M. (1978). Flight periods and ovarian maturation in Trichoptera in Iceland. In *Proceedings of the Second International Symposium on Trichoptera* (ed. M.I. Crichton), pp. 135–46. Junk, The Hague.

Grensted, L.W. (1939). Colonization of new areas by water beetles. *Entomologist's Monthly Magazine* **75**: 174–5.

Grodhaus, G. (1980). Aestivating chironomid larvae associated with vernal pools. In *Chironomidae. Ecology, Systematics, Cytology and Physiology* (ed. D.A. Murray), pp. 315–22. Pergamon Press, Oxford.

Grow, L. and H. Merchant. (1980). The burrow habitat of the crayfish *Cambarus diogenes diogenes* (Girard). *American Midland Naturalist* **103**: 231–7.

Hairston, N.G. and C.E. Cáceres. (1996). Distribution of crustacean diapause: micro- and macroevolutionary pattern and process. *Hydrobiologia* **320**: 27–44.

Hairston, N.G. and E.J. Olds. (1984). Population differences in the timing of diapause: adaptations in a spatially heterogeneous environment. *Oecologia* **61**: 42–8.

Hairston, N.G., R.A. Van Brunt, C.M. Kearns, and D.R. Engstrom. (1995). Age and survivorship of diapausing eggs in sediment egg banks. *Ecology* **76**: 1706–11.

Hall, F.G. (1922). The vital limit of desiccation of certain animals. *Biological Bulletin, Woods Hole* **42**: 31–51.

Hansen, A-M. (1996). Variable life history of a cyclopoid copepod: role of food availability. *Hydrobiologia* **320**: 223–7.

Harper, P.P. and H.B.N. Hynes. (1970). Diapause in the nymphs of Canadian winter stoneflies. *Ecology* **51**: 425–7.

Herbst, D.B. (1988). Comparative population ecology of *Ephydra hians* Say (Diptera: Ephydridae) at Mono Lake (California) and Abert Lake (Oregon). *Hydrobiologia* **158**: 145–66.

Hildrew, A.G. (1985). A quantitative study of the life history of fairy shrimp (Branchiopoda: Anostraca) in relation to the temporary nature of its habitat, a Kenyan rainpool. *Journal of Animal Ecology* **54**: 99–110.

Hildrew, A.G. (1986). Aquatic insects: patterns in life history, environment and community structure. *Proceedings of the European Congress of Entomology (Amsterdam)* **3**: 35–45.

Hinton, H.E. (1951). A new chironomid from Africa, the larva of which can be dehydrated without injury. *Proceedings of the Zoological Society of London* **121**: 371–80.

Hinton, H.E. (1952). Survival of chironomid larva after twenty months dehydration. *Transactions of the 9th International Congress of Entomology* **1**: 478–82.

Hinton, H.E. (1954). Resistance of the dry eggs of *Artemia salina* L. to high temperature. *Annals of the Magazine of Natural History* **7**: 158–60.

Hobbs, H.H. (1981). *The Crayfishes of Georgia*. Smithsonian Contributions to Zoology No. 318.

Hochachka, P.W. and G.N. Somero. (1984). *Biochemical Adaptation*. Princeton University Press, Princeton, NJ.

Horvath, G. and D. Varju. (1997). Polarization pattern of freshwater habitats recorded by video polarimetry in red, green and blue spectral ranges and its relevance for water detection by aquatic insects. *Journal of Experimental Biology* **200**: 1155–63.

Hornbach, D.J., T. Deneka, and R. Dado. (1991). Life-cycle variation of *Musculium partumeium* (Bivalvia: Sphaeriidae) from temporary and a permanent pond in Minnesota. *Canadian Journal of Zoology* **69**: 2738–44.

Horsfall, W.R. (1955). *Mosquitoes, their Bionomics and Relation to Disease*. Ronald Press, New York.

Horsfall, W.R. (1956). Eggs of floodwater mosquitoes. III. Conditioning and hatching of *Aedes vexans* (Diptera: Culicidae). *Annals of the Entomological Society of America* **49**: 66–71.

Hoy, M.A. and N.F. Knop. (1978). Development, hatch dates, overwintering success, and spring emergence of a 'non-diapausing' gypsy moth strain (Lepidoptera: Orgyiidae) in field cages. *Canadian Entomologist* **110**: 1003–8.

Husain, A. and W.E. Kershaw. (1971). The effect of filiriasis on the ability of a vector mosquito to fly and feed and to transmit the infection. *Transactions of the Royal Society of Tropical Medicine and Hygiene* **65**: 617–9.

Hynes, H.B.N. (1970). *The Ecology of Running Waters*. University of Liverpool Press, Liverpool.

Ingram, B.R. (1976). Life histories of three species of Lestidae in Northern Carolina, U.S. (Zygoptera). *Odonatologica* **5**: 231–44.

Innes, D.J. (1997). Sexual reproduction of *Daphnia pulex* in a temporary habitat. *Oecologia* **111**: 53–60.

Istock, C.A., K.J. Vavra, and H. Zimmer. (1976). Ecology and evolution of the pitcher-plant mosquito. 3. Resource tracking by a natural population. *Evolution* **30**: 548–57.

Jackson, D.J. (1956). Observations on water beetles during drought. *Entomologist's Monthly Magazine* **92**: 154–5.

James, H.G. (1969). Immature stages of five diving beetles (Coleoptera: Dytiscidae); notes on their habits and life histories, and a key to aquatic beetles of vernal woodland pools in southern Ontario. *Proceedings of the Entomological Society of Ontario* **100**: 52–97.

Johansson, A. and A.N. Nilsson. (1994). Insects of a small aestival stream in northern Sweden. *Hydrobiologia* **294**: 17–22.

Johansson, F. and L. Rowe. (1999). Life history and behavioural responses to time constraints in a damselfly. *Ecology* **80**: 1242–52.

Jones, R.E. (1975). Dehydration in an Australian rockpool chironomid larva (*Paraborniella tonnoiri*). *Journal of Entomology (A)* **49**: 111–19.

Kam, Y.C., C.F. Lin, Y.S. Lin, and Y.F. Tsal. (1998). Density effects of oophagous tadpoles of *Chirixalus eiffingeri* (Anura: Rhacophoridae): importance of maternal brood care. *Herpetologica* **54**: 425–33.

Kenk, R. (1949). The animal life of temporary and permanent ponds in southern Michigan. *Miscellaneous*

Publications of the Museum of Zoology, University of Michigan **71**: 1–66.

Kerney, M.P. (1999). *Atlas of the Land and Freshwater Molluscs of Britain and Ireland*. Harley, Colchester.

Khlebovich, V.V. (1996). The susceptibility to loss of diapause capacity in hydrobionts of ephemeral waterbodies. *Hydrobiologia* **320**: 83–4.

Keil, T.A. (1999). Morphology and development of the peripheral olfactory organs. In *Insect Olfaction* (ed. B.S. Hansson), pp. 5–47. Springer-Verlag, Berlin.

Kjellberg, G. (1973). Growth of *Leptophlebia vespertina* L., *Cloeon dipterum* L. and *Ephemera vulgata* L. (Ephemeroptera) in a small woodland lake. *Entomologica Tidskrift* **94**: 8–14.

Knowlton, G.F. (1951). A flight of water boatmen. *Bulletin of the Brooklyn Entomological Society* **46**: 22–3.

Kriska, G., G. Horvath, and S. Andrikovics. (1998). Why do mayflies lay their eggs en-masse on dry asphalt roads: water-imitating polarized-light reflected from asphalt attracts Ephemeroptera. *Journal of Experimental Biology* **201**: 2273–86.

Kushlan, J.A. (1973). Differential responses to drought in two species of *Fundulus*. *Copeia* **1973**: 808–9.

Lake, P.S. (1977). Pholeteros, the faunal assemblage found in crayfish burrows. *Newsletter of the Australian Society of Limnology* **15**: 57–60.

Lanciani, C.A. (1970). Resource partitioning in species of the water mite genus *Eylais*. *Ecology* **51**: 338–42.

Landin, J. (1968). Weather and diurnal periodicity of flight by *Helophorus brevipalpis* Bedel (Coleoptera: Hydrophilidae). *Opuscula Entomologica* **33**: 28–36.

Landin, J. (1980). Habitats, life histories, migration and dispersal by flight of two water beetles *Helophorus brevipalpis* and *H. strigifrons* (Hydroptilidae). *Holarctic Ecology* **3**: 190–201.

Lees, A.D. (1955). The physiology of diapause in arthropods. *Cambridge Monographs in Experimental Biology* **4**: 1–151.

Lehmkuhl, D.M. (1973). A new species of *Baetis* from ponds in the Canadian Arctic, with biological notes. *Canadian Entomologist* **10**: 343–6.

Lewontin, R.C. (1965). Selection for colonizing ability. In *The Genetics of Colonizing Species* (ed. H.G. Baker), pp. 79–93. Academic Press, New York.

Lincoln, R., G. Boxshall, and P. Clark. (1998). *A Dictionary of Ecology, Evolution and Systematics*. Cambridge University Press, Cambridge.

Lytle, D.A. (1999). Use of rainfall cues by *Abedus herberti* (Hemiptera: Belostomatidae): a mechanism for avoiding flash floods. *Journal of Insect Behaviour* **12**: 1–12.

Macan, T.T. (1939). Notes on the migration of some aquatic insects. *Journal of the Society for British Entomology* **2**: 1–6.

MacArthur, R.H. and E.O. Wilson. (1967). *The Theory of Island Biogeography*. Princeton University Press, Princeton, New Jersey.

Maguire, B. (1963). The passive dispersal of small aquatic organisms and their colonization of isolated bodies of water. *Ecological Monographs* **33**: 161–85.

McCafferty, W.P. and C.R. Lugo-Ortiz. (1995). *Cloeodes hydation*, n. sp. (Ephemeroptera: Baetidae): an extraordinary drought tolerant mayfly from Bazil. *Entomological News* **106**: 29–35.

McLachlan, A. and R. Ladle. (2001). Life in the puddle: behaviour and life-cycle adaptations in the Diptera of tropical rain pools. *Biological Reviews* **76**: 377–88.

McLachlan, A.J. (1983). Life-history tactics of rain-pool dwellers. *Journal of Animal Ecology* **52**: 545–61.

McLachlan, A.J. (1985). What determines the species present in a rain-pool? *Oikos* **45**: 1–7.

McLachlan, A.J. (1986). Sexual dimorphism in midges: strategies for flight in the rain-pool dweller *Chironomus imicola* (Diptera: Chironomidae). *Journal of Animal Ecology* **55**: 261–7.

McLachlan, A.J. (1993). Can two species of midge coexist in a single puddle of rain-water? *Hydrobiologia* **259**: 1–8.

Meryman, H.T. (1974). Freezing injury and its prevention in living cells. *Annual Review of Biophysics and Bioengineering* **3**: 341–63.

Miller, A.M. and S.W. Golladay. (1996). Effects of spates and drying on macroinvertebrate assemblages of an intermittent and a perennial prairie stream. *Journal of the North American Benthological Society* **15**: 670–89.

Moericke, V. (1969). Hostplant specific colour behaviour by *Hyalopterus pruni* (Aphididae). *Entomologia Experimentalis et Applicata* **1**: 524–34.

Morris, R.F. (1967). Factors inducing diapause intensity in *Hyphantrea cunea*. *Canadian Entomologist* **99**: 522–8.

Morse, J.C. and A. Neboiss. (1982). *Triplectides* of Australia (Insecta: Trichoptera: Leptoceridae). *Memoirs of the National Museum of Victoria* **43**: 61–98.

Murphy, P.J. (2003). Does reproductive site choice in a neotropical frog mirror variable risks facing offspring? *Ecological Monographs* **73**: 45–67.

Nielsen, A. (1950). On the zoogeography of springs. *Hydrobiologia* **2**: 313–21.

Nilsson, A.N. and B.W. Svensson. (1994). Dytiscid predators and culicid prey in two boreal snowmelt pools differing in temperature and duration. *Annales Zoologiae Fennici* **31**: 365–76.

Nilsson, A.N. and B.W. Svensson. (1995). Assemblages of dytiscid predators and culicid prey in relation to environmental factors in natural and clear-cut boreal swamp forest pools. *Hydrobiologia* **308**: 183–96.

Nolte, U., R.S. Tierböhl, and W.P. McCafferty. (1996). A mayfly from tropical Brazil capable of tolerating short-term dehydration. *Journal of the North American Benthological Society* **15**: 87–94.

Nurnberger, B. (1996). Local dynamics and dispersal in a structured population of the whirlygig beetle *Dineutus assimilis*. *Oecologia* **106**: 325–36.

Okamura, B. and J.R. Freeland. (2002). Gene flow and the evolutionary ecology of passively dispersing aquatic invertebrates. In *Dispersal Ecology* (eds J.M. Bullock, R.E. Kenward, and R.S. Hails), pp. 194–216. Blackwell Publishing, Oxford.

Pajunen, V.I. (1986). Distributional dynamics of *Daphnia* species in a rock-pool environment. *Annales Zoologici Fennici* **23**: 131–40.

Pajunen, V.I. and A. Jansson. (1969). Dispersal of rock-pool corixids *Arctocorixa carinata* (Sahler) and *Callicorixa producta* (Reut) (Heteroptera: Corixidae). *Annales Zoologici Fennici* **6**: 391–427.

Pearce, E.J. (1939). Colonization by water beetles. *Entomologist's Monthly Magazine* **75**: 208.

Pennak, R.W. (1953). *Freshwater Invertebrates of the United States*. Ronald Press, New York.

Pianka, E.R. (1970). On *r*- and K-selection. *American Naturalist* **104**: 592–7.

Popham, E.J. (1953). Observations on the migration of corixids (Hemiptera) into a new aquatic habitat. *Entomologist's Monthly Magazine* **89**: 124–5.

Pritchard, G., L.D. Harder, and R.A. Mutch. (1996). Development of aquatic insect eggs in relation to temperature and strategies for dealing with different thermal environments. *Biological Journal of the Linnean Society* **58**: 221–44.

Proctor, M.C.F. and V.C. Pence. (2002). Vegetative tissues: bryophytes, vascular resurrection plants and vegetative propagules. In *Desiccation and Survival in Plants: Drying Without Dying* (eds M. Black and H.W. Pritchard), pp. 207–37. CAB International, Wallingford, Oxford.

Proctor, V.W. (1964). Viability of crustacean eggs recovered from ducks. *Ecology* **45**: 656–8.

Rankin, M.A. (1974). The hormonal control of flight in the milkweed bug *Oncopeltus fasciatus*. In *Experimental Analysis of Insect Behaviour* (ed. L. Barton Browne), pp. 175–82. Springer, New York.

Readshaw, J.L. and G.O. Bedford. (1971). Development of the stick insect *Didymuria violescens* with particular reference to diapause. *Australian Journal of Zoology* **19**: 141–58.

Roff, D.A. (1986). The evolution of wing dimorphism in insects. *Evolution* **40**: 1009–20.

Roff, D.A. (1992). *The Evolution of Life Histories: Theory and Analysis*. Chapman and Hall, New York.

Rothschild, L.J. and R.L. Mancinelli. (2001). Life in extreme environments. *Nature* **409**: 1092–101.

Rundle, S.D., A. Foggo, V. Choiseul, and D.T. Bilton. (2002). Are distribution patterns linked to dispersal mechanisms? An investigation using pond invertebrate assemblages. *Freshwater Biology* **47**: 1571–81.

Rzoska, J. (1984). Temporary and other waters. In *Sahara Desert* (ed. J.L. Cloudsley-Thompson), pp. 105–14. Pergamon Press, Oxford.

Salt, R.W. (1961). Principles of insect cold-hardiness. *Annual Review of Entomology* **6**: 55–74.

Saski, A. and S. Ellner. (1995). The evolutionarily stable phenotype distribution in a random environment. *Evolution* **49**: 337–50.

Schneider, C.W. and D.C. McDevit. (2002). Are earthworms a possible mechanism for airborne dispersal of the alga *Vaucheria*? *Northeastern Naturalist* **9**: 225–34.

Schwartz, S.S., G. Cosyleon, and S. Yaselusky. (2002). A comparison of bet-hedging strategies among species in temporary aquatic habitats. *Abstract*, American Society of Limnology and Oceanography Meeting, Albuquerque, 2001.

Schwind, R. (1995). Spectral regions in which aquatic insects see reflected polarized light. *Journal of Comparative Physiology, Series A* **177**: 439–48.

Service, M.W. (1977). Ecological and biological studies on *Aedes cantans* (Meig.) (Diptera: Culicidae) in southern England. *Journal of Applied Ecology* **14**: 159–96.

Solbreck, C. (1978). Migration, diapause and direct development as alternative life histories in a seed bug, *Neacoryphus bicrucis*. In *Evolution of Insect Migration and Diapause* (ed. H. Dingle), pp. 195–217. Springer-Verlag, New York.

Southwood, T.R.E. (1977). Habitat, the templet for ecological strategies? *Journal of Animal Ecology* **46**: 337–65.

Southwood, T.R.E. (1988). Tactics, strategies and templets. *Oikos* **52**: 3–18.

Stagliano, D.M., A.C. Benke, and D.H. Anderson. (1998). Emergence of aquatic insects from two habitats in a small wetland in the southeastern USA: temporal patterns of numbers and biomass. *Journal of the North American Benthological Society* **17**: 37–53.

Statzner, B., K. Hoppenhaus, M.-F. Arens, and P. Richoux. (1997). Reproductive traits, habitat use and templet theory: a synthesis of world-wide data on aquatic insects. *Freshwater Biology* **38**: 109–35.

Stearns, S.C. (1976). Life-history tactics: review of the ideas. *Quarterly Review of Biology* **51**: 3–47.

Stearns, S.C. (1992). *The Evolution of Life Histories*. Oxford University Press, Oxford.

Stearns, S.C. (2000). Daniel Bernoulli (1738): evolution and economics under risk. *Journal of Bioscience* **25**: 221–8.

Strandine, E.J. (1941). Effect of soil moisture and algae on the survival of a pond snail during periods of relative dryness. *Nautilus* **54**: 128–30.

Svensson, B.W. (1972). Flight periods, ovarian maturation and mating in Trichoptera at a south Swedish stream. *Oikos* **23**: 370–83.

Talling, J.F. (1951). The element of chance in pond population. *Naturalist* **4**: 157.

Tauber, C.A. and M.J. Tauber. (1981). Insect seasonal cycles: genetics and evolution. *Annual Review of Ecology and Systematics* **12**: 281–308.

Taylor, D.J., T.L. Finston, and P.D.N. Hebert. (1998). Biogeography of a widespread freshwater crustacean: pseudocongruence and cryptic endemism in the North American *Daphnia laevis* complex. *Evolution* **52**: 1648–70.

Templeton, A.R. (1980). Modes of speciation and inferences based on genetic distances. *Evolution* **34**: 719–29.

Thienemann, A. (1954). *Chironomus*. Leben, Verbreitung und wirtschaftliche Bedeutung der Chironomiden. *Binnengewässer* **20**: 1–834.

Thiéry, A. (1997). Horizontal distribution and abundance of cysts of several large branchiopods in temporary pool and ditch sediments. *Hydrobiologia* **359**: 177–89.

Thomas, G.I. (1963). Study of population of sphaeriid clams in a temporary pond. *Nautilus* **77**: 37–43.

Turner, D. and D.D. Williams. (2000). Invertebrate movements within a small stream: density dependence or compensating for drift? *Internationale Revue gesamten Hydrobiologie* **85**: 141–56.

Ushatinskaya, R.S. (1959). Origin of insect diapause in the zone of temperate climate and its role in the formation of biological cycles. *Proceedings of the 15th International Congress of Zoology* (London, 1958), pp. 1051–53.

Venette, R.C., R.D. Moon, and W.D. Hutchinson. (2002). Strategies and statistics of sampling for rare individuals. *Annual Review of Entomology* **47**: 143–74.

Vepsalainen, K. (1978). Wing dimorphism and diapause in *Gerris*: determination and adaptive significance. In *Evolution of Insect Migration and Diapause* (ed. H. Dingle), pp. 218–53. Springer-Verlag, New York.

Vinogradova, E.B. (1960). An experimental investigation of the ecological factors inducing imaginal diapause in bloodsucking mosquitoes. *Entomological Review* (translation: *American Institute of Biological Sciences*) **39**: 327–40.

Wagner, R. and F.O. Gathmann. (1997). Long-term studies on aquatic dance-flies (Diptera, Empididae) 1983–1993: distribution and size patterns along the stream, abundance change between years, and the influence of environmental factors on the community. *Archiv für Hydrobiologie* **137**: 385–410.

Watson, J.A.L. and G. Theischinger. (1980). The larva of *Antipodophlebia aesthenes* (Tillyard): a terrestrial dragonfly (Anisoptera: Aeschnidae). *Odonatologica* **9**: 253–8.

Williams, D.D. (1977). Movements of benthos during the recolonization of temporary streams. *Oikos* **29**: 306–12.

Williams, D.D. (1983). The natural history of a nearctic temporary pond in Ontario with remarks on continental variation in such habitats. *Internationale Revue gesamten Hydrobiologie* **68**: 239–54.

Williams, D.D. (1991). Life history traits of aquatic arthropods in springs. In *Arthropods of Springs, with Particular Reference to Canada* (eds D.D. Williams and H.V. Danks), pp. 63–87. Memoir of the Entomological Society of Canada, No. 155.

Williams, D.D. and B.W. Feltmate. (1992). *Aquatic Insects*. CAB International, Wallingford, Oxford.

Williams, D.D. and I.D. Hogg. (1988). Ecology and production of invertebrates in a Canadian coldwater spring-springbrook system. *Holarctic Ecology* **11**: 41–54.

Williams, D.D. and H.B.N. Hynes. (1974). The occurrence of benthos deep within the substratum of a stream. *Freshwater Biology* **4**: 233–56.

Williams, D.D. and H.B.N. Hynes. (1976a). The ecology of temporary streams I. The faunas of two Canadian streams. *Internationale Revue gesamten Hydrobiologie* **61**: 761–87.

Williams, D.D. and H.B.N. Hynes. (1976b). The recolonization mechanisms of stream benthos. *Oikos* **27**: 265–72.

Williams, D.D. and H.B.N. Hynes. (1977a). The ecology of temporary streams II. General remarks on temporary streams. *Internationale Revue gesamten Hydrobiologie* **62**: 53–61.

Williams, D.D. and H.B.N. Hynes. (1977b). Benthic community development in a new stream. *Canadian Journal of Zoology* **55**: 1071–6.

Williams, D.D. and N.E. Williams. (1975). A contribution to the biology of *Ironoquia punctatissima* (Trichoptera: Limnephilidae). *Canadian Entomologist* **107**: 829–32.

Williams, D.D., N.E. Williams, and H.B.N. Hynes. (1974). Observations on the life history and burrow construction of the crayfish *Cambarus fodiens* (Cottle) in a temporary stream in southern Ontario. *Canadian Journal of Zoology* **52**: 365–70.

Williams, D.D., N.E. Williams, and I.D. Hogg. (1995). Life history plasticity of *Nemoura trispinosa* (Plecoptera: Nemouridae) along a permanent–temporary water habitat gradient. *Freshwater Biology* **34**: 155–63.

Williams, W.D. (1985). Biotic adaptations in temporary lentic waters with special reference to those in semi-arid regions. In *Perspectives in Southern Hemisphere Limnology* (eds B.R. Davies and R.D. Walmsley), pp. 85–110. Junk, The Hague.

Williamson, I. and C.M. Bull. (1999). Population ecology of the Australian frog *Crinia signifera* larvae. *Wildlife Research* **26**: 81–99.

Wissinger, S.A., W.S. Brown, and J.E. Jannot. (2003). Caddisfly life histories along permanence gradients in high-altitude wetlands in Colorado (USA). *Freshwater Biology* **48**: 255–70.

Wood, D.M., P.T. Dang, and R.A. Ellis. (1979). *The Mosquitoes of Canada (Diptera: Culicidae).* The Insects and Arachnids of Canada, Part 6. Agriculture Canada Publication 1686.

Wright, J.C. (1989). Desiccation tolerance and water-retentive mechanisms in tardigrades. *Journal of Experimental Biology* **142**: 267–92.

Yaron, Z. (1964). Notes on the ecology and entomostracan fauna of temporary rainpools in Israel. *Hydrobiologia* **24**: 489–513.

Yee, W.L. and J.R. Anderson. (1995). Tethered flight capabilities and survival of *Lambornella clarki*-infected, blood-fed, and gravid *Aedes sierrensis* (Diptera: Culicidae). *Journal of Medical Entomology* **32**: 153–60.

Young, F.N. (1960). The water beetles of temporary ponds in southern Indiana. *Proceedings of the Indiana Academy of Science* **69**: 154–64.

Zahiri, N., M.E. Rau, and D.J. Lewis. (1997). Starved larvae of *Aedes aegypti* (Diptera: Culicidae) render waters unattractive to ovipositing conspecific females. *Environmental Entomology* **26**: 1087–90.

Zaslavski, V.A. (1996). Essentials of the environmental control of insect seasonality as reference points for comparative studies in other invertebrates. *Hydrobiologia* **320**: 123–30.

Zera, A.J. and L.G. Harshman. (2001). The physiology of life history trade-offs in animals. *Annual Review of Ecology and Systematics* **32**: 95–126.

Chapter 6

Abell, D.L. (1984). Benthic invertebrates of some California intermittent streams. In *Vernal Pools and Intermittent Streams* (eds S. Jain and P. Moyle), pp. 46–60. University of California, Davis Institute of Ecology Publication 28, Davis, CA.

Anholt, B.R. (1990). An experimental separation of interference and exploitative competition in a larval damselfly. *Ecology* **71**: 1483–93.

Armbruster, P., R.A. Hutchinson, and P. Cotgreave. (2002). Factors influencing community structure in a South American tank bromeliad fauna. *Oikos* **96**: 225–34.

Arts, M.T., E.J. Maly, and M. Pasitschniak. (1981). The influence of *Acilius* (Dytiscidae) predation on *Daphnia* in a small pond. *Limnology and Oceanography* **26**: 1172–5.

Baker, R.L. (1981). Behavioral interactions and use of feeding areas by nymphs of *Coenagrion resolutum* (Coenagrionidae: Odonata). *Oecologia* **49**: 353–8.

Bärlocher, F., R.J. Mackay, and G.B. Wiggins. (1978). Detritus processing in a temporary vernal pool in southern Ontario. *Archiv für Hydrobiologie* **81**: 269–95.

Beaver, R.A. (1985). Geographical variation in food web structure in *Nepenthes* pitcher plants. *Ecological Entomology* **10**: 241–8.

Bengtsson, J. (1989). Interspecific competition increases local extinction rate in a metapopulation system. *Nature* **340**: 713–5.

Bilton, D., A. Foggo, and S. Rundle. (2001). Size, permanence and the proportion of predators in ponds. *Archiv für Hydrobiologie* **151**: 451–8.

Black, A.R. (1993). Predator-induced phenotypic plasticity in *Daphnia pulex*: life history and morphological responses to *Notonecta* and *Chaoborus*. *Limnology and Oceanography* **38**: 986–96.

Blaustein, A.R., E.L. Wildy, L.K. Belden, and A. Hatch. (2001). Influence of abiotic and biotic factors on amphibians in ephemeral ponds with special reference to long-toed salamanders (*Ambystoma macrodactylum*). *Israel Journal of Zoology* **47**: 333–45.

Blaustein, L. (1990). Evidence for predatory flatworms as organizers of zooplankton and mosquito community structure in rice fields. *Hydrobiologia* **199**: 179–91.

Blaustein, L. (1997). Non-consumptive effects of larval *Salamandra* on crustacean prey: can eggs detect predators? *Oecologia* **110**: 212–7.

Blaustein, L. (1998). Influence of the predatory backswimmer, *Notonecta maculata*, on invertebrate community structure. *Ecological Entomology* **23**: 246–52.

Blaustein, L. and J. Margalit. (1996). Priority effects in temporary pools: nature and outcome of mosquito larva–tadpole interactions depends on order of entrance. *Journal of Animal Ecology* **65**: 77–84.

Boix, D., J. Sala, X.D. Quintana, and R. Moreno-Amich. (2004). Succession of the animal community in a Mediterranean temporary pond. *Journal of the North American Benthological Society* **23**: 29–49.

Boulton, A.J. (1988). Composition and dynamics of macroinvertebrate communities in two intermittent streams. Ph.D. Thesis, Monash University, Melbourne.

Boulton, A.J. and P.S. Lake. (1992). The ecology of two intermittent streams in Victoria, Australia. III. Temporal changes in faunal composition. *Freshwater Biology* **27**: 123–38.

Boulton, A.J. and P.J. Suter. (1986). Ecology of temporary streams—an Australian perspective. In *Limnology in Australia* (eds P. DeDeckker and W.D. Williams), pp. 313–27. CSIRO/Junk Publishers, Melbourne and The Netherlands.

Boulton, A.J., C.G. Peterson, N.B. Grimm, and S.G. Fisher. (1992). Stability of an aquatic macroinvertebrate community in a multiyear hydrologic disturbance regime. *Ecology* **73**: 2192–207.

Chamier, A.C. (1985). Cell-wall degrading enzymes of aquatic hyphomycetes: a review. *Botanical Journal of the Linnean Society* **91**: 67–81.

Chan, K.L. (1997). Variation in communities of dipterans in *Nepenthes* pitchers in Singapore: predators increase prey community diversity. *Annals of the Entomological Society of America* **90**: 177–83.

Chen, X. and J.E. Cohen. (2001). Global stability, local stability and permanence in model food webs. *Journal of Theoretical Biology* **212**: 223–35.

Church, S.C. and T.N. Sherratt. (1996). The selective advantages of cannibalism in a neotropical mosquito. *Behavioural Ecology and Sociobiology* **39**: 117–23.

Closs, G.P. (1996). Effects of predatory fish (*Galaxias olidus*) on the structure of intermittent-stream pool communities in southeastern Australia. *Australian Journal of Ecology* **21**: 217–23.

Closs, G.P. and P.S. Lake. (1994). Spatial and temporal variation in the structure of an intermittent-stream food web. *Ecological Monographs* **64**: 1–21.

Cohen, J.E. (1978). *Food Webs and Niche Space*. Princeton University Press, Princeton, NJ.

Cohen, J.E. (1989). Food webs and community structure. In *Perspectives in Ecological Theory* (eds J Roughgarden, R.M. May and S.A. Levin), pp. 181–202. Princeton University Press, Princeton, NJ.

Diamond, J.M. (1975). The assembly of species communities. In *Ecology and Evolution of Communities* (eds M.L. Cody and J.M. Diamond), pp. 342–444. Harvard University Press, Cambridge, MA.

Donald, D.B. (1983). Erratic occurrence of anostracans in a temporary pond: colonization and extinction or adaptation to variations in annual weather? *Canadian Journal of Zoology* **61**: 1492–8.

Ebert, T.A. and M.L. Balko. (1982). Vernal pools as islands in space and time. In *Vernal Pools and Intermittent Streams* (eds S. Jain and P. Moyle), pp. 90–101. Institute of Ecology, University of California, Davis, Publication No. 28.

Enfield, M.A. and G. Pritchard. (1977). Estimates of population size and survival of immature stages of four species of *Aedes* (Diptera: Culicidae) in a temporary pond. *Canadian Entomologist* **109**: 1425–34.

Erman, N.A. and D.C. Erman. (1995). Spring permanence, Trichoptera species richness, and the role of drought. *Journal of the Kansas Entomological Society* **68**: 50–64.

Fairchild, G.W., J. Cruz, A.M. Faulds, A.E. Short, and J.F. Matta. (2003). Microhabitat and landscape influences on aquatic beetle assemblages in a cluster of temporary and permanent ponds. *Journal of the North American Benthological Society* **22**: 224–40.

Fauth, J.E. (1999). Identifying potential keystone species from field data: an example from temporary ponds. *Ecology Letters* **2**: 36–43.

Fincke, O.M. (1994). Population regulation of a tropical damselfly in the larval stage by food limitation, cannibalism, intraguild predation and habitat drying. *Oecologia* **100**: 118–27.

Fox, B.J. (2000). The genesis and development guild assembly rules. In *Ecological Assembly Rules: Perspectives, Advances, Retreats* (eds E. Weiher and P. Keddy), pp. 23–58. Cambridge University Press, New York.

Hanski, I. (1987). Carrion fly community dynamics: patchiness, seasonality and coexistence. *Ecological Entomology* **12**: 257–66.

Hanski, I. (2001). Population dynamic consequences of dispersal in local populations and in metapopulations. In *Dispersal* (eds J. Clobert, E. Danchin, A.A. Dhondt, and J.D. Nichols), pp. 281–98. Oxford University Press, Oxford.

Harvey, I.F. and P.S. Corbet. (1985). Territorial behaviour of larvae enhances mating success of male dragonflies. *Animal Behavior* **33**: 561–5.

Higgins, M.J. and R.W. Merritt. (1999). Temporary woodland ponds in Michigan. In *Invertebrates in Freshwater Wetlands of North America* (eds D.P. Batzer,

R.B. Rader, and S.A. Wissinger), pp. 279–97. John Wiley & Sons, New York.

Holyoak, M. (2000). Habitat subdivision causes changes in food web structure. *Ecology Letters* **3**: 509–15.

Jeffries, M. (1994). Invertebrate communities and turnover in wetland ponds affected by drought. *Freshwater Biology* **32**: 603–12.

Jeffries, M. (1996). Effects of *Notonecta glauca* predation on *Cyphon* larvae (Coleoptera: Scirtidae) populations in small, seasonal ponds. *Archiv für Hydrobiologie* **136**: 413–20.

Johansson, F. and L. Samuelsson. (1994). Fish-induced variation in abdominal spine length of *Leucorrhinia dubia* (Odonata) larvae? *Oecologia* **100**: 74–9.

Juliano, S.A. (1998). Species introduction and replacement among mosquitoes: interspecific resource competition or apparent competition? *Ecology* **79**: 255–68.

Kenk, R. (1949). The animal life of temporary and permanent ponds in southern Michigan. *Miscellaneous Publication of the Museum of Zoology, University of Michigan* **71**: 1–66.

Kerfoot, W.C. and M. Lynch. (1987). Branchiopod communities: associations with planktivorous fish in space and time. In *Predation* (eds W.C. Kerfoot and A. Sih), pp. 367–78. University of New England Press, Boston, MA.

Kiesecker, J.M. and A.R. Blaustein. (1999). Pathogen reverses competition between larval amphibians. *Ecology* **80**: 2442–8.

Kingsley, K.J. (1985). *Eretes sticticus* (L.) (Coleoptera: Dytiscidae): life history observations and an account of a remarkable event of synchronous emigration from a temporary desert pond. *Coleopterists Bulletin* **39**: 7–10.

Kitching, R.L. (2000). *Food Webs and Container Habitats*. Cambridge University Press, Cambridge.

Larson, D.J. (1985). Structure in temperate predaceous diving beetle communities (Coleoptera: Dytiscidae). *Holarctic Ecology* **8**: 18–32.

Law, R. and J.C. Blackford. (1992). Self-assembling food webs. A global view-point of coexistence of species in Lotka-Volterra communities. *Ecology* **73**: 567–78.

Law, R. and R.D. Morton. (1993). Alternative permanent states of ecological communities. *Ecology* **74**: 1347–61.

Lawton, J.H. and P.H. Warren. (1988). Static and dynamic explanations for patterns in food webs. *Trends in Ecology and Evolution* **3**: 242–5.

Levins, R. (1969). Some demographic and genetic consequences of environmental heterogeneity for biological control. *Bulletin of the Entomological Society of America* **15**: 237–40.

Ljungdahl, L.G. and K.E. Eriksson. (1985). Ecology of microbial cellulose degradation. In *Advances in Microbial Ecology*, Vol. 8 (ed. K.C. Marshall), pp. 237–99. Plenum, New York.

MacArthur, R.H. and E.O. Wilson. (1967). *The Theory of Island Biogeography*. Princeton University Press, Princeton, NJ.

Maly, E.J., S. Schaenholtz, and M.T. Arts. (1980). The influence of flatworm predation on zooplankton inhabiting small ponds. *Hydrobiologia* **76**: 233–40.

Martinez, N.D. (1992). Constant connectance in community food webs. *American Naturalist* **139**: 1208–18.

Martinez, N.D., B.A. Hawkins, H.A. Dawah, and B. Feifarek. (1999). Effects of sampling effort on characterization of food-web structure. *Ecology* **80**: 1044–55.

McLaughlan, A.J. (1985). What determines the species present in a rain-pool? *Oikos* **45**: 1–7.

McLaughlan, A.J. and M.A. Cantrell. (1980). Survival strategies in tropical rain pools. *Oecologia* **47**: 344–51.

Menge, B.A. and J.P. Sutherland. (1976). Species diversity gradients: synthesis of the roles of predation, competition, and temporal heterogeneity. *American Naturalist* **110**: 350–69.

Molla, S., L. Maltchik, C. Casado, and C. Montes. (1996). Particulate organic matter and ecosystem metabolism dynamics in a temporary Mediterranean stream. *Archiv für Hydrobiologie* **137**: 59–76.

Montoya, J.M. and R.V. Solé. (2003). Topological properties of food webs: from real data to community assembly models. *Oikos* **102**: 614–22.

Moore, N.W. (1964). Intra- and interspecific competition among dragonflies (Odonata). *Journal of Animal Ecology* **33**: 49–71.

Morin, P.J. (1999). *Community Ecology*. Blackwell Science Inc., Oxford.

Mouquet, N., G.S.E.E. Mulder, V.A.A. Jansen, and M. Loreau. (2001). The properties of competitive communities with coupled local and regional dynamics. In *Dispersal* (eds J. Clobert, E. Danchin, A.A. Dhondt, and J.D. Nichols), pp. 311–26. Oxford University Press, Oxford.

Nalewajko, C. (1977). Extracellular release in freshwater algae and bacteria: extracellular products of algae as a source of carbon for heterotrophs. In *Aquatic Microbial Communities* (ed. J Cairns, Jr.), pp. 589–624. Garland, New York.

Newman, R.A. (1989). Developmental plasticity of *Scaphiopus couchii* tadpoles in an unpredictable environment. *Ecology* **70**: 1775–87.

Nilsson, A.N. (1986). Community structure in the Dytiscidae (Coleoptera) of a northern Swedish seasonal pond. *Annals Zoologica Fennici* **23**: 39–47.

Paine, R.T. and S.A. Levin. (1981). Intertidal landscapes: disturbance and the dynamics of pattern. *Ecological Monographs* **51**: 145–98.

Park, T. (1962). Beetles, competition, and populations. *Science* **138**:1369–75.

Peacor, S.D. and E.E. Werner. (2001). The contribution of trait-mediated indirect effects to the net effects of a predator. *Proceedings of the National Academy of Science* **98**: 3904–8.

Peterson, C.G. and A.J. Boulton. (1999). Stream permanence infuences microalgal food availability to grazing tadpoles in arid-zone springs. *Oecologia* **118**: 340–52.

Pimm, S.L. (1982). *Food Webs.* Chapman and Hall, London.

Poff, N.L. and J.V. Ward. (1989). Implications of streamflow variability and predictability for lotic community structure: a regional analysis of streamflow patterns. *Canadian Journal of Fisheries and Aquatic Sciences* **46**: 1805–18.

Price, P.W. (1984). Alternative paradigms in community ecology. In *A New Ecology. Novel Approaches to Interactive Systems* (eds P.W. Price, C.N. Slobodchikoff, and S. Gaud), pp. 353–83. John Wiley & Sons, New York.

Reckendorfer, W., H. Keckeis, G. Winkler, and F. Schiemer. (1996). Water level fluctuations as a major determinant of chironomid community structure in the inshore zone of a large temperate river. *Archiv für Hydrobiologie* **115** (Suppl.): 3–9.

Richardson, B.A. and G.A. Hull. (2000). Insect colonization sequences in bracts of *Heliconia caribaea* in Puerto Rico. *Ecological Entomology* **25**: 460–6.

Rosenberger, A.E. and L.J. Chapman. (1999). Hypoxic wetland tributaries as faunal refugia from an introduced predator. *Ecology of Freshwater Fish* **8**: 22–34.

Sanders, H.L. (1968). Marine benthic diversity: comparative study. *American Naturalist* **102**: 243–82.

Schluter, D. and R.E. Ricklefs. (1993). Species diversity: an introduction to the problem. In *Species Diversity in Ecological Communities* (eds R.E. Ricklefs and D. Schluter), pp. 1–12. University of Chicago Press, Chicago, IL.

Schneider, D.W. (1997). Predation and foodweb structure along a habitat duration gradient. *Oecologia* **110**: 567–75.

Schneider, D.W. (1999). Snowmelt ponds in Wisconsin: influence of hydroperiod on invertebrate community structure. In *Invertebrates in Freshwater Wetlands of North America* (eds D.P. Batzer, R.B. Rader, and S.A. Wissinger), pp. 299–318. John Wiley & Sons, New York.

Schneider, D.W. and T.M. Frost. (1996). Habitat duration and community structure in temporary ponds. *Journal of the North American Benthological Society* **15**: 64–86.

Schoener, A. (1974). Colonization curves for planar marine islands. *Ecology* **55**: 818–27.

Schoener, T.W. (1983). Field experiments on interspecific competition. *American Naturalist* **122**: 240–85.

Sherratt, T.N., S.E. Ruff, and S.C. Church. (1999). No evidence for kin discrimination in cannibalistic tree hole mosquitoes (Diptera: Culicidae). *Journal of Insect Behaviour* **12**: 123–32.

Skelly, D.K. (2001). Distributions of pond-breeding anurans: and overview of mechanisms. *Israel Journal of Zoology* **47**: 313–32.

Smith, R.E.W. (1983). Community dynamics of the pool fauna in an intermittent stream. *Australian Society of Limnology Newsletter* **21**: 18.

Spencer, M. and L. Blaustein. (2001). Hatching responses of temporary pool invertebrates to signals of environmental quality. *Israel Journal of Zoology* **47**: 397–417.

Stav, G., L. Blaustein, and Y. Margalit. (2000). Influence of nymphal *Anax imperator* (Odonata: Aeshnidae) on oviposition by the mosquito *Culiseta longiareolata* (Diptera: Culicidae) and community structure in temporary pools. *Journal of Vector Ecology* **25**: 190–202.

Stoks, R., M.A. McPeek, and J.M. Mitchell. (2003). Evolution of prey behaviour in response to changes in predation regime: damselflies in fish and dragonfly lakes. *Evolution* **57**: 574–85.

Tavares-Cromar, A.F. and D.D. Williams. (1996). The importance of temporal resolution in food web analysis: evidence from a detritus-based stream. *Ecological Monographs* **66**: 91–113.

Therriault, T.W. and J. Kolasa. (2001). Desiccation frequency reduces species diversity and predictability of community structure in coastal rock pools. *Israel Journal of Zoology* **47**: 477–89.

Velasco, J. and A. Millan. (1998). Feeding habits of two large insects from a desert stream: *Abedus herberti* (Hemiptera: Belostomatidae) and *Thermonectus marmoratus* (Coleoptera:Dystiscidae). *Aquatic Insects* **20**:85–96.

Vermeij, G.T. (1977). The Mesozoic marine revolution: evidence from snails, predators and grazers. *Paleobiology* **3**: 245–58.

Vidondo, B., Y.T. Prairie, J.M. Blanco, and C.M. Duarte. (1997). Some aspects of the analysis of size spectra in aquatic ecology. *Limnology and Oceanography* **42**: 184–92.

Walton, W.E. (2001). Effects of *Triops newberryi* (Notostraca: Triopsidae) on aquatic insect communities in ponds in the Colorado Desert of southern California. *Israel Journal of Zoology* **47**: 491–511.

Ward, D. and L. Blaustein. (1994). The overriding influence of flash floods on species-area curves in ephemeral Negev Desert pools: a consideration of the value of island biogeography theory. *Journal of Biogeography* **21**: 595–603.

Wellborn, G.A., D.S. Skelly, and E.E. Werner. (1996). Mechanisms creating community structure across a freshwater habitat gradient. *Annual Review of Ecology and Systematics* **27**: 337–63.

Wiggins, G.B., R.J. Mackay, and I.M. Smith. (1980). Evolutionary and ecological strategies of animals in annual temporary ponds. *Archiv für Hydrobiologie* **58** (Suppl.): 97–206.

Wilbur, H.M. (1987). Regulation of structure in complex systems: experimental temporary pond communities. *Ecology* **68**: 1437–52.

Williams, D.D. (1980). Temporal patterns in recolonization of stream benthos. *Archiv für Hydrobiologie* **90**: 56–74.

Williams, D.D. and H.B.N. Hynes. (1976). The ecology of temporary streams I. The faunas of two Canadian streams. *Internationale Revue der gesamten Hydrobiologie* **61**: 761–87.

Williams, D.D. and H.B.N. Hynes. (1977). Benthic community development in a new stream. *Canadian Journal of Zoology* **55**: 1071–6.

Williams, D.D., N.E. Williams, and H.B.N. Hynes. (1974). Observations on the life history and burrow construction of the crayfish *Cambarus fodiens* (Cottle) in a temporary stream in southern Ontario. *Canadian Journal of Zoology* **52**: 365–70.

Wissinger, S.A., A.J. Bohonak, H.H. Whiteman, and W.S. Brown. (1999). Subalpine wetlands in Colorado—habitat permanence, salamander predation, and invertebrate communities. In *Invertebrates in Freshwater Wetlands of North America* (eds D.P. Batzer, R.B. Rader, and S.A. Wissinger), pp. 757–90. John Wiley & Sons, New York.

Wissinger, S.A., H.H. Whiteman, G.B. Sparks, G.L. Rouse, and W.S. Brown. (1999). Foraging trade offs along a predator permanence gradient in subalpine wetlands. *Ecology* **80**: 2102–16.

Zahiri, N. and M.E. Rau. (1998). Oviposition attraction and repellency of *Aedes aegypti* (Diptera: Culicidae) to waters from conspecific larvae subjected to crowding, confinement, starvation, or infection. *Journal of Medical Entomology* **35**: 782–7.

Zimmer, K.D., M.A. Hanson, and M.G. Butler. (2000). Factors influencing invertebrate communities in prairie wetlands: a multivariate approach. *Canadian Journal of Fisheries and Aquatic Sciences* **57**: 76–85.

Chapter 7

Armbruster, P., R.A. Hutchinson, and P. Cotgreave. (2002). Factors influencing community structure in a South American tank bromeliad fauna. *Oikos* **96**: 225–34.

Ashmole, N.P., J.M. Nelson, M.R. Shaw, and A. Garside. (1983). Insects and spiders on snowfields in the Cairngorms, Scotland. *Journal of Natural History* **17**: 599–613.

Barrera, R. and V. Medialdea. (1996). Development time and resistance to starvation of mosquito larvae. *Journal of Natural History* **30**: 447–58.

Barton, D.R. and S.M. Smith. (1984). Insects of extremely small and extremely large aquatic habitats. In *The Ecology of Aquatic Insects* (eds V.H. Resh and D.M. Rosenberg), pp. 456–83. Praeger Scientific, New York.

Beadle, L.C. (1981). *The Inland Waters of Tropical Africa. Longman, London.*

Beaver, R.A. (1983). The communities living in *Nepenthes* pitcher plants: fauna and food webs. In *Phytotelmata: Terrestrial Plants as Hosts for Aquatic Communities* (eds J.H. Frank and L.P. Lounibos), pp. 129–60. Plexus Publishing, Medford, NJ.

Benzing, D.H. (1980). *The Biology of the Bromeliads.* Mad River Press, Eureka, CA.

Blinn, D.W. and G.A. Cole. (1991). Algal and invertebrate biota in the Colorado River: comparison of pre- and post-dam conditions. In *Colorado River Ecology and Dam Management: Proceedings of a Symposium,* May 24–25, 1990. Sante Fe, New Mexico, pp. 102–23. National Academy of Sciences. http://books.nap.edu/books/0309045355/html.

Buxton, P.A. and G.H.E. Hopkins. (1927). Researches in Polynesia and Melanesia. *Memoirs of the London School of Tropical Medicine and Hygiene* **1**: 1–260.

Calvert, P.P. (1911). Studies on Costa Rican Odonata. II. The habits of the plant-dwelling larva of *Mecistogaster modestus*. *Entomological News* **22**: 402–11.

Cameron, R.E. and G.B. Blank. (1966). Desert algae: soil crusts and diaphanous substrata as algal habitats. *Jet Propulsion Laboratory, California Institute of Technology, Pasadena, Technical Report* **32–971**: 1–41.

Cameron, R.E., F.A. Morelli, and H.P. Conrow. (1970). Survival of micro-organisms in desert soil exposed to five years of continuous very high vacuum.

Jet Propulsion Laboratory, California Institute of Technology, Pasadena, Technical Report **32–1454**: 1–11.

Carpenter, S.R. (1982). Stemflow chemistry: effects on population dynamics of detritivorous mosquitoes in tree-hole ecosystems. *Oecologia* **53**: 1–6.

Cheng, L. (ed.). (1976). *Marine Insects*. Elsevier, New York.

Copeland, R.S., W. Okeka, and P.S. Corbet. (1996). Tree-holes as larval habitat of the dragonfly *Hadrothemis camarensis* (Odonata: Libellulidae) in Kakamega Forest, Kenya. *Aquatic Insects* **18**: 129–47.

Corbet, P.S. (1983). Odonata in phytotelmata. In *Phytotelmata: Terrestrial Plants as Hosts for Aquatic Communities* (eds J.H. Frank and L.P. Lounibos), pp. 29–54. Plexus Publishing, Medford, NJ.

Evans, K.W. and R.A. Brust. (1972). Induction and termination of diapause in *Wyeomia smithii* (Diptera: Culicidae), and larval survival studies at low and subzero temperatures. *Canadian Entomologist* **104**: 1937–50.

Evans, R.D. and J.R. Johansen. (1999). Microbiotic crusts and ecosystem processes. *Critical Reviews in Plant Sciences* **18**: 183–225.

Farkas, M.J. and R.A. Brust. (1986a). Phenology of the mosquito *Wyeomia smithii* (Coq.) in Manitoba and Ontario. *Canadian Journal of Zoology* **64**: 285–90.

Farkas, M.J. and R.A. Brust. (1986b). Pitcher-plant sarcophagids from Manitoba and Ontario. *Canadian Entomologist* **118**: 1307–8.

Fashing, N.J. (1975). Life history and general biology of *Naiadacarus arboricola* Fashing, a mite inhabiting water-filled treeholes (Acarina: Acaridae). *Journal of Natural History* **9**: 413–24.

Fish, D. (1983). Phytotelmata: flora and fauna. In *Phytotelmata: Terrestrial Plants as Hosts for Aquatic Insect Communities* (eds J.H. Frank and L.P. Lounibos), pp. 1–28. Plexus Publishing Inc., Medford, NJ.

Fish, D. and D.W. Hall. (1978). Succession and stratification of aquatic insects inhabiting the leaves of the insectivorous pitcher plant, *Sarracenia purpurea*. *American Midland Naturalist* **99**: 172–83.

Flechtner, V.R., J.R. Johansen, and W.H. Clark. (1998). Algal composition of microbiotic crusts from the central desert of Baja California, Mexico. *Great Basin Naturalist* **58**: 295–311.

Folkerts, D. (1999). Pitcher plant wetlands of the southeastern United States. In *Invertebrates in Freshwater Wetlands of North America* (eds D.P. Batzer, R.B. Rader, and S.A. Wissinger), pp. 247–75. John Wiley & Sons, New York.

Forest, H.S. and C.R. Weston. (1966). Blue-green algae from the Atacama desert of Northern Chile. *Journal of Phycology* **2**: 163–4.

Forsyth, A.B. and R.J. Robertson. (1975). Reproductive strategy and larval behaviour of the pitcher plant sarcophagid fly, *Blaesoxipha fletcheri*. *Canadian Journal of Zoology* **53**: 174–9.

Frank, J.H. (1983). Bromeliad phytotelmata and their biota, especially mosquitoes. In *Phytotelmata: Terrestrial Plants as Hosts for Aquatic Communities* (eds J.H. Frank and L.P. Lounibos), pp. 101–28. Plexus Publishing, Medford, NJ.

Frank, J.H. and G.A. Curtis. (1981). Bionomics of the bromeliad-inhabiting mosquito *Wyeomia vanduzeei* and its nursery plant *Tillandsia utriculata*. *Florida Entomologist* **54**: 491–506.

Friedmann, E.I. (1964). Xerophytic algae in the Negev Desert. *Abstract, Proceedings of the 10th International Botanical Congress*, Edinburgh, UK, pp. 290–1.

Friedmann, E.I., Y. Lipkin, and R. Ocampo–Paus. (1967). Desert algae of the Negev (Israel). *Phycologia* **6**: 185–96.

Ganning, B. (1971). Studies on chemical, physical and biological conditions in Swedish rockpool ecosystems. *Ophelia* **9**: 51–105.

Giberson, D.J., B. Bilyj, and N. Burgess. (2001). Species diversity and emergence patterns of nematocerous flies (Insecta: Diptera) from three coastal salt marshes in Prince Edward Island, Canada. *Estuaries* **24**: 862–74.

Giberson, D.J. and M.L. Hardwick. (1999). Pitcher plants (Sarracenia purpurea) in eastern Canadian peatlands. In *Invertebrates in Freshwater Wetlands of North America* (eds D.P. Batzer, R.B. Rader, and S.A. Wissinger), pp. 401–22. John Wiley & Sons, New York.

Giesel, J.T. (1976). Reproductive strategies as adaptations to life in temporally heterogeneous environments. *Annual Review of Ecology and Systematics* **7**: 57–79.

Hamilton, R., M. Whitaker, T.C. Farmer, A.A. Bern, and R.M. Duffield. (1996). A report of *Chauliodes* (Megaloptera: Corydalidae) in the purple pitcher plant, *Sarracenia purpurea* (Sarraceniaceae). *Journal of the Kansas Entomological Society* **69**: 257–9.

Hamilton, R., R.L. Petersen, and R.M. Duffield. (1998). An unusual occurrence of caddisflies (Trichoptera: Phryganeidae) in a Pennsylvania population of the purple pitcher plant *Sarracenia purpurea*. *Entomological News* **109**: 36–7.

Hammer, O. (1941). Biological and ecological investigations on flies associated with pasturing cattle and their excrement. *Videnskabelige Meddeleser Dansk Naturhistorik Forening. Kobenhaven* **105**: 1–257.

Henry, W.A. and F.E. Morrison. (1923). *Foods and Feeding*. Henry-Morrison Company, Madison, WI.

Hoham, R.W., J.E. Ullet, and S.C. Roemer. (1983). The life history and ecology of the snow alga *Chloromonas polyptera*, new combination Chlorophyta: Volvocales. *Canadian Journal of Botany* **61**: 2416–29.

Hunt, P.C. and J.W. Jones. (1972). The effect of water level fluctuations on a littoral fauna. *Journal of Fish Biology* **4**: 385–94.

Husby, J.A. and K.E. Zachariassen. (1980). Antifreeze agents in the body fluid of winter active insects and spiders. *Experientia* **36**: 963–4.

Hynes, H.B.N. (1961). The effect of water-level fluctuation on littoral fauna. *Verhandlungen Internationale Vereinigung für Theoretische und Angewandte Limnologie* **14**: 652–5.

Istock, C.A., S.S. Wasserman, and H. Zimmer. (1975). Ecology and evolution of the pitcherplant mosquito. I. Population dynamics and laboratory response to food and population density. *Evolution* **29**: 296–312.

Istock, C.A., K.J. Vavra, and H. Zimmer. (1976). Ecology and evolution of the pitcherplant mosquito. III. Resource tracking by a natural population. *Evolution* **30**: 548–57.

Itamies, J. and E. Lindgren. (1985). The ecology of *Chionea* species (Diptera, Tipulidae). *Notulae Entomologica* **65**: 29–31.

Janetzky, W. and E. Vareshi. (1993). Phytotelmata in bromeliads as microhabitats for limnetic organisms. In *Animal–Plant Interactions in Tropical Environments* (ed. W. Barthlott), pp. 199–209. *Proceedings of the Annual Meeting of the German Society for Tropical Ecology*. Zoologisches Forschungsinstitut und Museum Alexander Koenig, Bonn.

Janetzky, W., P.M. Arbizu, and J.W. Reid. (1996). *Attheyella (Canthosella) mervini* sp. n. (Canthocamptidae, Harpacticoida) from Jamaican bromeliads. *Hydrobiologia* **339**: 123–35.

Kaisila, J. (1952). Insects from arctic mountain snows. *Annales Entomologica Fennica* **18**: 8–25.

Kaufman, M.G., E.D. Walker, T.W. Smith, R.W. Merritt, and M.J. Klug. (1999). Effects of larval mosquitoes (*Aedes triseriatus*) and stemflow on microbial community dynamics in container habitats. *Applied and Environmental Microbiology* **65**: 2661–73.

Keilin, D. (1944). Respiratory systems and respiratory adaptations in larvae and pupae of Diptera. *Parasitology* **36**: 1–36.

Kingsolver, J.G. (1979). Thermal and hydric aspects of environmental heterogeneity in the pitcher plant mosquito. *Ecological Monographs* **49**: 357–76.

Kitching, R.L. (1971). An ecological study of water-filled treeholes and their position in the woodland ecosystem. *Journal of Animal Ecology* **40**: 281–302.

Kitching, R.L. (1983). Community structure in water-filled treeholes in Europe and Australia—comparisons and speculations. In *Phytotelmata: Terrestrial Plants as Hosts for Aquatic Insect Communities* (eds J.H. Frank and L.P. Lounibos), pp. 205–22. Plexus Publishing Inc., Medford, NJ.

Kitching, R.L. (2000). *Food Webs and Container Habitats*. Cambridge University Press, Cambridge.

Kitching, R.L. and A.G. Orr. (1996). The foodweb from water-filled treeholes in Kuala Belalong, Brunei. *The Raffles Bulletin of Zoology* **44**: 405–13.

Kneitel, J.M. and T.E. Miller. (2002). Resource and top-predator regulation in the pitcher plant (*Sarracenia purpurea*) inquiline community. *Ecology* **83**: 680–8.

Koste, W., W. Janetzky, and E. Vareshi. (1995). Zur Kenntnis der limnischen Rotatorienfauna Jamaikas (Rotifera). Teil II. *Osnabrüker Naturwissenschaftliche Mitteilungen* **20/21**: 399–433.

Lampman, R., S. Hanson, and R. Novak. (1997). Seasonal abundance and distribution of mosquitoes at a rural waste tire site in Illinois. *Journal of the American Mosquito Control Association* **13**: 193–200.

Landheim, R., C.P. McKay, E.I. Friedman, and D. Andersen. (2004). Hypolithic algae at Johnson Canyon: Death Valley sample collection of March 5–7, 1997. http://cmex-www.arc.nasa.gov/ExtremeEnvironments/.

Laurence, B.R. (1954). The larval inhabitants of cow pats. *Journal of Animal Ecology* **23**: 234–60.

Leewenhoek, A. (1701). Cited in Hall, F.G. (1922). The vital limit of exsiccation of certain animals. *Biological Bulletin, Woods Hole* **42**: 31–51.

Leinaas, H.P. (1981a). Activity of Arthropoda in snow within coniferous forest, with special reference to Collembola. *Holarctic Ecology* **4**: 127–38.

Leinaas, H.P. (1981b). Cyclomorphosis in the furca of the winter active Collembola *Hypogastrura socialis* (Uzel). *Entomologica Scandinavica* **12**: 35–8.

Lichti-Federovich, S. (1980). Diatom flora of red snow from Isbjorneo, Carey Oer, Greenland. *Nova Hedwegia* **33**: 395–420.

Little, T.J. and P.D.N. Hebert. (1996). Endemism and ecological islands: the ostracods from Jamaican bromeliads. *Freshwater Biology* **36**: 327–38.

Lounibos, L.P. (1980). The bionomics of three sympatric *Eretmapodites* (Diptera: Culicidae) at the Kenya coast. *Bulletin of Entomological Research* **70**: 309–20.

Lounibos, L.P., C.V. Dover, and G.F. O'Meara. (1982). Fecundity, autogeny, and the larval environment of the pitcher-plant mosquito, *Wyeomyia smithii*. *Oecologia* **55**: 160–4.

Lounibos, L.P., R.L. Escher, N. Nishimura, and S.A. Juliano. (1997). Longterm dynamics of a predator used for biological control and decoupling from mosquito prey in a subtropical treehole ecosystem. *Oecologia* **111**: 189–200.

Lozovei, A.L. (1999). Dendricolous mosquitoes (Diptera: Culicidae) in the internodes of bamboo in the Atlantic Forest of the Serra do Mar and the First Plateau (Parana, Brazil). *Brazilian Archives of Biology and Technology* **41**: 501–8.

Machado-Allison, D.J., R. Rodriguez, R. Berrera, and C.G. Cova. (1983). The insect community associated with inflorescences of Helicona caribaea Lamark, in Venezuela. In *Phytotelmata: Terrestrial Plants as Hosts for Aquatic Insect Communities* (eds J.H. Frank and L.P. Lounibos), pp. 247–70. Plexus Publishing Inc., Medford, NJ.

Magadza, C.H.D. (1995). Special problems in lakes/reservoir management in tropical Southern Africa. Keynote Address, pp. 1–6. *Proceedings of the 6th International Conference on the Conservation and Management of Lakes Kasumigaura*, Tsukuba, Japan.

Maguire, B. (1959). Aquatic biotas of teasel waters. *Ecology* **40**: 506.

Marchant, H.J. (1982). Snow algae from the Australian snowy mountains. *Phycologia* **21**: 178–84.

Mattingly, P.F. (1969). *The Biology of Mosquito-borne Diseases*. George Allen and Unwin, London.

McKay, C.P. (1993). Relevance of Antarctic microbial ecosystems to exobiology. In *Antarctic Microbiology* (ed. E.I. Friedmann), pp. 593–601. Wiley-Liss, New York.

Merritt, R.W. and J.R. Anderson. (1977). The effects of different pasture and rangeland ecosystems on the annual dynamics of insects in cattle droppings. *Hilgardia* **45**: 31–71.

Mestre, L.A.M, J.M.R. Aranha, and M.P. Esper. (2001). Macroinvertebrate fauna associated with the bromeliad *Vriesa inflata* of the Atlantic Forest (Paraná State, southern Brazil). *Brazilian Archives of Biology and Technology* **44**: 89–94.

Miles, D.H., U. Kokpol, and N.V. Mody. (1975). Volatiles in *Sarracenia flava*. *Phytochemistry* **14**: 845–6.

Mogi, M. and D.T. Sembel. (1996). Predator-prey system structure in patchy and ephemeral phytotelmata: aquatic communities in small aroid axils. *Researches on Population Ecology* **38**: 95–103.

Mogi, M., T. Sunahara, and M. Selomo. (1999). Mosquito and aquatic predator communities in ground pools on lands deforested for rice field development in central Sulawesi, Indonesia. *Journal of the American Mosquito Control Association* **15**: 92–7.

Moore, C.G. (1999). *Aedes albopictus* in the United States: current status and prospects for further spread. *Journal of the American Mosquito Control Association* **15**: 221–7.

Moran, J.A. and A.J. Moran. (1998). Foliar reflectance and vector analysis reveal nutrient stress in prey deprived pitcher plants (*Nepenthes rafflesiana*). *International Journal of Plant Sciences* **159**: 996–1001.

Moss, B. (1990). Engineering and biological approaches to the restoration from eutrophication of shallow lakes in which aquatic plant communities are important components. *Hydrobiologia* **200/201**: 367–77.

Müller, F. (1880). Wasserthiere in Baumwipfeln *Elpidium bromeliarum*. *Kosmos* (Leipzig) **6**: 386–8.

O'Meara, G.F., L.F. Evans, and M.L. Womack. (1997). Colonization of rock holes by *Aedes albopictus* in the southeastern United States. *Journal of the American Mosquito Control Association* **13**: 270–4.

Paradise, C.J. (1999). Interactive effects of resources and a processing chain interaction in treehole habitats. *Oikos* **85**: 529–35.

Paradise, C.J. and W.A. Dunson. (1997). Insect species interactions and resource effects in tree holes: are helodid beetles bottom-up facilitators of midge populations? *Oecologia* **109**: 303–12.

Parker, B.C., N. Schanen, and R. Renner. (1969). Viable soil algae from the herbarium of the Missouri Botanical Gardens. *Annals of the Missouri Botanical Garden* **56**: 113–9.

Paterson, C.G. (1971). Overwintering ecology of aquatic fauna associated with the pitcher plant *Sarracenia purpurea* L. *Canadian Journal of Zoology* **49**: 1455–9.

Paterson, C.G. and C.J. Cameron (1982). Seasonal dynamics and ecological strategies of the pitcher plant chironomid, *Metriocnemus knabi* Coq. (Diptera: Chironomidae) in southeast New Brunswick. *Canadian Journal of Zoology* **60**: 3075–83.

Prat, N., M. Real, and M. Rieradevall. (1992). Benthos of Spanish lakes and reservoirs. *Limnetica* **8**: 221–9.

Prus, M., P. Bijok, and T. Prus. (2002). Trophic structure of the benthic invertebrate community in the littoral zone of a mountain cascade system. *Web Ecology* **3**: 12–9.

Reid, J.W. and W. Janetzky. (1996). Colonization of Jamaican bromeliads by *Tropocyclops jamaicensis* n. sp. (Crustacea: Copepoda: Cyclopoida). *Invertebrate Biology* **115**: 305–20.

Rieradevall, M. and J. Cambra. (1994). Urban freshwater ecosystems in Barcelona. *Verhandlungen Internationale*

Vereinigung für Theoretische und Angewandte Limnologie **25**: 1369–72.

Robert, V., H.P. Awonoambene, and J. Thioulouse. (1998). Ecology of larval mosquitoes, with special reference to *Anopheles arabiensis* (Diptera: Culicidae) in market-garden wells in urban Dakar, Senegal. *Journal of Medical Entomology* **35**: 948–55.

Rohnert, U. (1950). Wasserfullte Baumhohlen und ihre Besiedlung. Ein Beitrag zur Fauna Dendrolimnetica. *Archiv für Hydrobiologie* **44**: 472–516.

Schreiber, E.T., C.F. Hallmon, K.M. Eskridge, and G.G. Marten. (1996). Effects of *Mesocyclops longisetus* (Copepoda: Cyclopida) on mosquitoes that inhabit tires: influence of litter type, quality, and quantity. *Journal of the American Mosquito Control Association* **12**: 688–94.

Schwabe, G.H. (1963). Blaualgen der phototropen Grengschicht. Blaualgen und Lebensraum. VII. *Pedobiologia* **2**: 132–52.

Service, M.W. (2000). *Medical Entomology*. Cambridge University Press, Cambridge.

Sota, T. (1996). Effects of capacity on resource input and the aquatic metazoan community structure in phytotelmata. *Researches on Population Ecology* **38**: 65–73.

Sota, T. and M. Mogi. (1996). Species richness and altitude variation in the aquatic metazoan community in bamboo phytotelmata from North Sulawesi. *Researches on Population Ecology* **38**: 275–81.

Steenis, C.G., G.J. van and M.M.J. van Balgooy. (1966). Pacific plant areas, Vol. 2. *Blumea* **5** (Suppl.): 1–312.

Sugden, A.M. and R.J. Robins. (1979). Aspects of the ecology of vascular epiphytes in Colombian closed forests I. The distribution of the epiphytic flora. *Biotropica* **11**: 173–88.

Sunahara, T. and M. Mogi. (1997). Can the tortoise beat the hare: a possible mechanism for the coexistence of competing mosquitoes in bamboo groves. *Ecological Research* **12**: 63–70.

Sutcliffe, D.W. (1960). Osmotic regulation in the larvae of some euryhaline Diptera. *Nature* **187**: 331–2.

Sutcliffe, D.W. (1961). Salinity fluctuations and the fauna in a salt marsh with special reference to aquatic insects. *Transactions of the Natural History Society of Northumberland and Durham* **14**: 37–56.

Thienemann, A. (1932). Die Tierwelt der *Nepenthes*-Kannen. *Archiv für Hydrobiologie*, **11** (Suppl.): 1–54.

Thompson, V. (1997). Spittlebug nymphs (Homoptera: Cercopidae) in Heliconia flowers (Zingiberales: Heliconiaceae): preadaptation and evolution of the first aquatic Homoptera. *Revue Biologica Tropica* **45**: 905–12.

Torres, J.R., B.M. Goebel, E.I. Friedmann, and N.R. Pace. (2003). Microbial diversity of cryptoendolithic communities from the McMurdo Dry Valleys, Antarctica. *Applied and Environmental Microbiology* **69**: 3858–67.

Trainor, F.R. (1962). Temperature tolerance of algae in dry soil. *Phycological Society of America, News Bulletin* **15**: 3–4.

Trpis, M. (1972). Seasonal changes in the larval populations of *Aedes aegypti* in two biotopes in Dar es Salam, Tanzania. *Bulletin of the World Health Organization* **47**: 245–55.

Turner, T.S., J.L. Pittman, M.E. Poston, R.L. Petersen, M. Mackenzie, C.H. Nelson, and R.M. Duffield. (1996). An unsual occurrence in West Virginia of stoneflies (Plecoptera) in the pitcher plant, *Sarracenia purpurea* (Sarraceniaceae). *Proceedings of the Entomological Society of Washington* **98**: 119–21.

Varga, L. (1928). Ein interessanter Biotop der Biocönose von Wasserorganismen. *Biologische Zentralblatt* **48**: 143–62.

Victor, R. and C.H. Fernando. (1978). Systematics and ecological notes on Ostracoda from container habitats of some South Pacific Islands. *Canadian Journal of Zoology* **56**: 414–22.

Vogel, S. (1955). Niedere Fensterpflanzen in der sudafrikanischen. *Wuste Beiträge Biologischen Pflanzen* **31**: 45–135.

Walker, E.D., G.F. O'Meara, and W.T. Morgan. (1996). Bacterial abundance in larval habitats of *Aedes albopictus* (Diptera: Culicidae) in a Florida cemetery. *Journal of Vector Ecology* **21**: 173–7.

Ward, J.V. and J.A. Stanford. (eds). (1979). *The Ecology of Regulated Streams*. Plenum Publishers, New York.

Watts, R.B. and S.M. Smith. (1978). Oogenesis in *Toxorhynchites rutilus* (Diptera: Culicidae). *Canadian Journal of Zoology* **56**: 136–9.

Weiss, R.L. (1983). Fine structure of the snow alga *Chlamydomonas nivalis* and associated bacteria. *Journal of Phycology* **19**: 200–4.

Williams, D.D. and N.E. Williams. (1976). Aspects of the ecology of the faunas of some brackishwater pools on the St. Lawrence North Shore. *Canadian Field-Naturalist* **90**: 410–5.

Winterbourn, M.J. and N.H. Anderson. (1980). The life history of *Philanisus plebeius* Walker (Trichoptera: Chathamiidae), a caddisfly whose eggs were found in a starfish. *Ecological Entomology* **5**: 293–303.

Yair, A. (1990). Runoff generation in a sandy area of the Nizzana Sands, western Negev, Israel. *Earth Surface Processes and Landforms* **15**: 597–609.

Yozzo, D.J. and R.J. Diaz. (1999). Tidal freshwater wetlands: invertebrate diversity, ecology, and functional significance. In *Invertebrates in Freshwater Wetlands of North America* (eds D.P. Batzer, R.B. Rader, and S.A. Wissinger), pp. 889–918. John Wiley & Sons, New York.

Zalewski, M., B. Brewinska-Zaras, P. Frankiewicz, and S. Kalinowski. (1990a). The potential for biomanipulation using communities in a lowland reservoir: concordance between water quality and optimal recruitment. *Hydrobiologia* **200/201**: 549–56.

Zalewski, M., B. Brewinska-Zaras, and P. Frankiewicz. (1990b). Fry communities as a biomanipulating tool in a temperate lowland reservoir. *Archiv für Hydrobiologie* **33**: 763–74.

Zettel, J. (1984). The significance of temperature and barometric pressure changes for the snow surface activity of *Isotoma hiemalis* (Collembola). *Experienta* **40**: 1369–72.

Chapter 8

Abbasi, A.A. (1997). *Wetlands of India: Ecology and Threats. Vol. 1: The Ecology and the Exploitation of Typical South Indian Wetlands.* Discovery Publishing House, New Delhi.

Abernethy, V.J. and N.J. Willby. (1999). Changes along a disturbance gradient in the density and composition of propagule banks in floodplain aquatic habitats. *Plant Ecology* **140**: 177–90.

Ackermann, W.C., G.F. White, and E.B. Worthington. (1973). *Man-Made Lakes: Their Problems and Environmental Effects.* Geophysical Monograph 17. American Geophysical Union, Washington DC.

Awachie, J.B.E. (1981). Running water ecology in Africa. In *Perspectives in Running Water Ecology* (eds M.A. Lock and D.D. Williams), pp. 339–66. Plenum Press, New York.

Baptista, D.F. (1999). Relation between flood pulse and functional composition of the macroinvertebrate benthic fauna in the lower Rio Negro, Amazonas, Brazil. *Amazoniana Limnologia et Oecologia Regionalis Systemae Fulminis Amazonas* **15**: 35–50.

Bird, G.A. and N.K. Kaushik. (1981). Coarse particulate organic matter in streams. In *Perspectives in Running Water Ecology* (eds M.A. Lock and D.D. Williams), pp. 41–68. Plenum Press, New York.

Bren, L.J. (1988). Effects of river regulation on flooding of a riparian Red Gum forest on the Murray River, Australia. *Regulated Rivers* **2**: 65–78.

Brinson, M.M. (1990). Riverine forests. In *Forested Wetlands* (eds A.E. Lugo, M.M. Brinson, and S. Brown), pp. 87–142. *Ecosystems of the World*, Vol. 15. Elsevier, New York.

Brinson, M.M., A.E. Lugo, and S. Brown. (1981). Primary productivity, decomposition and consumer activity in freshwater wetlands. *Annual Review of Ecology and Systematics* **12**: 123–61.

Cruz, E.M. and C.R. de la Cruz. (1991). Production of common carp (*Cyprinus carpio*) with supplemental feeding in ricefields in North Sumatra, Indonesia. *Asian Fisheries Science* **4**: 31–9.

de Graaf and N.D. Chinh. (2004). Floodplain fisheries in the southern provinces of Vietnam. http:www.nefisco.org/downloads/Vietnam.PDF.

Drobney, R.D. and L.H. Fredrickson. (1979). Food selection by wood ducks in relation to breeding status. *Journal of Wildlife Management* **43**: 109–20.

Dudley, R.G. (1976). Status of major fishes of the Kafue Floodplain, Zambia five years after completion of the Kafue Gorge dam. Final Report submitted to the National Science Foundation Scientists and Engineers in Economic Development Programme. University of Athens, Georgia Grant No. OIP75-09239.

Dussart, B.H. (1974). Biology of inland waters in humid tropical Asia. In *Natural Resources of Humid Tropical Asia*, pp. 331–53. Natural Resources Research No. 12, UNESCO, Paris.

Elphick, C.S. (2000). Functional equivalency between rice fields and seminatural wetland habitats. *Conservation Biology* **14**: 181–91.

Ezenwaji, H.M.G. (1998). The breeding biology of *Clarias albopunctatus* Nichols and LaMonte, 1953 in semi-intensively managed ponds in the floodplain of the River Anambra, Nigeria. *Ecology of Freshwater Fish* **7**: 101–7.

Fily, M. and F. D'Aubenton. (1966). *Cambodge: Grand Lac-Tonle Sap. Technologies des Pêches 1962–1963.* Ministére des Affaires Estrangeres Service de Coopération Technique, Paris.

Food and Fertilizer Technology Centre. (1994). Corn production in paddy fields during the off season. *FFTC Workshop, January 31–February 6, 1994*, Philippines. http://www.fftc.agnet.org/library/article/ac1994b.html.

Fugi, R., A.A. Agostinho, and N.S. Hahn. (2001). Trophic morphology of five benthic-feeding fish species of a tropical floodplain. *Reviews of Brazilian Biology* **61**: 27–33.

Gill, M.A. and H.M. Reisenauer. (1993). Nature and characterization of ammonium effects on wheat and tomato. *Agronomy Journal* **85**: 874–9.

Grant, I.F. and R. Seegers. (1985). Tubificid role in soil mineralization and recovery of algal nitrogen by lowland rice. *Soil Biology and Biochemistry* **17**: 559–63.

Gregory, S.V., F.J. Swanson, W.A. McKee, and K.W. Cummins. (1991). An ecosystem perspective of riparian zones. *BioScience* **41**: 540–52.

Grigarick, A.A., W.H. Lange, and D.C. Finlock. (1961). Control of the tadpole shrimp, *Triops longicaudatus*, in California rice fields. *Journal of Economic Entomology* **54**: 36–40.

Harper, D., J. Mekotova, S. Hulme, J. White, and J. Hall. (1997). Habitat heterogeneity and aquatic invertebrate diversity in floodplain forests. *Global Ecology and Biogeography Letters* **6**: 275–85.

Hora, S.L. and T.V.R. Pillay. (1962). *Handbook of Fish Culture in the Indo-Pacific Region*. FAO Fisheries Biology, Technical Publication 14, FAO, Rome.

Huat, K.K. and E.S.P. Tan. (1980). Review of rice-fish culture in Southeast Asia. In *Integrated Agriculture Farming Systems* (eds R.S.V. Pullin and Z.H. Shehadeh). *ICLARM Conference Proceedings 4*. International Centre for Living Aquatic Resource Management, Manila and the Southeast Asian Centre for Graduate Study and Research in Agriculture, College Los Baños, Laguna, Philippines.

Huffman, R.T. and S.W. Forsythe. (1981). Bottomland hardwood forest communities and their relation to anaerobic soil conditions. In *Wetlands of Bottomland Hardwood Forests* (eds J.R. Clark and J. Benforado), pp. 187–96. Elsevier, Amsterdam.

Huryn, A.D. and K.E. Gibbs. (1999). Riparian sedge meadows in Maine: a macroinvertebrate community structured by river-floodplain interactions. In *Invertebrates in Freshwater Wetlands of North America* (eds D.P. Batzer, R.B. Rader, and S.A. Wissinger), pp. 363–82. John Wiley & Sons, New York.

Husain, Z. (1973). Fish and fisheries of the lower Indus Basin (1966–1967). *Agriculture Pakistan* **24**: 297–322.

Hyslop, E.J. (1986). The food habits of four small-sized species of Mormyridae from the floodplain pools of the Sokoto-Rima river basin, Nigeria. *Journal of Fish Biology* **28**: 147–51.

Hyslop, E.J. (1988). A comparison of the composition of the juvenile fish catch from the Sokoto-Rima floodpalin, Nigeria in years preceding and immediately after upstream dam completion. *Journal of Fish Biology* **32**: 895–9.

Ibañez, M. do Socorro. (1998). Phytoplankton composition and abundance of a central Amazonian floodplain lake. *Hydrobiologia* **362**: 79–83.

International Centre for Aquaculture and Aquatic Environments. (2004). *Introduction to Fish Culture in Rice Paddies*. ICAAE Publication Series A, Number 14, Auburn University, Alabama.

Ita, E.O. (1993). *Inland Fishery Resources of Nigeria*. CIFA Occasional Paper 20. Food and Agricultural Organization, Rome.

Jacobson, P.J., K.M. Jacobson, and M.K. Seely. (1995). *Ephemeral Rivers and their Catchments: Sustaining People and Development in Western Namibia*. Desert Research Foundation of Namibia, Windhoek.

Junk, W.J. (1997). *The Central Amazon Floodplain*. Ecological Studies 126. Springer-Verlag, Berlin.

Junk, W.J., P.B. Bailey, and R.E. Sparks. (1989). The flood pulse concept in river-floodplain systems. In *Proceedings of the International Large River Symposium* (ed. D.P. Dodge), pp. 110–27. Special Publication 106, *Canadian Journal of Fisheries and Aquatic Sciences*.

Junk, W.J., M.G.M. Soares, and U. Saint-Paul. (1997). The Fish. In *The Central Amazon Floodplain* (ed. W.J. Junk), pp. 385–408. Ecological Studies 126. Springer-Verlag, Berlin.

Kirk, G.J.D and H.J. Kronzucker. (2000). Nitrogen uptake by rice roots. In *Carbon and Nitrogen Dynamics in Flooded Soils* (eds G.J.D. Kirk and D.C. Olk), pp. 147–62. International Rice Research Institute, Los Baños, Philippines.

Knollenberg, W.G., R.W. Merritt, and D.L. Lawson. (1985). Consumption of leaf litter by Lumbrcus terrestris (Oligochaeta) on a Michigan woodland floodplain. *The American Midland Naturalist* **113**: 1–6.

Kronzucker, H.J., A.D.M. Glass, M.Y. Siddiqi, and G.J.D. Kirk. (2000). Comparative kinetic analysis of ammonium and nitrate acquisition by tropical lowland rice: implications for rice cultivation and yield potential. *The New Phytologist* **145**: 471–6.

Lowe-McConnell, R.H. (1977). *Ecology of Fishes in Tropical Waters*. Edward Arnold, London.

Lowe-McConnell, R.H. (1991). *Ecological Studies in Tropical Fish Communities*. Cambridge University Press, Cambridge.

Lugo, A.E. (1990). Introduction. In *Forested Wetlands* (eds A.E. Lugo, M.M. Brinson, and S. Brown), pp. 1–14. *Ecosystems of the World*, Vol. 15. Elsevier, New York.

Mandima, J.J. (1995). Household fish sources, supply and consumption patterns in Mutasa and Chivhu districts, Zimbabwe. *Zimbabwe Science News* **29**: 32–34.

Matsunaka, S. (1976). Tadpole shrimps: a biological tool of weed control in transplanted rice fields. *Proceedings of the 5th Asian-Pacific Weed Science Society Conference*, Tokyo, 1975, pp. 439–43.

Matthes, H. (1977). The problem of rice-eating fish in the Central Delta, Mali. *Contributions to FAO/CIFA Symposium on River and Floodplain Fisheries*, November, Burundi.

Merritt, R.W. and D.L. Lawson. (1992). The role of leaf litter macroinvertebrates in stream-floodplain dynamics. *Hydrobiologia* **248**: 65–77.

Meschkat, A. (1975). Aquacultura e pesca em aguas interiores no Brasil. *Documente Téchnicale SUDEPE* **9**: 1–35.

Mitsch, W.J., C.L. Dorge, and J.R. Wiemhoff. (1979). Ecosystem dynamics and a phosphorus budget of an alluvial cypress swamp in southern Illinois. *Ecology* **60**: 1116–24.

Mozley, A. (1944). Temporary ponds, neglected natural resource. *Nature* **154**: 490.

Native Fish Australia. (2004). *Golden Perch*. www.nativefish.asn.au.

Patrick, W.H. (1981). Bottomland soils. In *Wetlands of Bottomland Hardwood Forests* (eds J.R. Clark and J. Benforado), pp. 177–85. Elsevier, Amsterdam.

Petry, A.C., A.A. Agostinho, and L.C. Gomes. (2003b). Fish assemblages of tropical floodplain lagoons: exploring the role of connectivity in a dry year. *Neotropical Ichthyology* **1**: 111–9.

Petry, P., P.B. Bayley, and D.F. Markle. (2003a). Relationships between fish assemblages, macrophytes and environmental gradients in the Amazon River floodplain. *Journal of Fish Biology* **63**: 547–53.

Raimondo, P. (1975). Monograph on operation fisheries, Mopti. *CIFA Occasional Papers* **4**: 294–311.

Rodríguez, M.A. and W.M. Lewis. (1997). Structure of fish assemblages along environmental gradients in floodplain lakes of the Orinoco River. *Ecological Monographs* **67**: 109–28.

Ruffino, M.L. (1996). Towards participatory fishery management on the Lower Amazon. *EC Fisheries Cooperation Bulletin* **9**: 15–8.

Sharitz, R.R. and W.J. Mitsch (1993). Southern floodplain forests. In *Biodiversity of the Southeastern United States* (eds W.H. Martin, S.G. Boyce, and A.C. Echternacht), pp. 311–72. John Wiley & Sons, New York.

Smock, L.A. (1994). Movements of invertebrates between stream channels and forested floodplains. *Journal of the North American Benthological Society* **13**: 524–31.

Smock, L.A. (1999). Riverine floodplain forests of the southeastern United States: invertebrates in an aquatic-errestrial ecotone. In *Invertebrates in Freshwater Wetlands of North America* (eds D.P. Batzer, R.B. Rader, and S.A. Wissinger), pp. 137–65. John Wiley & Sons, New York.

Soa-Leang and Dom Saveun. (1955). Aperçu général sur la migration et la reproduction des poissons d'eau douce du Cambodge. *Proceedings of the Indo-acific Fishery Commission* **5**: 138–62.

South Australia Fisheries Research Advisory Board. (2004). *Inland Fisheries—the Murray River, and the Lakes and Coorong Fisheries*. www.fishresearch.sa.gov.au/industry/inland.htm.

South Australian Research and Development Institute. (2001). *Inland Waters Fishery*. www.sardi.sa.gov.au/pages/aquatics/fish_stats.

Tappin, A.R. (2004). *Kakadu*. http://members.optushome.com.au/chelmon/Kakadu.htm

Truchelut, A. (1982). *Coutumes et Usages des Etangs de la Dombes et de la Bresse* (par C. Rivoire). Editions de Trevoux, Paris.

von Brandt, A. (1984). *Fish Catching Methods of the World*. Fishing News Books Ltd., London.

Welcomme, R.L. (1979). *Fisheries Ecology of Floodplain Rivers*. Longman, London.

Xiuzhen, F. (2003). Rice–fish culture in China. *Aquaculture Asia* **8**: 44–6.

Yonekura, M. (1979). Weeding efficiency of the tadpole shrimp (*Triops* spp.) in transplanted rice fields. *In Proceedings of the 7th Asian-Pacific Weed Science Society Conference*, Tokyo, pp. 237–40.

Chapter 9

Adhami, J. and P. Reiter. (1998). Introduction and establishment of *Aedes* (*Stegmyia*) *albopictus* (Diptera: Culicidae) in Albania. *Journal of the American Mosquito Control Association* **14**: 340–3.

Alphey, L., C.B. Beard, P. Billingsley, M. Coetzee, A. Cristani, C. Curtis *et al.* (2002). Malaria control with genetically manipulated insects. *Science* **298**: 119–21.

Amerasinghe, F.P., N.G. Indrajith, and T.G. Ariyasena. (1995). Physico-chemical characteristics of mosquito breeding habitats in an irrigation development area in Sri Lanka. *Ceylon Journal of Science (Biological Science)* **24**: 13–29.

Amerasinghe, F.P., F. Konradsen, K.T. Fonseka, and P.H. Amerasinghe. (1997). Anopheline (Diptera: Culicidae) breeding in a traditional tank-based village ecosystem in north-central Sri Lanka. *Journal of Medical Entomology* **34**: 290–7.

Andreadis, T.G. (1999). Epizooticology of *Amblyospora stimuli* (Microsporida: Amblyosporidae) infections in field populations of a univoltine mosquito, *Aedes stimulans* (Diptera: Culicidae), inhabiting a temporary vernal pool. *Journal of Invertebrate Pathology* **74**: 198–205.

Anon. (2005). Editorial: Reversing the failures of Roll Back Malaria. *Lancet* **365**: 1439.

Awono-Ambene, H.P. and V. Robert. (1999). Survival and emergence of immature *Anopheles arabiensis* mosquitoes in market-garden wells in Dakar, Senegal. *Parasite: Journal de la Societe Francaise de Parasitologie* **6**: 179–84.

Bannister, B.A., N.T. Begg, and S.H. Gillespie. (1996). *Infectious Disease*. Blackwell Scientific, Oxford.

Barbosa, F.S. and I. Barbosa. (1958). Dormancy during the larval stages of the trematode *Schistosoma mansoni* in snails aestivating on the soil of dry natural habitats. *Ecology* **39**: 763–4.

Beadle, L.C. (1981). *The Inland Waters of Tropical Africa*. Longman, London.

Becker, N. and M. Ludwick. (1993). Investigations on possible resistance in *Aedes vexans* field populations after a 10 year application of *Bacillus thuringiensis israelensis*. *Journal of the American Mosquito Control Association* **9**: 221–4.

Blaustein, L. (1992). Larvivorous fishes fail to control mosquitoes in experimental rice plots. *Hydrobiologia* **232**: 219–32.

Blaustein, L. and R. Karban. (1990). Indirect effects of the mosquitofish *Gambusia affinis* on the mosquito *Culex tarsalis*. *Limnology and Oceanography* **35**: 767–71.

Blaustein, L. and J. Margalit. (1991). Indirect effects of the fairy shrimp, *Branchipus schaefferi* and two ostracod species on *Bacillus thuringiensis* var *israelensis*-induced mortality in mosquito larvae. *Hydrobiologia* **212**: 67–76.

Blaustein, L. and J. Margalit. (1994). Differential vulnerability among mosquito species to predation by the cyclopoid copepod, *Acanthocyclops viridis*. *Israel Journal of Zoology* **40**: 55–60.

Blaustein, L., M. Kiflawi, A. Eitam, M. Mangel, and J.E. Cohen. (2004). Oviposition habitat selection in response to risk of predation in temporary pools: mode of detection and consistency across experimental venue. *Oecologia* **138**: 300–5.

Brown, M.D., J.K. Hendriks, J.G. Greenwood, and B.H. Kay. (1996). Evaluation of *Mesocyclops aspericornis* (Cyclopoida, Cyclopidae) and *Toxorhynchites speciosus* as integrated predators of mosquitoes in tire habitats in Queensland. *Journal of the American Mosquito Control Association* **12**: 414–20.

Canada. (1996). *The State of Canada's Forests, 1995–1996*. Forestry Department, Natural Resources Canada, Government of Canada.

Carter Centre. (1999). The global 2000 river blindness. *River Blindness News* **11**: 1–8.

Chaniotis, B. and Y. Tselentis. (1996). Water wells as a habitat of sandfly (Diptera: Psychodidae) vectors of visceral leishmaniasis in Greece. *Journal of Medical Entomology* **33**: 269–70.

Charlwood, J.D. and D. Edoh. (1996). Polymerase chain-reaction used to describe larval habitat use by *Anopheles gambiae* complex (Diptera: Culicidae) in the environs of Ifakara, Tanzania. *Journal of Medical Entomology* **33**: 202–4.

Chimbale, M. (1993). *Schistosomiasis Control Measures for Small Irrigation Schemes in Zimbabwe*. HR Wallingford Report OD 128. Wallingford, UK.

CDC (Centre for Disease Control and Prevention). (2004a). *Eradication of Malaria in the United States (1947–1951)*. http://www.cdc.gov/malaria/history/eradication_us.htm.

CDC (Centre for Disease Control and Prevention). (2004b). *Eradication Efforts Worldwide: Success and Failure (1955–1978)*. http://www.cdc.gov/malaria/history/.

Chui, V.W.D., K.W. Wong, and K.W. Tsoi. (1995). Control of mosquito larvae (Diptera: Culicidae) using Bti and teflubenzuron: laboratory evaluation and semi-field test. *Environment International* **21**: 433–40.

Collins, F.H. and S.M. Paskewitz. (1995). Malaria: current and future prospects for control. *Annual Review of Entomology* **40**: 195–219.

Collins, F.H. and R.K. Washino. (1985). Insect predators. In *Biological Control of Mosquitoes* (ed. H.C. Chapman), pp. 25–41. *American Mosquito Control Association Bulletin* 6.

Collins, L.E. and A. Blackwell. (2000). The biology of *Toxorhynchites* mosquitoes and their potential as biocontrol agents. *Biocontrol/News and Information* **21**: 1–12.

Comiskey, N.M., R.C. Lowrie, and D.M. Wesson. (1999). Role of habitat components on the dynamics of *Aedes albopictus* (Diptera: Culicidae) from New Orleans. *Joural of Medical Entomology* **36**: 313–20.

Croll, N.A. (1966). *Ecology of Parasites*. Heinemann, London.

Dale, P.E.R., S.A. Ritchie, B.M. Territo, C.D. Morris, A. Muhar, and B.H. Kay. (1998). An overview of remote sensing and GIS for surveillance of mosquito vector habitats and risk assessment. *Journal of Vector Ecology* **23**: 54–61.

Dreyer, G., J. Noroes, J. Figueredo-Silva, and W.F. Piessens (2000). Pathogenesis of lymphatic disease in bancroftian filariasis: a clinical perspective. *Parasitology Today* **16**: 544–8.

Edman, J.D., T.W. Scott, A. Costero, A.C. Morrison, L.C. Harrington, and G.G. Clark. (1998). *Aedes aegypti* (Diptera: Culicidae) movement influenced by availability of oviposition sites. *Journal of Medical Entomology* **35**: 578–83.

Focks, D.A., S.R. Sackett, and D.L. Bailey. (1982). Field experiments on the control of *Aedes aegypti* and *Culex quinquefasciatus* by *Toxorhynchites rutilus rutilus* (Diptera: Culicidae). *Journal of Medical Entomology* **19**: 336–9.

Food and Agriculture Organization (2004). *Disease Ecology*. http://www.fao.org/docrep/V8350E/v8350e0c.htm.

Frank, J.H. (1983). Bromeliad phytotelmata and their biota, especially mosquitoes. In *Phytotelmata: Terrestrial Plants as Hosts for Aquatic Insect Communities* (eds J.H. Frank and L.P. Lounibos), pp. 101–28. Plexus Publishing Inc., Medford, New Jersey.

Fry, L.L., M.S. Mulla, and C.W. Adams. (1996). Field introductions and establishment of the tadpole shrimp, *Triops longicaudatus* (Notostraca: Triopsidae), a biological control agent of mosquitoes. *Biological Control* **4**: 113–24.

Gerold, J.L. (1977). Evaluation of some parameters of house-leaving behaviour of *Anopheles gambiae s.l. Acta Leiden* **45**: 79–90.

Goldberg, L.J. and J. Margalit. (1977). A bacterial spore demonstrating rapid larvicidal activity against *Anopheles sergentii, Uranotaenia unguiculata, Culex univitattus, Aedes aegypti* and *Culex pipiens. Mosquito News* **37**: 355–8.

Gorrochoteguiescalante, N., I. Fernandezsalas, and H. Gomezdantes. (1998). Field-evaluation of *Mesocyclops longisetus* (Copepoda: Cyclopoidea) for the control of larval *Aedes aegypti* (Diptera: Culicidae) in northeastern Mexico. *Journal of Medical Entomology* **35**: 699–703.

Gubler, D.J. and G. Kuno. (eds). (1997). *Dengue and Dengue Haemorrhagic Fever*. C.A.B. International, Wallingford.

Guerin, P.J., P. Olliaro, F. Nosten, P. Druilhe, R. Laxminarayan, F. Binka, W.L. Kilama, N. Ford, and N.J. White. (2002). Malaria: current status of control, diagnosis, treatment, and a proposed agenda for research and development. *The Lancet Infectious Diseases* **2**: 564–73.

Hubbard, S.F., S.L.C. O'Malley, and R. Russo. (1988). The functional response of *Toxorhynchites rutilus rutilus* to changes in the population density of its prey, *Aedes aegypti. Medical and Veterinary Entomology* **2**: 279–83.

Hurlburt, H.S. (1938). Copepod observed preying on first instar larva of *Anopheles quadrimaculatus* Say. *Journal of Parasitology* **24**: 281.

Jones, A.W. (1967). *Introduction to Parasitology*. Addison-Wesley, London.

Jordan, P. and G. Webbe. (1969). *Human Schistosomiasis*. Heinemann, London.

Keeton Industries (2005). *Aquaculture Equipment and Designs for the 21st Century*. www.keetonaqua.com.

Kerwin, J.L., C.A. Simmons, and R.K. Washino. (1986). Oosporogenesis by *Lagenidium giganteum* in liquid culture. *Journal of Invertebrate Pathology* **47**: 258–70.

Kerwin, J.L., D.D. Dritz, and R.K. Washino. (1994). Pilot scale production and application in wildlife ponds of *Lagenidium giganteum* (Oomycetes: Lagenidiales). *Journal of the American Mosquito Control Association* **10**: 451–5.

Kidson, C. and K. Indaratna. (1998). Ecology, economics and political will: the vicissitudes of malaria strategies in Asia. *Parasitologia* **40**: 39–46.

Lainson, R. and J.J. Shaw. (1971). Epidemiological considerations of the leishmanias with particular reference to the New World. In *Ecology and Physiology of Parasites* (ed. A.M. Fallis), pp. 21–57. University of Toronto Press, Toronto.

Marshall, L. (1913). *The Story of the Panama Canal*. L. T. Myers Publisher, New York.

Marten, G.G. (1990). Elimination of *Aedes albopictus* from tyre piles by introducing *Macrocyclops albidus* (Copepoda: Cyclopoida). *Journal of the American Mosquito Control Association* **6**: 689–93.

Marten, G.G., E.S. Bordes, and M. Nguyen. (1994). Use of cyclopoid copepods for mosquito control. *Hydrobiologia* **292/293**: 491–6.

McKenzie, E., J.W. Barnwell, G.M. Jeffery, and W.E. Collins. (2002). *Plasmodium vivax* blood-stage dynamics. *Parasitology* **88**: 521–35.

Mitchell, C.J., C.D. Morris, G.C. Smith, N. Karabatsos, D. VanLandingham, and E. Cody. (1996). Arboviruses associated with mosquitoes from nine Florida counties during 1993. *Journal of the American Mosquito Control Association* **12**: 255–62.

Mitchell, C.J., L.D. Haramis, N. Karabatsos, G.C. Smith, and V.J. Starwalt. (1998). Isolation of La Crosse, Cache Valley, and Potosi viruses from *Aedes* mosquitoes (Diptera: Culicidae) collected at used-tyre sites in Illinois during 1994–1995. *Journal of Medical Entomology* **35**: 573–7.

Mokany, A. and R. Shine. (2002). Pond attributes influence competitive interactions between tadpoles and mosquito larvae. *Austral Ecology* **27**: 396–404.

Nannini, M.A. and S.A. Juliano. (1998). Effects of the facultative predator *Anopheles barberi* on population performance of its prey *Aedes triseriatus* (Diptera: Culicidae). *Annals of the Entomological Society of America* **91**: 33–42.

Nasci, R.S. and C.G. Moore. (1998). Vector-borne disease surveillance and natural disasters. *Emerging Infectious Diseases* **4**: 1–2. http://www.cdc.gov/ncidod/EID/vol4no2/nasci.htm.

Neilsen-LeRoux, C. and M.H. Silva–Filha. (1996). Bacteria: a tool in the integrated control of mosquito vectors of diseases. *Bulletin of Pest Management* **7**: 9.

Neng, W., L. Guohou, L. Yulin, and Z. Gemei. (1995). The role of fish in controlling mosquitoes in ricefields. In *Rice-Fish Culture in China* (ed. K.T. MacKay), Section III. International Development Research Centre, Ottawa, Canada.

Neng, W., W. Shusen, H. Guangxin, X. Rongman, T. Guangkun, and Q. Chen. (1987). Control of *Aedes aegypti* larvae in household water containers by Chinese catfish. *Bulletin of the World Health Organization* **65**: 503–6.

New Jersey Mosquito Control Association. (1998). *Environmental Protection Agency's Pesticide Environmental Stewardship Programme*. Partnership Strategy Document. http://www-rci.rutgers.edu/~insects/psd.htm.

Olsen, O.W. (1974). *Animal Parasites: Their Life Cycles and Ecology*. University Park Press, Baltimore, MD.

Omer, S.M. and J.L. Cloudsley-Thompson. (1968). Dry season biology of *Anopheles gambiae* Giles in the Sudan. *Nature* **217**: 879–80.

Ottesen, E.A. (2000). The global programme to eliminate lymphatic filariasis. *Tropical Medicine and International Health* **5**: 591–4.

Pajot, F. (1983). Phytotelmata and mosquito vectors of sylvatic yellow fever in Africa. In *Phytotelmata: Terrestrial Plants as Hosts for Aquatic Insect Communities* (eds J.H. Frank and L.P. Lounibos), pp. 79–100. Plexus Publishers Incorporated, Medford, NJ.

Pates, H. and C. Curtis. (2005). Mosquito behaviour and vector control. *Annual Review of Entomology* **50**: 53–70.

Pennisi, E. (2001). Malaria's beginnings: on the heels of hoes? *Science* **293**: 416–7.

Perich, M.J., P.M. Clair, and L.R. Boobar. (1990). Integrated use of planaria (*Dugesia dorotocephala*) and *Bacillus thuringiensis* var. *israelensis* against *Aedes taeniorhynchus*: a laboratory bioassay. *Journal of the American Mosquito Control Association* **6**: 667–71.

Pike, E.G. (1987). *Engineering Against Schistosomiasis/Bilharzia*. MacMillan, London.

Pont, D., E. Franquet, and J.N. Tourenq. (1999). Impact of different *Bacillus thuringiensis* variety *israelensis* treatments on a chironomid (Diptera: Chironomidae) community in a temporary marsh. *Journal of Economic Entomology* **92**: 266–72.

Pradeep Kumar, N., S. Sabesan, and K.N. Panicker. (1992). The resting and house frequenting behaviour of *Mansonia annulifera, Ma. uniformis* and *Ma. indiana*, the Vectors of Malayan filariasis in Kerala State, India. *Southeast Asian Journal of Tropical Medicine and Public Health* **23**: 324.

Rai, K.S. (1991). *Aedes albopictus* in the Americas. *Annual Review of Entomology* **36**: 459–84.

Rai, K.S. (1999). Genetics of mosquitoes. *Journal of Genetics* **78**: 163–9.

Reiter, P., M.A. Amador, R.A. Anderson, and C.G. Clark. (1995). Dispersal of *Aedes aegypti* in an urban area after blood feeding as demonstrated by rubidium-marked eggs. *American Journal of Tropical Medicine and Hygiene* **52**: 177–9.

Rejmankova, E., K.O. Pope, D.R. Roberts, M.G. Lege, R. Andre, J. Greico, and Y. Alonzo. (1998). Characterization and detection of *Anopheles vestitipennis* and *Anopheles punctimacula* (Diptera: Culicidae) larval habitats in Belize with field survey and spot satellite imagery. *Journal of Vector Ecology* **23**: 74–88.

Richards, C.S. (1967). Aestivation of *Biomphalaria glabrata* (Basommatophora: Planorbidae). *American Journal of Tropical Medicine and Hygiene* **16**: 797–802.

Ritchie, S.A. and C. Laidlaw-Bell. (1994). Do fish repel oviposition by *Aedes taeniorhynchus*? *Journal of the American Mosquito Control Association* **10**: 380–4.

Riviere, F., B.H. Kay, J.M. Klein, and Y. Sechan. (1987). *Mesocyclops aspericornis* (Copepoda) and *Bacillus thuringiensis* var. *israelensis* for the biological control of *Aedes* and *Culex* vectors (Diptera: Culicidae) breeding in crab holes, tree holes and artificial containers. *Journal of Medical Entomology* **24**: 425–30.

Rosenberg, R. (1982). Forest malaria in Bangladesh. III. Breeding habits of *Anopheles dirus*. *American Journal of Tropical Medicine and Hygiene* **31**: 192–201.

Rowe, S.G. and M. Durand. (1998). Blackflies and whitewater: onchocerciasis and the eye. *International Ophthalmology Clinics* **38**: 231–40.

Russell, R.C. (1999). Constructed wetlands and mosquitoes; health hazards and management options: and Australian perspective. *Ecological Engineering* **12**: 107–24.

Scholte, E.-J, B.G.J. Knols, R.A. Samson, and W. Takken. (2004). Entomopathogenic fungi for mosquito control: a review. *Journal of Insect Science* **4**: 19–43.

Service, M.W. (2000). *Medical Entomology for Students*. Cambridge University Press, Cambridge.

Sjogren, R.D., D.P. Batzer, and M.A. Juenemann. (1986). Evaluation of methoprene, temephos and *Bacillus thuringiensis* var. *israelensis* against *Coquillettidia perturbans* larvae in Minesota. *Journal of the American Mosquito Control Association* **2**: 276–9.

Smith, L.B. (1953). Bromeliad malaria. *Annual Report of the Smithsonian Institution* (1952): 385–98.

Soltes-Rak, E., D.J. Kushner, D.D. Williams, and J.R. Coleman. (1995). Factors regulating *cryIVB* expression in the cyanobacterium *Synechococcus* PCC 7942. *Molecular and General Genetics* **246**: 301–8.

Spielman, A. and J.J. Sullivan. (1974). Predation on peridomestic mosquitoes by hylid tadpoles on Grand Bahama Island. *American Journal of Tropical Medicine and Hygiene* **23**: 704–9.

Staal, G.B. (1975). Insect growth regulators with juvenile hormone activity. *Annual Review of Entomology* **16**: 417–60.

Sturrock, R.F. (1974). Ecological notes on habitats of the freshwater snail *Biomphalaria glabrata*, intermediate host of *Schistosoma mansoni* on St. Lucia, West Indies. *Caribbean Journal of Science* **14**: 149–62.

Sukumar, K., M.J. Perich, and L.R. Boobar. (1991). Botanical derivatives in mosquito control: a review. *Journal of the American Mosquito Control Association* **7**: 210–37.

Takagi, M., A. Sugiyama, and K. Maruyama. (1995). Effect of rice culturing practices on seasonal occurrence of *Culex tritaeniorhynchus* (Diptera: Culicidae) immatures in three different types of rice-growing areas in central Japan. *Journal of Medical Entomology* **32**: 112–8.

Turell, M.J., M.R. Sardelis, D.J. Dohm, and S. O'Guinn. (2001). Potential North American vectors of West Nile virus. *Annals of the New York Academy of Sciences* **951**: 317–24.

United Nations Educational, Scientific and Cultural Organization. (2004). *UNESCO Courier Archives*. http://www.unesco.org/courier/archives/2004uk.htm.

USAID. (1975). *Water, Engineers, Development and Disease in the Tropics*. U.S.AID, Washington DC.

Vossbrinck, C.R., T.G. Andreadis, J. Vavra, and J.J. Becnel. (2004). Molecular phylogeny and evolution of mosquito parasitic microsporidia (Microsporida: Amblyosporidae). *Journal of Eukaryotic Microbiology* **51**: 88–95.

Walter Reed Biosystematics Unit. (1997). *Key to the Mosquito Genera of North America, North of Mexico, Version 1.0: Diseases*. Smithsonian Institution, Washington, DC, USA. http://wrbu.si.edu/www/projects/cdvik/mosgend.html

Womack, M. (1993). The yellow fever mosquito, *Aedes aegypti*. *Wing Beats* **5**: 4.

Wood, D.M., P.T. Dang, and R.A. Ellis (1979). *The Insects and Arachnids of Canada, Part 6: The Mosquitoes of Canada (Diptera: Culicidae)*. Biosystematics Research Institute Publication 1686. Agriculture Canada, Ottawa, Ontario.

World Health Organization. (1993). Lymphatic filiariasis: diagnosis and pathogenesis. *Bulletin of the World Health Organization* **71**: 135–41.

World Health Organization. (1999). *Prevention and Control of Dengue and Dengue Haemorrhagic Fever. Comprehensive Guidelines*. WHO Regional Publication, SEARO No. 29, WHO Regional Office for South-East Asia, New Delhi.

World Health Organization. (2004a). *Environmental Management for Vector Control*. http://www.who.int/docstore/water_sanitation_health/vectcont1/ch3.htm.

World Health Organization. (2004b). Dengue and Dengue Haemorrhagic fever. http://www.who.int/mediacentre/factsheets/fs117/en/.

World Health Organization. (2004c). *Disease Information*. http://www.who.int/tdr/diseases/.

World Health Organization/South-East Asia Regional Office. (1997). *Insecticide Resistance of Mosquito Vectors*. Intercountry Workshop on Insecticide Resistance of Mosquito Vectors, Salatiga, Indonesia, August.

Worthington, E.B. (1983). *Arid Land Irrigation in Developing Countries: Environmental Problems and Effects*. Pergamon Press, London.

Chapter 10

Abbasi, S.A. (1997). *Wetlands of India: Ecology and Threats. Vol. 1. The Ecology and the Exploitation of Typical South Indian Wetlands*. Discovery Publishing House, New Delhi.

Abdullahi, B.A. (1990). The effect of temperature on reproduction in three species of cyclopoid copepods. *Hydrobiologia* **196**: 101–9.

Anon. (1988). *The Murrumbidgee River Corridor Policy Plan*. National Capital Development Commission, Canberra.

Berrill, M. (1978). Distribution and ecology of crayfish in the Kawartha Lakes region of southern Ontario. *Canadian Journal of Zoology* **56**: 166–77.

Biggs, J., A. Corfield, D. Walker, M. Whitfield, and P.J. Williams. (1994). New approaches to the management of ponds. *British Wildlife* **5**: 273–87.

Boulton, A.J. (2000). Limnology and conservation of rivers in arid inland Australia. *Verhandlungen*

Internationale Vereinigung für Theoretische und Angewandte Limnologie **27**: 655–60.

Bratton, J.H. (1990). Seasonal pools. An overlooked invertebrate habitat. *British Wildlife* **2**: 22–9.

Bratton, J.H. and G. Fryer. (1990). The distribution and ecology of *Chirocephalus diaphanus* Prévost (Branchiopoda: Anostraca) in Britain. *Journal of Natural History* **24**: 955–64.

Calhoun, A.J.K. and M.W. Klemens. (2002). Best development practices: conserving pool-breeding amphibians in residential and commercial developments in the Northeastern United States. Technical Paper 5, Metropolitan Conservation Alliance, Wildlife Conservation Society, Bronx, New York.

Campbell, D.A., C.A. Cole, and R.P. Brooks. (2002). A comparison of created and natural wetlands in Pennsylvania, USA. *Wetlands Ecology and Management* **10**: 41–9.

Carpenter, S.R. and J.F. Kitchell. (1984). Plankton community structure and limnetic primary production. *American Naturalist* **124**: 159–72.

Caspers, H. and C.W. Heckman. (1981). Ecology of orchard drainage ditches along the freshwater section of the Elbe Estuary. Biotic succession and influences of changing agricultural methods. *Archiv für Hydrobiologie* **43** (Suppl.): 347–486.

Clarke, A.H. (1981). Introduction. In *Proceedings of the Second American Malacological Union Symposium on the Endangered Molluscs of North America, Bulletin of the American Malacological Union, Incorporated*, pp. 41–2.

Cohen, J.E., F. Briand, and C.H. Newman. (1990). *Community Food Webs, Data and Theory*. Springer-Verlag, New York.

Collinson, N.H., J. Biggs, A. Corfield, M.J. Hodson, D. Walker, M. Whitfield, and P.J. Williams. (1995). Temporary and permanent ponds: an assessment of the effects of drying out on the conservation value of aquatic macroinvertebrate communities. *Biological Conservation* **74**: 125–33.

Coope, G.R. (1995). The effects of Quaternary climatic changes in insect populations: lessons from the past. In *Insects and a Changing Environment* (eds R. Harrington and N.E. Stork), pp. 29–48. Academic Press, London.

Cordery, I. (1976). Some effects of urbanization on streams. *Civil Engineering Transactions of the Australian Institute of Engineers* CE18: 7–11.

Council on Environmental Quality. (1995). *Environmental Quality*, 25th Anniversary Report, pp. 271–2. The Executive Office of the President, Washington DC.

Covich, A.P., S.C. Fritz, P.J. Lamb, R.D. Marzolf, W.J. Matthews, K.A. Poiani, E.E. Prepas, M.B. Richman, and T.C. Winter. (1997). Potential effects of climate change on aquatic ecosystems of the Great Plains of North America. *Hydrological Processes* **11**: 993–1021.

Davies, C.P., M.A. Simovich, and S.A. Hathaway. (1997). Population genetic structure of a California endemic branchiopod, *Branchinecta sandiegoensis*. *Hydrobiologia* **359**: 149–58.

D'Cruz, R. (1997). Tasek Bera. In *Wetlands, Biodiversity and the Ramsar Convention*. Chapter 4: *The Asian Region* (ed. A.J. Hails), pp. 15–16. http://www.ramsar.org/lib_bio_5.htm.

Duffey, E. (1968). Ecological studies on the large copper butterfly *Lycaena dispar batavus* Obth. at Woodwalton Fen National Nature Reserve, Huntingdonshire. *Journal of Applied Ecology* **5**: 69–96.

Endler, J.A. (1995). Multiple-trait coevolution and environmental gradients in guppies. *Trends in Ecology and Evolution* **10**: 22–9.

European Environment Agency. (1999). *Environment in the European Union at the Turn of the Century: Environmental Assessment Report, No. 2*. European Environment Agency, Copenhagen, Denmark.

Evans, R.R. (2000). California Farm Bureau Federation, ag groups: fairy shrimp not threatened. Agriculture Alert. http://www.cfbf.com/archive/aa-1029c.htm.

Federal Register (1996). Final notice of issuance, re-issuance and modification of nationwide permits. Army Corps of Engineers, DOD, Vol. 61. p. 65916.

Fretwell, S.D. (1987). Food chain dynamics: the central theory of ecology? *Oikos* **50**: 291–301.

Fry, L.L. and M.S. Mulla. (1992). Effect of drying period and soil moisture on egg hatch of the tadpole shrimp (Notostraca: Triopsidae). *Journal of Economic Entomology* **85**: 65–9.

Galatowitsch, S.M. and A.G. van der Valk. (1996). The vegetation of restored and natural prairie wetlands. *Ecological Applications* **6**: 102–12.

Gallaway, M.S. and W.D. Hummon. (1991). Adaptations of *Cambarus bartoni cavatus* (Hay) (Decapoda, Cambaridae) to acid mine polluted water. *Ohio Journal of Science* **91**: 167–71.

Garvey, J.E. and R.A. Stein. (1993). Evaluating how chela size influences the invasion potential of an introduced crayfish (*Orconectes rusticus*). *American Midland Naturalist* **129**: 172–81.

Goettle, B. (2000). A 'living fossil' in the San Francisco Bay Area? http://www.rl.fws.gov/sfbnwr/tadpole.html.

Greenway, M. and A. Woolley. (1999). Constructed wetlands in Queensland: performance efficiency and nutrient bioaccumulation. *Ecological Engineering* **12**: 39–55.

Grillas, P., P. Gauthier, N. Yavercovski, and C. Perennou. (2004). *Mediterranean Temporary Pools. Vol. 1: Issues Relating to Conservation, Functioning and Management*. Station Biologique de la Tour du Valat, La Sambuc, Arles, France.

Hengeveld, H.G. (1990). Global climate change: implications for air temperature and water supply in Canada. *Transactions of the American Fisheries Society* **119**: 176–82.

Higler, L.W.G. (1984). Caddis larvae in ditches. *Series Entomologie* **30**: 175–8.

Higler, L.W.G. and F.F. Repko. (1981). The effects of pollution in the drainage area of a Dutch lowland stream on fish and macroinvertebrates. *Verhandlungen Internationale Vereinigung für Theoretische und Angewandte Limnologie* **21**: 1077–82.

Higler, L.W.G. and P.F.M. Verdonschot. (1989). Macroinvertebrates in the Demmerik ditches (The Netherlands): the role of environmental structure. *Hydrobiological Bulletin* **23**: 143–50.

Higler, L.W.G. and P.F.M. Verdonschot. (1993). Stream valleys as wetlands. *Hydrobiologia* **265**: 265–79.

Hobbs, H.H. and E.T. Hall. (1974). Crayfishes (Decapoda: Astacidae). In *Pollution Ecology of Freshwater Invertebrates* (eds C.W. Hart and S.L.H. Fuller), pp. 195–241. Academic Press, New York.

Hogg, I.D. and R.H. Norris. (1991). Effects of runoff from land clearing and urban development on the distribution and abundance of macroinvertebrates in pool areas of a river. *Australian Journal of Marine and Freshwater Research* **42**: 507–18.

Hogg, I.D. and D.D. Williams. (1996). Response of stream invertebrates to a global-warming thermal regime: an ecosystem-level manipulation. *Ecology* **77**: 395–407.

Hogg, I.D., D.D. Williams, J.M. Eadie, and S.A. Butt. (1995). The consequences of global warming for stream invertebrates: a field simulation. *Journal of Thermal Biology* **20**: 199–206.

Hussain, Z. (1997). The Sundarbans. In *Wetlands, Biodiversity and the Ramsar Convention*. Chapter 4: *The Asian Region* (ed. A.J. Hails), pp. 4–6. http://www.ramsar.org/lib_bio_5.htm.

Key, K.H.L. (1978). The conservation status of Australia's insect fauna. *Australian National Parks and Wildlife Service Occasional Paper* **1**: 1–24.

Kirkman, L.K., S.W. Golladay, L. Laclaire, and R. Sutter. (1999). Biodiversity in southeastern, seasonally ponded, isolated wetlands: management and policy perspectives for research and conservation. *Journal of the North American Benthological Society* **18**: 553–62.

Korneck, D., W. Lohmeyer, H. Sukopp, and W. Trautmann. (1977). Rote Liste der Farund Blütenpflanzen (Pteridophyta et Spermatophyta). In *Rote Liste der gefährdeten Tierre und Pflanzen in der Bundesrepublik Deutschland* (eds J. Blab, E. Nowak, W. Trautmann, and H. Sukopp), IUCN, Morges, Switzerland.

Korpelainen, H. (1986). Genetic variation and evolutionary relationships within four species of *Daphnia* (Crustacea: Cladocera). *Hereditas* **105**: 245–54.

IPCC (Intergovernmental Panel on Climate Change) (1990). *Scientific Assessment of Climate Change*. Report of Working Group 1. World Meteorological Organisation, Geneva, Switzerland.

LaCaze, C. (1970). Crawfish farming. *Louisiana Wildlife and Fisheries Commission Fisheries Bulletin* **7**: 1–27.

Lehman, J.T. and C.E. Caceres. (1993). Food-web responses to species invasion by a predatory invertebrate: *Bythotrephes* in Lake Michigan. *Limnology and Oceanography* **38**: 879–91.

Levins, R. (1970). Extinction. *Lectures on Mathematics in Life Sciences* **2**: 77–107.

Löfroth, M. (1991). *Vatmarkerna och deras betydelse*. Report 3824, Swedish Environmental Protection Agency, Stockholm.

Loiselle, S.A., S. Bastianoni, L. Bracchini, and C. Rossi. (2004). Neotropical wetlands: new instruments in ecosystem management. *Wetlands Ecology and Management* **12**: 587–96.

Maguire, B. and D. Belk. (1967). Paramecium transport by land snails. *Journal of Protozoology* **14**: 445–7.

Magurran, A.E. (1999). Population differentiation without speciation. In *Evolution of Biological Diversity* (eds A.E. Magurran and R.M. May), pp. 160–83. Oxford University Press, Oxford.

Martin, C.W. and R.S. Pierce. (1980). Clearcutting patterns affect nitrate and calcium in streams of New Hampshire. *Journal of Forestry* **78**: 268–72.

Maude, S.H. (1998). A new Ontario locality record for the crayfish *Orconectes rusticus* from West Duffins Creek, Durham Region Municipality. *Canadian Field-Naturalist* **102**: 66–7.

Maude, S.H. and D.D. Williams. (1983). Behaviour of crayfish in water currents: hydrodynamics of eight species with reference to the distribution patterns in southern Ontario. *Canadian Journal of Fisheries and Aquatic Sciences* **40**: 68–77.

Mawdsley, N.A. and N.E. Stork. (1995). Species extinctions in insects: ecological and biogeographical considerations. In *Insects and a Changing Environment* (eds R. Harrington and N.E. Stork), pp. 321–69. Academic Press, London.

May, R. (1999). Unanswered questions in ecology. *Philosophical Transactions of the Royal Society of London B* **354**: 1951–9.

Merritt, R.W. and K.W. Cummins (1988). *An Introduction to the Aquatic Insects of North America, 2nd edn*, Kendall/Hunt, Dubuque, Iowa.

Meyer, J.L., M.J. Sale, P.J. Mulholland, and N.L. Poff. (1999). Impacts of climate change on aquatic ecosystem functioning and health. *Journal of the American Water Research Association* **35**: 1373–86.

Mills, E.L., J.H. Leach, J.T. Carlton, and C.L. Secor. (1993). Exotic species in the Great Lakes: a history of biotic crises and anthropogenic introductions. *Journal of Great Lakes Research* **19**: 1–54.

Moore, N.W. (1976). The conservation of Odonata in Great Britain. *Odonatologia* **5**: 37–44.

Moore, N.W. (1980). *Lestes dryas* Kirby—a declining species of dragonfly (Odonata) in need of conservation: notes on its status and habitat in England and Ireland. *Biological Conservation* **17**: 143–8.

Müller, J., E. Partsch, and A. Link. (2000). Differentiation in morphology and habitat partitioning of genetically characterised *Gammarus fossarum* forms (Amphipoda) across a contact zone. *Biological Journal of the Linnean Society* **69**: 41–53.

Neckles, H.A., H.R. Murkin, and J.A. Cooper. (1990). Influences of seasonal flooding on macroinvertebrate abundance in wetland habitats. *Freshwater Biology* **23**: 311–22.

Norris, R.H. (1991). Agriculture and modified temperate environments. The ecosystem and human impact: problems for assessing water quality. *Australian Biologist* **4**: 22–7.

Olive, L.J. and P.H. Walker. (1982). Processes in overland flow-erosion and production of suspended material. In *Prediction in Water Quality* (eds E.M. O'Loughlin and P. Cullen), pp. 87–119. Australian Academy of Science, Canberra.

Patala, K. (1975). The crustacean plankton communities of 14 North American Great Lakes. *Verhandlungen Internationale Vereinigung für Theoretische und Angewandte Limnologie* **19**: 504–11.

Penn, G.H. Jr. (1943). A study of the life history of the Louisiana red-crawfish, *Cambarus clarkii* Girard. *Ecology* **24**: 1–18.

Peters, R.H. (1991). *A Critique for Ecology*. Cambridge University Press, Cambridge.

Pianka, E.R. (1970). On *r*- and K-selection. *American Naturalist* **104**: 592–7.

Pimm, S.L., J.H. Lawton, and J.E. Cohen. (1991). Food web patterns and their consequences. *Nature* **350**: 669–74.

Putman, R.J. (1994). *Community Ecology*. Chapman and Hall, London.

Samways, M.J. (1996). Insects on the brink of a major discontinuity. *Biodiversity and Conservation* **5**: 1047–58.

Savage, A.A. (1989). *Adults of the British Aquatic Hemiptera, Heteroptera: A Key with Ecological Notes*, FBA Scientific Publication, No. 50, Freshwater Biological Association, Windermere.

Söderqvist, T. and T. Lindahl (2003). Wetland creation: socio-economic and institutional conditions for collective action. In *Managing Wetlands: An Ecological Economics Approach* (eds R.K. Turner, J.C.J.M. van den Bergh, and R. Brouwer), pp. 223–49. Edward Elgar, Cheltenham, UK.

Stansbery, D.H. (1971). Rare and endangered molluscs in the Eastern United States. In *Proceedings of a Symposium on Rare and Endangered Molluscs (Naiads)* (eds S.E. Jorgensen and R.W. Sharp), pp. 117–24. US. Department of the Interior, Fish and Wildlife Service.

Stevenson, A.C., J. Skinner, G.E. Hollis, and M. Smart. (1989). The El Kala National Park and Environs, Algeria: an ecological evaluation. *Environmental Conservation* **16**: 335–45.

Sweeney, B.W., J.K. Jackson, J.D. Newbold, and D.H. Funk. (1992). Climate change and the life histories and biogeography of aquatic insects in eastern North America. In *Global Climate Change and Freshwater Ecosystems* (eds P. Firth and S.G. Fisher), pp. 143–76. Springer-Verlag, New York.

Tavares-Cromar, A.F. and D.D. Williams. (1996). The importance of temporal resolution in food web analysis: evidence from a detritus-based stream. *Ecological Monographs* **66**: 91–113.

Turner, R.K., J.C.J.M. van den Bergh, and R. Brouwer. (2003a). Introduction. In *Managing Wetlands: An Ecological Economics Approach* (eds R.K. Turner, J.C.J.M. van den Bergh, and R. Brouwer), pp. 1–16. Edward Elgar, Cheltenham, UK.

Turner, R.K., R. Brouwer, S. Georgiou, I.J. Bateman, I.H. Langford, M. Green, and H. Voisey. (2003b). Management of a multi-purpose, open access wetland: the Norfolk and Suffolk Broads, U.K. In *Managing Wetlands: An Ecological Economics Approach* (eds R.K. Turner, J.C.J.M. van den Bergh, and R. Brouwer), pp. 250–70. Edward Elgar, Cheltenham, UK.

Watson, G.F., M. Davies, and M.J. Tyler. (1995). Observations on temporary waters in northwestern Australia. *Hydrobiologia* **299**: 53–73.

Wells, S.M., R.M. Pyle, and N.M. Collins. (1983). *The International Union for Conservation of Nature and Natural Resources Invertebrate Red Data Book.* IUCN, Gland, Switzerland.

Williams, D.D. and H.B.N. Hynes. (1977a). Benthic community development in a new stream. *Canadian Journal of Zoology* **55**: 1071–6.

Williams, W.D. (1981a). Running water ecology in Australia. In *Perspectives in Running Water Ecology* (eds M.A. Lock and D.D. Williams), pp. 367–92. Plenum Publishers, New York.

Wrubleski, D.A. (1987). Chironomidae (Diptera) of peatlands and marshes in Canada. In *Aquatic Insects of Peatlands and Marshes in Canada* (eds D.M. Rosenberg and H.V. Danks), pp. 141–61. Memoir of the Entomological Society of Canada, No. 140.

Young, L. (1997). Mai Po Marshes: conserving wetland biodiversity through shrimp farming. In *Wetlands, Biodiversity and the Ramsar Convention.* Chapter 4: *The Asian Region* (ed. A.J. Hails), pp. 8–9. http://www.ramsar.org/lib_bio_5.htm.

Index